# Environmental Issues in India

# Environmental Issues in India

## A Reader

*Edited by*

Mahesh Rangarajan

ISBN 978-81-317-0810-1

First Impression, 2007
**Current Impression, 2026**

Published by Pearson India Education Services Pvt. Ltd, CIN: U72200TN2005PTC057128.

Head Office: 1st Floor, Berger Tower, Plot No. C-001A/2, Sector 16B, Noida - 201 301, Uttar Pradesh, India.

Registered Office: 6th Floor, Tower A, Unit A and B, International Tech Park, CapitaLand Chennai,
200 Feet Radial Road,Zamin Pallavaram, Old Pallavaram Chennai – 600117, Tamil Nadu, India

Website: in.pearson.com; Email: companysecretary.india@pearson.com

Digitally printed in India by Trinity Academy for Corporate Training Ltd, New Delhi in the year of 2026

# Contents

## Section II: Colonial India



## Section III: Independent India's Environment

## Section IV: Movements and Alternatives



## Section V: Global Issues

# Tables

# PREFACE AND ACKNOWLEDGEMENTS

The environment as an issue is ubiquitous in the world we live in. There has been a consequent explosion of work that reflects on the reasons for and responses to ecological dilemmas. The last decade and a half in particular have seen a quickening of the pace of research and scholarship among both social and natural scientists, as well as among policy makers and activists. The range and scope of the issues these works address cannot be covered in one volume, but it is possible to put together a selection of papers that informs, stimulates and provokes thought.

This volume has its genesis in a workshop convened at the Department of History at Delhi University in September 2005. Environmental issues not only figure in a new paper being studied by undergraduates, but also constitute a topical issue in the times we live in. Professor B.P. Sahu, Head of the Department of History encouraged me to put together a volume that consolidated on the readings put together for the workshop. The 50 odd teachers drawn from across the vast network of affiliated colleges of the University were enthusiastic both in participation and in support of the idea. Professor Shahid Amin and Dr Prabhu Mahapatra, who were also instrumental in organizing the workshop, were equally forthcoming with their support.

Most papers in this volume have been reproduced from journals or books. This would have been impossible if not for the assistance and cooperation of the authors and the publishers. Each of them is acknowledged elsewhere in this reader but their support has been invaluable.

I am indebted to three scholars who took the time to write pieces especially meant for this volume: Darryl D'Monte, Professor Sumit Guha and Dr Mayank Kumar.

An associate of many years, Pallava Bagla, a photographer and journalist of wide repute, graciously agreed to lend a selection of his photographs for the book. I am grateful to him.

I do hope that those who read the extracts and pieces in this book will also venture further and sample the extensive and growing corpus of literature on environmental issues by these and other scholars, thinkers and activists.

Several people whose own works do not figure in this volume have gone out of their way to enrich my understanding. Their intellectual and personal support has been invaluable and I owe them a debt that cannot be repaid. These include Dr Rohan Arthur, Dr Madhulika Banerjee, Dr Amita Baviskar, Professor Ranjan Chakrabarti, Professor Indrani Chatterjee, Dr Aparajita Datta, Dr Rohan D'Souza, Professor Richard Grove, Professor Ronald Herring, Professor Barbara Lynch, Dr Jagdish Krishnaswamy, Dr Neema Kudva, Dr M.D. Madhusudan, Dr Subhadra Menon, Nitin Sethi, Dr Ghazala Shahabuddin, Professor K. Sivaramakrishnan, Professor Nandini Sundar, Dr V. 'Chellam' Suri, Professor Romila Thapar and Valmik Thapar.

A very special word of thanks to contemporaries whose work on environmental policies and futures is known far and wide: Sunita Narain and Ashish Kothari. Professor Ramachandra Guha, whose works figure in this volume made several valuable suggestions, as did Professor Nayanjot Lahiri.

The editorial team at Pearson has been unstinting in their support. I am especially grateful to Kamini Mahadevan for her fine sense of copy-editing and the way she adjusted to my habit of writing (or re-writing) down to the wire.

My in-laws, Wing Commander and Visalakshi Venkataraman provided a conducive atmosphere in Jalvayu Vihar, Bangalore where the introductory notes were finally written up. Chellam and Harini were ideal hosts for a fortnight, providing hospitality and acting as sounding boards for ideas. Geetha and Uttara tolerated yet another intrusion on their time and my absence on a trip to see the flowering of the *kurinji* in Kodaikanal.

I am also grateful to those institutions that have given me the chance as a Visiting Faculty to interact with students: The Mario Einaudi Center, Cornell University, Ithaca; the Department of History, Jadavpur University; the Centre for Ecological Studies, Indian Institute of Science; the Centre for Wildlife Studies, Bangalore; and the National Centre for Biological Studies, Bangalore.

Among those who read through the drafts of the introduction was Dr Heather Goodall. I am grateful to her for persistent questions about what exactly I meant; I hope the final product meets the high standards her own work has set as an environmental historian.

Two people, Dr Nayanjot Lahiri and Dr Tripta Wahi, who were active participants in the workshop at the Department of History in September 2005 played a formative role in my undergraduate years. Each has been an outstanding teacher and an original thinker, pushing the frontiers of knowledge even while inculcating new generations in the historical craft. I might also mention that in a country with sharp disciplinary divides between 'Arts' and 'Sciences', they encouraged us to cross chasms. Divisions among historians in India are slightly more complex: we divide into pens marked 'ancient', 'medieval' and 'modern'.

Dr Nayanjot Lahiri, who is now Professor at the Department of History, South Campus, Delhi University, is a historian of ancient India. Dr Tripta Wahi researches early modern and medieval India and continues to teach at the Department of History, Hindu College. Both are former teachers but are also colleagues and friends.

It is a privilege to dedicate this volume to Nayanjot and Tripta. I am not in touch with everyone in Hindu College's batch of 1985 from History (Honours) but I cannot think of a single one of them who would not but approve.

As Nayanjot and Tripta always reminded us, the quest for knowledge is an unending one and any work is only a spur for further thought and reflection. This is even more so for such a volume that seeks to place contemporary issues of environment in India against a larger social and historical backdrop. There is much more to learn than we know of the interactions of nature and cultures, of peoples and the environments that encompass them. This Reader hopes to draw on critical inputs and responses. Any suggestions and criticisms are most welcome.

**Mahesh Rangarajan**

# Introduction

## A Word in Advance

It is difficult to avoid mention of environmental issues in India today. Newspapers warn of impending water wars. Television stations beam news broadcasts of the state of the tiger or the rights of tribals. Big dams divide people, from film stars to commoners, into groups that deify and those that oppose them. Over two decades ago, environment seemed an idea whose time had come. *Time* magazine even made earth 'The Planet of the Year' in 1989, in place of their usual choice of a man or a woman. Today, environment is still a critical issue. What drives the salience of such issues? Just what do they mean? This Reader is an attempt to help find clues to the answer, or rather, to suggest ways to pose this question.

Environmental issues gained in importance when two different trends intersected each other. One was the capacity of humans to transform in a relatively short span of time their natural surroundings, with consequences that could be adverse for them. These adverse outcomes were not equally distributed across all societies or peoples, but they required and evoked responses from them. Here, there was a second critical change: it was precisely their concern about the environment that made such responses possible at all. The ideas of people, or rather of people who chose to or tried to act to overcome the adverse situations remain as crucial to environmental dilemmas as the material conditions themselves. Such a pithy summary can hardly do justice to the complexity of issues that concern humans and their home, the earth. But they will do for starters.

Human imprints on earth's ecology may be as old as our species itself, but the last century witnessed change to an unprecedented degree. The number of people in the world grew four-fold between 1900 and 2000 of the Christian Era (CE). It is but commonplace to attribute environmental impact to this demographic factor.

The addition in human numbers did not, however, proceed on its own but interacted with other, secular shifts. There was a concurrent expansion in the size of the world economy, with the Gross World Product growing fourteen-fold. But the kind of growth that took place was often more intensive in use of energy and raw materials, and generated more waste per unit of wealth. The carbon dioxide emissions rose thirteen-fold and energy use increased sixteen-fold. As a rule of thumb, economic growth and technological change have had much more to do with the human imprint on the earth than just the numbers of people (McNeill 2000:xiii–xiv, 3–17).

The intensity of resource use and the number of people had no precedent in the past. It marked a new phase of human existence on the planet. We are still coming to terms with the fact it may have marked a turning-point for the planet itself.

While the increases—whether of more people or of more wealth—were not distributed evenly across nations, the ecological consequences could and did cross borders. In the early part of the century, the United States of America (USA) and then the Union of Soviet Socialist Republics (USSR), which disintegrated in 1991, experienced ascendancy in part due to faster economic growth than older imperial powers. By the end of the century, the Asia-Pacific region was emerging as an engine driving global growth. Again, the spread of impacts was and is uneven. Just two countries out of nearly 200, China and India, account for four out of every 10 people on earth.

Nor was the proportion of people to land and resources equitably spread out across the globe. Russia has more land mass but far fewer people. India has one in six of world's people but only 5 per cent of the fresh water. Nature's endowment is not evenly distributed among countries. The consequences of growth in numbers, increase in wealth or waste, change in land, air or water use or abuse, can mean very different things in different nations. China, for instance, has 22 per cent of the earth's people but less than 10 per cent of its cultivable land area.[1]

## Changing Present, Changing Pasts

Of course, the environment has never been static, unchanging and in equilibrium. Even earlier, changes in land were as critical a part of human history as the rise and fall of empires, the patterns of trade or wars of conquest.

Had the earth's average temperatures not risen around 10,000 to 13,000 years ago, vast areas that are now habitable would have been under ice sheets. This is not only true of large parts of Europe and North America but possibly even of regions in the Himalayas and Asia's highlands. But this change in temperature took place over a long period, spanning over 8,000 years.

In the subsequent period, humans developed the capacity to transform the face of the earth. Fire was known to humans long before the end of the Ice Age. Domestication of animals and early agriculture became widespread over the last 10,000 years. Over the last two centuries, industrial production multiplied the human capacity to tap energy, use raw materials, move people and commodities over huge distances and generate enough waste to make its management a pressing and troubled question. The dates and timings and the sequence of these changes varied enormously across space and time (Mithin 2003; Williams 2000).

It is tempting to imagine that in our modern era, the wider environment does not matter all that much. It is an even more widely held view that whatever does matter can be 'fixed' with better technology. There is no doubt that advanced technologies have had a major role in abatement of pollution, in making inhabitable lands productive and in working out what to do with waste. Such views though common do less than justice to the complexity or immensity of environmental challenges. They represent not just a technical but much more so a social dilemma.

Environmental questions, whether of safety and risk, of resource depletion or access, of pollution or extinction of biodiversity, are more crucial today than ever before. Again, there have been attempts to grapple with them in the past. Some initiatives came from those with power and privilege. Supplies of timber became vital for navies in Europe from about five centuries ago: this led to regulations about who could use the forest and when. This was initially the case in the Mediterranean and then the Atlantic coastal countries. Venice preceded Germany, France and Britain. The herds of wild elephants in Asian kingdoms had attracted similar rule-making 3200 years ago. Rulers like the Mauryas in India set down harsh penalties for elephant poaching, the animals being more valuable alive than dead.[2]

Over the last two centuries and a little earlier, such concerns about specific resources, about who should secure their control and for what, acquired a new dimension. There were concerns about wider global impacts of the way resources were used. In the Indian Ocean region, early French and

British scientists pondered over possible linkages of deforestation, the wholesale denuding of the native tree cover and the cycles of water replenishment (Grove 1995). In America, there were similar fears by the late nineteenth century. The creation of the Forest Department in India in 1864 and of the US Forestry Service just over a 100 years ago were partly in response to such fears. They were also a bid to retain control of a critical and strategic economic asset: timber (Guha 2000).

Despite early warning signals and beginnings, great and small, the cluster of concerns we today label as 'environmental' were marginal to most decision makers till about three decades ago. The twentieth century saw two great World Wars (1914–18 and 1939–45) and the end of these conflicts led to an era of unprecedented growth. As the historian John McNeill has calculated, the energy used by humans in that 100-year run was more than that used by all our ancestors in the past 10,000 years. In the face of such huge advances, it was but natural that all problems looked solvable. Environmental concerns were marginal in the period of rapid growth of the global economy roughly from 1945 to 1973 (McNeill 2000: xv).

## Many Nations, One Earth

Yet, by the 1960s, there were signs that many things were not quite right with the way the earth was viewed or used. The publication of the book *Silent Spring* by the American marine biologist Rachel Carson warned of the unintended impacts of the overuse of pesticides. Her work was preceded and followed by the public espousal of causes such as the threat of radioactive contamination due to nuclear testing. Barry Commoner, once called the 'Professor with a class of millions', argued that technology could not solve ecological problems (Carson 1962; Commoner 1971; Lear 1997). Ecology was the science of the structure and function of life that would help abate larger consequences. In much of the western world and in the USSR, industrial and chemical pollution and the issues of wilderness came to be of central significance. In due course attention shifted to the differential impact on the poor and on those subject to racial discrimination.[3]

In the developing world, there had long been voices of caution. Mahatma Gandhi was an early exponent of less resource-intensive ways of living. A century ago, the Cuban nationalist Jose Marti worried about the implications of the rise in human numbers. Rabindranath Tagore commented extensively on the importance of rural crafts as a means for people to live in tune with nature. In much of Africa and colonial Asia (including India), there were major debates on how to evolve alternative patterns of development. In addition to bridging divides of town and country, class and gender, some among these early visionaries also took up ecological themes. Once freedom came, as in India in 1947, these voices were for the most part ignored.

It was around the time that Commoner's and Carson's work found resonance with a wider audience by the end of the 1960s and the early 1970s when the environmental movement came of age in countries like India. Indian ecologist Madhav Gadgil pointed to not one but several markers. In the western foothills of the Himalayas, villagers contested rights to forests with contractors. Trees like ash and hornbeam, oak and rhododendron were vital for their own livelihood. Profits clashed with livelihood also in the Jharkhand region where *Adivasi* or tribal peoples preferred sal to teak. On the western coast, Zuari in Goa and Mavoor on the river Chaliyar, Kerala, witnessed protest against industrial pollution (Gadgil 2001).

The list could be extended very easily. In Chikmangalur, Karnataka, landless labourers who ate crabs from rice fields were afflicted with illnesses. Pesticides travelled up the food chain and into their bodies. By the 1970s, debates on dams and the submergence of critical forest habitats came to the fore. The dams in Silent Valley, Kerala, and Moyar in Tamil Nadu were eventually stalled. By the early 1980s, a synthesis, a new paradigm of development where ecology and equity both mattered was in place.[4]

This made the battle lines—if not the common ground—in debates much clearer. Few realize today quite how exceptional *The State of India's Environment, the First Citizen's Report, 1982* was. The key driving force behind the report, Anil Agarwal would echo Barry Commoner in rejecting 'reductionism', the method of seeing the parts and not the whole. With a strong emphasis on prudence among the poor as contrasted to the ecological profligacy of the rich, this was but one of many approaches towards alternative models. The coming decades saw an explosion of journalistic and academic work on these themes, much of which has found its way into this volume. What was important is that these diverse strands addressed a range of issues (Agarwal 1985).

The gap between rich and poor nations was and remains a recurrent theme. It also presents a difficult dilemma for political leaders and citizens alike. Consumption in powerful countries could often offer strong, short-term incentive for environmentally destructive practises beyond their borders. Mexico and then parts of Central America degraded arid savannah and tropical forest respectively. The aim: to supply beef for American hamburger chains (Tucker 2000: 309–12; Ramakrishna and Woodwell 1993).

While consumer pressure could make a change here, there were more complex issues at hand once world leaders got down to negotiations. This was the case at the UN Conferences on the Environment held at Stockholm, Sweden (in 1972), Rio de Janeiro, Brazil (in 1992) and Johannesburg, South Africa (in 2002). Indira Gandhi, world's second ever woman Prime Minister, spoke of many such issues in her address at the first conference. 'The rich look askance at our continuing poverty,' she said, 'on the other they warn against our methods. We do not wish to impoverish the environment any further and yet we cannot for a moment forget the grim poverty of large numbers of people. Are not poverty and need the greatest polluters? The environment cannot be improved in conditions of poverty. Nor can poverty be eradicated without the use of science and technology.'[5]

Over the last three decades, in only one instance (the Montreal Protocol to halt the depletion of the ozone layer) have developed countries agreed to technology transfers. These were essential to reduce use of chlorofluorocarbons (CFCs). Yet, the issue of cutting down carbon dioxide emissions, which have risen sharply in an industrial age, has eluded agreement. The US emerged by the 1990s as a country that stood apart from successive efforts to make progress via multilateral agreements. But such isolationist nationalism though exceptional, pointed to a larger trend. The sense of unity that pervaded the run-up to the Stockholm Conference did not translate into effective action, save in exceptional cases.

## One Country, Many Voices

She may have been the voice of the developing world at Stockholm, but the country Indira Gandhi represented was deeply at odds with itself over similar issues. This should occasion no surprise. For few countries exemplified the diverse facets of the modern ecological dilemma as India. At the time, nearly 70 per cent of the work force was in agriculture, yet the country already had a substantial industrial base. Two years later, it would enter the exclusive nuclear club and also launch its first satellite, Aryabhata.

Contrary to usual claims, few societies have witnessed such a diverse array of effective or potentially effective responses. Several critical policy initiatives preceded and followed from the Stockholm visit. Project Tiger, launched in April 1973, was in its day the single largest wildlife conservation scheme in the world and soon secured a string of reserves across various habitats. The Wildlife (Protection) Act came in 1972 and the Forest Conservation Act in 1980, the year of Indira Gandhi's return to office. The Department of Environment at the Union government level, later upgraded to a Ministry was set up in November 1980. If these responses were significant, it still needs to be asked why there was so much strife around issues of the environment (Rangarajan 2006).

Just as the divide between nations rich and poor, there are basic divisions in Indian society on what environmental issues matter and how much. These were evident even at the time of Indira Gandhi's famous speech but by the 1980s there was a blossoming of grassroots and middle class groups with varying, even differing agendas. In more ways than one, it was such concerns that influenced, though not always shape, policies. Forest-based movements backed by scientific evidence put a halt to clear felling mixed forests for monocultures, long a standard practise for foresters eager to generate cash. In a *cause celebre*, the dam at Silent Valley, Kerala, was scrapped and a rain forest habitat saved. Mobilization in the Himalayan foothills and across *Adivasi* India played a key role in redirecting forest policy. Silent Valley was testimony to lobbying by scientists and even more so by the enlisting of school and college teachers and students into a campaign by a popular science organization in India's most literate state, Kerala.[6]

Even responses of those oriented to conservation can differ, and widely so. When Robert Frost famously wrote, 'Whose woods these are/I do not know', he could not have imagined how apt his words would sound at the turn of the century with regard to the Indian forests. Who ought to protect the forests were often tied up with the question of how one assessed where the threat to the forests came from. In a re-run of the colonial era, forest departments made a push for more powers. In turn, they repeatedly found resistance not only from tribal peoples but also marginal peasant groups. If the former saw the forest as a source of cash or commodities, as an ecological safety valve, the latter asserted the rights to livelihood. The woods themselves were and are imbued with many meanings.

The pieces in this volume will attest that these are far from being isolated in a landscape in turmoil. Contests over water and fisheries, grazing lands and commons, city spaces and mountains are a feature of twenty-first century India, as much as if not more than they were of the century just past. Often there is disagreement on what constitutes a resource and for whom. There are clashes and conflicts over whose interests should get priority and how.[7]

On one point, the scientists in our midst are right. Each of these choices has significance beyond our lifetimes. The immensity of power humans have gained over the globe may not be manifest in a country like India. Yet, life spans have more than doubled since 1947. There is more land under irrigation than ever before. By 2025, if not a little later than that, more Indians will reside in town, city and megalopolis than in the countryside for the first time ever in any millennium of history.

Will the coming century derive any lessons or rather insights from the past? No size can possibly fit all conditions. A rich vein of alternative approaches contests for priority in public policy. The ideas discussed range from more inclusive ways of achieving conservation without alienating resident peoples in and around parks to shifting cultivation to domestic stock-rearing.[8]

Mono-causal explanations are as abundant as they are prone to be misleading. Those who see demographic expansion as the prime mover of decline often down play the disparities in the size of the ecological footprint of different groups or sets of people. A stock broker in Surat, Gujarat, and a terrace rice farmer in the Apa Tani valley in Arunachal Pradesh may both command energy, land and raw materials. But the amount they use is not comparable. One treads on the earth less lightly than the other. Neither can be, however, unaffected by the course India takes in the coming decades.

## Opportunity or Pandora's Box?

Writing in her diary over a decade ago, a young urban Indian who forsook city life to work with Van Panchayats or village forest councils struck a note that mixed realism with optimism. Oona Mansingh (1962–1996) was working on rural reforestation of mixed oak forests in Mukteswar in the

Kumaon Himalaya at the time. She felt, 'I don't know if it is possible to save the Himalaya,' and went on, 'but I want to spend the rest of my life trying.' The short span of a decade cannot permit a definitive answer but there are encouraging signs that the kind of local grassroots initiatives she was working with can and will work, given the right conditions (Mansingh 2002).

The immensity of these issues and the range of powerful and entrenched interests cannot but make a confident reading more difficult for those of us who do not have her optimism or her courage of conviction. But there is little doubt on two matters.

The planet may be at a turning point in human–nature relations. What happens in India will have deep, long-range consequences for the entire planet. These will concern all who breathe air on the earth and walk its lands. The sheer biological endowment of India and the numbers of people in it would count in any environmental audit. The numbers, as they say, are stacked in our favour (or against our success) depending on your point of view (see Agrawal 2006).

But there is a second, better reason. Few countries outside of the western world have had such vibrant and deep rooted institutions of enquiry and debate, of representative government and an open media, of an independent (if not always wise) judiciary, of labour unions and women's committees, or farmer's associations and fish worker federations, of all those people who breathe life and soul into that magic word: democracy. Some feel democracy will doom the environment, that strong rules and authoritarian governments alone can secure the future. Yet, there is abundant evidence to the contrary in the public realm (Khilnani 1997; Sen 2006).

A small instance is the range of papers in this Reader. These form but a selection of the works of individuals, most though not all, based in India who have reflected on the patterns of our past, the issues of the present and the prospects of our future.

How the earth will fare may be a distant concern for a person who is unsure of the next meal. But how that meal and that person are made secure will have a lot to do with the fate of the earth. How poverty will be tackled may have little immediate relevance to any one with a credit card. But riches cannot fully guard against contaminated water and air any more than they can against the return of diseases once thought to have been eradicated. In contrast to 'purely' economic issues, those with an ecological edge or an environmental dimension can help trace the threads that bind us in a common future.

No place can be a bad place to begin. But surely in more ways than one, India is *the* place. Or is it?

Satellites in the sky show a denuded landscape where mangroves, wetlands and forests are vanishing. Negative trends seem to overwhelm positive ones. Yet, the signs of hope, the spirit of enquiry beckon equally strongly. In our capacity to master the means, ignorance is a luxury nobody can afford.

## II

## Of Ecology and Geography

Issues of the environment are bound up in various ways with our immediate surroundings. For example, even living in the city of Delhi makes you aware of not one but many ecologies.

The very location of the city is at the '*dahlij*', literally the doorstep of the fertile flood plains and alluvium deposits brought down by the Yamuna and Ganga from the foothills of the Himalayas. In times past, sandstone and marble from quarries in the Aravalis helped build the forts and palaces the city was famous for. Its most prominent physical feature, the Ridge, is the eastern edge of the range (Krishen 2006).

Delhi's other key feature, the Yamuna is a part of one of several river systems in North India that are fed year round by snow melt from the Himalayas, a privilege denied to the seasonal rivers of the peninsula that lies to the south. As the city grows outwards, it transforms fertile fields into urban living space, and already it has more motor vehicles than the other three metropolises of India combined.

Delhi nestles in a larger region, the diversity of whose ecologies constitutes an aspect of the heterogeneity of India. Riverine flood plain and scrub forest, crop land and pasture, motorway and industrial estate, urban slum and rural settlements, marshlands and rocky outcrops are but a sample of the environments in and around the city.

For a country that covers only 2.4 per cent of the earth's landmass, an array of landscapes is to be found in India. In the north-east, a land that shades off into South-East Asia, are gibbons and liana creepers and orchids, as also an ungulate, the *mithun* (*Bos frontalis*), kept for sacrifice and never put to work, being killed only on special occasions. In the north lies the great massif of the Himalaya, a geologically young mountain chain. Yaks are reared in its higher reaches and sheep in the foothills. With oak and pine, rhododendron and deodar, these ranges resemble that of Europe in their flora and fauna. And then there are lands of bamboo brakes and thorn forest, the India represented on the new 10-rupee note with its three large mammals: the rhino, the tiger and the elephant.

The array of ecologies from coral reef to cold desert, and the Thar Desert to rain forest in the Western Ghats are home to a billion people and more than 500 other mammals. They also are habitat to over 45,000 plant species, one of every ten on earth. These are spread across sixteen kinds of forest types and five types of savannah grasslands. Over 160 plant species are cultivated here with twice as many being their relatives in the wild. Around 4,000 species can be used for wood though only a handful of them are commercially valuable.[9]

Through history, the environment has been many things to different people. The mountains and forests were often barriers to movement of men, women and materials. These barriers were far from being complete and contact and slow diffusion of populations is a feature of the past. But, until as recently as 1941, there was no reliable land route eastward to Burma. To the west, the passes of the Bolan and Khyber were important for invaders or herders going back and forth. But they were even more so for horse traders who came across to sell Central Asian horses, always superior to Indian ones. These mares and stallions were sold to eager buyers who used them for rapid transport and in battle. Tall grasslands of the sort that now survive in patches like Kaziranga on the flood plains of the Brahmaputra could be traversed most easily on elephant back. The taming of these huge beasts (and that far more ill-tempered creature, the water buffalo) was a significant achievement several centuries before the birth of Christ. Yet in much of India, 'the forest was still barrier to overland movement'. This was true when Huen Tsang (or Xuan Xang), the prolific Chinese scholar, walked hill and jungle tracks in the seventh century CE and when Mughal armies foot-marched the same routes a thousand years later.

For centuries past, India was a land with islands of cultivation in a sea of forest. The latter was pushed back but often recovered, especially when disease depleted numbers or cultivators fled high taxes. Place names to this day attest to the traces of that past: Jharkhand in Persian meant 'the land of thickets' (Deloche 1993: 5–10).

Just over two thousand years ago, what is now India still had as few as 20 million people. This is an estimate for the second century BCE. Even in

1600 CE, during the reign of Akbar, the number was just over five times that many, at 114 million. More land was under forest and savannah, marsh and jungle than under the plough. Much of this land was cultivable: unlike China where only a tenth of the land is fit to plough, well over half of India can sustain agriculture, if fitfully. As Jean Deloche, the great French historian of transport in India suggests, where humans live depends both on geography and history.

The growth of numbers of people has been slower across the centuries but certain features persist. For two millennia, the centre of demographic gravity was and remains the Gangetic plain (Sumit Guha 2001: 24–34, 59). Elsewhere too, the valleys were more intensively cultivated: this was the case in Manipur in the north-east, and in the lowlands around that most majestic of rivers, the life giver to Assam, the Brahmaputra.

Much more than in the north, resource use in the peninsula depended on sea-borne contact. But the rivers of the south, like the Godavari or Kaveri, were seasonal, not perennial. Without snow melt all the year, they dried up in summer. The river valleys were narrower in expanse and the hinterland consisted of the Deccan plateau and numerous hill ranges.

Large extensive states were harder to create but ports enabled trade importing African ivory and coral from Roman times onwards. Ships off-loaded thorough-bred Arab horses in a land where they were paid for in silver and gold. Trans-ocean links to South-East Asia, Africa and West Asia linked destinies and ecologies together.[10]

For all the diversity, nature imposes on the subcontinent a basic unity of the water cycle. Eight of 10 buckets of rain that pours down from the heavens comes in a specific time of the year. It's a word with Arabic roots but one that is now part of any English dictionary: the monsoon (from *mausam*, the season). Most rainfall is between June and September, and even this is a gift of geography, the Himalayas blocking rain-bearing clouds that come up from the Bay of Bengal in the east. One part of India, Tamil Nadu gets most of its rain in winter from the north-east, much more than from the south-west monsoon.

Ecologies like societies have always been in flux but over the last two centuries, human-induced change has outpaced life in India. Human choices cannot override geography but they are reshaping the lands and waters, the soils and the plant life faster than ever before.

To learn how to moderate our impact, we have to turn not only to the future but the past. It is in assimilating insights from ecological pasts that we step towards a durable future. Knowing where we stand in relation to the land, a sense of geography is but the first step to remaking our histories.

## Notes

1. For an exposition of trends over time in China, see Elvin (2004).
2. On Europeans and timber see Williams (2000: 201–6, 259–70); Trautman (1982: 254–80).
3. The major source for this paragraph is Ramchandra Guha (2000); Shabecoff (2003: 103–117). On the Soviet Union see Weiner (2001).
4. Many of these themes in India and beyond are explored in Guha and Martinez-Alier (1997).
5. Indira Gandhi, 'Man and his world', Speech at the UN Conference on the Human environment, Stockholm, 14 June 1972, *On Peoples and Problems*, Hodder and Stoughton, London 1982, 2nd edition, 1983, pp. 60–67.
6. The Chipko movement has been the subject of several studies, of which the pioneering work was Ramachandra Guha (1989). On Silent Valley, see Zachariah and Suryamurthy (1994).
7. On the tiger, for instance, see the range of views in the debate in *Economic and Political Weekly*, 12 November 2005. For a biologist's perspective, see K. Ullas Karanth, 'Reconciling Conservation with Emancipatory Politics', pp. 4803–05. On displacement see Rucha Ghate, 'Relocation versus Wildlife Preservation', pp. 4807–09.
8. On wildlife, shifting cultivation and livestock respectively see Saberwal *et al.* (2000); Ramakrishna (1994); and Ghotge (2006).
9. Kalpavriksh and the Technical Support Group, *The National Biodiversity Strategy and Action Plan*, Delhi, 2005.
10. See the essay 'Landscapes and Peoples' in Romila Thapar (2002: 37–45).

## References


Agarwal, Anil, 1985, 'The Politics of Environment, II' in Anil Agarwal and Sunita Narain (ed.), *The State of India's Environment: The Second Citizen's Report, 1984–85*, Centre for Science and Environment, Delhi.

Agrawal, Arun, 2006, *Environmentality: Technologies of Government and the Making of Subjects*, OUP, Delhi.

Carson, Rachel, 1962 (1987), *Silent Spring*, special anniversary edition, Houghton Mifflin, Boston.

Commoner, Barry, 1971, *The Closing Circle*, Alfred Knopff, New York.

Deloche, Jean, 1993, *Transport and Communication in India Prior to Steam Locomotion, Volume I*, OUP, Delhi.

Elvin, Mark, 2004, *The Retreat of Elephants: An Environmental History of China,* Yale University Press, New Haven and London.

Gadgil, Madhav, 2001, *Ecological Journeys*, Permanent Black, Delhi.

Ghotge, Nitya, 2006, *Livestock and Livelihoods: The Indian Context,* Foundation Books and the Centre for Environmental Education, Delhi.

Grove, Richard, 1995, *Green Imperialism*, OUP, Delhi.

Guha, Ramachandra, 1989, *The Unquiet Woods*, OUP, Delhi.

———, 2000, *Environmentalism: A Global History*, OUP, Delhi.

——— and J. Martinez-Alier, 1997, *Varieties of Environmentalism, North and South*, OUP, Delhi.

Guha, Sumit, 2001, *Health and Population in South Asia: From Earliest Times to the Present*, Permanent Black, Delhi.

Khilnani, Sunil, 1997, *The Idea of India*, Penguin, Delhi.

Krishen, Pradeep, 2006, *The Trees of Delhi: A Field Guide*, Dorling Kindersley, Delhi.

Lear, Linda, 1997, *Nature's Witness: A Life of Rachel Carson*, Henry Holt, New York.

Mansingh, Jasjit, 2002, *Oona: Mountain Wind*, Bluejay, Delhi.

McNeill, Robert, 2000, *Something New Under the Sun: An Environmental History of the Twentieth Century*, Penguin, Allen Lane, London.

Mithin, John, 2003, *After the Ice: A Global Human History, 20,000–5,000 BC*, Weidenfield and Nicholson, London.

Ramakrishna, P.S., 1994, *Shifting Cultivation in North East India,* UNESCO, Paris and OUP, Delhi.

Rangarajan, Mahesh, 2006, 'Ideology, Policy and the Environment: Indira Gandhi, 1966-84', *India International Centre Quarterly,* Summer 2006.

Saberwal, Vasant K., Mahesh Rangarajan and Ashish Kothari, 2000, *Towards Coexistence: People, Parks and Wildlife,* Orient Longman, Delhi.

Sen, Amartya, 2006, *The Argumentative Indian: Writings on Indian Culture, History and Identity,* Penguin, Delhi.

Shabecoff, Philip, 2003, *A Fierce Green Fire: The American Environmental Movement*, revised edition, Island Press, Washington.

Thapar, Romila, 2002, *The Penguin History of Early India from the Origins to 1300 CE*, Penguin, Delhi.

Trautman, Thomas, 1982, 'Elephants and the Mauryas', in SN Mukherjea (ed.), *India: History and Culture, Essays in Honour of A.L. Basham,* Firma L Mukhopadhyay and Co., Calcutta.

Tucker, Richard, 2000, *Insatiable Appetite: The U.S. and the Ecological Degradation of the Tropical World*, University of California Press, Berkeley.

Weiner, Douglas, 2001, *A Little Corner of Freedom*, University of California Press, Berkeley.

Williams, Michael, 2000, *Deforesting the Earth, From Pre-history to Global Crisis,* University of Chicago Press, Chicago.

Zachariah, K. C., and S. Suryamurthy, 1994, *Science for Social Revolution*, Sage Publications, New Delhi.

# Section I

# Pre-colonial India

*Photo credit Pallava Bagla*

# Introduction

The British defeated Nawab Siraj-ud-Daulah's forces in a battle at Palashi, near Kolkata in 1757. Palash, of course, comes from the name of the flowering tree, *Palas*. Known to botanists under the inelegant name of *Butea frondosa*, its red flowers possess a rare beauty all of their own. Edible for monkeys, they can also be used to make *gulal*, the red colour powder used to smear friends in the spring festival of Holi.

Like the flowers gathered from *Palas* trees, over time humans had brought about changes in the land, long before the coming of the East India Company to these shores. Colonial rule and its multiple dispossessions, its subjugation of the waters, hills, lands, and peoples should not lead to a Golden Age view of the past that preceded it. The British were not the first European power in the land though they were the last and the most decisive: the Portugese who rounded Africa were by far the first to found settlements. In their wake they brought crops from South America. These soon became ingredients of dishes that make mouths water (like the potato or corn) or the eye (the chilli), or worse still endanger human health (tobacco). All were quickly taken up by peasants and grown extensively across much of India by 1600. These were not the first such infusions: sugar cane is from South-East Asia and the one-humped camel from lands to the west (Habib 1983).

Human imprints on the land were even older. Hominids may have come into India two million years ago. India ceased to be a pristine wilderness aeons ago. Even Stone Age peoples in the Vindhya hills hunted boars and deer, collected honey, and traded with plains folks 10,000 years ago. Their graphic rock-cut art in caves not far from Bhopal are witness to an era. Five millennia ago, herders in the Vindhyas were chopping palms to build stockades for sheep and lighting fires with dried dung to keep warm. They also had an animal central to much of South Asian life: the humped zebu cow (Mathpal 1984; Settar and Korisettar 1987). Hunts and herding could depress wild populations and fires lit by humans could change grassland micro ecologies.

Climates too have been in fluctuation for long. Glaciers, the moving rivers of ice in the high Himalaya, were at their maximum 20,000 years ago. By about 12,000 years ago they retreated and a warmer age known as the Holocene began. This was a larger global trend (Mithin 2003). Changes in rainfall patterns did not cease but could have huge impacts in a macro region where water, its scarcity or plenty, could make all the difference for plant, animal and human life.

Harappa and Mohenjo Daro were among South Asia's first cities. Their discovery of the 5,000-year-old civilization by archaeologists continues to cause ripples of excitement even today. This first wave of town-building in the Indus Valley fully came to light only in the century just past (Lahiri 2005). *Why* they ceased to be active centres has long been subject to controversy, with climatic change being *one* possible piece in the jigsaw puzzle. There was no one compelling reason. Climatic shifts evident via pollen analyses show very complex patterns that were not all in one direction. Misra's perceptive paper carefully sifts the evidence. Grains of pollen and other material remains are assembled detective-like to piece together patterns of a changing, in this case a collapsing past (see Chapter 1). Ecological change cannot be a simple substitute for larger explanations. But as we grow more conscious of our own ecological footprint, we cannot but look at the past for clues.

The ecologies of much of South Asia have undergone shifts but the broad pattern set then remains so to this day. What is striking is the continuity of different lifestyles even as innovation made new techniques of hunting, animal rearing or cultivation possible. This, as Makkhan Lal shows, remained the case even later. Iron axes helped clear forests along the Ganga valley and found new stable settlements but these were small and scattered. Compared with the present, the forest expanse was vast, the farms few and far between (see Chapter 2).

The material evidence can by the late centuries BCE be illumined with reference to literary evidence in a host of languages, Tamil, Pali, and Sanskrit being the chief ones. Thapar explores the forest-settlement interface taking on texts like the *Sakuntalam*, familiar to the reader. Rangarajan's extracts knit together evidence from the ancient and the medieval period to ask how humans and nature interacted. Divyabhanusinh focuses on sixteenth and seventeenth century Mughal hunts providing powerful, compelling evidence of landscapes that subsequently vanished (see Chapters 3, 4, and 5).

The vastness of India makes it inevitable that there will be gaps in time spans and regions. But the patterns are what matter: long centuries of interaction left the environment very different from say two millennia ago. But there were limits to these changes, imposed both by available technologies and by the vagaries of nature itself.

Colonial rule was consolidated after Palashi and came to dominate India till the mid-twentieth century. Its excesses should not obscure the fact of it not always being a land of plenty: paucity could be a recurrent pattern. Given the scarcity of water in a third of the land mass where the rainfall never totals more than 75 centimetres a year, centuries-old arrangements for saving water should come as no surprise. Rajasthan in the north-west is the subject of Kumar's enquiry. Even one region exhibits great diversity of ecology and also of human hierarchies (see Chapter 6).[1]

Equally important is how conflict and contest shaped the land. Every school child knows how kings were locked in combat over control of fertile river valleys, trade routes, and ports. But commoners could feel the negative impacts, as in eighteenth century western India. Powerful elements could corner fodder, armies shut off access to timber. Lower castes and women lost out the most. Sumit Guha shows how power and privilege could help a few to corner the wealth of nature more than most (see Chapter 7 by Sumit Guha in this volume).[2]

India before European rule had not one but many histories. Studies of the ecological interface with pre-colonial society, economy and state are still in their early days. A deeper understanding is still some time away. Little is agreed upon and there is much, much more to think about.

Yet the beginning holds promise. Like the *Palas* flowers that have but a short life as the sun beats down in an unforgiving Indian summer, the senses have awakened. The fact that the tree and the name survives must not delude us that all was static. This was a society with landscapes already in flux when colonial rule began.

## Notes

1. Also see his paper in *Conservation and Society,* 3,1 (2005), pp. 134–49. For its pioneering insights, see Anil Agarwal and Sunita Narain, *Dying Wisdom*, 1997, Centre for Science and Environment, Delhi.

2. Also see his 'Claims on the Commons, Political Power and Natural Resources in Pre-colonial India', *Indian Economic and Social History Review*, 39, 2 (2002), pp. 181–96.

## References

Habib, Irfan, 1983, *Atlas of Mughal India*, Oxford University Press, Delhi.

Lahiri, Nayanjot, 2005, *Discovering Forgotten Cities: How the Indus Civilization was Discovered*, Permanent Black, Delhi.

Mathpal, Yashodhar, 1984, *Prehistoric Rock Paintings of Bhimbetka*, Abhinar, Delhi.

Mithin, John, 2003, *After the Ice: A Global Human History, 20,000–5,000 BC*, Weidenfield and Nicholson, London.

Settar, S. and S. Korisettar (ed.), 1987, *Indian Archaeology in Retrospect*, Vol. I, Manohar, Delhi, pp. 151–257.

I

# CLIMATE, A FACTOR IN THE RISE AND FALL OF THE INDUS CIVILIZATION: EVIDENCE FROM RAJASTHAN AND BEYOND

**V.N. Misra**

About 50 years ago, Sir Aurel Stein (1931) and Sir John Marshall (1931), on the basis of their evaluation of the multiple archaeological evidence from Baluchistan and Sind, proposed that climate in the these regions during the Indus Civilization period was more wet than it is at present. This theory was accepted and supported by archaeologists like Stuart Piggott (1950) and Mommer Wheeler (1953), and it held unquestioned sway for three decades. Then in the 1950s, American archaeologists began taking an interest in the archaeology of the Indus Valley and neighbouring regions, and they brought the anthropological approach to bear on the archaeological problems of this region. Briefly put, this approach looked at cultural evolution in terms of cultural processes in contrast to the older British historical approach which laid more emphasis on events (Fairservis 1961: 51). The introduction of this new approach led to a questioning of several interpretations previously held more or less as facts (Dales 1964, 1966; Raikes 1964).

One of these interpretations was the theory of a more wet climate during the Indus Civilization. Robert L. Raikes, a hydrologist with keen interest in archaeology, and Robert H. Dyson, an archaeologist, teamed up in 1961 to critically examine various kinds of evidence adduced by Stein and Marshall in support of their climatic theory (Raikes and Dyson 1961). They came to the conclusion that alternative explanations were possible for each piece of evidence, and therefore, the theory of a more wet climate could

* Reprinted with permission from Nayanjot Lahiri (ed.), 2000, *The Decline and Fall of the Indus Civilization*, Permanent Black, Delhi, pp. 239–250.

not be accepted as proven fact. As an interesting coincidence, another American anthropologist, Walter A. Fairservis, Jr. (1961) also examined the same evidence at the same time and reached a similar conclusion. F.A. Durrani (1965) who did a similar exercise for the lower Indus Valley a few years later also arrived at an identical conclusion.

Except for a weak meteorological defence of the older theory by Ramaswamy (1968), the matter rested there for a decade until, in 1971, Gurdip Singh revived the theory of a more wet climate and rekindled interest in the subject (Singh 1971; Singh *et al*. 1974). Singh's theory differed from that of Stein and Marshall in three important respects:

1. The evidence for climatic change was based on palynological rather than on archaeological data.
2. The climatic change consisted of several fluctuations which covered almost the entire Holocene in north-west India and not just the Harappan period.
3. The fluctuations in rainfall were related not only to the growth and decline of the Harappan culture, but also to the origin of agriculture-based life in the eighth millennium BCE and to its expansion in the fourth and third millennia BCE in north-west India.

The palynological evidence for Singh's theory came from three salt lakes, namely, Sambhar (27° N; 75° E), Didwana (27° 20′ N; 74° 35′ E), and Lunkaransar (28° 30′ N; 73° 45′ E) and one fresh-water lake, Pushkar (26° 29′ N, 74° 33′ E) in Rajasthan. The first two lakes are situated in the semi-arid belt (25–50 cm average annual rainfall), the third in the arid belt (less than 25 cm rainfall), and the last in the semi-humid belt (50–60 cm rainfall). The climate evidence from the first three lakes only is relevant to archaeological history.

The climate sequence based on the pollen record from these three lakes can be briefly summarized as follows:

## Phase I (Before 8000 BCE)

Phase I, represented by wind-borne sand deposits at the base of lake sediments, was characterized by a severely arid climate which was unsuitable for habitation for a long time during at least the later part of the last glacial period.

## Phase II: Pollen Zone B (c. 8000 bce–c. 7500 bce)

Phase II is represented by the first sedimentation in the lakes. Rainfall at this time was at least 25 mm more than the present annual precipitation in western Rajasthan.

## Phase III: Pollen Zone B (c. 7500 bce–c.3000 bce)

A slight decline in rainfall is indicated at the beginning of Phase III, but it was not severe enough to substantially alter the ecological pattern established in Phase II. A noteworthy feature of this phase is an extra-ordinary rise in carbonized vegetable remains in the lake sediments at all sites. This rise is accompanied by the appearance of the *Cerealia* type of pollen. These two phenomena indicate scrub burning which probably formed the basis of a primitive cereal agriculture.

## Phase IV: Pollen Zone C (c. 3000 bce–c. 1000 bce)

Phase IV can be divided into three sub-phases: IVa (c. 3000 bce–c. 1800 bce); IVb (c. 1800 bce–c. 1500 bce); and IVc (c. 1500 bce–c.1000 bce).

### Sub-phase IVa: Pollen subzone CI: c. 3000 bce–c. 1800 bce

Sub-phase IVa is characterized by a rather sudden and considerable increase in rainfall. Annual average rainfall during this phase was probably at least 50 cm more than the present rainfall in the arid belt. Pollen of the *Cerealia* type, of the same size range as seen in the earlier phase, and the evidence of scrub burning in the form of carbonized remains, continue in this sub-phase. In chronological terms, sub-phase IVa sees the rise of pre-Harappan levels at Kalibangan together with the unbroken record of *Cerealia* type pollen in the pollen profiles and the evidence of forest burning from three different sites, leading one to believe that cereal cultivation perhaps does not start with the Indus Valley Civilization after all. It would seem, in fact, that the practise had existed in the region for a long time, indeed, as has been suggested, from the beginning of Phase III.

It can in fact be argued that the significant increase in rainfall at the beginning of the third millennium BCE attested by the palaeo-ecological evidence, played an important part in the sudden expansion of the Neolithic-Chalcolithic cultures in north-west India, ultimately leading to the prosperity of the Indus culture (Singh 1971: 188).

## Sub-phase IVb: Pollen subzone C2: c. 1800 BCE–1500 BCE

Sub-phase IVb is a short dry period. In the arid belt the Lunkaransar lake had started drying out while, in the semi-arid belt, the lakes began to turn saline. The beginning of aridity was not synchronous in the arid and semi-arid belts. At Lunkaransar it began around 2000 BCE while at Sambhat it began around 1000 BCE.

Commenting on the effect of aridity on Harappan culture, Singh says:

> The present evidence would suggest that the onset of aridity in the region around 1800 BCE probably resulted in the weakening of Harappan culture in the arid and semi-arid parts of north-west India but that the peripheral areas of the culture, such as in Gujarat and the Himalayan foot-hills were not affected to the same degree. The extinction of the Indus culture may thus have been initiated through gradual decline as a result of climatic change, but the process may yet have been completed by successive invasions from the north-west by the Aryans (Singh 1971: 188).

Further,

> This dry period (2000 BCE–1000 BCE) as already pointed out, was perhaps responsible for the wide cultural gap between the decline of the Harappan culture and the beginning of the succeeding Painted Grey Ware culture, generally put between 1000 and 600 BCE and tentatively associated with the colonization of impoverished land (Ghosh 1952). The weight of analytical and stratigraphic pollen evidence favours a climatic determination of the events of this period (Singh *et al*. 1974: 498–9).

## Sub-phase IVc: c. 1500 BCE—1000 BCE

Sub-phase IVc is represented by a slight reversal to a relatively weak wetter interval, lasting up to about 1000 BCE.

### Phase V: Early Centuries CE to the Present

Phase V is known only from Lunkaransar, and is dated, in the absence of $^{14}$C dates, on the basis of the rate of sedimentation. The pollen assemblage reflects more or less the present conditions in the area. There is hardly any evidence of either cereal cultivation, tree vegetation or the existence of any aquatic species in the area. The vegetation, on the whole, seems to comprise a typical 'Sand Formation' type, as presently seen in the Lunkaransar region (Singh *et al.* 1974: 496).

Singh's evidence for a wetter climate during the Harappan period has been questioned by scholars on palynological (Vishnu-Mittre 1972, 1978), archaeological (Pande 1977; Thapar 1977) and a combination of these and other grounds (Flam 1976; Seth 1978).

The purpose of the present paper is not to examine or question the evidence for climatic changes put forward by Singh. For one thing, not being a palynologist, I am not qualified to assess the pertinent data. For another, having examined the stratigraphic record at several salt lakes (Pachpadra, Sambhar, Kuchaman, Didwana, Tal Chhapar, and Malhar) and on numerous sand dunes in Rajasthan during the last five years, I am aware that the sedimentological evidence supports the palynological one for an increase in rainfall during the Holocene period though the degree and duration of this increase is a matter for debate.[1]

My purpose here is to examine the archaeological evidence in so far as it relates to the hypothesis that Singh has put forward: about decrease in rainfall around 1800 BCE which led to the decline of the Harappan Culture.

The following important points emerge from a survey of the geographical distribution of Harappan sites.

1. As in the case of the pre-Harappan farming-based settlements, Harappan sites are conspicuous by their total absence in west Rajasthan (except on the dry bed of the Ghaggar) where the lakes

that produced the climate sequence are located. This area receives an annual rainfall from less than 200 mm in the west to 400 mm in the east. Its southern part is drained by the Luni and its tributaries but not a single site of any phase of the Harappan culture has been found on any of them.

2. Similarly, Harappan sites (of all phases) are practically absent on the north Gujarat plain which is an extension of the west Rajasthan plain. This area receives an annual rainfall ranging from 400 mm in the west to 800 mm in the east and is drained by the Banas, the Sabarmati, the Mahi, and their tributaries.
3. In sharp contrast to the total absence of pre-Harappan and Harappan sites (of all phases) in these two areas is the great density of settlements of pre-Harappan, Early Harappan, Mature Harappan, Late Harappan, and even later sites (PGW and Medieval) in the Cholistan desert in Pakistan. This region has no active streams today. It receives less than 100 mm annual rainfall and is the most arid and inhospitable part of the Thar Desert (Mughal 1981).

   If increased rainfall was a causative factor in the emergence of agriculture, in the expansion of farming-based life, and in the development of Harappan culture surely one would expect a greater density of Neolithic-Chalcolithic and Harappan sites in the relatively congenial west Rajasthan and north Gujarat plans than in the hyper-arid Cholistan. But the archaeological evidence is quite contrary to that expected by Singh's hypothesis. It can therefore be safely asserted that increased rainfall was not a decisive factor in the rise and growth of the Harappan culture.
4. The densest distribution of Harappan sites is not on the Indus river and its tributaries but on the extinct Hakra-Ghaggar and its equally extinct tributaries. Of the over 800 Harappan sites (not including Degenerate Harappan or OCP known at present (Jansen 1980), more than 530 sites are located on the Hakra-Ghaggar system. If we add to this the nearly 200 Harappan sites from Kutch-Saurashtra and the nearly 70 Late Harappan sites from the Yamuna valley in Uttar Pradesh (U.P.), probably less than a 100 sites are left in the Indus valley proper and in Baluchistan.

The Harappan culture is, therefore, essentially a culture of the Hakra-Ghaggar valley, and any search for the cause or causes of the rise and fall of this culture must embrace an investigation into the fluctuations of the fortunes of this river as was rightly pointed out by Gupta (1978).

## Decline of Indus Civilization: A Fresh Look

Singh's hypothesis, namely, that a decrease in rainfall around 1800 BCE led to a decline of the Indus Civilization, can now be examined. Here it may incidentally be mentioned that the date of the decline in rainfall is not certain. While at Lunkaransar in the arid belt, the lake had started drying up around 2000 BCE, the same event at Sambhar, in the semi-arid belt, took place a thousand years later. It is difficult to explain such a long time difference between two points only 200 km apart. Singh's dating of this event therefore seems to be influenced by the need of reconciling the climatic evidence with archaeological data. But first let us examine the evidence for the decline of the Indus Civilization.

It is undeniable that in the eighteenth-nineteenth centuries BCE, Harappan cities declined, and some of them were even abandoned. In the lower Indus valley many causes for this event have been suggested: reduction in rainfall (Marshall 1931); exhaustion of the economic resources (Wheeler 1968; Fairservis 1961, 1967); excessive flooding (Raikes 1964; Dales 1966), and Aryan invasion (Wheeler 1953). Though some of these explanations have been questioned (Dales 1964, 1966; Lambrick 1967; Possehl 1967), the fact of decline and abandonment of the cities is accepted by critics. There is also an undeniable decline in material prosperity and in civic standards. But this decline did not lead to a decrease in population. It only forced the population to migrate from the lower Indus valley into Saurashtra and from the Hakra-Ghaggar valley into north Punjab, Haryana, and the upper Yamuna-Ganga doab. The sudden proliferation of Late Harappan sites in Saurashtra, north-east Punjab, and Haryana and the upper Yamuna-Ganga doab attests to this migration.

At least in the case of north Punjab and the upper Yamuna-Ganga doab, rainfall is not significantly higher than in the adjoining parts of Punjab and Haryana which had witnessed dense human settlements during the Early and Mature Harappan times. Therefore, these new regions of colonization could not have conferred any significant advantages to the immigrants. On the other hand, the shift of the courses of the Yamuna and Sutlej to the east and west, respectively, would have considerably reduced the availability of both surface and sub-surface water in the Ghaggar valley. This would, in due course, have adversely affected both natural vegetation and agriculture, and forced the population to shift to areas like north Punjab (Sutlej channel) and the upper Yamuna-Ganga doab where the rivers provided the ecological conditions the Harappans had long been accustomed to exploit. It is, therefore, quite unnecessary to invoke the deterioration of climate to explain this migratory phenomenon. In the case of the lower Indus valley, some or all of the various explanations suggested, could have accounted for the shift of the Harappan population into Gujarat.

The phenomenon of Degenerate Harappan (same as OCP or OCW) culture represents a continuation of the eastward migration of the Late Harappans. This would appear to be due to excessive population pressure in the limited land available in the upper doab. The density of Late Harappan sites in the districts of Ludhiana and Saharanpur is eloquent proof of this demographic pressure.

## Aridity in the Second Millennium: A Fresh Look

The final hypothesis of Singh is that aridity between 2000 and 1000 BCE is responsible for the 'hiatus' between the Harappan and PGW cultures. This too does not stand the scrutiny of new archaeological evidence. Recent excavations by the Archaeological Survey of India at Bhagwanpura, in the district of Kurukshetra (Haryana); Dadheri in the district of Ludhiana; Nagar and Katpalon, both in the district of Jullandur (Punjab), and Manda on the Chenab in Jummu have shown that the PG culture coexisted with the Late Harappan at these sites (Joshi 1977, 1978; Lal 1979). Gupta (Gupta and Ramachandran 1977) suggests 1300 BCE for this overlap phase. This new evidence eliminates the 'hiatus' between the Harappan and PGW cultures and effectively knocks the bottom out of the desiccation hypothesis of Singh. Even in the north Rajasthan part of the Ghaggar valley the new evidence available suggests that there was no 'hiatus' between the Harappan and the PGW cultures. Black and Red Ware has been found in profuse quantities in association with Painted Grey Ware both in surface explorations and in the excavation at Sardargarh in this region (Pande 1977: 56). Archaeological research during the last two decades has amply demonstrated that in north-east Rajasthan (Jodhpura and Noh) and in the central Yamuna-Ganga doab (Atranjikhera), Painted Grey Ware was preceded by an independent phase of Black and Red Ware culture which, in turn at many sites, was preceded by OCP or Degenerate Harappan. Fresh field research in the Ghaggar valley in Rajasthan and a re-evaluation of the older evidence is most likely to show a continuity of occupation in this area from Harappan to Painted Grey Ware.

However, there is evidence to show that population during the Painted Grey are period was much sparser in the Ghaggar valley. In the Pakistan part of this valley, only 14 sites of this culture have been found and all this occur in the eastern part (southeast of Bahawalpur) near the Indian border. This is in sharp contrast to the dense concentration of Mature Harappan sites further down in the same valley (southwest of Bahawalpur) (Mughal 1981: map). Further, Ghosh (1952) has shown that the Painted Grey Ware sites are often located on the Ghaggar bed. These two factors show that during the PGW period: (*a*) the Ghaggar (or Saraswati) no longer flowed along the way to the sea, but only in the upper part of its course; and (*b*) that its original bed had considerably shrunk. This must naturally have been the consequence of the diversion of the Yamuna from the Ghaggar channel to its own present channel.

## NOTE

1. Since 1977, the writer is directing a multi-disciplinary research project on 'Early Man and His Environment in North-West India'. Other members taking part in this project are S.N. Rajaguru (Deccan College, Pune), D.P. Agrawal (Physical Research Laboratory, Ahmedabad), R.P. Dhir (Central Arid Zone Research Institute, Jodhpur), and R.J. Wasson and Gurdip Singh of the Australian National University, Canberra. Besides conducting extensive explorations in western Rajasthan and excavation at Jayal and Didwana in the district of Nagaur, the research team has examined the stratigraphy of all the major salt lakes of Rajasthan. In January 1980, largely on the initiative of Wasson and Singh, we dug two wells in the Didwana lake to collect fresh samples for palynological, sedimentological, and radio-carbon analysis. The palynological data is being analysed by Gurdip Singh, sedimentological by Wasson and Rajaguru, and radiocarbon samples by Agrawal. Preliminary results of these studies show that the lake had already started carrying water in the terminal Pleistocene.

## References

Dales, G., 1964, 'The Mythical Massacre at Mohenjodaro', *Expedition* 6(3), pp. 36–43.

———, 1966 'The Decline of the Harappans', *Scientific American* 241(5), pp. 92–100.

Durrani, F.A., 1965, 'Climate of the Lower Indus Valley in Ancient Times', *Journal of the University of Peshawar* 10, pp. 33–7.

Fairservis, Jr. W.A., 1961, 'The Harappan Civilization: New Evidence and More Theory', *Novitates* 2055, New York, pp. 1–35.

———, 1967, 'The Origin, Character and Decline of an early Civilization', *Novitates* 2055, New York, pp. 1–35.

Flam, L., 1976. 'Settlement, Subsistence and Population: A Dynamic Approach to the Development of the Indus Valley Civilization', in K.A.R. Kennedy and G.L. Possehl (eds), *Ecological Backgrounds of South Asian Prehistory*, Cornell, pp. 76–93.

Ghosh, A., 1952, 'The Rajputana Desert—Its Archaeological Aspect', *Bulletin of the National Institute of Sciences of India* I, pp. 37–42.

Gupta, S.P., 1978, 'Origin of the Form of Harappa Culture: A new Proposition', *Puratattva* 8 (1976-76, pp. 141–6.

———, and Ramachandran, K.S., 1977, *Mahabharat: Myth and Reality*, Delhi.

Jansen, M., 1980 'Settlement Patterns in the Harappan Culture', in H. Hartel (ed.), *South Asian Archaeology 1979*, Berlin, pp. 251–69.

Joshi, J.P., 1977, 'Overlap of the Late Harappan Culture and Painted Grey Ware Culture in the Light of Recent excavation in Haryana, Punjab and Jammu' in a seminar on the 'Indus Civilization: Problems and Prospects', held at the IIAS, Simla.

———, 1978, 'Interlocking of Late Harappa Culture and Painted Grey Ware Culture in the Light of Recent excavations', *Man and Environment* 2, pp. 98–101.

Lal, B.B., 1979, 'West was West and East was East, But, When and How did the Twain Meet? The Role of Bhagwanpura as a Bridge between Certain Stages of the Indus and Ganges Civilizations', in the Conference on 'The Harappan Civilization: A Contemporary Perspective', Srinagar.

Lambrick, H.T., 1967, 'The Indus Flood Plain and the "Indus" Civilization', *The Geographical Journal* 133(4), pp. 483–95.

Marshall, J., 1931, *Mohenjodaro and the Indus Civilization*, 3 vols, London.

Mughal, M.R., 1981, 'New Archaeological evidence from Bahawalpur', in A.H. Dani (ed.), *Indus Civilization: New Perspectives,* Islamabad, pp. 33–42.

Pande, B.M., 1977, 'Archaeological Remains on the Ancient Saraswati', in D.P. Agrawal and B.M. Pande (eds), *Ecology and Archaeology of Western India*, Delhi, pp. 55–60.

Piggott, S., 1950, *Prehistoric India*, Middlesex.

Possehl, G.L., 1967, 'The Mohenjodaro Floods: A Reply', *American Anthropologist* 69(1), pp. 32–40.

Raikes, R.L., 1964, 'The End of the Ancient Cities of the Indus', *American Anthropologist* 63(2), pp. 284–99.

———, and Dyson, Jr. R.H., 1961, 'The Prehistoric Climate of Baluchistan and the Indus Valley', *American Anthropologist* 63(2), Part I, pp. 265–81.

Ramaswamy, C., 1968, 'Monsoon Over the Indus Valley During the Harappan Period', *Nature* 217 (5129), pp. 628–29.

Seth, S.K., 1978, 'The Dessication of the Thar Desert and Its Environs During the Protohistorical and Historical Periods', in W.C. Brace (ed.), *The Environmental History of the Near and Middle East,* London, pp. 279–305.

Singh, G., 1971, 'The Indus Valley Culture (Seen in the Context of Post-Glacial Climate and Ecological Studies in North-West India)', *Archaeology and Physical Anthropology in Oceania* 6(2), pp. 177–89.

———, Joshi, R.D. Chopra, S.K. and Singh, A.B., 1974, 'Late Quaternary History of Vegetation and Climate of the Rajasthan Desert, India', *Philosophical Transactions of the Royal Society of London* 267 (889), pp. 467–501.

Stein, A., 1931, *An Archaeological Tour of Gedrosia*, Memoirs of the Archaeological Survey of India 43, New Delhi.

Thapar, B.K., 1977, 'Climate During the Period of the Indus Civilization: Evidence from Kalibangan', in D.P. Agrawal and B.M. Pande (eds), *Ecology and Archaeology of Western India*, Delhi, pp. 67–73.

Vishnu Mittre, 1972, 'Palaeobotany and the Environment of Early Man in India', in S.B. Deo (ed.), *Archaeological Congress and Seminar Papers*, Nagpur, pp. 206–12.

———, 1978, 'Palaeoecology of the Rajasthan Desert During the Last 10,000 Years', *The Palaeobotanist* 25, pp. 549–58.

Wheeler, M., 1953, *The Indus Civilization*, Cambridge.

———, 1968, *The Indus Civilization*, 3rd edition, Cambridge (reprint 1979).

# 2

# Iron Tools, Forest Clearance and Urbanization in the Gangetic Plains

Makkhan Lal

## Introduction

In the context of urbanization in the Gangetic Plains during the middle of first millennium BCE it has been repeatedly argued by some archaeologists and historians that the urbanization was possible only because of wide-spread use of advanced iron tools which helped the people in clearing the then dense monsoonal forest of the Gangetic Plains and widespread use of iron tools was possible only when Bihar ore mines could be exploited in the middle of first millennium BCE (Agrawal 1969, 1971, 1973, 1983). However, some scholars argued that it was not the technology alone but the establishment of strong social and political institutions which brought the urbanization at home (Chakarabarti 1973a, 1973b; Ghosh 1973; Ray 1978). Ray (1978: 134) stated that slowly and steadily forests were cleared for the habitation and agricultural purposes and there is as yet no absolute cause for iron technology effecting the clearance of forests. Present writer's interest in the settlement patterns and systems in the Ganga Plains during the second and first millennium BCE, first hand knowledge of the setting of archaeological sites, excavated archaeological materials, and intensive explorations in the Ganga Plains lead to the conclusion that the so-called extensive use of iron tools and the large-scale forest clearing for human settlements and agricultural land is nothing but a myth.

The problem of clearing of forests for human settlement and agricultural land and a few references found in literature viz., *Satapatha Brahmana*, have been blown out of proportion. A careful analysis of archaeological data,

* Reprinted with permissions from *Man and Environment* Vol. 10, 1986, pp. 83–90.

location and density of settlements, an approximate estimate of population on the basis of size of settlements and an idea of land requirement for the agricultural purposes to support the population are bound to give us a fairly good idea about the problem of large scale use of iron tools, deforestation, and urbanization.

In the present paper, I propose to discuss these problems in four parts. Part I gives an idea of flora of the upper Ganga Plains as it is today and what it may have been in the past. In part II, iron technology and the frequency of tools in PGW (Painted Grey Ware) and NBPW (Northern Black Polished Ware) periods have been discussed. The part III deals with the colonization and the density of settlements during NBPW period in upper Ganga Plains. In part IV, taking a maximum population in the village or a settlement, its land requirement has been computed on the basis of need of foodgrains and the yield per acre and lastly the actual forest clearing that may have taken place during the middle of first millennium BCE. Taking into consideration above factors conclusions have been drawn about the urbanization in the Gangetic Plains.

## Forests of the Gangetic Plains

The vast alluvial plains of northern India as we see them today are the result of continuous deforestation for the last four millennia. In the traditional literature we have ample evidence for the Gangetic plain being covered with dense monsoonal forests. The *Satapatha Brahmana* (14.1.4.14, 15, 16) mentions that area between Saraswati and Sadanira (modern Gandak) river was dense forest and aryanization of this area was possible only after burning of the forests. According to the Mahabharata, Hastinapur, the Capital of Kurus was situated in a forested area. It also mentions that the kingdoms of Kurus and Panchalas—covering roughly the upper Gangetic Plains—were located in forested area. *Devipurana* (chapter 14) mentioned nine sacred forests of which Kuru Jungle, Nimisa (Naimisaranya) and Utpalaranya covered the upper Ganga Plains. Naimisaranya has been identified with modern Nimsar, 20 km northwest of Sitapur. This was the forest where several thousand sages lived and many Puranas were written. Utapalaranya was the forest where sage Valmiki lived and Sita gave birth to twins known as Lava and Kusha. This place is identified with modern Bithur, 20 km north of Kanpur where there is still a Valmiki Ashram.

Pollen records from Hastinapur excavations show the presence of *Pinus*, *Dalbergia sissoo*, and *Holorabera antidysentrica* (Chaudhary *et al.* 1977; Lal 1954–55: 120). Samples of charred wood from OCP to NBPW levels at Atranjikhera belong to *Pinus roxburghii*, *Dalbergia sissoo*, *Tectona grandis, Bochmeria platypyla*, *Cedrus deodara*, *Cupressus torulosa*, *Terminalia tomentosa, Tamarix articulata, Dendrocalamus strictus, Acacia nilotica,* and *Acacia arabica* (Chaudhury *et al.* 1977). The presence of these species some of which are now confined only to the *Tarai* region shows that the doab was then quite densely forested during the second and first millennium BCE. Pant says (Agrawal 1977: 2251) that in the past the fauna of Siwaliks included carnivores, monkeys, elephants, and ungulates suggesting that the Indo-Gangetic Plains had a thick cover of flora like that of *Tarai* and *Bhabhar* regions. In fact, the forest cover was so thick even in the fifteenth-sixteenth centuries CE that wild elephants, buffaloes, lions and tigers were hunted in the upper Ganga Plains (Abul Fazal, *Ain-I-Akbari*, pp. 283–93). But now the once rich and varied fauna has been reduced to small numbers of deer, antelopes, boars, hyenas, wolves, foxes, jackals, and monkeys.

The remnants of early forests can be seen in the patches and the vegetation includes *Butea frondosa, Caserica tomentosa, Alienthus excels, Woodfordia floribunda*, *Acacia laucophoria*, *Terminalia belerica*, *Tamarindus indica*, *Basia latifolia*, and species of *Bauhinia* and *Eugenia* (Puri 1960: 212). *Bombax malabaricum, Adina cordifolia, Lagerstroemia perviflora, Lannea grandis, Gamalina arborea*, *Casia fistula*, *Mallotus philippinensis,* and *Calicarpa microphyla* are now confined to the most part of the sub-Himalyan tract (Puri 1960: 251). In the central part of the Ganga–Yamuna doab thorny shrubs such as *Caparis aphylla, Prosopis spicigera, Tecoma undulate,* and *Acacia* which form an extension of semi arid vegetation of Rajasthan.

In fact no other part of India has undergone such a drastic change due to tillage of jungle land in the past two centuries. Considerable damage to the forests was done by the railways also which obtained their fuel supply directly from local timber until a century ago (Whitecombe 1972: 95). The intensive colonization and large scale deforestation has brought the balance between man and vegetation to a very delicate stage. Mukerjee (1938: 97–103) sums up:

> Nowhere is this interaction between man and vegetation so vividly illustrated over an extensive area than in the Gangetic plains. The balance between the

> progressive tendencies of vegetation and retrogressive influence of man is no where more evident than in the vegetation of this region. Throughout the Gangetic Plain may be said to have attained an apparent equilibrium in relation to variation in rain fall and temperature which at any moment is liable to be upset...On the plains the vegetation is rather delicately balanced against man at about thorn-scurb state. The soil over most of the Indo-Gangetic plains seems to be supporting all the human and bovine life that is possible under existing methods of exploitation.

Thus it is very clear that until during the beginning of centuries CE the forests were almost undisturbed except for small pieces for the land that might have been required for the agricultural purposes and for the settlements. The presence of lions, tigers, elephants etc., even uptil the sixteenth century indicates that forest was fairly intact well preserved. As a matter of fact the forests disappeared rapidly only in the last two centuries.

## II

After having an idea of the types of forests we had in the Gangetic plain in this section we shall consider the typology and the frequencies of iron tools found in GW and NBPW period to see if there is any difference between the two as it has been often postulated that it was with the use of new type of iron tools on a large scale that the NBPW people cleared the forests for extensive agricultural purposes which lead to urbanization.

The champions of the view that there has been fundamental changes in iron technology and tool typology from PGW to NBPW period and it was after the opening of Bihar ore mines the iron tools came to be extensively used never give any concrete evidence in support of their view. It is my feeling that most of such opinions have been expressed on incomplete information and many presumptions.

First of all let us accept the fact that after the publication of Hastinapur report not a single excavation report came out from where a comparative picture of PGW and NBPW periods could be drawn. Fortunately, now we have Atranjikhera excavations report with full details about every iron piece found in PGW and NBPW periods. We are really in a good position to draw a comparative picture of iron tools typology,. technology, and its extensiveness.

The table I shows the various tool types and other objects found in PGW and NBPW periods. The total area excavated for PGW is about 1,000 sq m (Gaur 3: 122) while that of NBPW is more than double (personal communication). In all there are 16 types of objects in PGW with a total of 135 pieces and there are three pieces in NBPW period. From this total if we remove slag pieces and the lumps then we are left with 16 pieces for PGW and 252 pieces for NBPW period. Fifteen new types that have been introduced in NBPW period mostly do not come in tools category and also number at 43 only. Also another fact is that most of these were introduced in Phase B and C which are dated between 500 and 200 BCE i.e. when the urbanization was already an established fact. It may be emphasized that all the forest clearing tool kit was very much there in the PGW period and there is no appreciable change in their number from PGW to NBPW period. Though the number of iron tools in NBPW is almost double but when we consider the fact that the excavated area was also almost double and the smaller objects like nails etc. have increased the number, we find that there has been neither any change in technology nor in the extensive use of tools; only its ghost has always haunted us.

Also the upper Ganga Plains did not have to look towards Bihar are mines to be opened as late as fifth–sixth centuries BCE for the extensive use of iron tools and implements. The presence iron objects in such profusion and the discovery of furnaces at Atranjikhera and, Noh and certain specific tools used by blacksmiths at Atranjikhera make it very clear that not only iron tools and other objects were manufactured here but also the smelting of ore was also done. The main source of iron ore was perhaps the region extending between Agra and Gwalior. The hills of this region are extremely rich in iron ore in which iron content is very high. The iron industry of Gwalior has been noted by Abul Fazal in his *Ain-i-Akbari*. Sir Alexander Cuningham (1864: 235) also commends the quality of Gwalior and Narwar iron and regrets the decline of the industry due to the scarcity of charcoal for smelting as the forests were being cut down.

Another fact toward which I would like to draw the attention of scholars is the location and frequency of NBPW settlement in U.P. and Bihar. In both the states combined there are nearly 500 settlements of which less than 100 are in Bihar and rest 400 in U.P. (Sharma 1983: 113). All the sites in Bihar are located in Arah, Chhapra, and Patna districts with a very few beyond this region. It is also evident that no NBPW site has been found in iron ore area. A detailed distribution map of NBPW sites according to their known size shows that there is no truth whatsoever, in the view that there was the nuclear zone of NBPW culture. Also it should not be forgotten that in Bihar from none of the NBPW site any appreciable number of iron tools have been found and neither the sites in Bihar are older than those found in the Upper Ganga Plains.

From the above discussion it is clear that the opinion that Bihar iron ore mines had a big role to play in urbanization does not stand the scrutiny of evidence.

## III

## Evidence for Intensive Colonization

To understand the settlement patterns and systems in the Upper Ganga plains over a period of nearly two thousand years i.e. in the second and first millennium BCE, a PhD project was taken up by the present author (Lal 1982, 1984a, 1984b). To understand the pattern of the change and also the density of settlements we should have, if not total, near total number of settlements of the period concerned. To explore whole of the Upper Ganga Plains is simply beyond the capacity and resources of a single explorer. Thus in such a situation a methodology was followed which is considered to be most reliable and ideal, i.e., it was decided to go for sampling and explore an area which could represent the plain on a miniature scale. As the methodology and the general results have already been published (Lal 1984a, 1984b), I shall confine here myself only with the results for the NBPW settlements which are directly relevant to us here.

In an intensive village to village survey in Kanpur district, 99 NBPW settlements were discovered. Of these 55 settlements are on the river and 34 away from the rivers. The settlements on the river banks are located on the elevated land, obviously to remain away from the range of annual flood. River-wise distribution shows that there are 20 settlements on the Ganga, one each on the Isan and Northern Non, four on the Pandu, 25 on the Rind, two on the southern Non, and three on the Yamuna.

The break-up of settlements according to size shows that of 99 settlements 41 are below 1.00 hect. and 40 between 1.00 and 2.00 hect. size. In other words there are as many as 81 settlements which were below 2.00 hect. size and each could have accommodated not more than 400 people (for details see Lal 1984c). There are only four settlements which were between 2–4 hect. size, i.e., having a population between 500 and 1,000 people. There are only four settlements which are above 4.00 hect. size and could have accommodated a population between 1200 and 1300 except in the case of Radhan there the resident population should have been more than 2000. This settlement is the biggest one found in Kanpur district. It is very apparent that most of the bigger settlements were located on the river Ganga along the banks of which open agricultural land, games, and other habitational facilities were more attractive and profitable than on the other rivers or away from the rivers. The settlements away from the river are mostly smaller in size.

One of the most important factors that we have to keep in mind is that the average spacing between the two settlements during NBPW period was 9 km though some times they were as close as 2 km and as far away as 12 km.

## IV

Now with a proper background of the location and density of settlements and the spacing between the two settlements we shall proceed to evaluate the land requirement for a settlement and what could have been a territorial limit for the settlements and the actual forest area that might have been cleared for human settlement and agricultural land.

### Catchment and Exploitation Area of a Settlement

The concept of effective area being exploited by a settlement was first formulated by C. Vita-Finzi and Eric Higgs. They used the term 'Site Catchment Area' and defined it as 'the study of relationship between technology and those natural resources lying within the economic range of individual sites' (Vita-Finzi and Higgs 1970). In other words the catchment area of an archaeological site is that area from which a site (more properly the inhabitants of a site) derive its resources. In broad outlines site catchment analysis delimits a territory or a set of concentric circles surrounding a site form where greatest quantity of resources were derived. The territory exploited from a site will vary according to a number of factors the most obvious among them being the nature of economy, the size of population, and the characteristics of terrain. However, the most important aspect of this exploited territory both in theory and practise is the capacity of the human population to exploit an area effectively.

This effective exploitation has another facet. 'Further the area is from the site, less likely to be exploited and less rewarding is its exploitation (unless it is particularly productive) since energy consumed in movement to and from the site will tend to cancel that derived from the resource' (Vita-Finzi and Higgs 1970:7). Chisolm's study of agricultural exploitation on the global level shows that 'as the average distance increased, per cent return declined' and it is exceptional for the farmers to travel more than 3–4 km to their fields' (Chisolm 1968 : 43–68 : Table 2).

In such a situation if we know the area and the land required for a particular size of settlement population it should not be difficult to compute the food grains needed and the agricultural land required.

In the case of district Kanpur on the basis of size of implements I have attempted to compute the size of population that might have been there (Lal 1984c). Here I shall take the site of Radhan (see site No. 115 in Lal: 1984c) as a model and try to compute the land requirement for the agricultural purposes.

At Radhan during NBPW period a maximum population that could have existed at a particular time is around 2400 people. This figure should not be out of reach for a site which was 8. 75 hect. in size.

## Requirement of Food Grains

A variety of cultivars have been found in archaeological excavations of protohistoric and early historic sites in the Ganga plains. They include rice, barley, eat, gram, lentils, millet, and peas (Chaudhury *et al.* 1977; Kajale 1974). Of these wheat, barley, and rice must be considered staple food. In addition cattle, sheep, and goat were the main source of milk and meat. Hunting and fishing must have also contributed to food economy.

Dhavalikar and Possehl (1974: 41) have estimated the diet of an individual as follows:

> Minimum requirement for long term survival are about 1, 700 calorie per day taken with reasonably balanced diet. Good health can be achieved with about 600 calories more. Based on this fact the ethnographic record of somewhat modern comparable cultures the estimate presented here is for 2, 000 calories. This figure has the advantage of being a mean estimate so that if it misses, the resulting distortion would be minimized. The 2,000 calories have distributed as follows:

| | |
|---|---:|
| All grains | 1,350 calories |
| Vegetables | 100 calories |
| Meat (eggs and fish etc) | 200 calories |
| Fruits and sweets | 50 calories |
| Dairy products | 300 calories |
| Total | 2,000 calories |

The content of 1 kg grains is approximately 3,350 calories. Within the diet worked out here this would feed 2.5 persons for one day. By multiplying this with 2,400 (the estimated population of settlement) the daily grain requirement some to 960 kg which is also the daily food requirement. Multiplying this figure by 365 days we get 346,750 kg, which will be the yearly food grain requirement.

## Land Requirement

It seems probable that there would be no spread of cultivation beyond the nucleus site before the evolution of cultivars capable of giving acceptable yield on less fertile and durable soils without seepage. This raises the question as to what may be an acceptable yield? If we are to answer this question and other related to quantitative analysis we must make some assumptions and drew analogies from modern comparable examples. Archaeology does not provide us with basic data on yields of early cultivars or on the systems of field management, seed rates etc.

Grains and vegetables are the two main features of this diet, which would require fair amount of farm land. For grains yields were computed for ten years from *fasali* 1305–1314 and taken from the *Gazetteer of Kanpur District* (1929). The average was combined and after deducting for seepage, regular seasonal disorder etc., the yield of 600 kg per acre was arrived at (for details see Lal 1984a: 240–42).

The area required for the cultivation can now be computed from the yield per acre and the total grain requirement for one year. This comes to 578 acres. To this estimate something must be added for seeds, spilage, spoils, loss due to rodents and disease, and processing. Vegetables, millet, and pulses would require additional land. A figure of 0.25 acre per person satisfies the grain requirements. If 25 per cent of this figure is added for vegetable crops (0.06 acre), 15 per cent for loss (0.04 acre) and 15 per cent (0.04 acre) for seeds the total land requirement comes to about 0.4 acre per person. By this estimate the total land requirement for 2,400 persons would be 960 acres. However, in this estimate no provision has been made for fallow land. If the land is left dormant every year it doubles the area required which is 1,920 acres and if the fallow period is of two years the land required would be 2, 880 acres.

Before we express this land requirement in terms of radius around the settlement we have to consider one fact. On one side of the settlement if the river Ganga, the bed of which is at places more than 3 km wide. Between the settlement and the regular course of the river a strip of land of 150–300 and at places more than this comes up after the monsoon flood. This strip is rich in organic matter and very fertile. If we draw a circle around the settlement one-third of it would go in the river and not more than two-third will be available for the cultivation. Taking into consideration this fact when we draw a circle of land requirement around the settlement it comes to 1.8 to 2.6 km in radius.

Similarly for a settlement of 1,000 population land requirement would be between 800 and 1,200 (1.3–1.6 km radius), for a settlement of 3,000 population 2,400 to 3,600 (2.2–2.7 km radius) and for a smaller settlement with 200 to 500 population the land requirement would be 250 to 1000 acres i.e., the land within a radius of 1 km radius.

It may be pointed out that in Kanpur district, the total area of which is 6,167 sq km, maximum population that might have existed at any given time during NBPW period on 99 settlements was nearly 4,000. To support this population the land requirement was never more than 122 to 184 sq km (Lal 1984a: Table 49, pp. 244–47) and most of this land was available on the soft alluvial soils along the rivers and lakes and other open areas in the forests.

## Actual Forest Clearings

Now we shall turn our attention towards the forest clearings that might have been done to acquire agricultural land. As we have already seen a good number of settlements are located on the rivers and lakes along which open spaces are available for the settlements and agricultural land. Even deeper into the forests we find that there are lakes and open spaces where settlements came up without disturbing the original forest.[1] Also in the case of NBPW we know that average spacing between the two settlements was 9 km and for 80 per cent settlements land requirement was rarely more than 600–700 acres i.e., within the radius of 1km from the land around the settlement. Thus between the two settlements minimum 8 km of forest remained before the requirement for the agricultural land arises for other settlement. Most of the agricultural land came from the open land and not by clearing the forest. It was absolutely not necessary to go for the clearing of forest, it was not required even at the climax of urbanisation period.[2] And I do not think that any community would go for destroying the forests unless it is absolutely needed. This is very much attested in the traditional literature. *Arthasastra* refers to the protection of forests by the State. The reference of *Satapatha Brahmana* should be taken a little more carefully like any other literary evidence has been taken in Indian context.

## Conclusion

We may you now conclude the results of the above analysis as follows:

1. That in the Ganga Plains thick forest cover remained as late as sixteenth and seventeenth centuries CE and it would be erroneous to believe that large scale forest clearing was done in the middle of first millennium BCE.
2. The there was no fundamental change in iron tool typology and technology and their extensive use as forest clearing tools from PGW to NBPW period.[3]
3. That whatever the new types of iron tools were introduced in NBPW period were in the mid and late phase when the urbanization was already an established fact.

4. That during the period of urbanization and intensive conlonization fairly good number of settlements were on the river banks and lakes where open land was available and the average spacing was nearly 10 km.
5. That nearly 80 per cent of the NBPW settlements were small villages having a population less than 500 people.
6. That the land requirements for the settlements indicate that rarely land was required beyond a radius of 1 km and most of it was available without forest clearing.
7. That there was never any need for the large scale forest clearings in any period. Whatever little forest clearing was needed could be done with technology available to PGW people.
8. That the extensive use of iron plough share should not he over emphasized to till the hard calcareous soil as there was very little of it to till.
9. That the urbanization of the Ganga plain was not due to the iron technology and the over exploitation to of the extensive tilling of the agricultural land for the surplus[4] but mainly due to the culmination of several social, political, and economic factors that were operating over many preceding centuries to channelize the energy and resources and generate a need and circulation of the surplus.[5]

## Notes

1. While carrying out fieldwork in Bastar and Koraput districts this was my experience. Large settlements are located deeper into the forests and the agricultural land was available to them near the marshy patches and the lakes. They have rarely gone for deforestation for the agricultural land. Whatever deforestation has been done is by the wood contractors who are supplying the wood for construction purpose and fuel.
2. While some of the proponents if iron technology and surplus like to argue in the favour of extensive agriculture I may point out that the surplus is always available by the intensive agriculture and not by extensive agriculture/where the input cancels out the output (Boserup 1965; Tiwari 1969).
3. It may be worthwhile here to mention that the experiments carried out to visualize the forest clearing with Neolithic stone axes, as the case during the Neolithic period, turned out to be surprisingly efficient. It was seen that

with the help of these axes which were obtained from National Museum, Copenhagen and not sharpened since the Neolithic age, two acres of jungle cold be cleared within a few days and the stone axes were quite effective for the purpose (Iverson 1971). Also it must be remembered that Harappans reached the stages of urbanization on a much greater scale without the help of iron.

4. To argue that technical feasibility was sufficient to start chain reaction which automatically led to state formation and urbanization would be too mechanical an interpretation of change (Thapar 1984: 77).
5. I think it is important to understand that surplus is not an event but a process as it has been rightly regard by Adam's (1965) and Chakrabarti (1973a, 1973b) to name just a few. The form of control is as important as the existence of surplus as is also the direction towards which it is channeled. It would be worthwhile to quote Romila Thapar. While discussing PGW culture she writes:

> It would seem feasible that this material culture provided the base for the emergence of state with a monarchical system of government in which the control of agricultural land and rights to succession would play a major role........ It is clear that the potentiality for urbanisation came with PGW culture (Thapar 1978: 91).

Thus it is very clear that the potentiality of urbanization was reached is PGW period but it could not be exploited simply because proper social, political, and economic infrastructure was insufficient to mobilize and channelize the surplus.

## References

Adams, R.M., 1966, *The Evolution of Urban Society,* Chicago.

Agrawal, D.P. 1969, Steps towards Urban Revolution in the Doab: Archaeological and Ecological Data. Cyclostyled paper.

———, 1971, *Copper Bronze Age in India,* Delhi.

Banerjee, N.R., 1965, *Iron Age in India*, Delhi.

Boserup, E. 1965, *The Conditions of Agricultural Growth,* Chicago.

Chakrabarti, D. K., 1973a, 'Concept of Urban Revolution and the Indian Context', *Puratattva*, 6, pp. 27–32.

———, 1973b, 'Beginning of Iron and Social Change in India', *Indian Studies: Past and Present,* XIV, pp. 329–38.

Chisolm, M., 1968, *Rural Settlement and Land Use*, London, pp. 50.

Chaudhury, K.A., K. S. Saraswati and G.M. Both, 1977, *Ancient Agriculture and Forestry in Northern India*, New Delhi.

Cunningham, A., 1864–65, *Archeological Survey of India Report III*, Delhi.

Deva, K., 1973, 'Comment on Chakrabarti's Paper', *Puratattva*, 6, pp. 33–34.

Dhavalikar, M.K. and G.L. Possehl, 1974, 'Settlement and Subsistence Pattern of an Early Farming Community in Western India', *Puratattva*, 7, pp. 39–46.

Gaur, R.C., 1983, *Excavations at Atranjikhera,* Delhi.

Ghosh, A., 1973, *City in Early Historical India,* Shimla.

Iverson, J., 1971, 'Forest Clearance in Stone Age', in *Readings from Scientific American* (eds), P.R. Ehrlich, J.P. Holdern, R.W. Holm, *Man and Ecosphere*, pp. 26–31.

Joshi, M.C., 1974, 'Early Historical Urban Growth in India: Some Observations', *Puratattva*, 7, pp. 90–91.

Kajale, M.D., 1974, 'Ancient Grains from India', *Bulletin of Deccan College Research Institute,* 34, pp. 55–57.

Lal, B.B., 1954–55, 'Excavations at Hastinapura and Other Exploration in the Upper Ganga and Sutlej Basins', *Ancient India,* 10–11, pp. 5–155.

Lal, M., 1982, Early Human Consolation Pattern in the Ganga, Yamuna Doab. Unpublished Ph.D Thesis. Poona University.

———, 1984a, *Settlement History and the Rise of Civilization the Ganga–Yamuna Doab,* New Delhi.

———, 1984b, 'Summary of Four Seasons of Exploration: Kanpur District, Uttar Pradesh', *Man and Environment* 8, pp. 61–80.

———, 1984c, *Archaeology of Population,* Varanasi.

Mukerjee, R.K., 1938, *The Regional Balance of Man: Ecological Theory of Population,* Madras.

Puri, G.S., 1960, *Indian Forest Ecology* 2 vols., New Delhi.

Ray, N.R., 1978, 'Technology and Social Changes in Early Indian History: A Note Posing Theoretical Questions', *Puratattva,* pp. 132–38.

Sharma, R.S., 1965, 'Material Background of the Origin of Buddhism', in M. Sen and M.B. Rao (eds), *Das Capital Centenary Volume: A symposium,* Delhi.

———, 1983, *Material Culture and Social Formations Ancient India,* Delhi.

Thapar, R., 1978, 'Puranic Lineages and Archaeological Culture', *Puratattva,* 8, pp. 86–98.

———, 1984, *From Lineage to State,* Delhi.

Tiwari, 1969, Impact of Irrigation on Resources and Levels of Income in Block Ghatampur, District Kanpur, Uttar Pradesh, Unpublished M.A Thesis, Department of Agricultural Economics, Government Agricultural University, Kanpur.

Vita-Finzi, C. and E. Higgs, 1970, 'Prehistoric Economy in Mount Carmel Area of Palestine: Site Catchment Analysis', *Proceedings of Prehistoric Society,* 36, pp. 1–37.

Whitecombe, E., 1972, *Agrarian Conditions in Northern India 1860–1900*, Berkeley.

# 3

# Forests and Settlements

**Romila Thapar**

It is possible to trace through Indian history a changing attitude on the part of the Indian towards his forests. This change is reflected in the texts in Sanskrit, Prakrit, Tamil and other languages as well as in the oral tradition. There is gradually a self-consciousness and a distancing from the forest, especially in the culture of high literacy. The distancing takes the form of romaniticising the forest at a time when some forests are being cleared and the forest people living there being forced to change their lifestyle. Often the oral tradition is more sympathetic to the forest.

The gradual clearing of forests over the centuries has changed the ecology of certain regions. The pace of change was not uniform. It was slower in earlier times and more limited in its geographical area, but as the demand for land increased, the clearing of the forests became faster and the area so cleared, larger. Seals from the Indus Civilization depict a variety of animals, of which the tiger, the rhinoceros, and the elephant are the most frequent. These are all animals which require a reasonable forest cover and it is not surprising that with the present desiccation of the Indus plain, it no longer remains the habitat of these animals. It has been suggested that in the past there were forest galleries where these animals made a home but which have disappeared in the last couple of centuries. The climate and ecology of the region are being further investigated by archaeologists, some of whom maintain that the climate was wetter at that time. The historical study of the ecology of the Indus region and Gujarat is especially important because it is being argued that the decline of the Indus cities was substantially due to their inability to check the degradation of their environment.

---

* Reprinted with permission from Romila Thapar, 1995, 'Forests and Settlements', in Paola Manfredi (ed.), *In Danger, Habitats, Species and People*, Ranthambhore Foundation, Delhi, pp. 112–19.

Some categories of literature are very aware of ecological differences and their links to culture. Thus the *tinnai* concept of Tamil Sangam texts, composed around the start of the Christian era, is a fascinating example of a detailed and early perception of the significance of eco-zones. The landscape has been classified by five types of eco-zones and these in turn are co-related to other activities, and to cultural articulation. The zones are listed as the littoral, the wetlands, the pastoral tracts, the dry zone, and the hilly backwoods. Occupations are said to differ in each of these. Fishing and the making of salt was important to the coastal area, the cultivation of rice in the wetlands, the breeding of livestock and practise of shifting cultivation in the pastoral tracts, which were subject to cattle-lifting from those who lived in the dry zone, and hunting and gathering was associated with those who lived in the backwoods. There are descriptions therefore of the way of life in each of these and their interactions. Wetlands, for example, were extremely limited as compared to the non-cultivated tracts, but over a period of time this changed. There was some interaction through the exchange of paddy and salt for other products, generally at places which remained constant and grew to become exchange centres. There have been some historical studies attempting to trace the process by which these micro eco-zones gradually evolved into macro-zones.

Early Sanskrit texts such as the Vedas make a distinction between what they call *grama* and *aranya*. These categories are generally seen as confrontational, but it is a confrontation which is sometimes used to highlight social perceptions. The *grama* is the settlement and is therefore orderly, disciplined, known, predictable, and the location of what came to be called civilization. It is the place where Vedic rituals can be performed. It is basic to agriculture, urban living, exchange, government, the arts, and the culture of elite groups (often referred to as high culture these days). The *aranya* is the forest, disorderly, unknown, unpredictable, and inhabited by predators and strange creatures, different from those living in the *grama*. This fantasy of associating the unknown with the dark depths of the forest is common to all societies which begin to view the forest from the settlement, as is evident from folk tales and children's stories. As the lifestyle of those who live in the settlement evolves into urban centres, the distancing from the people of the forest becomes greater. They are regarded as less civilized and because their social mores are different, they come to be treated as the backward peoples of society relegated to the lowest rungs of the social hierarchy.

This dichotomy between *grama* and *aranya* is generic to the narratives of the two epics, the *Mahabharata* and the *Ramayana*. The forest is the habitat of those who are sent into exile. But gradually the forest comes to be appropriated in the epics and there is to some degree an identifying with the forest. But this identity is kept distinct from that of the settlement in which the social mores of the *grama* can be and are, transgressed. The explanations given for these transgressions are fascinating. One wonders for example, whether the marriage of Draupadi to the five Pandava brothers would have been feasible in a setting untouched by the fact of exile.

Among the more graphic descriptions of the deliberate destruction of the forest and forest life are incidents narrated in the *Mahabharata*. Dusyanta goes on a hunt, deep into the forest which brings him to the *ashrama* where he meets Shakuntala. The hunt is a campaign against nature. He is accompanied by a large band of warriors and they indiscriminately kill the animals. Herds of deer, families of tigers and elephants are mowed down and trees are devastated in this process. The people of the settlement are demonstrating their power over nature. Elsewhere in the epic we are told that when the Pandavas were given half the kingdom, they had to build their capital, Indraprastha. This required the clearing of the forest and in order to do this effectively the god Agni came into action and the forest was burnt. Descriptions of the burning of trees and of animals attempting to flee the flames provide a glimpse of the destruction of the forest prior to establishing a settlement. The *grama* is again projected as triumphant, irrespective of what may have been involved in the building of the new city.

Exiles from the *grama* are different from the people of the forest, because for the former the forest remains a wild habitat which has to be tamed, whereas for the latter the forest is the natural habitat. Forest people are said to live on gathering roots and fruit and on hunting wild animals and this is a different culture from the cultivators and traders who constitute the settled society. The life of the forest people is seen as antithetical to the evolution of civilization if civilization is defined as the activities of the settled society.

But the concession to the forest was made in various ways. The most obvious is the continuing worship of trees. This was in the form of either the veneration of particular trees, such as *Ficus religiosa,* which has perhaps the longest continuity of a sacred tree in India. It occurs on the seals from the Indus Civilization, it is associated with Buddhism and is depicted in sculpture from Buddhist places of worship and it continues to be venerated in association with religious shrines of Hindus and Muslims.

Texts associated with Buddhism and Jainism refer to sacred groves, some maintained by the people of a city, others by a monastery, and still others by the wider community who lived on the edge of the forest. Thus the cities of Vaisali, Kushinara, and Champa, all maintained sacred groves. The importance of the forest had not disappeared as yet. The groves may even have been specially planted as they are often described as being of specific kinds of trees, such as the *banyan* or the *sala*. Individual trees under worship, sometimes referred to as a *chaitya* or a sacred enclosure, were cordoned off in various ways. The tree had platforms around it for offerings and the placing of ritual pitchers, the area being enclosed by a railing with doorways. The railings appear to have been decorated with sculpted relief. Sometimes the individual tree was located in the midst of a sacred grove. A platform, believed to be of Mauryan times, was excavated near the Bo-tree at Bodh Gaya, constructed of brick and stone and decorated with sculpted human and animal forms.

The worship of trees as part of a fertility cult has remained a constant feature of the religion of pastoralists, peasants, and those of lesser status among urbanites; an undercurrent which from time to time enters and suffuses even the more sophisticated expression of belief and practise. The frequency of sacred groves, some enchanted and magical and some inhabited by deities, is a common theme in Indian narratives. The tree was and is, personified as a deity and some trees are sacred in their association with deity. The spirit of the tree mingled with the spirit of the cults of water and mountains as well as those of animals—the *naga* and the tiger, for instance. These were not just mystical manifestations, but represented a sense of integration with the world of plant and animal life, where the sensitivity of the human merged awe with affection, very different from what has eventually become the insistence on dominating nature.

Ritual vessels used in the Vedic sacrifices had specific junctions and were only to be made of specific kinds of wood, which suggests a symbolism regarding trees and wood which was far more complex than in many other societies. The incorporation of the people of the forest into the society of the settlement required the accommodation of the deities of each to the other, in a process of acculturation. Of those that perform their rituals diligently, very few realize that there is much of the forest embedded in their rituals.

Pastoral groups form a kind of grey area, since they are people who live in the settled society but graze their animals in the forest. They are familiar with only the edge of the forest. But the fact that they have been so important to Indian culture means that the forest was not seen as altogether hostile or distant. The cycle of Krishna legends draw from both the pastoral and the more gentle forest associated with the grazing of cattle.

Not surprisingly when a person decided to opt out of the *grama* he became an ascetic and went to live in the forest which represented the opposite pole from the *grama*. For him the forest was not alien but was an avenue to the discarding of the mantle of civilization and the discovery of the self. Buddhist monks who were renouncers from society lived either near the *grama* from where they got alms, or the more respected ones lived in the forest away from civilization. The forest also curiously becomes the location for the salvaging of the highest ethical qualities which are associated with the settlement. The Chinese Buddhist pilgrim, Fa-hien, who visited India in the fifth century CE, tells us that one of the current beliefs concerned the future of the Buddha's teachings. He relates that when these teachings will decline, which they will because of the increase of evil in the world, then those who are still virtuous will flee into the forest. They will live there until the coming of the Buddha Maitreya—the future Buddha—who will restore the world to virtue and those who had escaped to the forest will return once more to live in the settlement.

The closeness of man to the forest through asceticism is also demonstrated in the idyllic picture which is often constructed of life in an *ashrama* in a forest. For example, the setting of the play on Shakuntala by Kalidasa emphasizes the dichotomy. The forest hermitage is a place of gentleness, peace, and closeness to nature, where even the plants and the deer reach out to Shakuntala. This is contrasted with the court in the capital and its hostility and violence. The romanticization of the forest is subsequent to its being the habitat of exiles and renouncers and reflects less a fear of the forest and more a familiarity with it. It could be argued that the romanticization of the forest begins when it is thought that Culture is superseding Nature. The forest of epic literature is a different kind of forest and has quite another place in the imagination from the forest of Kalidasa. Perhaps this image was exaggerated and the forest treated as a kind of imagined alternative, a forest paradise, when the complexities of civilized life became overwhelming. The symbolism of *grama* and *aranya* are marked in early Indian culture and possibly this enabled the two categories to live co-existentially for many centuries, where the characteristics of one did not impinge on the other. There was also in earlier times, enough forest available for it to remain a distant habitat. But today, the attempt that is being made is to insist that the *grama* subordinate the *aranya* to its needs, hence the exploitation of the forests.

The importance of forest wealth is recognized in the *Arthasastra* of Kautilya who states that no one is permitted to cut any part of the forest without the permission of the state. This was to ensure a control over revenue both from the produce of the forest and from the land cleared and brought under cultivation. Kautilya is also very precise in describing how the state should clear wasteland and settle families of agriculturalists on it as part of the process of extending agriculture and enhancing revenue. Emperor Ashoka took pride in the roads which were constructed by his administration and these he states were lined with shade-giving trees and with wells. The building of *baolis* at intervals along the major roads characterizes road construction at many points in history. These were mainly small structures with wells, often set in a garden. The more elaborate ones elsewhere were virtually underground places and a thick growth of trees in the vicinity added to keeping the place cool.

Sometime after the fourth century CE when agriculture was extended, both to bring in greater revenue and perhaps to support a larger population, there appears to have been a greater encroachment on the forests. There are references to violence against forest tribes in a Gupta period inscription and the need to subordinate the dwellers of the forest in inscriptions of a slightly later period—and inscriptions in those days were the official statements. Forest dwellers were brought under control, their forest resources—timber, mines, gem-stones—appropriated, and they were then converted into the lower castes of the area and established on the edges of towns or in separate settlements in the forest. This was a process of marginalizing those who lived and worked with the produce of the forest.

The clearing of wastelands which was substantially jungle and converting it to cultivated land, was necessary for the creation of the many small states which emerged at the time, dependent on agriculture for revenue, and which continued gradually to increase in number. Shifting cultivation lets the forest reclaim what is cleared. Settled cultivation means losing the forest altogether and it was revenue from the settled cultivation that maintained the new states. The expansion of trade and the opening up of new areas to commercial exchange also resulted in routes cutting through forests and the conversion of some forest settlements into markets. The establishing of monasteries and *ashramas* in the forest meant clearing the forest, and unless they were maintained as inaccessible to the rest of society (which was rare), there was gradually a building up of settlements around these institutions, particularly where they became the nucleus of a new settlement which worked to support the monastery or the *ashrama*. Religious interests, therefore, also sometimes coincided with economic interests in converting the forest into a settlement.

Significantly, the forest people play a central role in the origin of myths of many dynasties in every part of the subcontinent and this role cannot be ignored. Most dynasties were of humble or obscure origin and their mythologies in which they legitimised their coming to power provide clues to this origin. Yet there are some Brahmanical myths with a clear contempt for the forest dweller. One of the most powerful myths is that of the first ruler, Prithu. The original king, Vena, turned against the Brahmans and therefore, the Brahmans killed him. In the absence of a king there was the threat of chaos and disorder. Therefore, the Brahmans churned the left thigh of the body of Vena and out of it sprang a short, dark man, whom they banished to the forest and called him Nishada—which became the generic name for the forest people. They then churned the right arm of Vena and from it sprang a tall, fair, handsome man whom they made the king and who was responsible for introducing cattle-keeping and agriculture. So pleased was the earth with him that she bestowed her own name on him and he was called Prithu. Has there always been in Indian culture a contempt for the forest dweller? It was one thing to set up an *ashrama* in the forest but it was another to mingle with the forest people. The latter were looked upon as beyond the social pale, in spite of the romanticizing of life in the *ashrama*. Even to this day they are described as 'backward' and needing to be 'assimilated' into the mainstream of Hindu society. Even benign official policy gets vitiated by such attitudes. Yet it is the people of the forest who are closest to the forest and more in harmony with it than the others. Their knowledge of the forest is different from that of the officials who are supposed to be their benefactors or that of the environmentalists who are concerned about the forest and its people. The people living in the forests were in the past subordinated, but as yet, the forests were not destroyed. The pressure on the land was less since the pressure of population was not so great. Today, the problem is that not only are there few forest people left, for most are being brought into the net of 'modernization', but the forests themselves are being destroyed.

Another dichotomy lies in the terms *Prakriti,* that which is natural, and *Sanskriti,* that which is created by man, sell-consciously cultured and to that degree, artificial. This becomes apparent in all activities where things naturally existing are differentiated from those which are artificially created. Here the forest would be the natural unit and the settlement, the created one. Gradually over time, it is the latter, *Sanskriti,* which comes to be equaled with civilization. This dichotomy continues into later literatures in other languages of India, such as the regional languages, Turkish and Persian. The forest is the retreat of recluses and holy men and even the princes of the royal court have to go there to meet them—as many Mughal miniatures depict. This is again a form of turning away from the settlement.

The old ambiguity of using the forest for royal hunts also continues. The dichotomy of Nature and Culture is interestingly reflected in the activity of hunting. Those who live by hunting are treated as uncouth, looked down upon, and subordinated to outcaste status. Yet princes and kings frequently go on hunts, which are meant to be a sport but which can sometimes be described as a ferocious destruction of nature and animals. These activities do not affect the high status of royalty, even when they behave in a manner far more gruesome than the actions of professional hunters.

The earlier suspicion of the forest and contempt for the people of the forest was reinforced by colonial rule, when forests became an area to be exploited for their wealth—particularly timber—and the people were dismissed as backward and primitive in the worst nineteenth century sense of the word. Some were seen as wild tribes and some even as criminal tribes where their moral codes were different from those of the colonisers. The somewhat more ameliorating position of romanticizing the forest or treating it as the habitat of the renouncer, which softened the hostility to the forest, was no longer encouraged in colonial attitudes to the forest. The philosophy of colonialism was based on man's necessity to control nature at all costs, even if it meant destroying the forest. Hunting became a sport open to all and the target was the biggest and the best of the species. With the extensive availability of firearms, the sporting element of the hunt was changed to a carnage of animals.

A quantitative change came about with the colonial decision to build railway lines. This has been seen as one of the most destructive acts in the clearing of forests. Routes had to be cut through all kinds of terrain and if there were forests on the way, they were destroyed. The routes were originally located to serve the economic demands of carrying resources to the markets and the ports, to enhance the growth of industrialization. Inevitably the introduction of the railway had other results. Access to new lands through rail communication led to migrations and new settlements. These were not limited to land along the rail tracks but often took settlers into the interior of forests where land was cleared for cultivation. Railways brought communities closer together but they were also responsible for changing the environmental conditions in many areas. The building of an extensive canal network in the nineteenth century had much the same effect, for large tracts of jungle had to be cleared for canals to pass through and canal colonies to be established. What started as an attempt to improve and increase agriculture through providing irrigation facilities is now seen in many areas as terminating agriculture because of the ensuing salination of the soil.

In looking critically at these changes, the intention is not to suggest that all technical innovation has to be brushed aside. But now that there is a greater awareness of the crucial role of environmental conditions, innovations have to be examined more carefully to assess the alterations that they will introduce in the interaction of man, nature, and culture. The change is not even restricted to the environment, for, altitudes of mind change, as do attitudes to other human beings. If a technological innovation in an area is imperative, then the first concern should be that the least damage be done to those who live there and their environment. The point of having the capacity to plan and to project the effect of a plan is to ensure the minimum devastation. But it is this, which we have moved away from in the claim that development is ultimately all to the good.

The colonial interlude has taken us even further away from what might have been a holistic understanding of Nature and Culture. The pressure of population in the past has been encroachments into forest lands but the forests were plentiful enough to meet this pressure. Now this is coupled with the ruthless middle class determination to wrench the maximum reventue out of the forest. In the formidable alliance of the politician, the contractor, the bureaucrat and the industrialist, there is little hope for the forest and the people of the forest, for they are not permitted even to denounce, let alone stop, the desecration. We are all now silent witnesses to the holocaust of the forest in India.

4

# The Forest and the Field in Ancient India

Mahesh Rangarajan

When Rama, the prince of Ayodhya, is about to set out on his long exile in the forests south of the Gangetic plains, his mother, Kaushalya, expresses fears about his safety: 'May the huge elephants not harm you my dear son,' she says, 'nor the lions, tigers, bears, boars or ferocious horned buffalo.' Rama himself, in a bid to dissuade his wife, Sita, from following him into the woods, paints a similar portrait of the forest as a place of hidden menace. The roars of flesh-eating lions would assail the senses and the sharp kusa grass prick the soles of those who walk the tracks. Even the word *vana* or forest was only given to lands where pleasure gave way to hardship. But a very different picture of the same lands emerges when Sita finally has her way and joins her husband in exile. The forested lands are a source of pleasure for her. The poet conjures up vivid images of, 'lotus ponds in full bloom blanketed with flocks of geese and ducks.' Plenitude replaces pain, and beauty, menace. Sita is drawn to the honey-scented groves and the river's edge where herds of deer quench their thirst.

The Ramayana of Valmiki assumed the shape we read it in around 1500 years ago: its oral traditions are certainly older. The twin themes of the forest as a place of dangers to be confronted and of beauty to be enjoyed run like a thread through subsequent sections of the epic. By the third century CE, a fresh phase of town-building began in North India. The scribes in the city had their own conceptions of what the forest was like. Most often, it was seen as another land, culture, and space, set apart not only by its beasts and birds but also by its people. The forest was all that the city

---

* Reprinted with permission from Mahesh Rangarajan, 2000, *India's Wildlife History: An Introduction,* Permanent Black and The Ranthambhore Foundation, New Delhi, pp. 1–9.

was not. The forest hermitage of the brothers Rama and Lakshmana had the gleam of 'metal sword and axe' to warn any intruder. When building their home in the forest, they first sacrificed that most essential of sacrificial animals, the blackbuck. Lakshmana made sure the animal was killed but left with its limbs intact to be cooked, broiled and only then offered to the gods. The antelope, the *krisnasara*, had long represented the culture of the open spaces, of cereal-bearing dry land, that were the centre of civilization. Beyond its range lay the *anupa*, the marshy tracts where malaria struck and elephants abounded. In the texts from the Vedic times many centuries earlier, *Aryavarta*, the land of the Aryans, was often co-terminus with the land of the black antelope. Sometimes defined in terms of geography, these were areas to the north of the Vindhya mountain chain. At other times, it included lands to the south. The societies beyond were seen as different in culture and lifestyle: unruly, barbaric, and dangerous. Such simple images do little justice to the complex realities of the links between people and nature in early India. But they do give a sense of perceptions among the literate classes, whose life never moved far into the world beyond the city (Goldman 1986a: 131 6, 201, 278).

The range of landscapes could be seen in more ways than a simple two-way split between the *aranya* or forest and the *grama* or village. In the Tamil literature of the Sangam period dating back even further than Valmiki's epic, land is divided into as many as five eco-types, ranging from the littoral to the wet rice paddies. The ecology and nature of habitat changed over time. For instance, the dry land or *palai* where people lived by herding became a land of great scarcity and danger when the rains failed, a place of menace where wolves and thieves attacked people.

In a very different category were the medical treatises of Charaka and Susruta. Here, the land itself is seen in terms of the range of meats that it yielded. Creatures are graded in terms of when they may or may not be eaten and by whom. In an increasingly divided society, where the warrior caste or *kshatriya* was setting itself apart from the rest on hereditary lines, they asserted the right to eat certain animals. For instance, the meat of the lion and the tiger were to be eaten only on rare occasions by the king himself. This act of eating the flesh of 'eaters of flesh' was peculiar to the more privileged sections by the latter part of the first millennium CE.

In non-medical texts, composers see the animals or their habitat in a metaphorical and not a literal sense. This pointed to a 'self consciousness and distancing from the forest... in the culture of high literacy'. The lion is seen as a king of beasts, the wolf or *vrikka* as a glutton. The anger of an elephant in *musth* is often evoked as an image; a king in exile is referred to as 'a tiger among men'. Metaphors apply as much to seasons as to places or specific animals. The forest in the time of rains evoked deeply lyrical descriptions, summoning up images of fertility and bounty. At times, the creatures themselves are purely mythical like the *sarabha* or the *yali*. In Kalidasa's play, the *Shakuntalam*, a king slaughters families of tigers and elephants. But such evidence cannot be taken at face value. Court poets would celebrate royal hunts as kingly conquests of the forest, even as they looked down upon people who lived in or near the forest as having a less worthy lifestyle (Goldman 1986b: 365, 372; Watters 1961).

## Early Faunal Collapses

Texts give clues to sensibilities. Archaeological evidence in the form of artefacts and animal remains are a more reliable guide to the changes in the land in centuries past than literature is. There is little doubt that there were several sites in India where rice was being collected or cultivated several centuries before the birth of Christ. Hunting, the rearing of goats or sheep, and cereal-eating, often went together. The uncertain cycles of rainfall could push people relying on one occupation, towards another. There was no watertight division between hunter-gatherer, herder, and cultivator. Long before the times referred to by the Sanskrit texts, wild animals were a major source of meats in the various sites of the Harappan civilization. Over 1000 sites across north-western India dating before 1700 BCE have been located or studied. A range of bones of wildlife including the *chital*, hare, jackal, the great Indian one-horned rhino, wild ass, and elephant have been found. These make up upto a fifth of the animal remains in many sites in the Indus valley. In western Indian sites, most seeds found in the old dwellings are of wild plant species now extinct in the region. Some changes in faunal and floral distribution were probably the result of climatic shifts, such as increasing aridity in some tracts. Others may have been due to the impact of early humans. The swamp deer or *barasingha* was found in Mehrgarh in Baluchistan till around 300 BCE. Its local extinction was probably a result of over-hunting and cultivation of its riverside habitat. Its vulnerability to such changes hastened its disappearance, though it survived along the Indus river till about a century ago. More interesting was the case of a species of wild cattle, the *Bos primegenius nomadicus*, or zebu, a relative of the familiar humped cattle. The species vanished very early on, and hunting was perhaps not the cause. While people hunted a wide spectrum of wild animals, they herded only a few varieties. It is possible that the species interbred with domestic stock and vanished across much of its range, though it survived in some pockets of western India for longer than in other areas. Feral zebu and domestic cattle, both smaller sized than their wild relatives helped push the unique breed over the edge.

Such faunal collapses were still exceptional. One reason was the sheer immensity of the forest. Contrary to what was believed till recently, the 2000 years before the Christian era did not see extensive denudation across much of the Indo-Gangetic plain. Iron tools and fire are often celebrated in the Sanskrit texts as being responsible for replacing jungle with farmland, and nature with culture. But such celebratory accounts find little proof in material evidence. The whole of Kanpur district at that time had less than 4000 people, a density of less than one person to a square kilometre. There is no doubt that cultivation, domestication, the taming of animals like the elephant and the rooster, the water buffalo, and the zebu cow were major landmarks. Even small numbers of humans with techniques we would now consider primitive could be the decisive factor in the local extinction of herbivores. Still, it does not follow that the forest was on the retreat (Thapar 1997: 112–19; Zimmerman 1987: Ch.1, 107–73).

## Forests for Elephants

What did happen was that around the fourth and third centuries BCE there was for the first time a unified political entity, the Mauryan empire, that included vast stretches of the eastern Ganga plain as well as the central Indian forests. From now on, in varying degrees, military might would rest not only on horses and weapons but on that most remarkable of animals used in war—the elephant. Its domestication by the Harappan times is certain, but it was only later that it began to be put to wider use. It was at one and the same time, a siege engine, a mount in war, a sign of status, and a mobile platform to hunt in tall grass country. The animal was so vital for warfare across centuries that the control of elephant supplies was a critical (though not the sole) factor in successful military campaigns. Elephants were easier to catch and tame than to breed partly because of their long gestation period of over 20 months and the sheer time and trouble it took to rear calves before they could enter the workforce.

The protection of elephants became serious business by the time of Mauryan rulers such as Ashoka. The *Arthasastra*, containing maxims of ancient statecraft, lays down the duties of the Protector of the Elephant Forests with no room for doubt:

> On the border of the forest, he should establish a forest for elephants guarded by foresters. The Superintendent should with the help of the guards . . . protect the elephants whether on the mountain, along a river, along lakes or in marshy tracts . . . They should kill anyone slaying an elephant.

Even the tusks of animals dying from natural causes in the forest were to be handed over to the government. As many as eight *hasti-vanas* or elephant forests named in the book have been identified and mapped. Strangely, there was no such reserve in southern India, but it is possible this region was not a source of supplies for the Mauryas. The westernmost such reserve was in Saurashtra, today a semi-arid region. Even at the time, the animals from the latter were small in stature and not the most sought after. While the huge animals were more abundant and of better quality the further east one went, the picture with horses was the exact opposite. Horses were domestic animals and the best specimens were imported from lands to the west of the Indian subcontinent. All the more reason for rulers of the eastern Ganga plain to assert royal monopoly rights over elephants.

Elephants were not the only product of the forests to which the ruler laid claim. There were other forests where the aim was to secure timber, and tiger and lion skins. Elsewhere, the Protector of Animals worked to eliminate 'thieves, tigers, and other predators' from the woods to make them safe for cattle herders. Woodlands were also seen as the home of different kinds of peoples to be placated or suppressed. The writer of the *Arthasastra* regarded forest tribes with great suspicion: they were to be controlled through bribery and political subjugation. In contrast, the food-gathering tribes or Aranyacara were to be used to guard the borders and to trap animals. The empire relied on the continued cooperation of such groups, though the relationship could be tense and conflict-ridden. Even as rulers directed the colonisation of fresh lands, certain forested tracts were valued either in strategic or economic terms. Controls or curbs on the use of such tracts had a wider political and economic aim.

The Mauryas were only able to rule an extensive empire by drawing hunters or nomads into the wider networks of exchange. Elephants were taken as tribute from the forest chiefs. In the latter part of his reign, Ashoka brought about significant changes in his style of governance. Embracing Buddhism, he protected many birds and animals, even giving up the royal hunt. The Fifth Pillar Edict proclaimed how, 'Our king killed very few animals.' But its further claim that others followed suit and gave up killing is simply not credible. Ashoka was aware of the limits of his power over the forest-tribes, who were warned that his patience was not limitless. A ban on hunting, fishing, and forest fires was impossible to enforce. Not only the peoples in the forest, but even cultivators probably slew animals routinely. The fact that there was a fine of 100 panas or coins for poaching deer in forest set aside for the king shows that rule-breakers did exist. The edicts as well as the *Arthasastra* give us a sense of what was desired by the powerful rather than an accurate picture of what actually happened (Lal 1986: 83–90; Thomas and Joglekar 1994).

## The Sea of Forests

It was not easy to answer the simplest of questions: whose forests were these? Rulers claimed the woods, mountains, and forests as their own. This is what Rama told Bali, the monkey king, as the latter lay dying from a mortal wound: 'Men used traps, nooses, hid and hunted (and) killed for meat.' The epic is no guide to events.

But it does give some vital clues. The forests after all were so immense that even a few centuries after the composition of Valmiki's Ramayana in its final form, there were vast wooded areas even across the plains of north India. The Chinese traveller Hieun-Tsang in his travels across India during the seventh century CE refers repeatedly to the immensity of the forested spaces that made travel perilous and difficult. Not surprisingly, the fear of natural forces often went hand in hand with venerating various natural entities, whether in the worship of elements like water or the association of animals like the elephant or the tiger with divinity. This is of course a view from the city and the plains regions, but it drives home the point. Protection by rulers could extend to a chosen few varieties like the elephant and select patches or forest. But the jungles did matter: they re-grew where people abandoned cultivation due to a failure of the rains, disease or human-made calamity. Cultivated sites were still small islands in a sea of forest. The ecological map of South Asia was strikingly different from the one we know today. Epics and other narratives, and the shards of animal remains only give us a glimpse of what it was like (Parasher-Sen 1998; Rangarajan 1992).

## References

Goldman, R.P., 1986b, *The Ramayana of Valmiki, Volume IV*, *Kishkindakanda*, pp. 93.

———, (trans. and ed.), 1986a, *The Ramayana of Valmiki: An Epic of Ancient India, Volume II, the Ayodhyakanda*, Princeton University Press, Princeton, pp. 131–36, 201, 278.

Lal, Makkhan, 1986, 'Iron Tools, India's Wildlife History, Forest Clearance, and Urbanisation in the Gangetic Plains', *Man and Environment* 10, pp. 83–90.

Parasher-Sen, Aloka, 1998, 'Of Hunters, Tribes and Barbarians', *Studies in History* 14, pp. 173–92.

Rangarajan, L.N. (trans. and ed.), 1992, *The Arthashastra*, Delhi: Penguin, pp. 688–91, 320–3.

Thapar, Romila, 1997, 'Forests and Settlements', in P. Manfredi (ed.), *In Danger: Habitats, Species, Peoples*, Ranthambhore Foundation, Delhi.

Thomas, P.K., and P.K. Joglekar, 1994, 'Holocene Faunal Studies', *Man and Environment* 19, pp. 179–204.

Watters, T., 1904 & 1961, *On Yuan Chwang's Travels in India*, Delhi, Reprint Publishers, Vol. I, pp. 365, 372.

Zimmerman, F., 1987, *The Jungle and the Aroma of Meats: An Ecological Theme in Hindu Medicine*, University of Califonia Press, Berkeley.

5

# The Great Mughals Go Hunting Lions

**Divyabhanusinh**

The political turmoil of Central Asia had many a Timurid prince fighting for his piece of inheritance. Zahiruddin Muhammad Babar of Fargana was no exception, in spite of his illustrious lineage from Timur on the paternal side and Genghis Khan on the maternal. However, there was another unique quality in the man, his inexhaustible energy to conquer and ambition to find himself a permanent place in the sun, which ensured his success in his endeavour for political power. Born in 1483, Babar's life until 1526 was based in Central Asia and was nomadic in the extreme. Thrice he took Samarkand and thrice he had to leave; consequently his attention turned south, to Kabul and the ever-tempting plains of Hindustan. He made five excursions into Hindustan, the final one in 1526. As the successor ofTimur, Babar sent Ibrahim Lodi a goshawk (*qarchigha*) with a letter demanding his kingdom (Beveridge 1922: 384–85; Thackston 1996: 279). The Lodi king did not oblige and the rest, as they say, is history.

Babar lived barely four years after his success at Panipat, but these years and his death while still in India ensured that his son Humayun (1508–56) and later successors would stay and consolidate his conquests. His place in the annals of Asia is assured by his founding the great Mughal dynasty alone, but he is unique also in having maintained a diary and recording his memoirs, a feat unheard of during his time. Written in his native Chaghatai Turki, it was the forerunner of the imperial accounts commissioned by his grandson Akbar (1542–1605), and the memoirs of his great grandson Jahangir (1569–1627) who chose to write in his own hand in Persian, the court language of his time. Both Babar's and Jahangir's memoirs are remarkable for their

* Printed with permission from Divyabhanusinh, 2005, 'The Great Mughals Go Hunting Lions', in *The Story of Asia's Lions*, Marg Publications, Mumbai.

candour and for leaving behind a firsthand account of their times. Jahangir referred to the *Babarnama* in his own memoirs and lamented that though his ancestor described what he saw, he left no visuals for the record. The reigns of Shah Jahan (1627–58) and Aurangzeb (1658–1707) were recorded by their courtiers, apart from European travellers who left their own travelogues of the fabulous empire. While this is a firsthand corpus of information of the times, there is yet another, of equal importance, from which one can learn a great deal about the period.

With the return of Humayun from Persia to Kabul, miniature paintings of the Mughal school had their nascent beginning, for he recruited Mir Sayyid Ali, Abd as-Samad and Dust Muhammad, the principal artists from the court of Shah Tahmasp. The first two came to Hindustan when Humayun returned in 1555. With the continuous and plentiful supply of paper produced at Sialkot, Lahore and Daulatabad, paintings on a large scale became possible for the first time. Under Akbar who set up an autelier at his court, Mughal painting attained maturity, and under Jahangir it reached its zenith particularly as far as studies of animals and birds were concerned. However, Shah Jahan and Aurangzeb were preoccupied with numbers to give us a glimpse of their lifestyle including shikar. It is to these sources along with the written records that we must turn to find the lions' position under the Great Mughals.

Issuing forth from the northern vastness of Central Asia and Afghanistan, Babar found Hindustan different from the lands he had known. In his memoirs he describes the country he had conquered and its people. A keen hunter, he was intrigued by the animals and birds that he encountered which he had not known in his own homeland. His descriptions of Indian fauna and avifauna are unique and a worthy forerunner of his eclectic great-grandson Jahangir's descriptions penned over three-quarters of a century later. Babar's descriptions for instance of the peacock, the rhinoceros, and other animals or birds he saw are hard to beat.

Important to our story is the fact that Babar does not mention the lion and says little about the tiger; he was probably familiar with these animals as they were not peculiar to Hindustan. Though he records having killed a tiger (Beveridge 1922: 393; Thackston 1996: 285), there is no mention of lion being pursued.

Humayun's intermittent reign was fraught with many a calamity and was finally cut short by an untimely accident. The chronicles of his times were committed to paper years after his death by his daughter and also his ewer-bearer, at the command of Akbar. These throw no light on his encounters with lions. But before we come to Akbar we need to take a look at the size of the Mughal entourage which had grown to immense proportions.

Akbar's consolidation of his domains and his benevolent rule saw the emergence of a fabulously rich empire. The three centres of imperial governance at Agra and Fatehpur Sikri, Delhi and Lahore came to acquire a grandeur of their own, further embellished by his successors. The Mughal capital or centre of power moved with the emperor to one of the three cities where the emperor was in residence at any given time, or wherever the imperial cavalcade camped en route, or for that matter anywhere in their far-flung domains. Sir Thomas Roe, Ambassador of King James I of England, records that when Jahangir broke camp to move from Ajmer in 1616 after a stay of three years, his encampment had so acquired the trappings of permanence that 'The king gave order to fire all the *leskar* (camp followers) at Ajmer to compel the people to follow, which was gayly executed …' (Foster 1926: 297). In 1633 Peter Mundy saw the ruins of many a buildings as a result. Sir Thomas Roe estimated that the imperial camp on the move took 12 hours to pass a spot and when the tents were pitched, the process took only four hours. 'The circuit being little less then 20 English mile, the length some ways three course, comprehending the skirts …' (ibid.: 324). Jahangir himself estimated that it would require 100,000 buffaloes of Banjara tribesmen to keep the large army supplied with grain on its march from Multan to Kandahar since there was little cultivation in the regions of the army's march (Rogers and Beveridge 1909–14, vol. II: 233).

When the Mughal emperors moved with such large entourage, their pleasure spots had to be easily accessible from their camps, so it was natural that the imperial hunting grounds were located in the plains. Irfan Habib lists 16 such locations (Habib 1982: pls 4B, 6B, 7B and 8B), including such celebrated ones as Rupbas and Bari near Agra, Bhatinda and Sunam in the Punjab, and Jodhpur and Merta in Rajashtan. This meant that the Mughals were moving about in or near the most preferred habitat of the lion, the vast grasslands and scrublands of northern India. Francois Bernier who travelled through the Mughal Empire between 1656 and 1668 has left a graphic description of Mughal hunting grounds during the time of Aurangzeb:

I could never conceive how the Great Mughal could hunt with an army of one hundred thousand men; but there çertainly is a sense in which he may be said to hunt with two hundred thousand, or with any number of which his army may consist.

In the neighbourhood of Agra and Delhi, along the course of the *Gemna* [Yamuna], reaching to the mountains, and even on both sides of the road leading to Lahore, there is a large quantity of uncultivated land, covered either with copse wood or grasses six feet high. All this land is guarded with the utmost vigilance; and excepting partridges, quails, and hares which natives catch with nets, no person, be he who he may, is permitted to disturb the game, which is consequently very abundant.

Whenever the Monarch is about to take the field, every gamekeeper [*gardes chasse*] near whose district the army is to pass is called upon to apprise the Grand Master of the Hunt of the various sorts of game under his particular charge, and of the places where they are in the greatest plenty. Sentries are then stationed at the different roads of that district, to guard the tract of ground selected, which extends sometimes four or five leagues; and while the army is on the march on one side or the other, so as to avoid that tract, the king enters it with as many *Omrahs* [nobles] and other persons as have liberty to do so, and enjoys, leisurely and uninterruptedly, the sports of the field, varying them according to the nature of the game (Constable 1934: 374).

A favourite method of hunting was the Qamargha, the *battue,* an ever-decreasing circle of men, in this case the army, which would enable the emperor and his chosen nobles to hunt from horseback to their heart's content. A variation of this method was the Shakhbandh which was a stockade. In these methods of hunting the objects of the pursuit would be usually ungulates and lesser carnivores. In one such stockade at Lahore, Akbar hunted for five days in 1567 and the bag included chital, nilgai, hare, hyena, jackal, and blackbuck run down by trained cheetahs—as can be seen from the paintings commemorating the event (Beveridge 1904, vol. II: 422–24; Sen 1984: 39–40). Among the other favourite forms of sport at the time were coursing with trained cheetahs and hawking, both of these being pursued by the Mughals till the end of the dynasty's rule.

However, neither the Qamargha nor the Shakhbandh was suitable for hunting large carnivores. It was too dangerous to corner these animals in the chaos and confusion of other animals and men, for they could attack the hunters at close quarters. Lions were royal game and could be hunted only by the emperor and with his permission by the notables in empire. Thomas Roe had to take Jahangir's permission to tackle a troublesome lion (Divyabhanusinh 2005: 116) and Francois Bernier records from Aurangzeb's reign that

> … of all the diversions of the field the hunting of the hunting of the lion is not the most perilous but is peculiarly royal; for except by special permission, the king and princes are the only persons who engage in the sport (Constable 1934: 378).

Thomas Roe reports that the imperial standard sported a couchant lion and the rising sun (Welch 1986). A painting by Payag from the *Padshahnama,* the chronicle of Shah Jahan's reign by Abdul Hamid Lathery preserved in the Royal Library, Windsor Castle, depicts the siege of Qandahar in the Deccan in 1631 by Shah Jahan's army led by Nasiri Khan. The imperial standards are scarlet pennants with green borders with a passant lion and a rising sun behind it in each of them (Beach and Koch 1997: 18). Another *Padshahnama* painting by the Kashmiri painter, of Shah Jahan's royal procesors in 1655, clearly shows pennants which have a field of green with a couchant lion an the sun rising behind it in each (ibid.: pl. 34). This symbolism is ancient and was common in older Persian imagery; it also has echoes of the Hindu sun worship and the lion's association with Hindu royalty and divinity. No doubt the Mughals took these symbols from their own cultural past, but they must have been aware that the lion and the sun held a special place in the tradition of royalty preceding them in India.

Emperor Akbar maintained an account of the game he had shot and even the records of the guns used for the hunt were kept (Constable 1934: 379). Jahangir too was equally meticulous. In the eleventh year of his reign he records that in a span of 39 years, as many as 28,532 animals and birds were hunted in his presence of which he himself accounted for 17,167. Lions take pride of place for they appear at the top of the list the tally and he records that he had shot 86 of them (Thackston 1999: 216).[1] Not surprisingly, one of the finest studies of an animal in Mughal painting is a painting of the Jahangir period. 'Lion drinking water after devouring its kill' attributed to Nanha circa 1618, preserved at the Sawai Man Singh II Museum, Jaipur.[2] The painting is realistic to high degree in its depiction of the animal, its environment and its prey a bull, or a cow, the remnants of which are being sequestered by a jackal! But in the hunting field the animal was pursued with vigour and its successful shikar was a matter of good omen. Bernier records the following:

> You must know, then, that as it is considered a favourable omen when the king kills a lion, so is the escape of that animal portentous of infinite evil to the state. Accordingly the termination of the hunt is attended with much grave ceremony. The king being seated in the general assembly of the *Omrahs,* the dead lion is brought before him, and when the carcass has been accurately measured and minutely examined, it is recorded in the royal archives that such a king on such a day slew a lion of such a size and of such a skin, whose teeth were of such a length, and whose claws were of such dimensions, and so on down to the minutest detail (Constable 1934: 379).

Though the weighing and measuring of lions with extreme care was the established Mughal practise, as reported by Bernier, just one record made by Jahangir survive. When the emperor encountered and shot an exceedingly large lion near Rahimaba between Delhi and Agra on December 22, 1623, he took great care to have it weighed and its length measured from the top of its head to the tip of its tail: He noted that in his entire hunting career he had not come across such a large, magnificent, or proportioned animal. As was his practise he ordered the imperial painters to take its likeness (Thackston 1999: 411, 473), as he had done several times before when he came across a rare or magnificent bird or animal, such as the turkey, the zebra, or barbary falcon.

It is regrettable that no other such records are available, as much information about these cats could have been extracted from them. The lion shot by Jahangir weighed eight and a half Jahangiri maunds or 255 kilograms (1 Jahangiri maund = 30.10 kilograms) (Habib 1963: 368). It was three and a half cubits (*zar-i-shari*) and two *tasu* long, i.e. 9 feet 4 inches or about 2.85 metres (1 cubit = 31.7 inches and 1 *tasu* = 1.33 inches) (Thackston 1999: 473). The weight range of a lion is between 145 and 225 kilograms (Sunquist and Sunquist 2002: 288). Jahangir's lion was obviously heavier than usual whereas in length it would be among the larger trophies recorded. From Jahangir's description it is clear that he had the animal measured and not its skin which could have yielded a longer result.

Before we go further it is necessary to examine the words used for the lion and the tiger in the chronicles of the times. The Persian word for the lion is *shir* while it is *babr* for the tiger (Humphreys and Kahrom 1995: 75, 77). The two animals have distinct words, with no hint of confusion in the minds of Mughal emperors or their chroniclers. In India, however, by the time the Urdu language got round to recognizing the big cats, the issue got mixed up. By the nineteenth century therefore the word *babbar* or *babbar sher* came to mean lion in Urdu/Hindustani, and the word *sher* meant tiger—in any case by the second half of the nineteenth century the latter was largely the only one to survive in the Hindi-speaking belt. (In Sanskritized Hindi the word is *simha* for lion, *bagh* for tiger.) The translators of the Mughal texts who were working in north India in the second half of the nineteenth century or later paid more attention to the political, social, and other aspects of what the texts contained as far as the accuracy of the translations was concerned. They simply translated *shir* as tiger without much ado.[3] They were also misled by their familiarity with Hindustani in which the word *sher* is used both for lion and tiger, such usage being common even today. In contrast, the Mughal emperors and their chroniclers were closer to nature than their British translators living in presidency towns such as Calcutta or smaller 'stations' and they knew what they were seeing and recording.

If we look at the locations where Akbar or Jahangir came across the big cat the picture becomes clear. Akbar encountered it at Mathura and Mewat near Delhi; Jahangir records his encounters with it at Bak Bhal and Nag Thal near Bari, near Dholpur, Palam (the site of the international airport at Delhi), Rupbas near Agra, Ajmer, and Mandu where, as we shall see, Thomas Roe had his problems with a lion. These were all grassland and scrub at the time, which was the preferred habitat of the lion. There may have been stray tigers but they were not likely to have been encountered in groups as described by the Mughals. Of the 33 wild species of the family *Felidae* (Sunquist and Sunquist 2002: v–vi), the lion is the only gregarious cat; the tiger is a solitary creature as a rule except when a female is with cubs.

Another noteworthy fact is that there are no known Mughal paintings in the public domain depicting a tiger hunt or the study of the tiger as an animal per se, except for one which records the chance encounter of Akbar in 1561 near Narwar in central India when a family of five tigers, two of which were 'white', attacked his cavalcade. The event is recorded for all time with a double-page illustration in the *Akbarnama* and Abul Fazl uses the correct Persian word for them, *babri* in the text (Sen 1984: pl. 20).[4] Ashok Kumar Das has shown that there are more than thirty paintings of the Mughal period in the public domain depicting a lion, a lion hunt, or a dead lion. It is equally noteworthy that not one illustration in any of the paintings is of a hunt from a machan, unlike the depictions in Kotah paintings where the quarry, either lion or tiger, was hunted in dense jungle rather than grassland or scrub thus necessitating the perch for the safety of the royal sportsman. The open scrub jungles and grasslands were easily accessible from the huge Mughal camps and they usually contained lions, not tigers.

The absence of paintings of tigers does not suggest that the Great Mughals did not hunt them. The *Muraqqa-e-Gulshan* (Gulshan Album) was prepared for Emperor Jahangir between 1605 and 1610 and some folios were added in 1616 and 1617. The album is now dispersed, the bulk of it being preserved at the Gulistan Palace Library, Tehran, with some folios in the Staatsbibliothek, Berlin and the Nelson Atkins Museum of Art, Kansas City. Das who has examined the Tehran folios and studied others, points to the *hashiyas* or margins of the album's folios. They contain monochrome drawings in black and gold which depict hunting scenes. Among them are a tiger behind a hilly outcrop, a tiger killing a spotted deer, a Rajput nobleman shooting a tiger with a gun, a Mughal prince hunting one of a pair of tigers with bow and arrow from horseback. There are also two drawings of a dead tiger being loaded on a saddled elephant.[5] However, barring such exceptions, the absence of paintings per se of tiger hunts and of such descriptions in chronicles testify to the lion's central position in Mughal hunts, while the tiger remained marginal.

A miniature painting has recently been published for the first time, titled 'Akbar on a Tiger Hunt' from the *Akbarnama*. A close look at the feline being speared by Akbar's entourage and watched by the emperor from elephantback shows uniform stripes on its body, a leonine mane, a cheetah-like tearline on its face, and a tail which has no feline characteristics (Thapar 2004: 250). It is obviously a composite animal. Das feels that the painting is of recent origin.[6]

According to his memoirs, Jahangir came across a '*shir babar*' at Giri near 'Kaḷiyadaha'—Kaliyadoh is the current name—in the neighbourhood of Ujjain. Having shot it, he had it opened to find the location of the gall bladder of the animal and it occurred to him that its location within its liver may be a source of its ferocity (Rogers and Beveridge 1909–14, vol. I: 351). Both Rogers and Beveridge and Wheeler Thackston translate the

Persian word '*shir babar*' as lion but the accurate rendering would be tiger (ibid.; Thackston 1999: 207). Wheeler Thackston's translation of Jahangir's memoirs has otherwise consistently and correctly translated the word *sher* as lion. A person with Jahangir's extensive and detailed interest in nature was not likely to make a mistake in the name of the animal he encountered. It may be recalled that he describes at great length other animals and birds that caught his fancy.

There are two instances of Akbar hunting lions which have been recorded in the *Akbarnama* (Beveridge 1904, Vol. II: 294, 482–83).[7] In 1562 he went lion hunting near Mathura in UP. He chanced upon a pride of seven lions. Abul Fazl records that 'five were leveled with the dust by arrow and bullet and one that repository of courage caught alive and so was a subject of a thousand wonderment'. The seventh lion he mentions was caught by the combined efforts of Akbar's courtiers. However, he does not describe how the lions were caught. The second incident took place in the Mewat region near Alwar in present-day Rajasthan, in 1568 when Akbar was travelling from Ajmer to Alwar. While out hunting, a lion emerged from the reed bed and was quickly slain by a hail of arrows. Akbar was most upset and ordered that should another lion come out he would tackle it himself. At that very moment another lion did come out. Akbar shot it in the eye with an arrow, but still the animal advanced and sat in such a position that Akbar could not get a good shot even though he dismounted from his horse. In the excitement of the moment a courtier shot an arrow which infuriated the lion, who mauled him. The animal had to be finished off by the other courtiers (ibid.: 482–83).

In the *Ain-i-Akbari*, Abul Fazl records two more instances of Akbar hunting lions. One incident occurred in Beri where a man-eater attacked Akbar's elephant, and had to be killed by his men. In another instance near Mathura, a lion came out to attack when one of Akbar's courtiers, Shujaat Khan, approached too close. But Akbar was able to subdue the lion with just a furious look: 'The brute cowered down before that divine glance, and turned right about trembling all over. In a short time it was killed' (Blochman 1873, Vol. I: 294).[8]

There is only one painting of Akbar hunting a lion known to us. Preserved in the National Museum, Delhi, the painting titled 'Akbar on a hunt' is by an unknown artist and may be dated circa 1598–1600 (Daljeet 1999: 41). The middle of the painting has Akbar on a chestnut charger. He has just shot an arrow which has pierced the lion's abdomen and it is bleeding while it is in the act of killing a wild ass. On the lower left of the painting another wild ass is bolting from the scene of action. On the right-hand bottom of the painting area a group of courtiers that include Bairam Khan on a white horse, others carrying mace and gun, and falconer with a falcon on his gauntlet. In front of Akbar's horse a male and a female blackbuck are leaping away along with other small animals which appear to be foxes and a hare. In the top right-hand corner are a group of men with hawks and cheetah brought on horseback to the hunting grounds.

Though the painting has been dated circa 1598–1600, it pictures Akbar as a young man. From the foregoing passages we know that Akbar hunted lions with bow and arrow from horseback. The lion being killed while it is on its own kill of a wild ass is reminiscent of the painting of Bahram Gur hunting lions. While this may be a rendering of an actual occurrence, the presence of blackbuck, foxes and a hare in such proximity to the lion suggests that the painting represents various events of a day's hunting or it is a collage of several hunts depicting Akbar's prowess in the field rather than a single event.

Akbar's bravery and fearlessness in the field have been recorded several times by his chroniclers. It is not at all surprising that he chose to hunt lions from horseback with bows and arrows and sometimes even on foot though guns were in common use in his time. His successors, who inherited the grand empire, preferred the safer elephant and matchlocks for their lion hunts. (It is however interesting to note here that lions were speared from horseback by the British near Deesa in Gujarat in the nineteenth century.)

To Jahangir, the hunting grounds were a second home. His passion for the chase and his acute observations of flora and fauna from his memoirs has left us an incomparable record of this facet of his life and times. He could be as considerate as providing warm water for bathing the imperial elephants as they shivered in the cold or building a grand monument to his favourite blackbuck Mansraj at Shaikhupura near Lahore, and yet think nothing of ordering the execution of a servant for inadvertently interrupting his hunt. His records of game bagged show the variety of animals he shot, from the largest to the smallest. The lion hunt to him was special and it always featured at the top of his list. In February and March of 1610 he camped at Mandakar Garden near Rupbas. During his stay there of 56 days he accounted for seven lions, 70 nilgais, 51 blackbucks, 82 other animals,

129 birds, and 1,023 fish (Thackston 1999: 108)! Again he records that over a period of three months at the beginning of 1611 he had accounted for 12 lions (ibid.: 120). Here we get a clear glimpse of the lion's as well as their prey, nilgai and to a lesser extent blackbuck. It is interesting to note that he had hunted one lion per week in 1611, an indication of the abundance of these cats in the region. Of the several shikar events concerning lions recorded by him, some may be described here.

In February 1608, Jahangir was informed that lions were making a nuisance of themselves on the road between Panipat and Karnal in present-day Haryana. He went out to the spot, got onto a female elephant, and ordered the rest of the elephants to encircle the lions as in a Qamargha. Here was a quarry-specific variation of the large Qamarghas which used vast resources and in which all and sundry animals were taken. The method was successful and 'by God's grace' the emperor was able to 'eliminate the evil' of the pair of big cats that had harassed the people (ibid.: 91). Elephants rather than any other animal were used for encirclement as they provided a perch from which the quarry could be seen, they were large enough to be respected by lions, and finally, they ensured a measure of safety to the imperial person.

Jahangir abstained from hunting with a gun on Sunday and Thursdays; yet when provoked he broke his own rule and it did not take much for him to do so. He records that on Sunday, 2 March 1617 while he was at Mandu, he was informed of the presence of a lion 3 *Kos* away from the camp. He promptly rode out on an elephant and when he came to the spot he found the lion sitting under a tree. He fired from elephant back and killed it with one shot. The bullet entered its mouth leaving no external mark of injury. His retinue were baffled when they examined the animal's body, until Jahangir cleared up the mystery by pointing to its mouth (ibid.: 213).

Jahangir also showed off his hunting skills to the young Mewar prince Karan who was at his court at Ajmer in 1615. On hearing of a lioness in the vicinity of his camp he went out to shoot it. He asked Karan where he should shoot the quarry. Karan indicated its eye. Jahangir obliged successfully, to the astonishment of all present. Jahangir knew that firing from elephantback which is not a steady perch, that too on a windy day and with a smooth bore musket, it was miraculous that the ball should find its predetermined minute target. He was honest enough to record in his autobiography, 'God almighty did not allow me to be ashamed before the prince, and as I had agreed, I shot her in the eye' (Rogers and Beveridge 1909–14, Vol. I: 280).[9]

Even for a Mughal emperor and for a keen marksman, things could and did go horribly wrong in a lion shoot. In 1610 when Jahangir was near Bari in Agra *suba,* one of his courtiers Anup Rai came across a half-devoured cow. Suddenly a lion came out of the thicket. It was promptly surrounded and the emperor was informed. Jahangir, Khurram, and others rushed to the spot. Jahangir set his musket on the stand and fired twice but missed as is not uncommon when it comes to judging the 'elevation' of a bullet when firing from a height as Jahangir was doing. In the event, the lion charged, Jahangir's retinue bolted, and, according to him, some even trampled him in their flight! Anup Rai saved Jahangir's life by battling the lion almost with his bare hands, while Prince Khurram struck the animal with a sword. At the end of it the lion was dispatched and the emperor was saved. In recognition of his valour Jahangir gave Anup Rai the title of *Ani Rai Singhdalan* meaning commander of troops, lion crusher (Thackston 1999: 117–18). This incident shows the personal bonds between the Mughals and Rajputs which held the Mughal domains together until the early 1700s. The fact that Jahangir and Khurram were accompanied by a Rajput courtier in a hunt shows the trust and respect placed in him.

Hunting lions on foot was certainly not a recommended method for Mughal emperors and their family, unless of course the situation arose accidentally. There is an unpublished painting from the personal collection of Maharaja Sawai Bhavani Singh of Jaipur which has been examined by Ashok Kumar Das. He dates it circa 1600, it is by an unknown artist and it shows unmistakably a young Jahangir. It depicts a night scene where a remarkably accurately drawn male lion has attacked an attendant of Jahangir in the left hand side of the painting. Though the lion has pinned him down, the attendant is about to plunge a dagger into his tormentor. The night scene is lit by two torch bearers behind the lion while a third holds his torch down near the lion's body as if to illuminate it for Jahangir who stands slightly to the right of the lion. He is about to strike the lion—his arms are raised and he holds a sword. An attendant stands beside him with a mace-like staff in his hand. There is a hill in the background, and a few more torch bearers stand at the top right-hand corner of the painting. The incident is depicted in open country interspersed with tall grass. Hunting antelope, particularly blackbuck, at night with torches was known then.

Was Jahangir out hunting antelope that night when his party surprised a lion? The painting has the moon partially covered by clouds—did Jahangir purposefully go after the lion on a moonlit night? Or is the painting entirely apocryphal? We do not know. What is certain, according to Dr Das, is that the painting is contemporaneous.[10]

On occasion there was even a romantic twist to a lion hunt. In April 1617 Jahangir went out lion hunting with his harem near Mandu. The party was on elephantback when four lions came out towards the elephants—presumably they were driven towards the hunters. Nurjahan asked permission of the emperor to shoot, which was readily granted. Two lions were dispatched with one shot each and the other two were accounted for by two bullets each, thus four lions were killed with six bullets. Considering that Nurjahan fired from a covered howdah with obviously restricted maneuverability, from a mobile, and therefore none too steady, elephant perch, and using a smooth bore musket, it was a truly remarkable feat of marksmanship. Jahangir was so pleased with the empress's hunting skill that he scattered a thousand *ashrafis* over her head and gifted her a pair of pearls and a diamond worth a lakh of rupees (Thackston 1999: 219).

In his autobiography Jahangir referred to his father with the utmost reverence and yet if he had scored a point over him, he was quick to document it. For instance, he records that Akbar tried hard to get cheetahs to breed in captivity but failed though he had collected a thousand cheetahs. In the year 1613, however, a pair mated and three cubs were born at Jahangir's court. The emperor states that he has recorded the fact as it was strange.[11] In the same breath he goes on to record that in the benevolent reign of his times wildness had been eliminated, lions had become tame and were roaming in prides among people without the restraint of a chain, not harming them. This is far-fetched and if such was indeed the case he would hardly have pursued them so vigorously in the field. However, the fact is that lions can be tamed easily and there is a painting by the artist Padarath depicting a young lion—it has a partially grown mane—being taken for a walk by keeper, only lightly chained (Alvi and Rahman 1968: pl. vii) in the manner of a not too aggressive canid. Jahangir notes further that he had heard from physicians that a lioness's milk was very good for the eyes. He tried to have a lioness who had given birth to three cubs milked, but did not succeed. He concluded that the milk could not be obtained owing to the irascible nature of the animal (Thackston 1999: 144).

As mentioned earlier, Jahangir had ordered a painting of the very large lion he had shot at Rahimabad, but the whereabouts of this painting, if it has survived, are not known. We may however look at another very remarkable work of his period. It is painting entitled 'Animals' in the *St Petersburg Muraqqa* (Kostioukovitch 1996: pl. 146, f. 72 recto; pp. 97–98). It is a collage which Stuart Cary Welch calls a surrealistic *omnium gatherum*. The bottom of the painting has goats, sheep, and nilgai; in the lower centre portion a hound suckles one pup while the other two frolic about her. In the upper portion a cat suckles her three kittens. The artist is unknown but the cat with her kittens may have been painted by Abul Hasan according to Welch (ibid.: 97–98). The painting has small figures of other animals, including an elephant attacked by a dragon-like creature. Of abiding interest is a couchant lion with the tip of the tail held high. The animal has been accurately rendered, its skin being the exact shade between dark yellow and brown, and its mane in darker brown with lighter hair around its face and sparse over its head as it should be—a characteristic of Indian lions. The tip of the tail is dark and the tuft prominent—another characteristic of the lion from India. Its mouth is slightly open, giving it a truly regal aspect. Of equal interest are the scenes at the top left and right corners of the painting. On the left a lion is in the process of strangulating a nilgai with his hold on its neck. On the right a lion is in the process of bringing down a chital. These are both depictions of a lion's methods of killing its prey and in both cases the animals are its usual prey. The artist was not only a master of his craft but was also familiar with the hunting behaviour of the lion in the wild, which no doubt would have been appreciated by Jahangir's discerning eye.

Though Jahangir's autobiography is replete with instance of lion hunts, only two paintings can be identified with the incidents described there. One is the Anup Rai incident which has been painted several times, the most famous rendering being the one by Balchand circa 1640 in the *Padshanama*. It shows Anup Rai under the lion, while Prince Khurram is about to strike it. One member of the party strikes the lion on its head with a stick, another strikes it with a sword on its hindquarters, yet another courtier looks on as does Jahangir himself (Beach and Koch 1997: pl. 30). The other is the incident when Jahangir shot a lioness in the presence of Prince Karan of Mewar (Das 1978, frontispiece; Divyabhanusinh 1999).

From these paintings we learn little of the techniques of hunting lions, since one was a chance encounter while another tells us that the shikar took place from elephantback. For this we turn to other sources. Bernier has left a firsthand account of what he saw in Aurangzeb's time (1658–1707). When a lion was located,

> As a preliminary step, an ass is tied near the spot where the gamekeepers have ascertained the lion retires. The wretched animal is soon devoured, and after so ample a meal the lion never seeks for other prey, but without molesting either oxen, sheep or shepherds, goes in quest of water, and after quenching his thirst, returns to his former place of retirement. He sleeps until the next morning, when he finds and devours another ass, which the gamekeepers have brought to the same spot. In this way they contrive, during several days, to allure the lion and attach him to one place; and when information is received of the King's approach, they fasten at the space an ass where so many have been sacrificed, down whose throat a large quantity of opium has been forced. This last meal is of course intended to produce a soporific effect on the lion (Constable 1934: 378).

This time-honoured method, without the opium—Bernier himself states in the narrative that 'it is a tale of the vulgar, and the lion is sufficiently disposed to sleep without it when he has eaten to satiety'—has been used in India by maharajas and nawabs for shikar and even in our own times when lions were 'anchored' for census operations or for the 'lion shows' which were staged until not so long ago for the benefit of tourists in the Gir forest. Bernier continues:

> The next operation is to spread by the means of the peasantry of the adjacent villages, large nets, made on purpose, which are gradually drawn closer in the manner practised in hunting the nil-ghaux. Every thing being in this state of preparation, the king appear on an elephant protected in places with thin plates of iron, and attended by the Grand Master of the Hunt, some *Omrahs* mounted on elephants, and great numbers both of *gourze-berdar* on horse back and of gamekeepers on foot, armed with half pikes. He immediately approaches the net on the outside, and fires at the elephant, according to the invariable practise of lions, but is arrested by the net; and the king continues to discharge his musketoon until the lion is at length killed (ibid.).

A painting attributed to Daulat and dated circa 1635 in the *Padshahnama,* 'Shah Jahan hunting lions near Burhanpur' (Beach and Koch 1997: pl. 46) illustrates this technique. There are three elephants. Shah Jahan is seated on the middle one, the howdah is designed for the hunt as it is open on all sides to enable the hunter to manoeuvre the gun in all directions. Shah Jahan has rested his musket on the shoulder of the mahout; if he were shooting on the ground, it would have rested on a tripod or a bipod. A lion, a lioness, and two cubs have been encircled by nets though, unlike in Bernier's description, Shah Jahan and the lions are on the same side of the net.

In the incident of Jahangir showing his hunting skills to Prince Karan, there is no sign of the net or for that matter any obstruction between the lioness and the imperial elephant. Colonel Kesri Singh, head of the Shikar Khana of Gwalior state in the 1930s and a renowned tiger hunter of his time, describes a tiger hunt with nets. The occasion was staged for the Columbia Broadcasting Corporation in 1958. In this event the tiger was driven to the nets by a Land Rover whereupon the animal was speared by the people surrounding the net from the outside (Singh 1959: pl. opp. p. 161, 1965: 116–19). Bernier's observation therefore may not have been a common practise, as when the lion was hunted from elephantback with a musket it could be dispatched from a safe distance, whereas keeping the net between the hunter and the hunted would be necessary while using a bow and arrow or using a long spear from elephantback. There are paintings however, of lions being speared without the use of a net by imperial Mughal hunters, as we have noted earlier in the case of a painting of Akbar.

In the hunting field methods of hunting had to be adapted to the circumstances pertaining to the hunting grounds, the quarry, and naturally the wishes of the hunter. In a painting of circa 1660 by an unknown artist, Emperor Shah Jahan and his four sons are depicted. The painting may be titled 'Shah Jahan hunting lions with the four princes' and is in the collection of Edmund de Unger. The whole hunting party involving Shah Jahan and all his direct and nearest in the hunt is unknown but the painting is described by Robert Skelton as follows:

> Emperor Shah Jahan... Can be identified [on] the penultimate elephant in a line of five, advancing from the left. The emperor is shown preparing to aim his matchlock at a lion which is attacking the elephant. The howdahs of the three elephants to the right of Shah Jahan contain the figures of his sons Murad Baksh [next to his father], Dara Shikoh, and Shah Shuja. The figure on the right hand side of the nullah are three elephants, two taking part in the beat and the central one bearing a prince who can probably be identified as Aurangzeb,

> about to throw [sic] at a lion which has leaped onto his elephant. On the left bank, three buffaloes. Are advancing on the nullah. A lion spring from a thicket and another lion mauls a fallen shikari. In the background is a line of nets supported by poles enclosing the hunting field. The perimeter is patrolled by elephants and solitary horseman rides in the distance (Robinson 1976: pls 130–31, p. 264).

While the net is a feature of the hunt, here the hunters and the hunted are again on the same side of it. Also the presence of buffaloes is not an accident. These animals are used as 'stops' to confine the lions in a more or less predetermined area. The use of 'stop' in driving game in or towards the desired ground or on the flanks of a beat has been known in India well into our own times. This method was not unique to lion hunts alone, it was used by Mughals for other game as well such as confining nilgai on the hunting ground, as in the painting of 'Dara Shikoh shooting nilgais' (Koch 1998: 23).

Bernier records that Aurangzeb put his son Sultan Muazzam to a strange test to check out his obedience and courage. He ordered him to attack and kill a lion which had been tracked down without the use of nets—that is without the help of confining the dangerous prey in an area which would give the advantage to the hunter. The prince obeyed, and in the event two or three men lost their lives, some horses were mangled, and the wounded lion jumped on Prince Muazzam's elephant before meeting its end (Constable 1934: 182–83). Here was a peculiar Mughal variant of baptism by fire!

Bernier was also witness to a hunt in which a lion escaped Aurangzeb's gun. The whole army was deployed to locate the escaped beast, it was 'subjected to great inconveniencies and thrown into a considerable degree of confusion' as a result and the march had to be halted for three or four day. There were no bazaars pitched at this place and many went hungry (ibid.: 379). Bernier does not mention what happened to the lion, but its escape was considered a bad omen, just as the king's success in killing it would have been a favourable one. Was this a foreshadow of coming events? Aurangzeb's policies left a fractured legacy and the end of his reign saw the beginning of the decline of the Mughal empire.

Abul Fazl describes another method of hunting lions wherein a hunter rides a male buffalo and makes it attack a lion, the buffalo tossing the lion with its horns and killing it (Blochman 1873: 294).[12] This is surely not a method recommended for the emperor or princes, but a variation of it was certainly in vogue. A painting titled 'Aurangzeb hunting lions' circa 1670–1700 by an unknown artist, preserved in the Chester Beatty Library (Acc. No. CBL IN 11A.28), shows Aurangzeb and his entourage on three male elephants on the right-hand side of the painting. The elephants are behind a phalanx of 10 buffaloes in front of which are a lion and lioness. Towards the top left is an elephant which has been attacked by a lion from the back; another elephant is crushing a lion with its tusks. The emperor is aiming his gun at the lion while sitting on an open howdah, with the stock of the matchlock resting on its frame. Almost an exact reproduction of the picture is found in a painting from Bikaner by Rashid dated circa 1693. The crucial difference is that the hunters are Maharaja Anup Singh of Bikaner and his three brothers (Goetz 1950: pl. 77, p. 170) instead of Mughal party! In both these paintings the hunters and the hunted are both within the nets.

In many ways the Great Mughals were no different from their predecessors. However, unlike the earlier sultans of Delhi, the lion to them was royal game. Though it was not spared in the hunting fields by them, no one else could meddle with it without their permission. It symbolized regal power and it sat proudly on their imperial standards. As was the case with everything they did, their shikar expeditions were larger than life. Their successors in the eighteenth century continued hunting but the pomp and circumstance progressively waned, until the last of the Mughals, Bahadur Shah Zafar, was exiled to Rangoon, Burma in 1858. The later Mughals were but a shadow of their fabulous predecessors, they could not be expected to add any new dimension to the courtly pursuit of hunting.

Jahangir did not describe a single lion in detail in his memoirs as he did, for example, the turkey, zebra, and barbary falcon. This is not surprising, as lions were plenty in his time, they were commonplace. It is estimated that the Agra *suba* at the heart of the Mughal empire had only 27.5 per cent land under cultivation in about 1600, land other *subas* in the plains had even less (Trivedi 1998). The rest of the land was either uncultivable or cultivable wasteland. In other words the Mughal empire had large tracts which were suitable habitat for lions. They were numerous enough to be found almost everywhere the Mughal emperors went. They did not have to think of the Gir, the lion's present and only home in India. To them it was a far off corner in their domains and they estimated its area to be no more than 3×3 *Kuroh* that is just about 90 sq km! (Habib 1982a, Sheet 7A: 25)

But 200 years after Jahangir, lions had almost died out in India. The reasons for this are not far to seek. Unlike their Sultanate predecessors, the Great Mughals restricted the hunting of game animals and birds, and for the smaller creatures which could be snared they set aside large hunting grounds as well. With their eclipse, protection no longer remained effective. The lion was an easy target because of its gregarious nature and diurnal habits. The constant pursuit of this great cat in the field, along with steady encroachment into its habitat, took a heavy toll. The human population within the bounds of India had grown from an extrapolated/estimated figure of 116 million in 1600 to 159 million in 1800 (Guha 2001: 34, 58).[13] Just before the last Mughal was to live out the final scene of the great drama, a single British officer is reported to have shot as many as 300 lions. Even if this figure was an exaggeration, which it probably was, we may recall that Emperor Jahangir with all his enthusiasm for the hunt and all the imperial resources at his command had shot only 86 lions in a span of 39 years! What was left behind by the Mughals was about to be decimated by their successors.

## Notes

1. Rogers and Beveridge (1909–14, Vol. I: 369) give the same figure, but identify the animal as a tiger. We shall come to this matter presently.
2. I am grateful to Ashok Kumar Das for bringing this painting from the Maharaja Sawai Man Singh II Museum, Jaipur, Acc. No. 863, to my attention and dating it.
3. I am grateful to Professor Muzaffar Alam who checked the Sayeed Ahmud text of *Tuzuk-i-Jahangiri* and found the word *shir* in all relevant cases on pp. 65, 78, 80, 89, 92, 117, 140, 175, 179, 182, 185, 221, 366 and 375. See Ahmud (1864).
4. For a detailed analysis of the event, see Divyabhanusinh (1986 and 1999).
5. Ashok Kumar Das, personal communication, 2003.
6. Ashok Kumar Das, personal communication, 2004.
7. Professor Muzaffar Alam checked the *Akbarnama* text edited by Maulvi Abdul Rahim and found the word to be *sher* in both cases on pp. 189 and 238. See Rahim (1879).
8. Professor Muzaffar Alam checked the *Ain-i-Akbari* text edited by Saiyad Ahmad Khan and found that the word is *shir* on p. 145. See Khan (1882).
9. See Divyabhanusinh (1999) for a detailed analysis of the incident.
10. The description of the painting is based on my discussions with Ashok Kumar Das.

11. See Divyabhanusinh (1987) for a detailed analysis of the incident.
12. The translated word for *shir* is tiger here.
13. Earlier estimates by Habib (1982b), which have now been paired down, were a population of 145 million in 1600 and less than 200 million by 1800.

## References


Ahmud, Sayeed, 1864, *Toozuk-i-Jehangeeree*, Ally Gurh: published by author.

Alvi, M.A. and A. Rahman, 1968, *Jahangir—The Naturalist*, National Institute of Sciences of India, New Delhi.

Beach, Milo Cleveland and Ebba Koch, 1997, *King of the World: The Padshahnama: An Imperial Mughal Manuscript from the Royal Library*, Windsor Castle, Washington, DC, Azimuth Editions/Sackler Gallery.

Beveridge, Annette Susannah (tr.), 1922, *Babur Namah (Memoirs of Babur)*, vols I and II, repr. 1979, Oriental Books Reprint Corporation, New Delhi.

Beveridge, H. (tr.), 1904, *The Akbar-Nama of Abul-Fazl*, vols I and II, repr. 1979, Es Es Publication, New Delhi.

Blochman, H. (tr.), 1873, *The A'in-I Akbari by Abu'l Fazl*, vol. 1, 2nd ed., rev. and ed. in 1977, by Lt. Col. D.C. Phillot, Oriental Books Reprint, New Delhi.

Constable, Archibald (tr.), 1934, *Travels in the Mughal Empire by Francois Bernier A.D. 1650–1668*, 2nd ed. rev. by Vincent A. Smith, repr. 1983, Oriental Reprint Delhi.

Daljeet, 1999, *Mughal and Deccani Paintings of the National Museum*, Prakash Book Depot, Delhi.

Das, Ashok Kumar, 1978, *Mughal Painting during Jahangir's Time*, Asiatic Society, Calcutta.

Divyabhanusinh, Chavada, 1986, 'Earliest Record of a White Tiger', *Journal, Bombay Natural History Society*, 83 (supplement): 163–65.

———, 1987, 'Record of Two Unique Observations of the Indian Cheetah in Tuzuk-i-Jahangiri', *Journal of Bombay Natural History Society*, 84 (2): 269–74.

———, 1999, 'Hunting in Mughal Painting', in Som Prakash Verma (ed.), *Flora and Fauna in Mughal Art*, Marg Publications, Mumbai.

———, 2005, *The Story of Asia's Lions*, Marg Publications, Mumbai.

Foster, Sir William, 1926, *The Embassy of Sir Thomas Roe to India 1615–19 as narrated in his Journal and Correspondence*, repr. 1990, Munshiram Manoharlal Publishers, New Delhi.

Goetz, Hermann, 1950, *The Art and Architecture of Bikaner State*, Bruno Cassirer, Oxford.

Guha, Sumit, 2001, *Health and Population in South Asia from Earliest Times to the Present*, Permanent Black, New Delhi.

Habib, Irfan, 1963, *The Agrarian System of Mughal India (1556–1707)*, Aligarh Muslim University and Bombay, Asia Publishing House, Aligarh.

———, 1982a, *An Atlas of Mughal India: Political and Economic Maps with Detailed Notes, Bibliography and Index*, repr. 1986, Oxford University Press, Delhi.

———, 1982b, 'Population', in Tapan Raychaudhuri and Irfan Habib (eds), *The Cambridge Economic History of India, vol. I: c1200–c1700*, repr. 1984, Orient Longman, Hyderabad, and Cambridge University Press, Cambridge, pp. 163–71.

Humphreys, P.N. and E. Kahrom, 1995, *The Lion and the Gazelle: The Mammals and Birds of Iran*, Comma International Biological Systems, Lower Coed Morgan, Nr., Abergavenney, Gwent, UK.

Khan, Saiyad Ahmad, 1882, *Ain-i-Akbari*, Nawal Kishore, Lucknow.

Kostioukovitch, Elena (ed.), 1996, *The St Petersburg Muraqqa: Album of Indian and Persian Miniatures from the 16th through the 18th Century and Specimens of Persian Calligraphy by Imad Al-Hasani*, Lenardo Arte srl, and Lugano: ARCH Foundation, Milan.

Rahim, Maulvi Abdul (ed.), 1879, *Akbarnama*, Asiatic Society, Calcutta.

Rogers, Alexander and Henry Beveridge, 1909–14, *The Tuzuk-i-Jahangiri or the Memoirs of Jahangir*, repr. 1980, Munshiram Manoharlal Publishers, New Delhi.

Sen, Geeti, 1984, *Paintings from Akbar Nama*, Lustre Press, Varanasi.

Singh, Col. Kesri, 1959, *The Tiger of Rajasthan*, Robert Hale Ltd, London.

———, 1965, 'Hints on Tiger Shooting (Tigers by Tiger)', *The Hindustan Times Press*, New Delhi.

Sunquist, Mel and Fiona Sunquist, 2002, *Wild Cats of the World*, The University of Chicago Press, Chicago.

Thackston, Wheeler M. 1996, *The Baburnama: Memoirs of Babur Prince and Emperor*, Freer Gallery of Art/Arthur M. Sackler Gallery, Oxford University Press, New York.

———, 1999, *The Jahagirnama: Memoirs of Jahangir, Emperor of India,* Freer Gallery of Art/Arthur M. Sackler Gallery, Washington, DC; Oxford University Press, New York.

Thapar, Valmik, 2004, *Tiger: The Ultimate Guide*, CDS Books/Two Brothers Press, New York.

Trivedi, K.K., 1998, 'Estimating Forests, Wastes and Fields, c1600', *Studies in History*, 14 (2): 301–11.

Welch, Stuart Carey, 1986, 'A Lion-King's Lion', in Lokesh Chandra and Jyotindra Jain (eds), *Dimensions of Indian Art: Pupul Jayakar at Seventy*, vol. I, Agam Kala Prakashan, Delhi.

6

# Ecology and Traditional Systems of Water Management: Revisiting Medieval Rajasthan

Mayank Kumar

Despite its central role in sustaining agriculture, the political and social relations of water use and abuse in Indian society have only recently been scrutinized by historians. Recent writings on Indian environmental history have examined the role of state and society with special reference to environmental degradation. Most writings have analysed the situation in the terms of decay during colonial rule and in turn tended to romanticize the pre-modern period.[1] These works have extensively catalogued the nature of social participation, argued the relative unconcern of state towards the agricultural production process and suggested that states were concerned almost solely with revenue realization. Few writers have probed the significance of pre-colonial water systems;[2] this is especially true of north and north-west India (but see Fisher 1997; Hardiman 1995: 185–209; Wahi 1997). Most scholars have stressed the role of traditional village community in the construction and maintenance of irrigation technologies. David Hardiman suggests that 'small-dam systems of irrigation existed in the past which were sustained over long periods of time... by community based control' (Hardiman 1995: 209).[3] There have been several excellent works on pastures, fields and forests of colonial and contemporary Rajasthan. But except for passing references in studies of agrarian production, and revenue, few have examined the dynamics of water management in Rajasthan prior to 1800 (Gold and Gujjar 2002; Jodha 2001).[4]

* This paper was especially written for this volume. The author is grateful to Mahesh Rangarajan and the reviewers for their comments.

This essay explores and examines various systems for water appropriation and agricultural practises adapted by society in the arid and semi-arid conditions of medieval Rajasthan (between 1500 and 1850 CE). It also tries to question the prevalent notions of pre-colonial state where it was presumed that the state became more intrusive only with the increased need to move towards social engineering (Scott 1998: 185). There is ample evidence of state intervention in the pre-colonial Rajasthan, where peculiar environmental conditions forced states to deeply penetrate into the hinterland to ensure good and continuous supply of revenue, both agricultural and non-agricultural. It was a necessity as the agricultural production in this arid and semi-arid region greatly depended upon the possibilities of water appropriation and management. It holds true for the survival of the population-subjects and the state. The nature of intervention is evident in the detailed mapping of the physical landscape, and also of the social landscape. This is evident from the seventeenth century administrative manual, *Marwar Pargana Ri Vigat*, written by royal Diwan, Munhot Nainsi, and other archival records available for the period.

Most writings on medieval Rajasthan have focused on social and politico-economic factors. Scholars like G.D. Sharma, S.P. Gupta, and Dilbagh Singh have examined the issue of agrarian relations with respect to revenue appropriation and distribution in an attempt to analyse the nature of political apparatus (Sharma 1966; Sharma 1977; Chandra 1982; Devra 1981; Gupta 1986; Singh 1990; Bhadani 1999). Apart from the process of state formation, these works have examined agrarian structure with reference to state and social stratification, namely, the caste system. However, these works have not dealt with the environmental constraints, related social concerns and indigenous mechanism of water appropriation. They have centrally focused on revenue realization and related issues. The neglect of issues like, the role of 'common property resources', water management systems, forest conservation, etc., can be solely attributed to the difference in their points of references.

In the recent past, few fieldwork-based researches have examined issues like, man-made water shortage, declining forest coverage and implications of 'development brought about by industry-induced capitalism.[5] Most of the studies have examined these issues with respect to colonialism and suggested existence of a better man–nature relationship in the pre-colonial period. Ann Grodzins Gold has questioned the pre-colonial period being the 'golden age', by examining the popular perceptions and provides a vantage point. She argues that, 'the narrative says that in the old days the court's rule was harsh, exploitative, bitter; ... But the *commons* [forests, etc] was preserved, the forest was dense and alive,' (Gold and Gujjar 2002: 317). The examination of the concerns for the 'commons' shared among the masses with respect to modern democratic tendencies is very significant. This essay exposes the fallacies of the approach adopted by present day advocates of the 'golden age'.[6] Reducing the role and domains of pre-colonial state in Southeast Asia, James C. Scott has suggested:

> The demography of pre-colonial Southeast Asia was such that control of land per se, unless it was a strategically vital estuary, strait, or pass, was seldom decisive in state building. Control of the population—roughly five persons per square kilometre in 1700—mattered far more. The key to successful statecraft was typically the ability to attract and hold a substantial, productive population within a reasonable radius of the court. Given the relative sparseness of the population and the ease of physical flight, the control of arable land was pointless unless there was a population to work on it. The pre-colonial kingdom thus rode a narrow path between a level of taxes and coercive exactions that would precipitate wholesale flight. Pre-colonial wars were more often about rounding up captives and settling them near the central court than asserting a territorial claim. A growing, productive population settled in the ambit of a monarch's capital was a more reliable indicator of a kingdom's power than its physical extent (Scott 1998: 185).[7]

Such generalizations are problematic as they attribute little significance to the role of local environmental conditions in modulating the socio-political apparatus. This general neglect of the environmental conditions has led to creation of stereotypical images of the pre-colonial societies where the state is seen as primarily an appropriator of agricultural surplus and virtually unconcerned with the very process of agricultural production.

The linkages between environment, social structure, statecraft and cultural values need to be examined to establish the necessity of state intervention in the appropriation and management of water. In pre-colonial period people of this region on their own initiative, and/or through community efforts, and/or by state support had developed several mechanisms of water appropriation and management to overcome the natural scarcity. It was attempted both by conservation of whatever water received during the rains, monsoons based and/or otherwise, and its prudent use in the lean season. In the absence of any perennial river, the agricultural economy depended on the monsoon as it is even today. Therefore the available water had to be used to secure maximum possible agricultural production in this semi-arid/arid region of Rajasthan. The availability was more constrained as the location of Rajasthan on the Indian subcontinent places it at the tail end of the monsoons. The annual rainfall scarcely averages more than 12 cms in the western region. The Aravallis functioned as a barrier to monsoon rains to the east and south-east but in turn formed a rain-shadow region on the other side, i.e., western and north western Rajasthan.

At the same time, relative absence of forests further contributed to the aridity, yet the land seems to be fertile. S. P. Gupta and Dilbagh Singh have pointed out that commercialization of agriculture was taking place in the seventeenth and eighteenth centuries with the growth of irrigation facilities (Gupta 1986; Singh 1990). As such, large part of India can be termed as semi-arid, but the relative absence of perennial rivers further worsened the conditions in Rajasthan. Rivers like the Looni, Chambal and Betwa are highly seasonal; no snow fed rivers existed unlike in Punjab, north India, or Assam.

Adaptation to these conditions was clearly manifested in the methods of human appropriation of the immediate environment, especially with reference to conservation of water, to be more precise, rainwater. Here, it is important to emphasize, that in the process of appropriation of naturally available water, various geographical features were integrated to secure maximum possible benefits without disturbing the regeneration capacity. It can be seen in the way natural cycles—monsoons, the moon's monthly cycle, the sun's annual cycle— were understood and put to constructive purposes in conjuncture with the nature of soil, the local topography and other related factors. This essay begins with the delineation of the environmental features of the area. The next section provides the description of methods of potable water management as practised in the late medieval period. The last section carries an analysis of water management systems for irrigation evolved in response to possibilities of agricultural production.

Most of the evidence for this paper has been located in the seventeenth and eighteenth century archival and literary records of Rajasthan. The documents preserved at state archives Bikaner are significant as these provide an insight into the functioning of the state apparatus. At the same time, these records also document the hardships faced by the peasantry in the very process of agrarian production in pre-colonial times. Rajasthan is rich in archival collection and we have used *Arzdashts* (correspondence between *pargana* officials and central administration of Jaipur State), *Sanad Parwana Bahin* [8] (primarily imperial directives issued to the pargana officials in response to various complaints and representations received by the court at Marwar, Jodhpur). Along with these, seventeenth century administrative treatise, *Marwar Pargana Ri Vigat*,[9] written by Diwan Munhot Nainsi (1610–1670 CE), has been illustrative for this study. These documents provide details of administrative mechanisms from pargana to the village level and illuminate the concerns of medieval states in the region. A careful reading provides means and methods for understanding the states (these documents are addressed to the king and his officials posted at the court, thus we have used term 'state' for convenience) and it represents the state policy on the issue of interference in agricultural production and delineates the wider concerns of the peasantry in general. Even the role of the Brahmins (priestly class) has been dealt with in these documents.[10]

## Settlement Geography

For a better comprehension of water conservation methods it is necessary to delineate the geographical features of Rajasthan. Geographically, Rajasthan is highly heterogeneous. It is difficult to classify the whole province as a singular environmental entity. The Aravalli range of hills running diagonally from southwest to northeast, divide the macro region broadly into two natural divisions, namely, the north-western, and the south-eastern, the former being arid and the latter being semi-arid. About 60 per cent of Rajputana lies to the northwest of this range of mountains and two-fifths on the south-east. Even these two regions are not homogenous.

As far as general weather is concerned, heat dominates the scene in the whole of Rajasthan except some hilly parts.[11] In the summers, the temperature rises above 40 degrees centigrade—the maximum occasionally reaching 50 degrees centigrade at places in the desert area (see, Lowe 1986: 239). Rainfall also varies greatly depending upon the route of monsoons and other related factors in the western part, i.e., Jaisalmer, Bikaner and greater part of Marwar, the annual rainfall scarcely averages more than

12 cms. Since the clouds have to pass extensive and heated tracts of sand before reaching these plains they are emptied of much of their moisture upon the high ranges in Kathiawar and the nearer slopes of the Aravallis (*The Rajputana Gazetteer*, Vol. I, p. 20). However, in south-west Rajasthan, which is more directly reached and with faces less intermediate evaporation due to periodical rains, rainfall is much more abundant; and at Mount Abu sometimes it even exceeds 250 cms. Except in the south-west highlands of the Aravallis, rain is most abundant in the south-east Rajasthan. In this part of the country, if the south-west rains fail early, the south-east rains usually come to the rescue later in the season; so that the region is rarely subjected to the extreme droughts of the northwestern tracts (*The Imperial Gazetteer* of India, Vol. XXI, p. 92). If the eastern winds are strong, they bring hard rains from the Bay of Bengal; whereas if the south-west monsoon prevails the rain is comparatively late and light. Sometimes a good bout of rainfall comes from both the seas, but rarely so. Rainfall varies substantially from year to year and even within a year.' [12] Similar variations are visible in the natural vegetation as well. North-western Rajasthan is bereft of any forest cover, and grass and bushes dominate the landscape (Habib 1982: Sheet 6, A&B), whereas the south-eastern part has greater forest cover. To this effect, we have references of several *shikarghahs* of the Mughal period.[13] This semi-arid bio-geographic zone was home to four large cats—lions, tigers, cheetahs and leopards. However, lions became extinct by 1870s and cheetahs by the 1900s (see Divyabhanusinh 1999; Pandey 1998: 36).[14]

Any conjecture about the number of people living in this region during the pre-census period is doubtful. Colonel Tod tried to give an estimate of the population of region for the first half of the nineteenth century. He estimated a population of 539,250 for the state of Bikaner at an average density of 10 persons per sq km, and 74,400 with an average density of 1 person per sq km for Jaisalmer. For Shaikhawati, he estimated an average density of 30 persons per sq km, leading to a population of 424,800. Hence, the total population for the arid region was about three million approximately with an average density of 14 persons per sq km. However, the most authentic record for the previous period, *Ain-i-Akbari* by Abul Fazl, does not give any idea about the number of people. Moreland remarked, 'Rajputana was in general sparsely inhabited' (Moreland 1920: 22), at the beginning of the seventeenth century.[15]

## Water Security: A Prerequisite to Habitation

It is necessary to point out that the delineation of various water appropriation systems should be examined with respect to the larger question of the necessity of states' intervention to supplement the efforts made at the individual and/or community level. The portrayal of the pre-colonial states, as suggested by James C. Scott, simply in terms of the revenue extraction machinery does not hold true for our region:

> If the state's goals are minimal, it may not need to know much about the society. Just as a woodsman who takes only an occasional load of firewood from a large forest need have no detailed knowledge of that forest, so a state whose demands are confined to grabbing a few carts of grain and the odd conscript may not require a very accurate or detailed map of the society. If, however, the state, is ambitious—if it wants to extract as much grain and manpower as it can, short of provoking a famine or a rebellion, if it wants to create a literate, skilled and healthy population, if it wants everyone to speak the same language or worship the same god—then it will have to become both far more knowledgeable and far more intrusive (Scott 1998: 184).

The most possible points of disagreements lie in the fact that state in pre-colonial Rajasthan emerged and survived in very peculiar environmental conditions where it was forced to develop detailed mapping of the region to offset the vagaries of nature/erratic monsoons and general paucity of water, and attempts toward social or political engineering were more an aberration than the dominant practise. Therefore, it is no surprise to find description of the nature and number of water-bodies in each village of the principality of Marwar by Nainsi.[16] Describing the politico-social landscape, he lists villages with single source of water (Bhati 1968: 207, 454, 455, 515; 1969: 145, 239, 329, 346) or without any source of water to sustain habitation round the year (Bhati 1968: 207, 454, 455, 515; 1969: 24, 25, 250, 259). Simultaneously he has also taken care to notify the name of the neighbouring village where they were supposed to get water (Bhati 1968: 207, 454, 455, 515; 1969: 145, 239, 329, 346). It is pertinent to point out at the very onset that unlike colonial states, pre-colonial political systems were often extension of kin-relations. This was manifested in the distribution of resources and social stratification (Sharma 1977). It was a society in which family ties were crucial in determining status and power. The polity and kinship systems coexisted and were intermixed implying empowerment based on family ties. The structure was hierarchical with the ruler as the head of the clan, who acquired territorial rights and established the principality; then came the *thikanadar*s (nobles) who were his *saga*s (blood relatives) brothers, sons and other close relatives. The dominant sections of village society comprised various strata of the rural aristocracy (like *jagirdar*s, *bhomia*s *and bohra*s) who owed their status partly to hereditary superior proprietary rights in land and partly to their position in the apparatus of the revenue administration (Gupta 1986:134–43; Singh 1990: 42–50). Moneylenders known as *Mahajan*s and *Bohra*s were the source of capital for the ordinary peasantry. The rulers offered a certain amount of patronage and protection to *mahajans*, as the state recognized their importance (Singh 1974). The paucity of natural resources necessitated greater ties of interdependence among the ruling class and the peasantry.

In Rajasthan, the water-table varied a lot and was brackish in the western part for most of the times. As one moves away from the Aravallis in the broad belt across the desert, the water table sinks rapidly. The depth of water in the west and north-west varies from 50 metres to 115 metres (*The Rajputana Gazetteer*, Vol. p. 174). At the same time, monsoons are confined to four months only, with some rains in winter due to western disturbances over the Arabian Peninsula. It was necessary to manage the water thus received during monsoons for drinking as well as irrigation purposes.

Confirming the importance of water for the growth of human settlements in pre-colonial times, B.L. Bhadani has pointed out that 'a relatively large number of villages per 100 sq miles were located in the *pargana*s: Sojhat, Merta, Jaitaran and some parts of Jalore, which comprise the most fertile zone of Marwar'. He further argues that 'from the irrigation pattern map … we again get a confirmation that the pattern of settlement depended largely on the character of water supply.' (Bhadani 1999: 7)

In this respect it is important to realize that depth at which ground water was available and the brackish nature made it non-feasible for drinking as well as irrigation purposes (ibid.: 51).[17] It seems that as a response to the given level of technology and such environmental conditions we witness the greater emphasis on the water storage mechanisms rather than on water procurement systems in pre-colonial Rajasthan.

We have evidence for various indigenous methods of water conservation prevalent in pre-colonial period and some of them are even used today. Although, most of the methods were practised at the individual level, some required larger community participation or the support from the political power. Wherever feasible, the most important method of water appropriation had been the wells. With the variation in the depth of the water-table, the structures and functioning of the wells also changed. Moreover, the decline in the water-table necessitated the deployment of water lifting devices. These devices differed according to the topography and the depth of the water-table. The selection of water lifting device varied also according to the nature of requirement, i.e., drinking or irrigation. Nainsi has described several types of wells, for example, the *chanch, kostia, dhimraor, dhibra, kohar, kuwa,* and *bera* (Bhati 1974: 132). The character of the subsoil, especially in the desert and mountainous parts of Rajasthan renders the construction of wells or tanks, labour-intensive, and therefore, expensive. At the same time, the cost of construction of brick lined wells must have been an important consideration (Powlett 1874: 90). It is evident in the way Nainsi has carefully distinguished between earthen and brick lined wells (Bhati 1969; 1974). Detailed mapping of the wells in Marwar region by the official (Diwan) of the kingdom shows that the state was concerned that irrigation ensure good agricultural production and rulers were aware of the necessity of providing sources of potable water for settlements. There is evidence that state also extended support for the maintenance of wells. In the village Dhadahanriyo of pargana, Jodhpur, a well which had became non-functional as it was covered with sand was restored by state support (Bhati 1968: 296).

Apart from wells, which were the most common method of water harvesting, several other methods of rainwater harvesting were practised. One such method of water harvesting was known as *kuin* or *par* or *teen*. In appearance it was similar to a well but with a different mechanism which evolved to adapt to the particular environmental niche (Bhati 1969: 96). It was difficult to construct a *talab* or tank in sandy soil of Churu, Bikaner, Jaisalmer and Barmer districts of modern Rajasthan. Sandy soils of the region made it difficult to construct a *talab* or tank as water could permeate deep and traditional wells could not be constructed. However, these districts had an underground layer of chalk or limestone, which restricted percolated rainwater from mingling with the brackish ground water. The presence of underground layer of limestone ensured that rainwater would remain potable and could be appropriated. Moreover, sand does not bind up like soil, hence restricting the possibilities of evaporation of trapped percolated water (see, Mishra 1995, for details on construction of such wells). Rainwater so retained by sand gradually seeps into the open space available at the bed of *kuin*. It was usually not more than three to four earthen pots in a day. Therefore, *kuin* was not meant for round the clock water availability (Bhati 1968: 313, 327; 1969: 96) and was used for drinking purposes only. The sheer fact that prevalence of such wells is reported by Nainsi suggests the significance of such devices as it ensured continuance of human settlement.[18] This technique speaks for the better appropriation of given environmental conditions and better management of limited natural resources.

Geo-physical conditions of the region suggest limited or negligible possibilities of agriculture; however, the prevalence of such methods of water harvesting clearly suggests human settlements in the region. References of such settlements and their methods of water appropriation which perhaps even did not contribute significantly to the state exchequer highlights the state's concerns for human habitations. It also strengthens Scott's submission that Southeast Asia has traditionally witnessed low population density and states actively guarded their human wealth (Scott 1998: 185). One must also keep in mind that the pre-colonial states in Rajasthan had enjoyed substantial incomes from the non-agricultural sector (see Devra 1976: 190–95; 1981: 178–82). Several important trade-routes passed through the region and perhaps such settlements supplied much needed human and draught power (De Laet 1928: 64–66; Verma 1978: 113).

Similarly, *bera* or *beri*, another method of water harvesting, reflects a social adaptation to the environmental constraints of the region. It is a variation of a well. *Bera* were constructed near a water-body or on the dry bed of a water body. The permeation from any water-body to the adjoining area provided an opportunity to extract trapped water (Boileau 1837: 33-34, 37, 49, 105). It is extensively practised even today in western Rajasthan. The village Godhali of pargana Phalodi had a *beri*, near seasonal streams (Bhati 1969: 21). Village Naberi of pargana Phalodi sustained itself on the dry bed of a seasonal stream, which flows only during rains (ibid.: 25). In the village Bhabhclai of pargana Jodhpur, the groundwater was brackish. The village was without any source of potable water, but a *beri* constructed on a nearby tank sustained the habitation (Bhati 1968: 263).

Similarly, village Mahev in pargana Sojhat had a *bera* in the *talab* of the village and provided water when the *talab* dried up during the summers (Bhati 1968: 457). Village Chanda Vasani also had a *bera* (ibid.). In Jaisalmer the bed of water tank—Amarsagar sustained several *beri*. Amarsagar tends to retain water for only six to eight months and for the remaining part of the year these *beri* were the only source of water (Mishra 1995: 56–57). It is interesting to note that it was prevalent in the regions with deep groundwater and limited possibilities of retention of water in the ponds for more than a few months. Low retention of water in these water-bodies was perhaps due to the limited availability of water due to scarce rains and more importantly due to the excessive heat in the region. The rate of evaporation in the open ponds was high and it was nominal in the case of *bera*. This technique, once again speaks for the comprehensive knowledge of the region and use of appropriate mechanism to extract maximum possible benefits.

Similarly, another important maneuvering of the given environmental niche was practised to manufacture salt in the region. In most part of this region of Rajasthan, ground water was brackish, thus providing options to produce salt. In pargana Pokaran, there was a *rin*, a salt land. There is evidence that the water procured through *beras* was also used to procure salt by spreading in fields to dry up (Bhati 1969: 36) and after several stages salt was obtained.

It implies that the same environmental conditions, which led to greater evaporation of water from open water bodies like, *talab*, etc, were fruitfully utilized through *teen*. Human ingenuity had put the harsh weather conditions of the region to constructive use whereby, brackish water was spread on the fields to increase the rate of evaporation and help procurement of salt. To further provide impetus to the naturally available ground water, in this region *bera* were dug to utilize percolated water of neighbouring water bodies, which became brackish due to the nature of soil (ibid.: 21, 34–36).

Another important method of water conservation was *kund* or *kundi*. It was a mechanism based on the direct accumulation of rainwater and was primarily used for drinking purposes. This method of accumulation of water was extensively used in the northwestern Rajasthan, where groundwater was either brackish or available at great depths or both (Boileau 1837: 13). *Kundis* were constructed to cater both for domestic and public purposes. The roof of the house was used as a catchment area to collect rainwater. Before the onset of rains the roof was cleaned and washed with the first showers. Subsequent rains were directed to the underground water tank. The underground tanks functioned as reservoirs and provided water in times of scarcity (Mishra 1995: 56–57). Even now there are innumerable *kundis* in houses all over Rajasthan. Locally available lime mortar was used for the construction of kundi/tanka because it was able to restrict seepage (Chand 1999: 203). Easy mechanism and minimal maintenance made it the most viable source of water. It seems that the cost of construction of *kundi*, although by the locally available material, was beyond the reach of the common man. It is possible that those sections of the society, who could afford pucca house, used the *kundi* and lower sections of the society depended upon the *kuin*.

Even today, all along the western region of Rajasthan one can find numerous *kundis* existing in open fields for public use.[19] The open field *kundi* also used a similar method to harvest rainwater. The roof of the underground tank itself functioned as catchment area. An opening was made to enable the move into the tank as well as to draw the water. At times, even the state promoted construction of such techniques (Chand 1999: 200). Numerous epigraphical evidences are available to suggest notion of religious merit being traditionally attached with the construction of water-bodies for public use (Sharma 1995).[20]

The bigger form of *kundi* is known as *tanka*. 'Occasionally where the wells were brackish, drinking water was supplied from *kundis* or pits in which rainwater was collected and preserved, but in such small quantities as to be used only in an emergency' (Boileau 1837: 13–14). The most important example of this mechanism is *Karor Pati Tanka* of Jaigarh fort near Amber constructed in the fifteenth century. The parapet of the fort was used to collect rainwater and channelised to the tank located in the fort. This technique also solved the problem of carrying water from the lakes situated far below the hillocks. Its efficiency can be gauged from the fact that even in present times, during the rainy season, water is carried through the same dilapidated channels and viaducts. The water is still drawn from the *tanka* to cater to the needs of tourists and inhabitants of the fort.[21]

Tanks were among other methods to harness rainwater. Tanks were constructed in a region, which had clay soils or hard ground surface, which allowed limited percolation and construction was easy and feasible. The construction of tanks required a thorough knowledge of the geography of the area. The site was chosen carefully. The direction of the slope, its gradient and topography of the catchment area were important considerations. Munhot Nainsi provides a long list of tanks constructed by individuals, communities and the state in the seventeenth century. *Ranas* or the rulers of Mewar or the kings constructed numerous water storage systems primarily for drinking purposes (Bhati 1968: 579–602). It is true for other regions as well. Few prominent tanks include, the Udaisagar (Skaria 1960: 30) built in 1564 CE, Rajsummand constructed around 1662–75 CE in Mewar (Das 1886: 401–49, 2204) and Manasagar built near Amber in the early eighteenth century (Khan and Kumar 1993). Maharana Raj Singh, the Sisodia Rajput ruler of the former Princely State of Mewar (1662–1680 CE), is credited with the construction of the Rajsummand as engraved on the eulogy pasted on the embankment (Vashistha 1993: 111).

Of singular significance were (the constructions of Garsisar in the fourteenth century and Jaitsar few generations later in the vicinity of city of Jaisalmer (Chand 1999: 49–51). The selection of the site exemplifies a thorough knowledge of the local topography. The tank of Garsisar was constructed by erecting a *bund* to block the drainage of the catchment area created by the slopes of the hill on which fort of Jaisalmer is located (Chand 1999; Bhati n.d.). The construction of Garsisar, could tackle the requirement of water for a capital city, it could also address the predicament of a permanent capital.[22] Furthermore seven subsequent reservoirs ensured utmost utilization of the overflow from Garsisar. One of the subsequent tanks was Amarsagar built in the fifteenth century CE and was provided with seven *beris* on its bed to supply water when the tank dried up after four to five months (Mishra 1995: 56–57). It ensured the continuous survival of the city of Jaisalmer in the arid region of Rajasthan (ibid.: 49-51).[23] It is interesting to note that dry beds of Jaitsar and Amarsagar were used to grow wheat, i.e., *rabi* crop (Chand 1999: 49–51).

We have enough evidence to safely argue that the state appreciating the peculiar environmental conditions extended financial and material support for the construction and maintenance of water appropriation mechanisms. Most of the investment was made to retain the human wealth so as to ensure the continuance of agricultural production. The ruling elite extended support for the construction of wells to encourage the habitation in the region to enlarge the revenue base of the states. Kings actively promoted the construction of mechanisms for water management. Construction of wells was promoted in the regions with relatively high water-table.[24] The

(village pastureland) (ibid.: 516). On these lands, crops could be sown especially the ones, which needed limited water.

*Rela*, a regional variation of *bahla* functioned on the same principle (Bhati 1968: 207, 454, 455, 515; 1969: 145, 239, 329, 346), *bahla* utilized rain-water, whereas *rela* used overflow from seasonal streams. Such a method, however, could not be used for irrigating fields other than those inundated by the floodwater of the seasonal streams. Lands irrigated by the *rela* method were generally located along the banks of rivers (ibid.: 353). These were generally, natural depressions and put into productive use by the village communities. The *rela* sustained *rabi/unali* crops. The village Peepaliyo Kapadiya Ro utilized the *rela* method of irrigation to generate revenue amounting to Rs 1091 for the year 1658, Rs 2250 for the year 1659, Rs 817 for the year 1662 and so on (ibid.: 516). Village Anandpur Khas and Lambia of pargana Merta were producing good quantities of wheat and gram due to availability of *rela* irrigated land (Bhati 1969: 120). It should be kept in mind that *rela* was the overflow water spread over fields and the crop grown on this field was known as *sewanj* (ibid.), and these were *rabi* crops. Usually wheat was grown due to low requirement of moisture and limited watering. Generally winter showers provided the watering required for the better growth of these crops.

Another method of rainwater harvesting for irrigational purposes was practised in Jaisalmer region and was known as *khareen* (Bhati 1969: 12; Chand 1999: 92). It is generally believed that Paliwal Brahmins settled in Jaisalmer during the early medieval times, developed *khareens*. The land irrigated through *khareens* supported a rich harvest of wheat (Mishra 1995: 62). *Khareens* were basically *bunds* created in the path of *mullah* or seasonal streams. The height of the *bund* was kept low so as to retain only limited water, which could dry up in two to three months. This dried tank bed was used to produce *rabi/unali* crops. Another advantage of keeping the height of the *bund* low was that the overflow water could be utilized to create another *khareen* downstream, which would in turn provide a larger cultivable land (Mosse 2003). [34] It is important to note that cultivation with the help of *khareen* could be achieved in the region adjoining the hills where soil was largely clay and it was irrelevant for the sandy soil.

Like other parts of India, tanks and reservoirs too played an important role in irrigation in pre-colonial Rajasthan. The rulers and nobles had constructed numerous small and large reservoirs and were concerned for their upkeep. In 1577 CE the Mughal Emperor Akbar (1556–1605 CE), ordered de-silting of a reservoir used for irrigational purposes at Merta (Beveridge 1972: 308–9).[35] Pillar inscriptions at Junjala tank in Nagore specifies that the income from the tank should only be used to maintain the tank by regular de-silting.[36] The 521 villages in the kingdom of Marwar had as many as 659 tanks. The highest number of tanks was in pargana Sojhat, followed by pargana Jaitaran and pargana Siwana (Bhadani 1999). Pokaran and Phalodi were prominent for having a number of tanks. The environmental features of the region suited for tank irrigation. Lakes and dams played an important role as sources of irrigation in the Mewat region also.[37]

A close look at the revenue collection based on the agricultural production from the various parts of the Kingdom of Marwar clearly establishes the correlation between investment made by the ruling elite towards development of irrigation facilities and the agricultural potential. The correlation can be seen in Table 6.1 based on the details compiled by B.L. Bhadani from the contemporary sources.

**TABLE 6.1: REVENUE COLLECTION AND NUMBER OF WELLS/TANKS (1658 CE)**

| *Pargana* | *Tanks*[38] | *Wells*[39] | *Jama per sq mile (in Dams)*[40] |
|---|---|---|---|
| Phalodi | 42 | 538–638 (including Pokaran) | 326.31 |
| Jodhpur | 36 | 4031–4065 | 617.51 |
| Siwana | 96 | 500–544 | 958.84 |
| Sojhat | 172 | 1441–1522 | 1807.55 |
| Jaitaran | 94 | 1789–1883 | 4083.35 |
| Merta | 16 | 6497–7782 | 4033.04 |

At the same time, rulers had a degree of control over irrigation. Rulers not only arbitrated on the issue of ownership of means of irrigation but also controlled the amount of water to be distributed.[41] At times the state had to arbitrate between claimants over the use of irrigational device.[42] It is interesting to note that, the state tried to increase the land to be cultivated after digging of new wells in the villages.[43] The state's concern for optimal use of irrigational facilities is also evident in the same document. Similar evidence is also available from eastern Rajasthan. An *arzdasht* written by Fateh Chand Ramji on *Paus vadi* 5, 1773 vs., informs the king Sawai Jaisingh that well water is being used for irrigation as directed.[44] It reflects that state is very much aware of the ecological context and therefore attempts to maximize utilization of the available water.

The dispute over use of water for irrigation purposes in the semi-arid climate cannot be ruled out and we follow cases of arbitration by the state. It was reported from village Dhamorki, pargana Khori, that Patel had diverted water from the river without permission. He was punished with a fine of Rs 20.[45] Rupa Patel of village Raitoli breached the embankment of river to divert water.[46] In another incidence, state punished a Gujar for diversion of water from other person's field in *qasba* Baswa, pargana Bahatri in 1774.[47] Similar dispute is reported from village Neembla, pargana Bahatri. Goverdhan Brahman fought with Tulani Brahman over the issue of diversion of water and hit him with an axe. He paid Rs 11 as fine.[48] The village pond had been used not only for washing clothes and ancillary activities but also by cattle for drinking purposes. Thus, any act of its pollution was punishable. We have evidence from *qasba* Malpura, pargana Malpura, that Lona dyer was fined Rs 5.50 for washing dyed cloth in the pond.[49]

We do not have sufficient evidence to comment on the use of water-bodies by different castes given the prevalence of caste system in pre-colonial Rajasthan.[50] It is equally important to analyse the role of the Turkish rulers with respect to the caste system in general and access to common sources of water in particular. An inscription at the step well located at Makrana carries an official order by Mirza Ali Beg, prohibiting the lower castes to not to fetch water along with the higher castes.[51] Were notions of pollution used to deny access to those of the lower castes and how did these relate to pactitioners of Islam, it is difficult to answer at present. Moreover, these are larger issue and further exploration of the issues shall provide insight to the socio-political reality.

In the agrarian economy of this semi-arid and arid region the failure of monsoons very often led to drought. These droughts could easily have assumed the form of a famine. Famines hit the material base of the peasantry crucially. High cost of seed and manure and dry wells restricted irrigation. Ploughing could not be done[52] thus affecting agricultural production in the region. The drought or infliction of crops, scarcity of fodder and dry ponds made cattle vulnerable, [53] affecting the pastoral communities of the region. Apart from expecting relief from the state, peasants tried to minimize the impact of drought by adapting various solutions. For example, if rains were delayed, peasants shifted to cultivation of *moth* as it has the shortest maturity period. Amar Chand and Sahib Ram, *vakil* of the Amber court inform that due to meager and delayed rains till the month of Bhadava,[54] only *moth* could be sown.[55] *Vakil* Ajitdas and Man Ram inform that due to drought, not even *moth* could be sown.[56] The statement that 'not even *moth* could be grown', testifies to the fact that *moth* was the last crop that could be sown in the case of delayed rains. During delayed or insufficient rainfall, land is ploughed extensively to make it soft and then seeds were sown. Kanwar Pal and Bhopat Ram inform the *sarcar* on *Kati Sudi* 9, 1698 CE, that in the regions, which have received insufficient rains, land is being worked upon to make it soft so as to sow the seeds.[57] Kanwar Pal and Hari Valabh Das supplement the information that wherever soil is hard, it needs one more shower and wherever rains have been insufficient, artificial irrigation through wells was arranged as well.[58]

## Conclusions

In recent times traditional systems of water management systems have been extensively examined in terms of their ecological relevance and wider issues of socio-political dimensions have been neglected. The availability of water in any region had been an important consideration in any pre-colonial political formation (Wahi 1997: 267–84) and it was more of a concern in Rajasthan due to peculiar environmental conditions. Contemporary records have suggested that the peasantry managed most of the water requirement both for the agricultural operations and potable purposes. Generally, the peasantry took care of the water needs by adapting various indigenous methods of water management, which evolved in response to the given environmental conditions. Better management of water along with drought resistant agricultural production had been a characteristic of pre-colonial Rajasthan. It is important to note that a tendency towards cultivation of water-intensive cash crops became evident only in the seventeenth century. However, it is only the wealthy sections of the society, with the capacity for greater capital investment, practised it. At the same time, in the times of water scarcity, even they went back to the cultivation of traditional drought resistant crops (Gupta 1986; Singh 1990).

The general perception that states in traditional societies played a very limited role with respect to water management for settlement purposes and agriculture, has been challenged. Our evidence have established that to a great extent agricultural production in Rajasthan necessitated financial assistance by the ruling class.[59] Rulers not only offered concessions but at times exempted revenue demand along with financial and material help (Singh 1974: 22). It has also highlighted the concerns of the state and the means and methods to attain its objectives suggest wider and deeper intervention of the state in the appropriation and regulation of water. Further, exploration of the subject may help in integrating it with the wider socio-cultural milieu of the period. The foregoing findings compel us to reconsider the significance of regional ecological niches to examine the socio-political structure. Traditional practise of placing environmental factors in the background has been questioned. It is suggested that the relationship between ruling elite and the ruled are necessarily to be mediated through the local environmental conditions, especially in any study of pre-colonial study (Scott 1998).[60]

## Notes

1. Early writings on the issue sought to establish a 'golden age' approach to Indian environmental history (Gadgil 1985). In the same vein Ramchandra Guha asserted that in the pre-British period there was little or no interference with the customary use of forest and forest produce (Gadgil and Guha 1992). The belief in the existence of pre-colonial common village property and communal customary forest use was based on the assumption of local self-sufficiency (Agarwal and Narain 1997; Misra 1995).
2. David Mosse has examined the interplay of 'developmental politics' to explain the level and process of state intervention. The role of community based programmes to tackle contentious issues like management and allocation of 'common property resources' like water bodies, etc., have also been examined (Mosse 2003: 1–27) See also, Stein (1994; 1980); Ludden (1978: 1–13).
3. Hardiman has attributed the decline of the system to the 'colonial disinterest, resettlement of certain regions by shifting cultivators...and the decay in the old system of community-based control as a result of colonial taxation... In the same vein Elizabeth Whitecombe (1995: 237–259) has argued that irrigation 'works were financed by loan capital.' Hence, in the sanctioning of constructions the emphasis was necessarily placed on the prospect of their remunerativeness; see Rohan D'Souza, 'Capitalist Property in Land and the

Making of the "Calamity of Season" in Colonial Orissa', presented at *Contemporary Studies Series*, Nehru Memorial and Museum Library, July 8, 2(X)3 p. 15.

4. Primary concern of Jodha has been to examine 'the changing status and usage pattern of natural resources... and the possibilities of arresting their negative trends characterizing these changes'. Also see Kavoori (1999). He has explored the issue of 'common property resources' by examining the conditions of the pastoralists in the contemporary era. Relative shortage of the 'common grazing land' and the contemporary stress over the sedentary lifestyle has reduced the opportunities for the pastoralists. By the same token, it has also reduced the opportunities available with the peasantry in times of drought and famine (Rosin 1987).
5. Lyla Mehta has argued that the 'water scarcity is both "real" and "constructed."' She suggests that '[T]he external "essentialised" notions of scarcity generated by state discourse and state programmes are often quite different from local people's knowledge systems and livelihood strategies that allows them to adapt to the unpredictability and temporary scarcity of water', Mehta (2003: 5066).
6. Even valuable works like *Dying Wisdom* and *Aaj Bhi Khare Hai Talab* share this oversight and often neglect the autocratic nature of medieval rule.
7. However, Sumit Guha argues that Southeast Asia has always had greater human densities since the ancient period (Guha 2001).
8. The subject matter of these documents ranges from routine complaints of undue collection of revenue against the state officials to the mutual disputes related even to the amount of water of a well to be shared among them. These pertain to the seventeenth–nineteenth century.
9. It is basically an administrative manual providing detailed description of the territory, soils, nature of habitation, agricultural types, nature and number of wells, other sources of water, caste and class composition of population, etc. It is such an informative document that details of villages, which were deserted, and later rehabilitated, also find mention here. These pertain to seventeenth–nineteenth century.
10. However, these documents seem to be of limited use with respect to the examination of role of lower classes/castes. It requires further research to examine such aspects.
11. On his way Babur had realized that north-east Rajasthan is very hot (see Beveridge 1970: 577); Abul Fazl and Jahangir have placed it in the second of the seven categories in which whole world was traditionally divided. The second category implied equitive heat and winter, see Jarrett (1978: 273); Beveridge (1968: 340–41).

12. *Arzdasht, Asad Vadi* 1, 1638 vs. and *Bhadva Sudi* 6, 1638 vs. Historical Section, Jaipur Records, Rajasdian State Archives, Bikaner. Henceforth, HS, JR, RSAB.
13. *Arzdasht, Migsar Vadi* 13, 1640 vs., HS, JR, RSAB; Devra (1999: 97–107).
14. Some of the famous protected areas here are Ranthambore, Sariska, Mount Abu and Sitamata.
15. Sumit Guha holds that South Asia has always had a greater density in pre-colonial times than other comparable societies (Guha 2001).
16. Similar details are available for Jaisalmer in Chand (1891, 1999).
17. For Marwar region, Bhadani concludes that the sparse distribution of wells in the Pokaran/Phalodi was in line with the scarcity of underground water and its availability at great depth. This was more so probably because the brackish water made only one out of twenty wells useful for irrigation.
18. Village Bhojavas of pargana Jodhpur had a *kuin*, which could supply water only twice in a single day (see Bhati 1968: 313–315).
19. The religious merit attached with the construction of water storage systems for public seems to be one reason for the greater proliferation.
20. For example, Pratapgarh stone inscription (946 CE), Eulogy at the Lahar stepwell, Basangarh (Sirohi) (1042 CE). Indragarh Inscription 'A' (1683 CE), Inscription at Bhuvana village stepwell inscription (1660 CE), Eulogy at Trimukhi Stepwell (1675 CE), Eulogy at Gujar Stepwell (1715 CE), Surtan Stepwell inscription at Bedla Village (1717 CE), Sagdogra Slepwell. Inscription (1801 CE). Udaivav Inscription at Dungarpur (1880 CE), etc. The notion of religious merit was shared even by the followers of Islam also. For example, Sambhar Stepwell Inscription (Amber) (1363 CE), Gaurishankar Tal. Narayana Inscription (1437 CE). Makrana Stepwell Inscription (1651 CE), Bakalia Inscription Dist. Nagore, 1670 CE), etc.
21. 'The most peculiar feature of this fort appears to be its hydraulic system which would not only have added to its basic strength, but also took care of it's daily needs.' (Sharma and Rezavi 1993: 129–31)
22. Here it is very important to recall the history of Jaisalmer. The ancient capital was situated at Lodurva and later on at Kuldara. However both these sites were short lived, perhaps due to its inability to sustain a large urban population.
23. However with the growth of urbanisation, the catchment area of these tanks is being encroached leading to gradual decay.
24. *Sanad Parwana Bahi* No 1707 vs., Jodhpur Records, RSAB; Wherever wells occurred more frequently in c. 1660, the water table was nearer the surface.
25. *Arzdasht, Kartik Sudi* 13, 1629 vs.; Bhadva Sudi 11, 1629 vs.; *Asad Sudi* 15, 1629 vs.; *Sawan Sudi* 13, 1626 vs. and *Asoj Vadi* 12, 1638 vs. HS, JR, RSAB.
26. *Sanad Parwana Bahi*, No.2, 1708 vs. Jodh. Rec. RSAB.
27. *Sanad Parwana Bahi*, No. 8, 1711 vs. Jodh. Rec. RSAB.

28. *Arzdasht*, *Kartik Vadi* 3, 1742, vs. HS, JR. RSAB.
29. *Yaddashti Pargana Patiari Mauzjn- Darobastki*, 1784 vs. HS. JR. RSAB.
30. *Arzdasht Asoj Vadi* 12, 1752 vs.; *Asoj Vadi* 1, 1751 vs.; *Kartik Vadi* 3. 1751 vs. HS, JR, RSAB.
31. *Arzdasht*, *Sawan Sudi* 11, 1659 vs. HS, JR, RSAB.
32. *Arzdasht*, *Sawan Sudi* 1, 1647 vs. HS, JR, RSAB
33. An *arzdasht* written by Mouji Ram dated *Kati sudi* 15. 1774 vs. HS. JR. RSAB. informs us about meagre rainfall leading to drought. This resulted in the migration of the peasantry. In his *arzdasht* dated Jeth sudi 1. 1762 vs. HS. JR, RSAB., Lai Chand Daja Ram informs about migration of peasantry due to drought and resultant decline in revenue collection. Similarly another *arzdasht* by Ajit Das, Man Ram dated Chait vadi 3, 1752 vs. HS. JR. RSAB., informs that Ullages were deserted.
34. Mosse has examined extensively the functioning of tank-system of irrigation as practised in the coastal regions of Tamil Nadu. It was an extensive network of tanks and required inter-village regulations and management. However, we do not have such references for our region perhaps due to meagre rains.
35. We have similar evidence from the ancient period also where King Rudradaman ordered repair of Mauryan dam in 1300 CE. See Thapar (1966: 98). Similarly, Habib (1992: 295) cites an evidence 'of an ambitious proposal to advance Rs 40,000 to 50.000 as '*Taqavi*' from the imperial treasury for financing the construction of dams (*bunds*) in the province of Khandesh and Berar.'
36. Junjala Tank pillar inscription (1679 CE) cited in Sharma (1995: 230).
37. Mughal Emperor Babur extensively describes the Koda Lake and noticed its utility for irrigation. It lies partly in the district Nuh, partly in Gurgaon, where the two tracts join at the foot of the Alwar hills (see Beveridge 1970).
38. See Bhadani (1999: 52, Table no. 10)
39. Ibid., p. 46; Table No. 7.
40. Ibid., p. 268; Table No. 3.
41. *Sanaad Parwana Bahi*, No. 2, 1708 vs., Jodh. Rec. RSAB.
42. *Sanad Parwana Bahi*, No. 2, 1708 vs., No. 8, 1711 vs. Jodh. Rec. RSAB.
43. *Sanad Parwana Bahi*, No. 2, 1708 vs., Jodh. Rec. RSAB.
44. *Arzdasht*, *Paus Vadi* 5, 1659 vs., HS. JR. RSAB.
45. Ibid., *Pargkna Khori,* 1801 vs., HS. JR. RSAB.
46. Ibid., *Pargana, Dausa*, 1825 vs., HS. JR. RSAB.
47. Ibid., *Pargana Bahatri*, 1774 vs., HS. JR. RSAB.
48. Ibid.
49. *Pargana Malpura*, 1791 vs., HS. JR. RSAB.

50. Any description of Indian society is not without the distinct references of caste system. Abul Fazl (see Beveridge 1972); Nainsi (see Bhati 1968; 1969); *Tawarikh* (see Chand 1999), while providing the administrative classification have based their descriptions on caste categories.
51. Makran Stepwell Inscription (1651 CE) cited in Sharma (1995: 227).
52. *Arzdasht, Paus Sudi*, 4, 1647 vs. HS, JR, RSAB.
53. *Sanad Parwana Bahi*, *Falgun Vadi* 10, 1672 vs. Jodh. Rec. RSAB.
54. Traditionally *Bhadva* is the third month of rainy season and usually the last one also.
55. *Arzdasht, Bluidva Vadi* 7, 1660 vs. and *Arzdasht Sawan Sudi* 9, 1638 vs., HS, JR, RSAB.
56. *Arzdasht*, *Falgun Sudi* 11, 1638 vs. HS, JR, RSAB.
57. *Arzdasht, Kartik Sudi* 9, 1641 vs. HS, JR, RSAB.
58. *Arzdasht*, *Asoj Vadi* 7, 1642 vs. HS, JR, RSAB.
59. Dilbagh Singh argues that, 'The ecological limitations of the region made the need for agricultural inputs [irrigation] considerably higher than in other more favorably located areas...'. Singh (1974: 51–53)
60. Scott has very aptly demonstrated that the negligence of local environmental conditions have been primary cause of the failure of major modern-age projects.

## References


Agarwal, Anil and Sunita Narain (eds), 1997, *Dying Wisdom*, Centre for Science and Environment, New Delhi.

Arnold, David and Ramachandra Guha (eds), 1995, *Nature, Culture, Imperialism: Essays on the Environmental History of South Asia*, OUP, Delhi.

Beveridge, H. (ed. and tr.), 1968, *Tuzuk-i-Jahangiri* by Jahangir, Munshiram Manoharlal Publishers, Delhi.

Beveridge, H., (tr.), 1970, *Babur-Nama* by Zahir-ud Din Muhammad Babur, Oriental Books Reprint Corporation, New Delhi.

Beveridge, H., (tr.), 1972, *Akbarnama* by Abul Fazl, Vol. III, Rare Books, Delhi.

Bhadani, B.L., 1999, *Peasants, Artisans and Entrepreneurs: Economy of Marwar in the Seventeenth Century*, Rawat Publications, Jaipur.

Bhati, Narain Singh (ed.), 1968, *Marwar ra Pargana ri Vigat* by Munhot Nainsi, Vol. I, Rajasthan Praehyavidya Pratishthan, Jodhpur.

———, 1974, *Marwar ra Pargana ri Vigat*, Vol III, Rajasthan Prachyavidya Pratishthan, Jodhpur.

Boileau, A.H.E., 1837, *Personal Narrative: Tour through the Western States of Rajwara in 1835 Comprising Beekaner, Jesulmer and Jodhpur*, Baptist Mission Press, Calcutta.

Chand, Lakhmi, (1891, 1999), *Tawarikh Jaisalmer*, Rajasthani Granthagar, Jodhpur.

Chandra, Satish, 1982, *Medieval India: Society, The Jagirdari Crisis and the Village*, Macmillan, New Delhi.

Das, Shayamal, 1886a, *Vir Vinod, Vol. I*, Motilal Banarsidas, Delhi.

———, 1886b, *Vir Vinod, Vol. II*, Motilal Banarsidas, Delhi.

De Laet, J., 1928, *The Empire of the Great Mughals*, tr. J.S.Hoyland, Bombay.

Devra, G.S.L., 1976, 'Nature and Incidence of Rokad Rakam (non-agricultural taxes) in the Land Revenue System of the Bikaner State (1650–1700 CE)', *Proceedings of Indian History Congress*, Calcutta.

———, 1981, *Rajasthan ki Prashashnik Vyavastha*, Dharti Prakashan, Bikaner.

———, 1999, 'Desertification and Problem of Delimitation of Rajputana Desert during the Medieval Period', *Human Ecology, Special Issue*, No. 7, pp. 97–107.

Divyabhanusinh, Chavada, 1999, *The End of a Trail: The Cheetah in India*, OUP, New Delhi.

Fisher, R.J., 1997, *If Rain Doesn't Come: An Anthropological Study of Drought and Human Ecology in Western Rajasthan*, Manohar, Delhi.

Gadgil, M., 1985, 'Towards an Ecological History of India', *Economic and Political Weekly*, 20: 910–18.

———, and Ramchandra Guha, 1992, *This Fissured Land: An Ecological History of India*, OUP, Delhi.

Gold, Ann Grodzins and B.R. Gujjar, 2002, *In the Times of Trees and Sorrows: Nature, Power, and Memory in Rajasthan*, Duke University Press, Durham.

Guha, Sumit, 2001, *Health and Population in South Asia from the Earliest Times to the Present*, Permanent Black, New Delhi.

Gupta, S.P., 1986, *Agrarian System of Eastern Rajasthan*, Manohar, New Delhi.

H.S. Jarrett (tr.) 1978, *Ain-i-Akbari* by Abul Fazl, Vol. II, Bibliotheca Indica, New Delhi.

Habib, Irfan, 1982, *Atlas of the Mughal Empire*, OUP, Delhi.

———, 1992, *The Agrarian System of Mughal India* 1556–1707 (2nd revised edition) OUP, Delhi.

Hardiman, David, 1995, 'Small Dam Systems of the Sahyadris', in David Arnold and Ramachandra Guha (eds), *Nature, Culture, Imperialism: Essays on the Environmental History of South Asia*, OUP, Delhi.

Jodha, N.S., 2001, *Life on the Edge: Sustaining Agriculture and Community Resources in Fragile Environments*, OUP, New Delhi.

Kavoori, P.S., 1999, *Pastoralism in Expansion: The Transhumant Herders of Western Rajasthan*, OUP, Delhi.

Khan, M.I.A. and Ravindra Kumar, 1993, 'The Mansagar Dam of Amber', in Ahsan Jan Qaisar and Som Prakash Verma (eds), *Art and Culture in Medieval India*, Jaipur.

Lowe, W.H. (tr.), 1986, *Muntakhab ut Tawarikh* by Abdul Qadir Badauni, Vol. II, Renaissance Publishing House, Delhi.

Ludden, David, 1978, 'Ecological Zones and the Cultural Economy of Irrigation in Southern Tamilnadu', *South Asia*, 1(1): 1–13.

Mehta, Lyla, 2003, 'Contexts and Constructions of Water Scarcity', *Economic and Political Weekly*, November 29, p. 5066.

Misra, Anupam, 1995, *Rajasthan Ki Rajat Boonden*, Gandhi Shanti Pratishthan, New Delhi.

Moreland, W.H., 1920, *India at the Death of Akbar: An Economic Study*, A.Ram & Sons, Delhi.

Mosse, David, 2003, *The Rule of Water: Statecraft, Ecology and Collective Action in South India*, OUP, New Delhi.

Narain Singh Bhati (ed.), 1969, *Marwar ra Pargana ri Vigat*, Vol II, Rajasthan Prachyavidya Pratishthan, Jodhpur.

Pandey, Deep Narayan, 1998, *Beyond Vanishing Woods: Participatory Survival Options for Wildlife, Forests and People*, Himanshu Publications, Udaipur.

Powlett, P.W., 1874, *Gazetteer of Ulwar*, Sallantyne Press, London.

Rosin, R. Thomas, 1987, *Land Reforms and Agrarian Change: Study of a Marwar Village from Raj to Swaraj*, Rawat Publications, Jaipur.

Scott, James C., 1998, *Seeing like a State: How Certain Schemes to Improve the Human Condition Have Failed*, Yale University Press, New Haven and London.

Sharma, Dashratha, 1966, *Rajasthan through the Ages, Vol. 1*, Dharti Prakashan, Bikaner.

Sharma, G. D., 1977, *Rajput Polity: A Study of Politics and Administration of the State of Marwar*, 1638–1749, Manohar Publications, Delhi.

Sharma, Gopinath (ed.), 1995, *Rajasthan ke Itihas ke Srota* (in Hindi), Rajasthan Hindi Grantha Academy, Jaipur.

Sharma, Rajiv and Syed Nadeem A.H. Rezavi, 1993, 'Aspects of Hydraulic Engineering in Medieval Rajasthan: A Case Study of Water System in Jaigarh Fort', Ahsan Jan Qaisar and Som Prakash Venna (eds), *Art and Culture*, Publications Scheme, Jaipur.

Singh, Bhati Narain (ed.), n.d., *Jaisalmer Ri Khyat*, published in, Rajasthan Prachyavidya Pratishthan, Pamnipara, Jodhpur.

Singh, Dilbagh, 1974, 'The Role of the Mahajanas in the Rural Economy in Eastern Rajasthan during the 18th Century', *Social Scientist*, Vol.2, No. 10.

———, 1990, *State, Landlords and Peasants*, Manohar Publications, Delhi.

Skaria, Badri Prasad (ed.), 1960a, *Munhot Nainsi ri Khyat* by Munhot Nainsi, Vol. 1, Rajasthan Prachyavidya Pratishthan, Jodhpur.

———, 1960b, *Munhot Nainsi ri Khyat*, Munhot Nainsi, Vol. II, Rajasthan Prachyavidya Pratishthan, Jodhpur.

Stein, Burton, 1980, *Peasant*, *State and Society in the Medieval South India*, OUP, Delhi.

———, 1994, *The New Cambridge History of India*, *Vol. 1*, *No. 2*, Vijayanagar Empire, Cambridge University Press, Cambridge.

Thapar, Romila, 1966, *A History of India*, Penguin, London.

*The Imperial Gazetteer* of India, 1909, Vol. XXI, Clarendon Press, London.

*The Rajputana Gazetteer*, 1879, Vol. I, Office of the Superintendent of Government Printing, Calcutta.

Vashishatha, Neelima, 1993, A Note on the Sculptures of the Raj Samudra Lake in Rajasthan', in Ahsan Jan Qaisar and Som Prakash Vernal (eds), *Art and Culture*, Jaipur.

Verma, H.C., 1978, *Medieval Routes to India*, Naya Prokash, Calcutta.

Wahi, Tripta, 1997, 'Water Resources and Agricultural Landscape: Pre-colonial Punjab', in Indu Banga (ed.), *Five Punjabi Centuries: Polity, Economy, Society and Culture, c. 1500–1990*, Manohar, Delhi.

# 7

# A Historical Study of the Control of Grass and Fodder Resources in Eighteenth-Century Maharashtra

**Sumit Guha**

## Introduction

The great majority of the population of India has for long lived in the countryside, largely deriving its subsistence by cultivating crops and rearing animals as well as gathering plants and catching wild animals. Most people were quite poor and crop failures sometimes caused famines in which tens of thousands perished of hunger or disease. While historians have studied many aspects of life in that time, it is only recently that many scholars have turned to the study of environmental history. Historical research is an effort to reconstruct the past by using the evidence *from that time*. This is necessary because even people who were witnesses to an event may have distorted or confused recollections of it. Consider how even today witnesses can give completely contradictory evidence in a court of law. And historians often discuss things that happened hundreds of years ago!

Many scholars have found it easier to work with papers and records from the period of British colonial rule because these were carefully organized in government offices (archives) and written in English. Many of the records of Indian rulers before the British conquest were lost and they are all in different languages and scripts. As a result they have been less studied than those of the last 200 years.

Many of the government departments and policies that we have in India today were established by the British. This includes the structure of the

---

* This paper was especially written for this volume.

services (IAS, IPS, etc.) and departments like Revenue, Forests, Health, Irrigation, PWD, and so on. It is well-known that India has suffered serious environmental problems in the last 30 or 40 years. Forests have shrunk, wildlife is vanishing, the air and water have become laden with dangerous pollutants. Nor are we the only people to have such problems: people all over the world experienced similar difficulties. Many people decided that the problems arose because of the new political and economic structure that followed the Industrial Revolution. Others thought that traditional societies in Asia had never experienced such problems and so the answer was to restore the practises that existed before the era of western colonial rule. Vandana Shiva for example, writes that

> [f]or centuries, vital natural resources like land, water and forests had been controlled and used collectively by village communities, thus ensuring a sustainable use of these renewable resources ... Colonial domination systematically transformed the common vital resources into commodities for generating profits and the growth of revenues (Shiva 1991).

Shiva makes this claim without any tangible historical material of how earlier communities were constituted, or how they actually managed their resources. This paper will consider actual evidence of environmental conditions and environmental institutions before 1800 or so. It will focus on the western and central parts of modern-day Maharashtra. In the eighteenth century most of this area was ruled by the Peshwas and we have many papers and letters from that time. We are therefore able to describe how 'land, water and forests' were actually used and controlled in this period.

Local village communities may be left to manage resources in certain circumstances. For example, when the population of a country is very small and the villages lie far apart. There is a lot of vacant land and large areas of forest. For this reason, there is enough for everybody and so kings and chiefs leave management in the hands of local people. By the 1700s, India had a large population—nearly 200 million people or 25 times the number that lived in England at the time (Guha 2001; for the historical background see, Kulkarni 2000). Regions like Maharashtra were inhabited by millions of people and governed by thousands of officials. I will introduce you to some documents from that time that enable us to understand the effects that this had on the environment.

The region discussed here extends in a long belt down the western side of the Indian peninsula, from the high Sahyadri mountains east to the Deccan plateau. We have heavy rainfall on the coast, mountains which intercept the monsoon clouds, and a semi-arid regime in the plains east of the Sahyadri ranges. As early as the eighteenth century this rainfall regime, combined with the activity of a fairly dense human population, had led to the creation of a landscape of open savanna and cultivated plains in the east with fairly dense forest on the mountains and in their narrower valleys and woodlands, rough grazing, and rice paddy-fields along the eastern mountains. Major cities like Satara, capital of Chhatrapati Shahu, and Pune, the seat of the Peshwas, were located in the transitional zone between the wide, dry plains to the east and the narrow moist valleys to the west.

## How were Biotic Resources Used?

Indian agriculture, and indeed, Indian society, has long depended on domesticated cows and bullocks: the major source of power for agriculture and transport. Oxen were used for ploughing land, sowing, weeding and even threshing the grain. These animals ate grasses, straws and leaves indigestible by man, provided animal power and their dung served for manure and fuel. Finally, the major burden of transporting goods was borne by animals since road conditions usually precluded wheeled vehicles: such goods as did not move by water were carried overland by droves of bullocks. While the dairy yield of the village cattle was small, professional herdsmen reared both cows and buffaloes for milk and milk-products like clarified butter. The fodder and grazing needs of all these beasts made (as we shall see) significant demands on the environment, and led to active contests over its control.

Other animals were also numerous. Sheep and goats yielded fibre, meat, leather, milk and valuable fertilizer. They were sometimes herded over considerable distances to take advantage of seasonal pastures. But if the bullock was the central animal economically, its political equivalent was the horse. The Maratha armies were centrally constituted of light cavalry, and their strength depended on their numbers and mobility. The care and feeding of horses were important matters of state and much attention is devoted to them in the administrative records of the time.

There is evidence that the demands of livestock were straining the limits of the sustainable output of the intensively settled lands of the western Deccan in the eighteenth century when population density in the tract may be estimated at perhaps 80 per sq mile or 32 per sq km (Sykes n.d.). To put it another way, the total density was about one-third of that rural population density as determined by the Census of 1961.

There were several effects of these shortages in an economy so profoundly dependent on horses and oxen. One of these was the development of an active market in fodder. This was not a completely novel development. In 1228, the potentate Kholesvara endowed the temple of Sakalesvara in Beed district of east-central Maharashtra with certain revenue sources: one of these was the right to levy 'one bundle on the sale of dry grass and fodder' (Shastri 1972).

This means that people were already buying and selling this product: this implies that it was relatively scarce and hard to find. Later papers confirm this reasoning. For example, in the month of Kartik, Shaka year 1709 (September 1787) we find nine landholding peasants of the village of Pimple, Chakan subdivision borrowing a total of 10,000 bundles of millet stalks (*kadba*) valued at 250 rupees from Balaji Shankar Sonavani of Pune with the undertaking to repay in kind within Mrugshirsha i.e., within two months. They were required to deliver good quality *kadba*, not less than one and a half cubits (70 cm) long to the lender's store in Pune city. If this was the village of Pimple Saudagar on the Chakan road, then this involved cartage over about 14 km. But then as the astute Thomas Coats observed in 1819, 'the riches of the cultivator, nay his existence, depend on his cattle, he always nurses them with great care'. But he also noted that towards the end of the dry season 'grass is always scarce, and if the rains are late in falling, as seldom any provision is made for this, the cattle become extremely thin and weak, and a *murrain* [a disease] not infrequently gets among them at this time, and destroys many; which reduces the cultivators to beggary'. At that time, *bajri* straw sold at prices ranging from 2.5 to 5 rupees 'per thousand bundles of three handfuls each'; while *jowari* straw at harvest time was 1.5 rupees a bundle of three handfuls but rose to 4 or 5 rupees 'when the green forage is late'. So valuable was it that Coats commented *jowari* 'near the city is cultivated more than *bajri*, in consequence of the high price its straw bears as a forage'. But Coats also adds that 'as grasses where there is no demand for animal food would not yield a profit sufficient to pay the land-tax, they are never cultivated on arable lands'. All the uncultivated land of the village was used as a common pasture; this was clearly land unfit for tillage, for Coats noted that nowhere in the common land was the soil more than a few inches deep (Coats 1823). How do we explain the incapacity of the market to stimulate investment in the production of fodder, given the scarcity of it? Why didn't the village communities try to grow enough fodder for their animals?

## POWER AND CONTROL

The answer is that the peasants often found that powerful chiefs and rulers simply seized such resources. The Maratha government fixed a standard ration of six bundles of fodder for oxen and ponies, eight to full-grown horses, and 12 to camels. Elephants received 125 bundles of fodder as well as four loads of sugarcane.[1] These supplies were often simply seized from the peasants. Great armies were gathering all over India in 1801–2. Their animals had to be fed. Ibrahim Khan who commanded a unit of the Maratha cavalry ordered all the village headmen in the region east of Pune, to reserve all grassland for his needs (Rajvade 1909). Cavalry units would obviously claim priority over local requirements: in another news-report we learn that men from a nearby military camp simply came and cut down all the standing crops, both rain-fed and irrigated, in the village of Karathi in Khandesh. The desperate villagers then thought of abandoning the village and settling elsewhere (Rajvade 1909). Thus peasants would have been helpless to protect their grasslands and so did not take active measures to protect them, even though scientific studies in the 1950s showed that closing grassland to free grazing resulted in a 10-fold increase of net yield.

> In the first year of protection the cut yield was 4,500 kg which, therefore, represented what the animals might have obtained by grazing (minus 40 per cent loss due to trampling) in the last grazing year. In the second year of protection the cut yield was 10,000 kg, and in the third year 27,000 kg. (Dabadghao p. 187)

Even in times of peace, powerful people like the important Senapati Dabhade could forcibly encroach on village resources. We see this in a letter from the administrator of the township of Kadus to the government, written in March 1736.

> The honourable Senapati's camp-followers go daily from the main camp to Talegaon. They turn elephants and camels into the fields, and they get into the irrigated lands and steal. The Lord (Peshwa) may command on this matter. Rajshri Mahadji Govind has been granted the village of Turakdi. He has just reserved its grazing lands; he beat (our) cowherds; to the north Rajshri Mahadji Govind has reserved the grazing, and that of Sayegaon is reserved by Rajshri Tryambakrao Mama. Where will the people of Kadus take their cattle to graze, from which forest will they fetch their wood? It is not possible to carry on the life of the settlement without touching the border tracts of the adjoining villages. The Lord is able to command. (Sardesai 1931–35)

This was not an unusual occurrence; thus in 1778 the headman of Kaloli, a village near Saswad, complained that a powerful noble Jiuba Chitnis had similarly closed some land in the adjoining village of Naloli. The rains had failed the previous year, and the only available water for the local cattle was in a ravine in the reserved lands. Chitnis's officer had beaten and threatened the local villagers when they took their animals there. The situation was aggravated by the arrival of the Peshwa's officer, Avji Kavde, with a large train, whose cattle also went to drink there. If this continued (the headman wrote) the peasants would be severely distressed (Oturkar 1950). These documents bring out how inequalities of power affected the control of biotic resources: the small needs of local villagers and townspeople could be met under an arrangement where people went out and collected what they needed, or simply sent their animals out to graze on uncultivated land. But rich and powerful people would try to seize exclusive control of grazing land and forest. Such closures might be temporary, or they might become permanent: in that case these lands would become the private or government meadows. These were called *kuran* in Maharashtra. Many of these formed the core of Forest Reserves in the nineteenth century (*Forest Reports* 1869). The creation of one such reserve was ordered by the Peshwa Bajirao in 1758. His order stated that the court often marched through the district of Karde-Ranjangaon and needed wood and fodder. So the local officer should find a village (preferably a partly-cultivated one) assessed at 400–500 rupees in tax, knock down most of it, allow a little cultivation to remain, and convert the remaining lands into a *kuran* (Vad 1903). It does not seem that the villagers were asked if this was acceptable to them.

Powerful people would sometimes authorise others to use village lands even if the villagers did not want it. Thus we find the village of Karaje in the Nira valley complaining that several bands of shepherds with twenty thousand sheep had descended on their village. These animals were ravaging the crops and if they were not restrained then the village would be obliterated. The shepherds had been rebuked but paid no heed; if the governor did not listen then death (*moksha*) was the villagers' lot.[2] It is likely that these were shepherds licenced by the state, or in charge of the flocks that supplied meat for the tables of the gentry. This is suggested by documents in the archives such as this order issued in 1752–3:

> Letter to Sivaji Salokhe—Baji Govere, Ravalji Manka, and Yesaji Manka are in charge of the King's shepherds in Miraj province. *They are permitted to graze everywhere; do not molest them*. Issue stern injunctions to the Mangs and Bedars of the province to see to their safety.[3]

Of course, villagers might ignore the commands of their rulers. For example, a meadow had been reserved for the state elephants near the village of Vade, Vandan subdivision. The bold headman of that village, Sakhoji Navlage, not only turned his cattle into the meadow to graze but assaulted its keeper when he protested. Sakhoji was summoned to the court. Another village headman was found to have ploughed up meadowland allocated to the herdsmen, and had to be warned to desist.[4] The royal hunting preserve near the village of Khopsi was tended by the villagers. But soldiers from Kalyangad fort disputed their control and refused to recognise it. The government ordered them to desist from interfering in the lands below the hill-crest.[5] Thus stealth and force were deployed to constantly threaten the valuable biotic resources preserved by king and commoner alike. So the picture that we find in actual historical records is very different from that imagined by Vandana Shiva. The situation was in many ways similar to what exists in much of rural India today as government officials, powerful landowners, poor peasants, and tribals all try to draw on the resources of the countryside. But modern technology—such as firearms and motor vehicles—allows much more destructive extraction even as the much larger population makes our dwindling resources all the more valuable to their users.

## Notes

1. Maharashtra Government Archives, Pune (MSA, hereafter Pune archives), Chitnisi rumal 57, Pudke 3, docs. 30122 and 30169.
2. Pune archives, Chitnisi Rumal 36, Pudke 1, doc. 18870.

3. Pune archives, Shahu Daftar Rumal 8, Pudke 4, doc. 8996; another example is Rumal 8, Pudke 5, doc. 9724.
4. Pune archives, Shahu Daftar, Rumal 8, Pudke 4, doc. 8909; Pudke 5, doc. 9739.
5. Pune archives, Shahu Daftar, Rumal 8, Pudke 4, doc. 8845.

## References

Coats, Thomas, 1823, 'Account of the Present State of the Township of Lony', in *Transactions of the Literary Society of Bombay*, Vol. III.

Dabadghao, P.M., and K.A. Shankaranarayan, 1973, *The Grass Cover of India*, ICAR, New Delhi.

Guha, Sumit, 2001, *Health and Population in South Asia from the Earliest Times to the Present*, Permanent Black, New Delhi, pp. 25–31.

Kulkarni, A.R., 2000, *Maharashtra: Society and Culture,* Books and Books, Delhi.

Oturkar, R.V. 1950, *Peshvekalin Samajik va Arthik Patravyavahara*, Pune, p. 7. *Forest Reports of the Bombay Presidency 1860-61 to 1867-8*, 'Forest Report of the Bombay Presidency 1863–4', p. 102, Bombay, 1869.

Rajvade, Vishvanath K. comp. and ed., 1909, *Marathyanchya Itihasanchi Sadhane—Khand Dahava*, Pune, pp. 449, 473.

Sardesai, Govind Sakharam, ed., 1931-35, *Selections from the Peshwa Daftar*, 46 Vols., Bombay , Vol. 30, p. 129.

Shastri, Ajay M., 1972, *Yadava Insriptions from Ambe Jogai*, Hoshiarpur, p. 10.

Shiva, Vandana, 1991, *Ecology and the Politics of Survival*, New Delhi, p. 14.

Sykes, W.H., n.d., 'Special Report', p. 254.

Vad, G.C. *et al*. (eds), 1903-11, *Selections from the Satara Raja's and Peswa Diaries*, 9 Parts, Pune, Pt. III, vol. 1, p. 282.

# Section II

# Colonial India

*Photo credit Pallava Bagla*

# Introduction

If you time-travelled to the India of the mid-eighteenth century and returned nearly two centuries after that, you would see a subcontinent where the land and waterscape had undergone an unprecedented change. Many diverse ways of living, on land and water, and of their produce, would have endured. Yet, the content and context would be found radically altered. The coincidence of new pressures with the new face of an alien power, the British, would be unmistakable. However, the outcome of this coincidence would be far from uniform. Different regions, ecosystems and peoples would show vastly varying responses and experiences. Change wouldn't be new: its pace and scope would be.

The British rule saw many changes and phases. Among its most far-reaching impacts were the new pressures, economic and political, that it unleashed. Teak from the Western Ghats was used for ships for the Royal Navy. Hill sides in Assam and Kerala were cleared for new plantations of tea and coffee. Jute and rice were grown for exports. In turn, new institutions managed natural resources: the Departments of Forests and Irrigation. The rulers hoped to get more revenue to fund a global empire, to enlist soldiers to guard and expand it. They changed not just the lives of people but the land itself.[1]

In this Section, we get a window on the life and works of three men of the British era. The three men chosen here exemplify very different facets of India's environment or indeed of the colonial encounter. The first, Mahatma Gandhi, was born in a princely state at a time when such territories covered one-third of the land. The second, Salim Ali, was from the great port city of Bombay where his ancestors had made a fortune that helped him pursue a passion for birds. The third, Jim Corbett, served in the Army and was a white man born in India, who left soon after India's independence, never to return.

Mahatma Gandhi (1869–1948) was an outspoken critic of the British Empire but one with a very different notion of development, more benign and inclusive of fellow humans and of nature itself. *Ahimsa* or non-violence was a personal creed, less successful as a practise for the millions of Indians. In the movements to mobilize people, he broke the salt laws at Dandi to assert rights of commoners. Further inland, where there was no sea shore, tribals entered government forest in 'Gandhi's name'. His ethical critique of modern development had space for recycling human and animal waste and a parable about the earth having enough for the need of all but the greed of none. Ramachandra Guha takes us to the Gandhi we do not know and leaves the reader with a better appreciation of his legacy (see Chapter 8).

Salim Ali, an educated Indian in British India, and an ornithologist (scholar of birds), was not a political leader, unlike Gandhi. He died in 1987, in his early nineties, but for almost half his life he lived in British India. His views on the world of nature were of profound importance to the new emerging India. Born in the late nineteenth century, in a world where science was white man's work, he rose to international eminence. By the 1970s, he and his American colleague, S. Dillon Ripley, had compiled the *Handbook of the Birds of India and Pakistan* in all of 10 volumes. His pictures might seem to show a bearded, old, tired-looking man but they are misleading to say the least. To cite one example, in the 1930s he drove to an international ornithology congress in Germany on a motorbike, all the way over land from India. That spirit never quite left him even as a Grand Old Man of Indian science (see Whitaker 2003).[2]

Unlike Ali, Jim Corbett (d. 1955) was a man of the foothills of the north, a famous hunter of man-eaters and cattle lifters. While the other two were Indian by origin, he was a domiciled Englishman. His family had lived away from Britain—in India—for more than one generation. This may explain why independent India named a national park after him even after Independence, a rare tribute to an Englishman in a newly free nation state. Though he shot many tigers, he later turned to the camera and labeled the big cat 'a gentleman of the forest' that harmed people only when driven to desperation (Corbett 1944; Kala 1999).

The imperial ecological impact was not all malignant in character. By the year 1904, British foresters had enclosed over 500,000 square kilometres in government forest. These were source of revenue and fodder, timber for the expanding railways and for the then new underground coal mines. Such husbandry of the forest resources alleviated scarcity of forest produce, and also engaged foresters with the regeneration of vegetation. Game birds and animals found sanctuary in forest reserves like Dachigam in Jammu and Kashmir and Kerala's Periyar. Finally, India and contemporary Pakistan gained from huge perennial canals. This was true not only in the north with the Indus Canal Colonies but in the south, with Krishna Raja Sagar on the river Kaveri. In the west, the idea of the Rajasthan canal was already discussed before 1947.[3]

British officers and Indian princes saw their rule as one of benevolent tutelage. However, the ecological costs are evident to this day. By 1947, a huge forest estate had been created and protected from axe and plough. But this 'jewel in the crown' remains for peasants and *Adivasis*, artisans and usufruct right holders, 'a running sore' that still hurts. Protection of timber and other produce for industrial raw materials was carried on at the cost of livelihoods. Rural, tribal peoples, whether artisans or peasants, found forestry by government a costly imposition on their already difficult lives (Williams 2003). Faunal extinctions were averted by forest reserves and princely hunting grounds, but the slaughter and captures of wild beasts took a toll on their numbers. Over 85,000 tigers were killed in half a century ending 1925: the species like many more has never really recovered. Huge canal systems in the Indus and Ganga valleys were the pride of the Raj, but both basins reported extensive salt deposits that endangered fertility (on tigers/elephants, see Rangarajan 1998: 114–116; on canals, see Whitcombe 1993).

The costs were uneven in distribution as were the gains. The colonial era saw concerted efforts to impose new styles of life on mobile herders, sustained pressure on shifting cultivators and a new intensity for control of ponds, lakes and woodlands. City-building also entailed new conflicts over access to water and land. Some of these tensions simmer to this very day (see Nair 2005).

Environmental dilemmas of today need careful engagement with the early visionaries. They often foresaw more than people around them realized. It is no surprise that today's concerned citizens often turn to their experiences for insight. The legacies of the British Empire were mixed, and the experience calls for reflection.

## Notes

1. For further elaboration of these points see D'Souza (2003: 317–22). Free access at <www.conservationandsociety.org>. Also see Sivaramakrishnan (2003: 25–30). Free access at <www.seminar-india.com>.
2. Also see Salim Ali's autobiography (1987).
3. The best overview of the process is still the one done by Madhav Gadgil and Ramachandra Guha (1989).

## References


Ali, Salim, 1987, *The Fall of a Sparrow*, OUP, Delhi.

Corbett, Jim, 1944, *The Man Eaters of Kumaon*, reprinted 1995, OUP, Bombay.

D'Souza, Rohan, 2003, 'Nature and the writing of environmental history', *Conservation and Society*, Vol.1, No. 1. pp. 317–322.

Gadgil, Madhav and Ramachandra Guha, 1989, *This Fissured Land, An ecological history of India*, OUP, Delhi.

Kala, ,D.C., 1999 (1979), *A life of Jim Corbett*, Ravi Dayal Publishers, Delhi.

Nair, Janaki, 2005, *The Promise of the Metropolis, Bangalore's twentieth century*, OUP, Delhi, especially pp. 28–32 and 38–62.

Rangarajan, Mahesh, 1998, 'The Raj and the natural world, The campaign against 'dangerous beasts' in colonial India, 1875-1925', *Studies in History*, 14 , pp. 167–199.

Sivaramakrishnan, Kalyanakrishnan, 2003, 'Nationalism and the writing of environmental history', *Seminar*, 522, pp. 25–30.

Whitaker, Zai, 1987, *Salim Ali for Schools*, Permanent Black, Delhi.

Whitcombe, Elizabeth, 1993, 'The costs of irrigation: water logging, salinity, malaria' in David Arnold and Ramachandra Guha (ed.), 1993, *Nature, Culture, Imperialism, Essays on the Environmental History of South Asia*, OUP, Delhi.

8

# Mahatma Gandhi and the Environmental Movement

Ramachandra Guha

There is a great tradition of critical enquiry and social reform associated with Pune, a tradition possibly unmatched by any other Indian city. Although I am revisiting Pune after more than a decade, in the intervening period my life has been touched at every point by this tradition. My closest intellectual relationship over the last 10 years has been with Madhav Gadgil, an ecologist with a civic conscience to match his wide learning. Madhav, who delivered the inaugural Parisar lecture in 1982, is himself the son of a remarkable figure of the Indian Renaissance, D.R. Gadgil, who lived most of his life in this city. Among the Pune intellectuals of the elder Gadgil's generation were D.D. Kosambi and Iravati Karve, two of the scholars to whom Madhav and I, as a mark of respect and admiration for their work, dedicated our ecological history of India, *This Fissured Land*. A third dedicatee of that book was Verrier Elwin, on whose life my current research revolves. Elwin also had close ties with Pune; he was based here during his early years in India, while his later work among *Adivasis* was keenly influenced by A.V. Thakkar of the Servants of India Society.

This paper asks and seeks to answer the question: was Mahatma Gandhi an early environmentalist? The life and work of Gandhi have had a considerable influence on the contemporary environmental movement in India. This movement truly began with the Chipko Andolan in April 1973; in one of the first printed accounts of Chipko, a breathless journalist announced that Gandhi's ghost had saved the Himalayan trees. Ever since, Mahatma Gandhi has been the usually acknowledged and occasionally unacknowledged patron saint of the environmental movement. From the Chipko

* This paper is based on the Parisar Lecture delivered in Pune on 1 October 1993.

Andolan to the Narmada Bachao Andolan, environmental activists have relied heavily on Gandhian techniques of non-violent protest, and have drawn abundantly on Gandhi's polemic against heavy industrialization. Again, some of the movement's better known figures, for example, Chandi Prasad Bhatt, Sunderlal Bahuguna, Baba Amte, and Medha Patkar, have repeatedly underlined their own debt to Gandhi.

One must of course not deny other influences; for under the broad umbrella of the Indian environmental movement are many groups with little connection to Gandhi. Think, for example, of an organization like the Kerala Sastra Sahitya Parishad; a group coming from a background of Marxism, but whose contribution to the environmental movement is second to none. Other voluntary groups in the environment field are variously influenced by socialism, liberation theology and traditions of self-help. All the same, it is probably fair to say that the life and practise of Gandhi are the single most important influence on the environmental movement.

I have myself had the privilege of studying, at close quarters, the work of two extraordinary Gandhian environmentalists of the present day, Chandi Prasad Bhatt and Sunderlal Bahuguna. My personal acquaintance with Bhatt and Bahuguna is slight; but I was to acquire some intimacy with their lifework through my research on the movement with which they are associated, the Chipko Andolan. As I have argued in my book, *The Unquiet Woods,* it is misleading to think of Chipko itself as a 'Gandhian' movement, for its roots lie more strongly in a century-old tradition of peasant resistance in defense of forest rights. However, Bhatt and Bahuguna, the Andolan's best-known leaders, themselves exemplify the highest traditions of Gandhian constructive work.

Urban admirers of Chipko can be usually identified as being either Chandi Prasad or Sunderlal supporters, but there is in fact ample reason to celebrate both men. Bhatt and his organization, the Dashauli Gram Swarajya Mandal (DGSM), played a seminal role in the origins of Chipko: the technique itself was suggested by Bhatt to the villagers of Mandal. Since co-ordinating those early protests against commercial forestry, the DGSM has focused increasingly on environmental restoration. Here it has taken the initiative in organizing women for afforestation work in villages of the Alakananda valley, where its tree planting and protection programmes have been a good deal more successful than the lavishly funded schemes of the Forest Department.

Whereas Chandi Prasad Bhatt must be reckoned the pioneer of Chipko (if we were to bestow that accolade on one individual), Sunderlal Bahuguna has a record of social work that goes back even further. He and his wife, Vimla, were among the first group of Sarvodaya workers trained in the hills by Sarla Devi (Catherine Mary Heilman), a remarkable English disciple of Mahatma Gandhi who moved to Kumaun in the 1940s. In the Bhageerathi valley, which is his home base, Bahuguna organized several important Chipko protests between 1977 and 1980. In the years since, where Bhatt and his colleagues have concentrated on ecological restoration within the Himalaya, Bahuguna has chosen to take the message of Chipko beyond the hills. An indefatigable *padayatri,* with the endurance of men half his age, he has travelled widely in India and abroad. He is a captivating speaker too, and in this capacity has done a great deal to alert the urban intelligentsia (in particular) to the dangers of unbridled materialism.

These two Chipko leaders are, in my book at any rate, among the greatest living Indians. The example of Gandhi animates the lives of both men, but I like to think that each has taken something distinctive from the life of the master. Here Bahuguna, in his uncompromising denunciation of industrial society, more closely follows the Gandhi of *Hind Swaraj* (1990), that slim volume which contains a massive indictment of modern civilization. As expressed through his walking tours and lectures, Bahuguna's is quintessentially an appeal to the conscience of individuals, urging them to abjure consumerism and return to a simpler way of life. In contrast, by working out in practise a sustainable economic alternative to centralized development, Bhatt and his group are more in line with the Gandhi of the Sabarmati and Wardha ashrams. Chandi Prasad's work has helped infuse a new ecological meaning to the Mahatma's ideal of *gram swaraj,* or village self-reliance.

I speak of Chandi Prasad Bhatt and Sunderlal Bahuguna only because my study of the Chipko movement brought me in close contact with their work. Those of you who have been associated with the Narmada Bachao Andolan might like-wise have seen something of the spirit of Gandhi in the lives and actions of its leaders. The Chipko and Narmada Andolans are outstanding, but by no means isolated examples of the living heritage of Gandhi, as it is embodied in the contemporary environmental movement. However, the environmentalists of today do not merely claim that they are following the example of Gandhi; they go on to argue that the Mahatma himself foresaw the ecological crisis of modern industrial society. This question, whether Gandhi was indeed an early environmentalist is usually answered in the affirmative by his admirers, but rarely with supporting evidence. That is, it is taken for granted that Gandhi anticipated our environmental concerns, but without demonstrating precisely where and in what ways he did so. If at all his writings are invoked for the purpose, it is almost always his work *Hind Swaraj* (published in 1909) which a distinguished Gandhian of the present day has claimed gives us an 'alternate perspective' on development while explaining how 'the current mode of development is exploitative of man by man and of Nature by man'. Rereading *Hind Swaraj* recently, I found myself unable to agree with this verdict. Despite its eloquent denunciation of modern western culture, the book has nothing to say about man's relationship with nature: still less does it offer an alternate perspective on development.

But perhaps *Hind Swaraj* is not the place to look. That book was, of course, written while Gandhi was still in South Africa. On his return to India in 1914, Gandhi began immediately to acquaint himself at first hand with economic and social conditions in the village. Through his travels in the Indian countryside and the organization of those early satyagrahas among the peasants of Champaran and Kheda (in 1917 and 1918), Gandhi was to come face to face with colonialism as a system of economic exploitation, not merely—as had been his experience in South Africa—of racial discrimination.

Through his immersion in village India and this deeper understanding of colonialism, Gandhi came to see that it would be impossible for India to emulate western patterns of industrial development. It must be acknowledged at once that he does not anywhere offer an alternate model of development for India—for one thing, he was not a systematic thinker; for another, he was preoccupied with more pressing questions of political mobilization and social reform. All the same, scattered through his writings of the 1920s 1930s, and 1940s are clues to such an alternate path. It is to these writings that I now turn.

Gandhi's reservations about the wholesale industrialization of India are usually ascribed to moral grounds—namely the selfishness and competitiveness of modern society—but they also had markedly ecological undertones. Take this remarkable passage, from *Young India* of 20 December 1928:

> God forbid that India should ever take to industrialization after the manner of the West. The economic imperialism of a single tiny island kingdom (England)

is today keeping the world in chains. If an entire nation of 300 million took to similar economic exploitation, it would strip the world bare like locusts.

Two years earlier, Gandhi had claimed that to 'make India like England and America is to find some other races and places of the earth for exploitation'. As it appeared that the western nations had already 'divided all the known races outside Europe for exploitation and there are no new worlds to discover', he pointedly asked: 'What can be the fate of India trying to ape the West?' (*Young India,* 7/10/1926).

The answer to his question is by now painfully obvious. For in the last few decades, we have attempted precisely to 'make India like England and America'. Without the access to resources and markets enjoyed by those two nations when they began to industrialize, India has had perforce to rely on the exploitation of its own people and environment. The natural resources of the countryside have been increasingly channelized to meet the needs of the urban-industrial sector while the diversion of forests, water, etc., to the elite having accelerated processes of environmental degradation even as it has deprived rural and tribal communities of their traditional rights of access and use. The Chipko and Narmada movements are outstanding, but by no means isolated examples of the living heritage of Gandhi. Meanwhile, the modern sector has moved aggressively into the remaining resource frontiers of India, the North-East and the Andaman and Nicobar islands.

Perhaps Gandhi would not have been surprised. As he recognized, the bias towards urban-industrial development could result only in a one-sided exploitation of the hinterland. In 1946, he had expressed this with characteristic lucidity: 'The blood of the villages is the cement with which the edifice of the cities is built' *(Harijan,* 23/6/1946). On an earlier occasion, Gandhi had, in his characteristically gentle yet forceful manner, alerted a gathering in Indore to the concentration of resources on which city life has come to rest. 'We are sitting in this fine *pandal* under a blaze of electric lights', he remarked, 'but we do not know we are burning these lights at the expense of the poor' (*Harijan,* 11/5/1935).

From this diagnosis of the ills of industrialism flowed Gandhi's preferred solution, wherein economic development would be centred on the village. He wished, above all, to see that 'the blood that is today inflating the arteries of the cities run once again in the blood vessels of the villages'. Preeminent here was the decentralization of political and economic power, so that villages could resume control over their own affairs. When he was accused of turning his back on the great scientific inventions, including electricity, Gandhi remarked (in words to inspire all proponents of decentralized energy systems): 'If we could have electricity in every village home, I should not mind villagers plying their implements and tools with the help of electricity. But then the village communities or the state would own power houses, just as they have their grazing pastures' (*Harijan,* 22/6/1935).

In 1937, some years after he had moved to Wardha to devote himself to rural reconstruction, Gandhi defined his ideal Indian village as follows:

> It will have cottages with sufficient light and ventilation, built of a material obtainable within a radius of five miles of it. The cottages will have courtyards enabling households to plant vegetables for domestic use and to house their cattle. The village lanes and streets will be free of all avoidable dust. It will have wells according to its needs and accessible to all. It will have houses of worship for all, also a common meeting place, a village common for grazing its cattle, a co-operative dairy, primary and secondary schools in which industrial (i.e. vocational) education will be the central fact, and it will have Panchayats for settling disputes. It will produce its own grains, vegetables and fruit, and its own Khadi. This is roughly my idea of a model village... (*Harijan* 9/1/1937).

There are many elements in this picture that would fit nicely into the Utopia of the environmentalist: local self-reliance, a clean and hygienic environment, the collective management and use of those gifts of nature so necessary for human life, water and pasture. But Gandhi himself had an uncanny knack of combining a Utopian vision with shrewdly practical means. Notable in this connection is the attention he paid to the crucial problem of soil. Towards the end of his life, he warned the proponents of the rapid mechanization of agriculture that trading in soil fertility for the sake of quick returns would prove to be a disastrous, short-sighted policy. It would result in virtual depletion of the soil (*Harijan,* 25/8/1946). He was an enthusiastic supporter of organic manure, which enriched the soil, improved village hygiene through the effective disposal of waste, saved foreign exchange, and enhanced crop yields—all this, as we now know, without the attendant pollution and resource exhaustion caused by modern chemical techniques. He greatly admired the work of Albert Howard, who had pioneered methods of organic agriculture at his Institute of Plant

Industry in Indore. In his own journal, *Harijan,* Gandhi described approvingly and in great detail, the methods developed by Howard and his associates to convert a mixture of cowdung, farm wastes, wood ash, and urine into invaluable fertilizer (*Harijan,* 17/8 and 24/8/1935).

Finally, Gandhi's philosophical critique of modern civilization also has profound implications for the way we live and relate to the environment today. For him, 'the distinguishing characteristic of modern civilization is an indefinite multiplicity of wants'; whereas ancient civilizations were marked by an 'imperative restriction upon, and a strict regulating of, these wants' (*Young India,* 2/6/1927). In uncharacteristically intemperate tones, he spoke of his 'wholeheartedly detesting this mad desire to destroy distance and time, to increase animal appetites, and go to the ends of the earth in search of their satisfaction. If modern civilization stands for all this, and I have understood it to do so, I call it satanic' (*Young India,* 17/3/1927).

At the level of the individual, Gandhi's code of voluntary simplicity offers a sustainable alternative to modern lifestyles. One of his best-known aphorisms, that the 'world has enough for everybody's need, but not enough for everybody's greed', is, in effect, an exquisitely phrased one line environmental ethic. This was an ethic he himself practised; for resource recycling, and the minimization of wants, were integral to his life.

His analysis of macro processes of economic development, his prescriptions for rural reconstruction, and his ethics for living; at all these levels Gandhi's writings, when reinterpreted in contemporary terms, offer acute insights into the environmental crisis. During his lifetime, this economic philosophy was elaborated and fleshed out by one of the Mahatma's close disciples, J.C. Kumarappa. Kumarappa has strong claims to being considered the first Gandhian environmentalist; as his work is largely unhonoured and forgotten today, a brief assessment is perhaps not out of place here.

Kumarappa was a Tamil Christian who had been trained in Accountancy in London. He had a flourishing practise as an auditor in Bombay, which he left temporarily to take a Master's degree at Columbia University in New York. There he embarked on a study of public finance, in the course of which he systematically uncovered the colonial exploitation of the Indian economy. He returned home in 1929, now a nationalist, and soon came into contact with Gandhi. His thesis on public finance was serialized in *Young India,* and Kumarappa himself abandoned his practise to join the ashram at Sabarmati. He was put in charge of Gandhi's schemes of village reconstruction, and over the next decade conducted important surveys of the agrarian economy and helped run two key Gandhian institutions, the All India Spinners Association and the All India Village Industries Association.

In a number of books written in the 1930s and 1940s, J.C. Kumarappa attempted to formalize Gandhian economics. As with his mentor, strewn through his writings are observations with profound ecological implications. This remark, for instance, could well serve as a basic condition for ecological responsibility: 'If we produce everything we want from within a limited area, we are in a position to supervise the methods of production; while if we draw our requirements from the ends of the earth it becomes impossible for us to guarantee the conditions of production in such places.'

Like his teacher, J.C. Kumarappa powerfully denounced industrial civilization. 'There can be no industrialization without predation', he observes, whereas agriculture is, and ought to be, 'the greatest among occupations', in which 'man attempts to control nature and his own environment in such a way as to produce the best results.' Notably, he expressed this contrast between agriculture and industry in terms of their impact on the natural world. As he put it:

> In the case of an agricultural civilization, the system ordained by nature is not interfered with to any great extent. If there is a variation at all, it follows a natural mutation. The agriculturist only aids nature or intensifies in a short time what takes place in nature in a long period... Under the economic system of (industrial society)... we find that variations from nature are very violent in that a large supply of goods is produced irrespective of demand, and then a demand is artificially created for goods by means of clever advertisements.

Like most Gandhians of his generation, however, Kumarappa was primarily interested not in theoretical reflection but in ameliorating the lost pride of the Indian peasant and artisan. A theme that runs through much of his work is the careful husbanding of natural resources in the agrarian economy. Thus he stressed the need to use night soil as manure, asking for subsidies to be given to individuals, as a means of overcoming caste prohibitions, for converting human excreta and village waste into organic fertilizer. At the same time, Kumarappa also dwelled on the importance of maintaining soil quality by checking erosion and water logging.

Water and forests are perhaps the two resource sectors that have most exercised the Indian environmental movement in recent years. In this connection, Kumarappa was not slow to criticize the poor maintenance of irrigation tanks under British rule, or to urge the conservation of water to augment the water table and reduce brackishness. And in a pithy comment on actual and preferred models of forest management, he says:

> The government will have to radically revise its policy of maintaining forests. Forest management should be guided, not by considerations of revenue but by the needs of the people... Forest planning must be based on the requirements of the villagers around. Forests should be divided into two man classes: (1) those supplying timber to the planned from the long range point of view, and (2) those supplying fuel and grasses, to be made available to the public either free of cost or at nominal rates. There are village industries such as palm gur, paper making, pottery, etc. which can flourish only if fuel and grass can be supplied to them at cheap rates.

Equally farsighted are Kumarappa's remarks on potential biomass shortages in the rural economy. He was particularly concerned about fodder availability, pointing out that cash crops like jute, tobacco, and sugarcane reduce food availability for men and their domestic animals. He also noted the widespread complaint of peasants that there was not enough grazing land, taking the colonial government to task for its reluctance to allow grazing on waste land without payment of a fee.

Soil maintenance and fertility, water conservation, recycling, village forest rights, biomass budgets—this is an agenda of rural environmental problems that are still very much with us. In setting firmly in its natural setting, Kumarappa could be said to have begun the task of building an ecological programme on Gandhian lines. Although they are for the most part unaware of his work, the environmentalists of today are only taking up where he left off.

Another associate of Mahatma Gandhi with environmental ideas far in advance of her time was Mira Behn (Madeline Slade), the daughter of an English Admiral who joined the Sabarmati Ashram in 1925. Mira Behn, like J.C. Kumarappa, was part of the Mahatma's inner circle, the core group of his followers: and like the Tamil economist, she too spent many years working for rural reconstruction, elaborating in practise the precepts of her teacher. In 1947 she set up an ashram near Rishikesh, at the foot of the Himalaya, shifting her base several years later to the Bhilangna valley in the interior hills. In articles written at the time, Mira Behn drew the attention of the public and of policy makers to the intimate links between Himalayan deforestation, soil erosion and floods. Years before the Chipko Andolan was to give popular force to these criticisms, she identified the lacunae of forest management as being, first, the lack of involvement of villagers; and second, the replacement in many areas of oak with pine, a species with much less capacity to absorb and retain rain water. She sent detailed reports with photographs to the Prime Minister, Jawaharlal Nehru; he passed them on to the forest officials concerned, but (as Mira Behn wryly noted many years later) the 'necessary changes were too fundamental' for the forest department to make.

In her years in rural Uttar Pradesh, Mira Behn also made some percipient comments on the chief ecological problems of Indian agriculture, problems that remain with us. These included the large scale water-logging that appears to be an almost inescapable feature of canal irrigation; the ploughing up of lands more suitable for growing pasture for cattle (this adversely affecting the quality of livestock); and rampant soil erosion. For Mira Behn, the rapidity of ecological change and disturbance was a distinguishing feature of modern life. While ancient civilizations in North Africa and the Middle East had collapsed due to their abuse of the natural environment, she wrote in the *Hindustan Times* of 5th June 1950, 'in those days it took centuries and centuries to reach complete destruction, but in these days of modern machinery and science, what took a thousand years or more in the past may be accomplished in a paltry hundred years!'

In common with Gandhi and Kumarappa, Mira Behn's primary concern was with rehabilitating the village economy of India. Yet her interest in the natural environment was not merely instrumental: at times she expresses a spiritual affinity with nature of a word worth a kind, straight out of the European romantic tradition. She called herself a devotee of the great primeval Mother Earth. As she wrote in April 1949:

> The tragedy today is that educated and moneyed classes are altogether out of touch with the vital fundamentals of existence—our Mother Earth, and the animal and vegetable population which she sustains. This world of Nature's planning is ruthlessly plundered, despoiled, and disorganized by man whenever he gets the chance. By his science and machinery he may get huge returns for a time, but ultimately will come desolation. We have got to study Nature's balance, and develop our lives within her laws, if we are to survive as a physically healthy and morally decent species.

I began my talk by recognizing and commenting upon the visible influence of Mahatma Gandhi on the Indian environmental movement. I then went back in time to investigate to what extent Gandhi himself had anticipated the distinctive ecological problems of the present day. The evidence in this respect does confirm that the ideas of Gandhi, as also of his followers J.C. Kumarappa and Mira Behn, do constitute an eminently usable past for the environmental movement.

It is time now to turn our attention to one widely prevalent myth that has its origins in the environmental movement's reclamation of Mahatma Gandhi. It is an unfortunate tendency, prevalent especially among the radical fringe of the movement, to identify good and evil with particular individuals. For the radical environmentalist, Gandhi is good in almost exact proportion to which Nehru is bad. While celebrating Gandhi as a model to honour and follow, he wishes simultaneously to demonize Jawaharlal Nehru, at whose feet he lays the blame for the ecological crisis of Indian society. Many environmentalists even believe that Gandhi had himself outlined a model of ecologically sound development; and that this Gandhian alternative was cast in to the dustbin by Nehru, who then imposed, on independent India, his own model of capital intensive, environmentally destructive economic development. A tale to illustrate this was recently told by an expatriate Indian environmentalist now based in Britain. Apparently Gandhi was staying with Nehru in the latter's family house at Allahabad, and asked for a bucket of water to wash up in the morning. Nehru sent two buckets, whereupon Gandhi sent one back. 'Why Gandhiji,' protested Jawaharlal, 'this is the city where the Ganga and Yamuna meet; there can be no shortage of water here.'

This incident is meant to exemplify the prudence of Gandhi, and the profligate ways of his host; ways which after 1947 are believed to have found expression in the path of destructive development followed by the new nation. No source is given for the story, which is almost certainly a figment of the environmentalist's imagination. Yet the beliefs underlying his tale are widely held by the Gandhian environmentalists of today. Let me give one more example, out of several I could have chosen, to illustrate this. In an essay published some years ago, a prominent Indian environmental writer and activist claimed that Mahatma Gandhi 'tried in vain to persuade Jawaharlal Nehru not to take India down the path of over-consumption.' This statement expresses in succinct fashion the two core elements of the myth: first, that ecologically speaking Nehru was as profligate as Gandhi was prudent; and second, that Gandhi had his own, alternate plan of development for India, which Nehru in his arrogance rejected out of hand. It is in this manner that the environmental debates of today have brought Gandhi and Nehru into fierce, if posthumous, public competition violating in spirit and letter the intimate relationship that actually exited between the two men.

The environmentalists' opposition of Gandhi to Nehru stems, in part, from the need to explain, a self-evident puzzle: that the development experience of independent India has been marked by a profound insensitivity to ecological considerations—and this despite what we have just demonstrated, namely that the 'Father of the Nation' was, in our terms, emphatically an 'early environmentalist'. The puzzle can be most conveniently explained by contrasting the prudent Gandhi with the profligate Nehru, and by putting forth a conspiracy theory whereby the younger man first took over the Congress in some kind of palace coup, then swiftly rid it of its Gandhian heritage.

That the puzzle exists I do not dispute; but I do wish to qualify, perhaps even challenge, the way it is usually explained by my friends in the environmental movement. To challenge their black and white portraits of Nehru and Gandhi is not, of course, to ignore the profound philosophical differences between the two men. Gandhi's vision of free India centred on village renewal; Nehru's vision, just as firmly, on rapid industrial development. The older man preferred stability to change; the restless Nehru, change to stability. These differences come out clearly in an exchange of letters between the two in October 1945. Following a Working Committee meeting on social and economic objectives after independence, Gandhi wrote to Nehru of his belief that India could 'realize truth and non-violence only in the simplicity of village life'. He went on to liken industrial society to the moth that whirls faster and faster around the light, only to perish in it. In his reply, Nehru disputed that the village, for him a mile backward both intellectually and culturally, could ever embody the principles of truth and non-violence. He identified, as the chief goal of economic planning, not over-consumption as the environmentalist quoted earlier would have us believe but rather a sufficiency of food, clothing, housing, education, sanitation, etc., for every Indian. This was a goal on which both Nehru and Gandhi agreed, but the younger man, in common with other intellectuals of the time, was convinced that it could be achieved only through rapid industrialization and the use of modern technology.

These differences notwithstanding, we must also recognize the deep and abiding love between Gandhi and Nehru. 'I cannot think of myself as a rival to Jawaharlal or him to me,' wrote Gandhi in July 1936. He continued:

> Or if we are, we are rivals in making love to each other in the pursuit of the common goal. And if, in the joint work for reaching the goal, we at times seem to be taking different routes, I hope the world will find that we had lost sight of each other only for the moment, and only to meet again with greater mutual attraction and affection.

I do not know how environmentalists reconcile this with the Gandhi/Nehru polarity they so fervently uphold; or indeed how they ignore the Mahatma's public anointing of Nehru as his heir in the early 1930s, a succession Gandhi repeatedly confirmed in later years. More substantively the environmentalists' interpretation of the crucial years before independence fails to recognize that by 1940 or there about, Gandhi's own economic ideas had been decisively rejected by the national movement. There had come about an overwhelming consensus, among politicians and intellectuals, that rapid industrialization was the only viable economic strategy in independent India—a strategy its proponents believed would go a long way in reducing poverty and unemployment, and in making for a strong, self-reliant, genuinely independent society. Nehru expressed this consensus in a particularly eloquent fashion; but behind him stood a solid phalanx of utterly sincere and deeply patriotic men.

Indeed, if the Gandhian model had indeed been adopted as the basis for economic policy in 1947, that would have been an undemocratic imposition in the face of strong, majority opinion to the contrary. The actual marginalization of the 'Gandhian alternative', such as it was, is well expressed in the career of J.C. Kumarappa. In 1937, he was appointed to the National Planning Committee (NPC) of the Congress as a representative of the All India Village Industries Association, but resigned when his fellow members of the NPC did not agree to put the village at the centre of planning. After Independence, Kumarappa was deputed by the Sarva Seva Sangh to represent it in the Planning Commission's Advisory Body. Again, the Gandhian economist quickly sensed that he was in a minority of one, and left the committee.

From our own vantage point, it is possible to celebrate the Mahatma and his disciple as environmentalists before the age of environmentalism. By contrast, India's first prime minister represented the majority intellectual opinion within the national movement, namely, that the revitalization of India could only come about through massive industrialization. One may justly honour Gandhi and Kumarappa for being ahead of their time; but it is grossly unhistorical, as well as unfair, to condemn Nehru for being, merely, a man of his time.

The great British socialist Edward Carpenter once remarked that the Outcast of one age is the Hero of another. Perhaps the converse, that the Hero of one age is the Outcast of another, is equally true. For no man was as greatly adored during his lifetime as Jawaharlal Nehru, yet no man has been more vilified since his death. It appears that Nehru was responsible for all that is wrong with India today. Thus the right holds Nehru's policies of pseudo-secularism and state planning squarely responsible for communal conflict and economic stagnation, while the left, just as effortlessly, traces the roots of economic inequality and environmental degradation to the same man's practise of pseudo-socialism and ecological arrogance.

This demonization of Nehru, within and outside the environmental movement, fails to allow for the possibility that times change, and men and ideas with them. Take for instance the controversy around the Sardar Sarovar Project, which environmentalists have found easy to represent in terms of the Gandhi/Nehru opposition. Thus a critic of the project wrote recently of a historic old temple being submerged by the rising waters of the dam, characterized by him as 'one of Jawaharlal Nehru's temples of modern India'. Here a man who died 30 years ago was being held guilty for the construction of a dam today, on account only of a phrase he had used to describe another dam built in the early years of Independence. But how can one be so sure that a man as generous and open minded, as Nehru would have held steadfast to a viewpoint, despite mounting evidence to the contrary? For myself, I have little doubt that were both Nehru and Gandhi alive today, on the Sardar Sarovar controversy they would have found themselves on the same side.

The urge to demonize Nehru comes from a Cowboys and Indians vision of history, wherein the world is effortlessly divided up into good and bad guys. These black and white portraits are especially congenial to social activists: they were once characteristic of the Marxist, and they now, sadly, appear to be characteristic of the radical environmentalist. But the ideas and actions of individuals must be set in context: that is the task of the historian, in doing which he might find himself qualifying, to lesser or greater degree, the beliefs of the activist. It is in this spirit that I have contested the environmentalists' portrayal of Nehru in uniformly dark colours, and it is in the same spirit that I now wish to qualify their portrayal of Gandhi in uniformly light ones. As I have argued, the historical figure of Gandhi provides a body of ideas, and a vocabulary of protest against unjust laws that have proved critical to the environmental movement. This much is indisputable—but perhaps it is now time to ask: are there ways in which the heritage of Gandhi might actually limit the movement? Or put more plainly, does Gandhi provide *all* the answers to those working for environmental and social renewal today? Some environmentalists are emphatic that he does, indeed—thus one friend recently claimed that for each and very environmental event or crisis or challenge one can find inspiration and guidance in Gandhi. This most emphatic statement notwithstanding, I think that Gandhiji does not provide all the answers; sometimes he does not even ask the right questions.

Let me clarify. I believe that the heritage of Mahatma Gandhi has limited the vision of the environmental movement in two crucial respects. First, it is striking how heavily focused on the countryside are the horizons of most environmentalists. Like Gandhi, his present day followers appear to have little understanding of the urban context and its distinctive social and environmental problems. In their angry denunciations of the urban–industrial way of life, Indian environmentalists, by and large, have yet to come to terms with the fact that by the turn of the century India shall have the largest urban population in the world. Speaking in this city, at the invitation of a group that has done much to alert us to these questions, I do not have to belabour the ecological problems associated with such rapid and unregulated urbanization: massive pollution, overcrowding, and the diseases associated with it, acute water shortages, grossly inadequate housing, and sanitation, and a system of transportation that is highly inefficient from an energy conservation and environmental point of view. In actively engaging with these problems, and in trying to make our cities and towns habitable, environmentalists can find no help from Gandhi, who in his own life and work simply turned his back on the city.

Like the city, the wilderness has no attraction for Gandhi. In last year's Parisar lecture, Bittu Sahgal perceptively noted that some environmental activists had no time for nature conservation, dubbing it an elitist fad. Ironically, Gandhi was, by all accounts, himself profoundly indifferent to the wild. It is true that his practise of vegetarianism and non-violence oriented Gandhi towards unspoilt nature. This might perhaps be attributed to his severely practical tempermanent, for there was nothing of the romantic in Gandhi. Intriguingly, Nehru, by far the more romantic of the two men, was deeply appreciative of the natural beauty of India. There is a near mystical quality to Nehru's invocation, in his last will and testament, of his affinity with the soil, the mountains, the rivers of India.

An anecdote to illustrate this contrast is told by Edward Thompson, the British educationist and writer who was a close friend of both Gandhi and Nehru. When the Congress Ministries were formed in different provinces of British India in 1937, Thompson tried hard to interest the nationalist leaders in the cause of saving India's disappearing wildlife, with (as he noted) 'animal after animal... either extinct or on the danger list'. When he confronted Gandhi with the problem, the Mahatma merely joked, saying, 'we shall always have the British lion.' But then, noting Thompson's disappointment, Gandhi asked him to speak to Jawaharlal, as one who might show more interest. Nehru did, indeed—he went on to speak of the issue to the Prime Ministers (as they were then called) of Congress-ruled states. Later, Nehru was able to report to Thompson, with some pride, that C. Rajagopalachari's last act, as Premier of Madras, was to put through the Periyar Nature Reserve.

Nature lovers and those with a focus on the urban environment would, therefore, will find little direct help from Mahatma Gandhi. But between the wilderness and the city lies a vast terrain, home to the seven hundred thousand villages Gandhi spoke of so often and so eloquently. It is here that his life and message admit of more direct application, in the resistance to environmentally destructive projects or in the restoration of the relationship between the agrarian economy and its natural environment. And all of us without exception—whether living in the city, the country or the wild—can try and simplify our lifestyles to the extent compatible with individual circumstance, taking our lead from a man who, in his own life, made remarkably few demands on the earth. And so it is that the environmental movement must perennially return to Mahatma Gandhi, while at the same time going beyond him.

## NOTE

The talk gave rise to a lively discussion, with a number of penetrating comments and critiques coming from the audience. As I felt it would be contrary to the spirit of my talk, I have not chosen to revise it in the light of these comments. However, as they take the debate several steps further a brief gist of the main points is provided below.

First, the question of population. It was suggested that Gandhi's rather rigid attitude to modern methods of birth control might have subtly influenced the Indian environmental movement's failure to lay adequate stress on the need to check population growth. This drew the response (from another member of the audience) that one might instead see Gandhi's opposition to contraception as flowing out of his innate respect for the human body and for the sanctity of life. He might very well have felt vindicated if he were to have seen the new invasive technologies of hormonal implants and fertility vaccines, technologies as invasive as major dams and utter violations of the rights of women.

Where population was a subject I had myself not touched upon, my use of the village/city dichotomy was criticized as a simplification of Gandhi's views. As one friend so beautifully put it, Gandhi might be seen not simply as an admirer of the village who turned his back on the city, but rather as wishing to transcend this dichotomy, by working towards a synthesis, an ideal state which would be both non-village and non-city.

On the differences between Nehru and Kumarappa regarding the path of planning in independent India, I was urged to take Vinoba Bhave (whom my talk had ignored) as the third point of reference in what was in essence a three-sided affair. It was suggested that Kumarappa was a purist, a Utopian who had a somewhat rigid understanding of Gandhi, from which flowed a unreserved hostility to the ideas of Nehru. But the more perceptive Vinoba recognized that, in the India of the 1950s, Nehru's ideas were very widely shared. It was thus that he did not reject or turn his back on the pattern of economic development followed at this time, but rather, tried both to spiritualize and soften it. In this respect, Vinoba could be seen as mediating between Gandhi and Nehru.

Finally, I was reminded that the overwhelming intellectual consensus around Nehru, which I had laid such stress upon, was a consensus not so much on the need of heavy industrialization as on the importance of science and the scientific worldview. It was believed (at any rate by the intelligentsia that in modern science lay the key to unlimited human knowledge; in modern technology the key to unlimited expansion of human welfare. The alliance of science and technology in the development process might be seen, therefore, as the distinguishing characteristic of what is sometimes called the Nehruvian model. Of course, from today's vantage point we can clearly see the fragility of these claims, for modern science and technology, far from providing unlimited human welfare, have, concentrated wealth and power in a ever narrower elite while leading to massive environmental degradation.

**Sources:** The *Collected Works of Mahatma Gandhi,* which run to more than 90 volumes, are the basic source for all scholarly studies of Gandhi. For my own limited purposes, namely, the reassessment of Gandhi as an 'environmentalist', I have relied heavily on three invaluable thematic anthologies of his writings: *Industrialize—And Perish*!, compiled by R.K. Prabhu (Navjivan, 1966); *Village Swaraj,* compiled by H.M. Vyas (Navjivan, 1962); and *My Picture of Free India,* compiled by Anand T. Hingorani (Pearl Publications, 1965).

J.C. Kumarappa's important works include *Why the Village Movement?* (second edition: Hindustan Publishing, 1938); and *The Economy of Permanence* (second edition: All India Village Industries Association, 1948). Kumarappa's ideas are examined in greater detail in my essay 'Prehistory of Indian Environmentalism: Intellectual Traditions', *Economic and Political Weekly,* 4–11 January 1992.

Finally, a fascinating selection of Mira Behn's writings compiled by her associate Krishna Murti Gupta, which I have drawn on here, has been published in a recent issue of the journal *Khadi Gramodyog* (Volume 39, Number 2, November 1992). A fuller selection of articles, as well as tributes to her work, can be found in Krishna Murti Gupta (ed.), *Mira Behn: Birth Centenary Volume* (Himalaya Seva Sangh, 1992).

9

# THE USES OF ECCENTRICITY: THE MAKING OF SALIM ALI

Madhav Gadgil

Salim Moizuddin Abdul Ali (1896–1987) was indisputably the greatest biologist of twentieth-century India. His meticulous observations of the bird life of the country spanned most of the century, for the earliest of these was recorded in 1906 when he was all of 10 years old. It read:

> The cock sparrow perched on the nail near the entrance to the hole while the female sat inside on the eggs. I ambushed them from behind a stabled carriage and shot the male. In a very short while the female acquired another male also, sat 'on guard' on the nail outside. I shot this male also, and again in no time the female had yet another male in attendance. In the next seven days I shot eight male sparrows from this perch; each time the female seemed to have another male in waiting who immediately stepped into the gap of the deceased husband.

These observations were to continue, entirely without break, for the next 80 years, through the shooting of birds—some for the pot, and most for scientific collection—was gradually to give way to just watching them through binoculars or netting and ringing them to understand their behaviour. These studies unravelled the fascinating breeding system of weaver birds and the role of sunbirds and flowerpeckers in pollinating and dispersing the seeds of mistletoe. Salim Ali investigated the famed Flamingo City in the Rann of Kutch. He undertook a series of regional bird surveys of Hyderabad, Travancore, Cochin, Afghanistan, Kailas Manasarovar, Kutch, Mysore, Goa, Sikkim, Bhutan, Arunachal Pradesh. He elucidated the migrations of waterfowl which, in a number of species, reached all the way to

---

* Reprinted with permission from Madhav Gadgil, 2001, *Ecological Journeys: The Science and Politics of Conservation in India*, Permanent Black, New Delhi.

Siberia. Throughout, he maintained the most meticulous records and distilled his knowledge into a series of superbly written and illustrated books; he began in 1941 with *The Book of Indian Birds* and followed this by *The Birds of Kutch, Indian Hill Birds, Birds of Kerala, Birds of Sikkim*, and finally his magnum opus, the ten volume *Hand-book of the Birds of India and Pakistan*. His last bird book, *The Field Guide to the Birds of Eastern Himalaya* was published in 1997. His last book, a most charming autobiographical account, *The Fall of a Sparrow* was published in 1985. In time, Salim Ail's passion for understanding the life of birds led to a passion for conserving them. His total dedication to science and to nature conservation was unique, unmatched by anyone else in the Indian scientific and conservation community.

Salim Ali's scientific work was grounded in a careful, painstaking observation and recording of events in the living world. This was something entirely novel in India. There had of course been other great Indian scientists before, Jagadish Chandra Bose (1858–1937) with his remarkable studies of plant physiology being a notable example. But J.C Bose was basically an experimentalist. The great field biologists who worked in India before Salim Ali were all Europeans like Sir Joseph Hooker (1817–1911) who not only produced a definitive work on Indian plants but was critical to advancing Charles Darwin's theory of evolution. Sir C.V. Raman (1888–1970), a contemporary of Salim Ali's was an experimental physicist; but most other luminaries of Indian science in his age such as the great number-theorist Ramanujam (1887–1920), S.N. Bose (1894–1974) of the famous Bose–Einstein distribution, and P.C. Mahalanobis (1893–1972) known for the $D^2$ statistics—all were theoretically oriented. Amongst all of these distinguished scientists Salim Ali was the one man to devote his life to scientific observation of the natural world. He was all alone because he came from a society in which men of learning had little wish to involve themselves with the natural world. After all Shankara had convinced them that the empirical world was an illusion. And in any case working with one's hands was not considered a fit occupation for people of literate castes.

The historian–mathematician D.D. Kosambi attributes the stagnation of the Indian science of Ayurveda to high-caste men of learning giving up the eating of meat. They then lost all touch with animals and in consequence, human anatomy. But even the study and documentation of herbal medicines was progressing haltingly. New Nighanthus (ayurvedic pharmacopoeas) continued to be written but none of them have careful, detailed descriptions of the plants, with excellent colour drawing, as early as the twelfth century, none of the ayurvedic Nighanthus have any illustrations. Perhaps the only Indian in pre-British times to write down careful descriptions of living organisms, accompanying them by excellent colour drawings was Emperor Jehangir. Jehangir devotes many pages in his diary to a meticulous recording of the breeding behaviour of the Sarus crane; he commissioned drawings of a large number of birds by Mansur, a learned painter of his court. When told that a lake in Ajmer was bottomless, he had its depth sounded by ropes and stones and concluded that it was only 22 metres deep.

Jehangir came of the stock of Mongols, who were nomadic herders, hunters and plunderers. For them, careful observation of the natural world as they traversed the vast cold deserts of Central Asia would have been a matter of life and death. But Jehangir's literate Indian contemporaries absorbed little of this. And it was to cost them dear. For instance the Peshwas, the learned rules of Pune, never understood the significance of the maps and magnetic compasses they captured from the British, and continued to plan their losing strategies on the basis of oral traditions concerning the terrain they were fighting in.

Salim Ali's accurate observation and recording of bird life have to be viewed against this backdrop. There were undoubtedly many Indians living close to earth, such as the Kanis of Kerala, who had as intimate a knowledge of the birds that surrounded them as Salim Ali. But this knowledge was need-based; the Kanis hunted most of these birds for the pot, and their grasp of bird life was quite naturally restricted to their own locality. Salim Ali was the only man whose careful observations of birds spanned the whole subcontinent and were made from the detached perspective of a scientist.

Such observation could simply not have been made by a traditional man of learning from the Indian society of his times. But fortunately for Indian biology, Salim Ali came from a most unusual family, that of one of the founders of the Indian National Congress, Badruddin Tyabji (1844–1906). (Incidentally, A.O. Hume [1829–1912], the British civil servant who took a lead in the founding of the Congress, was the foremost student of Indian birds in the period 1864–1898). The Tyabjis belonged to the trading community of the Bohras from Kutch. Badruddin's father, Tyab Ali (1803–1863), had amassed a large fortune as a trader. So theirs was a wealthy Muslim family that did not owe its prosperity to the ownership of vast acres of land. It was a rich family that was forward-looking, and valued learning, including modern British education. Many of the men, and even the women of the family, including Salim Ali's wife Tehmina, were educated in England and absorbed European manners and passions. These included a love for the outdoors, and of nature and a fondness for *shikar*—which is why Salim Ali started shooting birds for the pot from a young age. Shooting birds required careful observation of the habits of birds in the wild. At the age of 12, this landed Salim an unusual sparrow—a yellow-throated one. He was intrigued at this clearly distinctive bird. Fortunately for him his uncle Amiruddin was one of the few Indian members of the Bombay Natural History Society, the one association of people in India which combined an interest in natural history with that in modern science. Salim Ali's visit to the Society's bird collection was a turning point in his life. From that time onwards he made the scientific study of birds his life's calling.

The study of birds as one's life's work was certainly not something acceptable to an ordinary Indian family of those times. It was not apt to be financially rewarding. Indeed Salim Ali and Ramanujam provide two unusual examples of people who took to their chosen branch of learning—ornithology and mathematics—with total dedication, regardless of whether it was to earn them a living or not. Sundar Lal Hora (1896–1955), acontemporary of Salim Ali's was an outstanding zoologist. But his biography tells us that his interest in zoology was kindled rather late: in his early collage years he did badly in the subject. This disappointment spurred him on to study harder. That led to the excellent marks and the scholarship that he won to continue his honours course in zoology. He really began research after he got a job with Annandale at the Zoological Survey of India in Calcutta.

That was a very different route from Salim Ali's. He tried his hand at the family mining concern in Myanmar for a couple of years before giving it up as uninteresting, returning to biology. As a biologist he earned a living for just two years in 1923–24 as a Guide Lecturer in the Natural History section of the Prince of Wales Museum in Mumbai. He was able to devote the rest of his life to the study of birds simply because his wealthy relatives accepted what other Indian families would undoubtedly have dubbed as madness: instead they continued to support him.

While this happy acceptance of eccentric hobbies is a most un-Indian feature of the Tyabji clan, their acceptance of the responsibility for a kinsman is very much a part of Indian tradition. Salim Ali's wife Tehmina died young, when he was 43 years old. For the next 45 years, his sister Kumoo and brother-in-law Hassan Ali gave him a home, looked after his needs, and left him entirely free to do his lifework. After their death in 1984, he moved to their daughter Laeeq's sea-side cottage near Alibag. Indeed Laeeq and her husband Zafar Futehally did more than just help Kumoo and Hassan Ali in taking care of the domestic needs of Salim Ali. Zafar Futehally is himself an enthusiastic birdwatcher and one of India's leading conservationists; he served for many years as the Honorary Secretary of the Bombay Natural History Society. In that capacity he made sure that Salim Ali was assured of the support needed for his studies, and in particular for the completion of his great work—the 10 volume *Handbook of the Birds of India and Pakistan*. This total freedom from any demands of family or of having to earn a living, or even to struggle to obtain money to do research and writing, was of course a major factor behind Salim Ali's tremendous scientific productivity. It permitted him to lose himself in his work. Remarkably enough, he took all of this so much for granted that he barely talks of this family support in his autobiography. But it is abundantly clear that the life and work of this great Indian scientist was shaped, made possible, and fostered by the Tyabji clan with its unique blend of the traditions of the orient and the occident.

Salim Ali became a legend in his own lifetime, not only in India, but amongst students of birds and lovers of nature the world over. He achieved this distinction partly because he devoted his talents to a scientific discipline in which Indians are at a natural advantage. A tropical land with a tremendous variety of ecological habitat ranging from the coral island of Lakshadweep and the mangrove swamps of the Sunderbans to the hot desert of Rajasthan and the cold desert of Ladakh, India is one of the top 12 megadiversity countries of the world. Much of the study of the diversity of this life requires little sophisticated equipment. It therefore presents a great opportunity for Indians to make contributions of an international standard without being hampered by the many handicaps that students of experimental science such as molecular biology or solid state physics suffer from. It is a pity that India's social setting has not permitted us to make much use of these opportunities.

Salim Ali's example has begun to change this. An increasing number of talented biologists are now taking to field studies. Amongst the more notable example of these are M.K. Chandrashekaran's and Marimuthu's studies on bats; S.M. Mohnot's on the hanuman langurs; Uma Shaanker and

Ganeshaiah's on flowering plants; A.J.T. Johnsingh's on the wild dog and the lion; R. Sukumar's on the elephant; R. Gadagkar's on paper wasps; and Ranjit Daniel's on birds and frogs. There are then many hopeful signs that Indian biology is beginning to capitalise on the diversity of life so easily accessible to us. This will help us restore Indian science to its rightful place internationally, as well as contribute towards an understanding of how to conserve this rich heritage for future generations.

10

# Jim Corbett: A Life Story

D.C. Kala

The setting of this paper is two continents—Asia, which raised Lieutenant-Colonel Edward James Corbett, IARO, VD, OBE, CIE,[1] Kaisar-i-Hind; and Africa, which gathered him to its arms in death. His triumphs were distributed between two countries, India and Kenya. Except for the last seven years of his life, he lived in India, a tiger among men, lover of the underdog, a hero in war and pestilence, a model zamindar and employer, an ascetic, naturalist, and, above all, a hunter of man-eating tigers and leopards for 32 active years in the then three hill districts of Uttar Pradesh comprising Garwhal, Naini Tal and Almora. Others hunted, but he also wrote. Of his seven books, *Man-Eaters of Kumaon,* a breathless whirl through tigerland, got him millions of fans and a top niche as a narrator of true jungle stories. In the realm of high adventure, he is a man of the five continents where his fans are and will be.

Except for a lone triumph, Africa was a continent of sorrow for him. He stayed there, dogged by ill-health, as a refugee after the exit of the British Raj from India, unsure of the changing times, the White Sahib looking for another colony. True, his two closest relations were there and also his friends—we shall come to them later. The hero's exit from the Indian scene was silent. None of his tenants knew their landlord was going, nor did the hundreds of poor Whites and near-Whites who stayed behind in Kumaon. The secret was let out only to a few friends.

Corbett was admired in his home town, Naini Tal, an unassuming man greeting high and low, exuding bonhomie. As a city father for no less than 28 years, he is well remembered for keeping the forested area of the town

---

* Reprinted with permission from D.C. Kala, 2000, *Jim Corbett: A Life*, Ravi Dayal Publishers, Delhi.

intact from vandals. His tenants adored him. And all over the hill districts of Kumaon the legend of the hunter lives on in a dozen folk songs and a thousand tales of Corbett sahib. The Prime Minister to be of the province, G.B. Pant, was a crony of his with whom he spent hours (whenever Pant was in town and out of prison), swapping yarns of the Tarai, where both had worked and lived. He was the only Congressman Corbett liked. At least in one land deal he was Corbett's lawyer too.

Corbett's heart was always in Kumaon, in the village of Kaladhungi where he had his winter home, and in Naini Tal, high up 15 miles away, where he lived in summer. 'I'd like to be reborn in Kumaon,' he would often say. Ensconced in the Baden-Powell cottage of the Outspan Hotel, Nyeri, off the Aberdare National Park in the Kenya Highlands, he sat worrying about his former tenants—he had freed them when he left—giving advice to a friend suffering from beriberi, sending a fountain pen to a friend's son on his passing the Bachelor of Arts examination, wanting to take back the house (now the Corbett Museum) he sold at Kaladhungi for the benefit of his tenants and dutifully sending till the end the land rent of Rs 910 a year on behalf of his tenants to the government.

From Kenya, he swapped notes with E.P. Gee, a brother naturalist (*Wild Life of India*), on the tigers of Hailey National Park, which Gee visited in 1954. And he compared notes with his friends in India on the Mau Mau and the Indian freedom movement. He was spared the indignity of seeing the exit of the Raj from Kenya though. Kenya became free in 1963 after 68 years of White rule. Corbett died on 19 April 1955. But his sister, Margaret Winifred Corbett, who stuck to him till the end, saw the exodus.

India remembered Corbett by renaming its first national park after him in 1957. This was Hailey National Park, set up in 1935. In the first flush of freedom, while the tide of nationalism crested, it was found necessary to erase the memory of the builders of the Raj, and the Park was harmlessly renamed the Ramganga National Park in 1955 after the river which formed its northern boundary. The government would have perhaps immortalized Corbett's friend, who became the Chief Minister of UP, but Pant stepped aside and helped in renaming it after Corbett.

I am sure Corbett would have been embarrassed to take over the Park, once named after Malcolm Hailey, Governor of UP in 1928–30 and 1931–34, a friend of his and a keen conservationist. Hailey outlived Corbett by at least 15 years. Corbett was dead when the renaming came. And what could be a more fitting memorial than a tract of 324 square kilometres (now 528.8) in the Himalayan foothills for the tiger to survive in, I hope forever? Pant too got several memorials, including the renaming of the Ramsay Hospital at Naini Tal after him. This was the great Henry Ramsay who came to Kumaon in 1837 and ruled it for 28 years as its sixth Commissioner. An early builder of the Raj, he was responsible for hundreds of projects and a thousand good deeds.

Corbett and his Premier friend, two of Naini Tal's best-known sons, share 1,000 square yards of land on the Flats, as the upper end of the lake is known. Pant stands on a pedestal. His statue is black and larger than life. Corbett stands by the water unobtrusively and unknown in the form of a bandstand for which he donated Rs 7,300 in 1919 for the 'benefit of Naini Tal'. The hunter and angler is where the water is. The politician has his back on half the town and faces the other half where the road begins downhill to the dusty plains. Few know the history of the bandstand, but Gurney House on Ayarpata Hill, which Corbett's sister Margaret Winifred (Maggie) inherited from her mother, is well known. He lived in this house before he left for Kenya. A third son of Naini Tal who left his mark but no monument was Gen. Order Charles Wingate, the Chindit, who died in an air crash in Burma in 1944.

The 'Corbett Museum' at his village of Choti Haldwani, near Kaladhungi, came later, a pitiable effort of the state Forest Department to perpetuate the hunter's memory by displaying some photographs and letters he wrote to Indian friends. This too was done in the Premier's days. The odds and ends displayed there are what the villagers have not rifled from the untenanted house. The government acquired the house, restored it, and set up the museum. Our man was generous with his minor trophies, and the Forest Department has not been able to collect the major ones, for these were auctioned by the executors of his will in Nairobi and are scattered all over the globe.

World recognition from brother naturalists came in 1968 with the naming of a subspecies of tiger after Corbett. This is *Panthera tigris corbetti,* found in Indo-China and extreme South China. It is slightly shorter than the Indian tiger and characterized by a 'darker ground coloration and more numerous, rather short, narrow and rarely doubled stripes'. It was named after Corbett by Dr Vratislav Mazak to honour the 'excellent naturalist who devoted his life to the study and the protection of Indian wildlife, particularly the tigers'.[2]

I spoke earlier of a personal triumph of Corbett's in Africa. That came in 1952, a few years before his death. He was privileged to escort Princess Elizabeth, who later ascended the British throne as Queen Elizabeth II, to Tree Tops Hotel in the Aberdare Mountains, near Nyeri, on a sightseeing safari a day before her father's death. Right from 1919, when he finished with the railways and practically retired to be a gentleman of leisure and businessman, ending up with a considerable fortune by the prevailing standards, he was in the foothill forests every winter with governors, collectors and at least one viceroy, hobnobbing with them, gun or fishing rod in hand.

Every collector of the district or the forest officer who shot and fished became a friend. So did the maharajas, Jind for one. Young White rookies posted to the district turned to him for their first lessons in tiger hunting. Because of his association with the top district officers he commanded considerable influence in the area. A Corbett complaint was always heard. To Kaladhungi he had a museum approach. He wanted it unspoiled, with himself as patriarch to whom all could turn for help.

Born on 25 July 1875, Corbett spent his childhood at Kaladhungi and Naini Tal. He was educated at Naini Tal. After school, he went straight to Bihar to work on the railways for 23 years. In 1914, when World War I broke out, he returned to Kumaon to raise a labourer contingent for the British Army and served in France and Waziristan. After World War I, he settled down at Naini Tal, where he lived almost continuously till 1947, except for several trips to Tanganyika (modern Tanzania) and a stint during World War II training Allied troops in jungle war as a Lieutenant-Colonel. As a city father of Naini Tal, he stepped into his father's shoes. The rest of the time he kept watch on all the bad tigers and leopards of the high hills and the adjoining plains.

These big, bad cats, man-eaters to be precise, were Corbett's extra charge. When one was proclaimed a maneater by the district authorities, they turned to him for help to rid them of it. Sometimes the call came from the stricken villagers themselves. Several requests even reached Mokameh Ghat, Bihar, where he worked during his railway days. He always agreed. Tracking and killing man-eaters needs considerable skill and daring. The man-eating leopard of Rudraprayag he killed was responsible for the death of 125 persons in a reign of terror of eight years in northern Garhwal. Another less-publicized leopard, the man-eater of Panar, took 400 human lives.

When the call came, the 40-pound tent, the suitcase and the bedroll were hurriedly packed by sister Maggie, the porters were collected, and the hunter set out in forced marches of 20 to 40 miles a day—depending on the urgency—to the dak bungalow nearest to the last reported kill. Often, even after the last day's long march, he denied himself rest, left kit and porters at the dak bungalow and made a beeline to the kill or checked the lay of the land. If the kill was fresh enough, he would select a tree overlooking it, seek out a fork or comfortable branch and spend the night in it waiting for a shot at the marauder.

For weeks and sometimes months the strenuous hunt would go on in high tension up the hills and down the valleys as each new kill was reported, for every day gone meant more lives lost. Breakfasts, lunches, and dinners were skipped. A man-eating tiger or leopard acquires a special cunning by its long association with humanity. It loses all fear, finding the biped the most defenceless creature in nature's kingdom. When this happens, the hunter is also the hunted and gives himself at best a 50: 50 chance in spite of his shooting iron. Corbett himself admits: 'There is no more terrible thing than to live and have one's being under the shadow of a man-eater' (Corbett, 1993).

## The Naturalist

And the last of the seven tigers passed within ten feet of the camera lens.

'What would you have done, Colonel Corbett, if it had seen and attacked you?'

'I should have thrown it a khaki cushion which I always carried in case of emergency.'

'But...a cushion?...how would that have helped?'

'A friend of mine once saved his life by throwing a rolled khaki blanket to a tiger who was about to spring at him. The tiger stopped to inspect the blanket long enough for my friend to shoot it.'

'And you hoped the cushion might help you do the same?'

'Well, I never carried a gun when I went out to photograph the animals. Only my camera...and the cushion.'

Then what...?'

'I hoped the tiger might be interested in the cushion long enough for me to get away' (Ward 1951).[3]

The day was 10 October 1951, at Amen House, then headquarters of Oxford University Press in London. Corbett had just finished screening his wildlife films, one of them *Seven Tigers,* and the viewers had been asked to put questions. Corbett was 'inordinately proud' of the tiger film. After Corbett's death the films went first to the National Film Archive in London and were eventually given to the British Natural History Museum, where they are now with the education section

> wound on to bobbins, about 20 of them, each length having a running time of ten minutes or so. They have not been edited, and the scenes are in random order. They are of course all black and white and silent. They are becoming rather fragile with age. About half the film was shot in India and about half in Africa.[4]

With every hunter comes a time when he starts questioning whether it is worth his while cluttering up a house with hide and horn. Besides, seeing the decimation of wildlife over the years, he pauses and asks himself the question: Will anything be left for the coming generations? F.W. Champion, a pioneer wildlife photographer and a former forest officer of Kumaon (*With a Camera in Tiger-Land*), gave Corbett the idea of having trophies that could be shared by all. Corbett had received a 16-millimetre cine camera as a gift from a friend, Lord Strathcona, in 1928. Champion's still photography had been by night with the aid of trip wire and flashlight, but Corbett now went one better, getting his photographs in broad daylight.

For 10 years he tried in vain to get decent pictures of tigers. Either it was the weather that ruined the film or plain inexperience. Finally, he decided on setting up his own jungle studio near Kaladhungi in 1938 and succeeded in drawing seven tigers to within 10 to 16 feet of the studio and filming them. To drown the whirr of the cine camera, he dammed the stream in the centre of the studio. Cascading water did the trick. On other occasions, he drowned the noise with bird and animal calls. The filming took four and a half months' hard work and countless hours of waiting at the studio, to which he retired before first light. Before this, Corbett had been toting an 11-pound camera for years, unsuccessfully trying to get tiger pictures in broad daylight, part of the time in the Patli Dun valley of the Ramganga which was later to become the Corbett National Park.

Reviewing the films screened at Amen House, A.C. Ward wrote in *The Lantern:*

> But what films they were! No studio sets, no million-candlepower lights, no makeup, no props: not even a continuity girl. No cameraman or director but Jim Corbett; no scenery but the Indian and African landscape, no 'stars' but leopards, tigers, elephants, hippopotamuses, rhinoceroses, and nearly every kind of buck. The star of stars, however, was none of these. It was a goat . . .
>
> This was in the first spool that Col Corbett screened—mainly about leopards.... When setting out to film a leopard it is, of course, first necessary to produce the leopard. 'So I called up a leopard,' said Col Corbett, as any one of us might say 'So I rang up for a taxi'—and, 'unlike some taxis we know, a leopard came. Then the film "cut" to a herd of browsing or nibbling goats. Watch the goat on the extreme right. We shall see him again in a few minutes.' We did. Separated from the rest he was left alone with the leopard: which was what the leopard had played for.
>
> After a brief exhibition of ringcraft, the leopard went for the goat. The goat countered by meeting the leopard horns first. The leopard stepped back to think things over, decided that its technique was sound, and went for the goat again. Again those confounded horns! When the leopard came up for the fourth round the goat went into action on its own account with the goatesque equivalent of an uppercut. The contest ended in the sixth round. As the leopard came in once more, the goat appeared to stop the assailant in its tracks with horns and all four hoofs together. What any properly equipped sports writer would call the 'local derby' or 'needle contest' or even 'homeric battle' was over. The leopard threw in a spectral towel, muttered 'Goats is hell' and slunk away to tell 'why I lost' for the next Sunday's *News of the Jungle*. The goat, with an air of bored modesty, as though brushing off leopards was mere kid's play, resumed its nibbling or browsing. Anyway, that's one picture that Hollywood didn't get.
>
> The next two films were comparatively peaceful, being concerned with the Indians' method of collecting honey from wild bees who locate their hives in the treetops, and with the day's work of elephants.
>
> In the film which might be called *The Tale of the Seven Tigers* we saw the pictorial result of four months' patience and persistence by Jim Corbett. He explained that tigers are by nature solitary, preferring to roam alone or at most in couples. He was determined to film a number of tigers in company, however. Damming a small stream, to form a pool in a spot suitable for a drinking place, he then took steps to collect the tigers. Going a field to various

points of the compass where tigers were likely to be, he put down food to lure them. When this was taken, he put down further meals nearer to the pool he had made, and in this way enticed the tigers into a progressively smaller area until at length seven were together by the pool. One of the seven was an albino, its whiteness showing up queerly against the dark rocks and undergrowth. Another had the manlike trait of being in a constant bad temper about nothing in particular—growling and snarling with the venom of a politician of any party out of office. Although Col. Corbett made it clear that neither the camera nor the photographer must be seen by the animals, three of those tigers appeared as lens conscious as Marlene Dietrich. They posed themselves couched on a shelf of rock with their paws (or should it be pugs?) curving negligently over the edge: three great cats.

The last film was of the birds in the Corbett garden in Kenya and of a large herd of elephants, also made in Kenya. During the nearly two hours that the films took to project, *only one dead animal was seen*—a man-eating leopard—and Col Corbett apologized for showing even that one.... Though he was warmly and spontaneously thanked for his generosity in providing a unique experience, thanks seemed something of an impertinence and an irrelevance, for—as a member of the audience said privately afterwards—'Jim Corbett is the nearest to a saint that any of us is likely to see.'

According to Maggie, Jim always went unarmed to take his jungle pictures.

I felt he was taking risks in so doing, especially during the time he spent in filming the seven tigers he had managed to get together. He always sat in the same place, about eight feet from the ground on the branch of a very small tree under which the tigers had to pass on the way to the place where they fed. So as not to disturb them, Jim took up his position very early in the morning. By putting out his hand, he could have touched the tigers on their backs as they passed below him.[5]

Wildlife photography has its hazards too. Maggie recalls how Jim was once brought home 'in great pain and hardly aware of what he was doing or saying', having unaccountably fallen off a tree he was perched in while attempting to photograph a tiger. He was diagnosed by a doctor as having 'a broken back, severe concussion, and internal haemorrhage', which, not surprisingly kept him bedridden for some months.[6]

In later years, knowing well that Stripes was on his way out, Corbett became a saviour to the dozen tigers within a radius often miles of his village. Whenever a delegation called to seek his assistance in killing a tiger which had turned cattle-lifter, he would open his purse-strings and offer to compensate for the cattle killed. Critics in his own village, however, imputed motives to him—he wanted them for his friends or for his camera studies.

A villager of Choti Haldwani approached Corbett once to kill a tiger which had taken his best bullock. Persuasion and compensation failed to mollify the villager. Fetching his own shotgun and shouting obscenities at Corbett's gate against the tribe of 'photographers'—that was Corbett's photography phase—the villager made a beeline for the kill. Corbett reluctantly followed and killed the tiger from a tree after calling it to the kill. The villager ruefully remembered, after the shot had been fired, that Corbett's only regret was that he had not brought the camera to take a nice picture!

This, incidentally, was one of his last tigers and 'shot after Hitler's war'. About the very last he killed, Maggie had doubted Corbett's capacity to hold a rifle steady after years of sickness during the war. So it had to be called rather close and shot through the eye. To justify the killing, Corbett said: 'It resisted all my attempts to drive it away' and 'it was difficult to replace the animals it killed because of a shortage of farm animals brought about by the war' (Corbett 1999: 49).

A lot of people today think that the preservation of wildlife and other ecological matters are new-fangled ideas. If that is so, Corbett was a pioneer. He was not a killer sportsman and all his writings plead for the preservation of wildlife. What he himself destroyed, in terms of tigers, were mostly those that were dangerous to man.

On 21 May 1955, the Delhi edition of the *Statesman* reproduced a letter Corbett wrote in 1948 to an unidentified friend on the tigers of India. It read:

> For 20 years I have fought in defence of wildlife and my opponents have invariably been people one would have expected to help, and not to oppose.
>
> Men, and in some cases women, with a blood lust are always ready with an excuse—a potential man-eater, possible cattle-killer, and so on—and the excuses they have made are now being made by their successors.
>
> Until India realizes that wildlife is an asset, the killing will go on.
>
> Two years ago, Lord Wavell asked me the same question about tigers that you have done, and I told him that in my opinion there were 3,000 tigers in India.

When he asked me how long I thought tigers would survive, I said that except in sanctuaries and one or two Indian states tigers would be wiped out in ten years.

That situation has nearly come about. An earlier warning came in 1944 in the author's note in *Man-Eaters of Kumaon,* where he wrote:

> There is, however, one point on which I am convinced that all sportsmen—no matter whether their point of view has been a platform on a tree, the back of an elephant, or their own feet—will agree with me, and that is, that a tiger is a largehearted gentleman with boundless courage and that when he is exterminated—as exterminated he will be unless public opinion rallies to his support—India will be the poorer by having lost the finest of her fauna.

He never sat in judgment even over the maneating tigers he shot, for he fully believed 'a man-eating tiger is a tiger that has been compelled, through stress of circumstances beyond its control, to adopt a diet alien to it' (Corbett 1997a: 9). Man was responsible for this mostly. A tiger left wounded and incapable of preying on its legitimate game would seek easier pastures. Sometimes it was the tiger's own folly when porcupine quills got embedded in its flesh while killing one, and old age was a third reason for its turning to man-eating.

Perhaps the biggest factor that has gone against the tiger, apart from wanton killing and the encroachment on its habitat by population pressure, is the commercially oriented forest policy which believes in planting one species of trees to the exclusion of others for convenient exploitation. The eucalyptus plantations in the Tarai and the Nilgiris and the conifers in the hills illustrate this.

A plantation is not a forest. It is an ecological desert for wildlife. A natural forest is not trees alone. It is the mulch (decayed vegetation), ground herbs, tubers, undergrowth, and creepers too. The exclusion of all these in favour of the tree alone has wrought havoc with nature's delicate balance. The hard truth is that a 'standardization of forests', as Corbett called the plantations, provides neither shelter nor feed to wildlife.

This is how modern forestry works to the detriment of wildlife. A patch of natural forest is selected for planting commercial timber. The trees are felled and sold as fuel. The stumps and the undergrowth are systematically burnt. Then come the pits or tractor furrows—if the terrain allows it—and the saplings are planted. Barbed wire strands then go up to protect the young plants from stray cattle and deer. Watch towers are erected for forest guards, and such deer as manage to clear some four strands of barbed wire to forage in the plantation are systematically shot. Sambar, alas, have the habit of rubbing the itch off their hides on saplings and damaging them.

Modern forestry also demands that, for quicker growth and as protection from fire, the grass and undergrowth should be burnt regularly. The end result is evenly spaced rows of one family of trees, be it the eucalyptus, chir pine or teak, with a dry, bare forest floor which does not provide cover even to a scurrying rat. Deer may get grass here but not cover, and they desert these plantations. The undergrowth and ground vegetation, which provide the cover and supplementary feed, are not there. The pheasant and other ground feeders, who thrive on the worms that mulch produces, have nothing to eat here, for the mulch has gone. The natural forest untouched by man has all the things that sustain wildlife, a nature's garden of trees and shrubs fruiting, flowering, and foliaging all the time. Alas, this forest is *kukath* (bad timber) for the forester and has to go.

The substitution of oak forest by conifer plantations in the hills is another tragedy. The worm in the oak leaf-mulch sustained the pheasant. The young leaf of the oak and the lush undergrowth sustained the deer. And the acorn sustained all the herbivora—deer, porcupine, bear, and a multitude of forest birds. With the cutting down of the oak all the wildlife sustained by it is vanishing. The conifer forest with no undergrowth is hostile to wildlife.

In Corbett's Tarai, once a hospitable home for the tiger, the pattern is the same. The semal trees, whose fleshy red flowers sustained monkeys, langur and deer, which in turn sustained the tiger, have largely gone, swallowed up by a match factory at Bareilly started about 80 years ago. Vast tracts have been cleared for farming and new forest plantations, and the tiger, driven from the deerless forest, perforce goes out to feed on domestic cattle in villages along the periphery of the forest and gets shot. Unless a semblance of wilderness is restored to the forest—in other words no interference—nothing can be done to save wildlife.

The modern ecologist here faces a challenge. Can he fight plantation commercialism? Where are the biostudies of all the wildlife-sustaining trees and shrubs that season by season in an unbroken cycle provide for the needs of wildlife? What exactly is a wilderness? It is what begins where the road ends. The natural forest, undoctored by the messy hands of *Homo faber,* man the maker of things, is the friend of wildlife. Study it, restore it, sustain it, and you have the solution.

For Corbett the tiger always remained a largehearted gentleman with boundless courage. He had reason to reach this endearing conclusion, for how else could one explain that two children, one of 2 and the other of 3, lost for 77 hours in Kaladhungi were rescued safe and sound, with not even a scratch, in a forest which to Corbett's 'certain knowledge (Corbett 1997b: 67)' held five tigers, eight leopards, a family of four sloth bears, two Himalayan black bears and a number of hyenas?

Hailey, who was a frequent hunting and angling companion of Corbett in the 1930s, writes of delegations of villagers visiting their camp and asking Corbett to rid them of a tiger turned man-eater or cattle-lifter. 'But the rubric that Corbett applied to the inquisition which was now opened was strict, however friendly and considerate in its terms. It was no use for them to plead their losses in cattle or goats. The tiger was lord of the jungle and must have its dues. Not until he himself was convinced that a tiger had been killing human beings, not by chance or in anger, but because it sought them as food, would he agree to come to their help', he wrote (Hailey's Introduction in Corbett, 1998).

What exactly turned Corbett into a naturalist? According to the Rev. A.G. Atkins,[7] it happened one day on a duck shoot. In the 1930s, Atkins was the pastor of the Union Church at Naini Tal for two summers. This was the church of Philander Smith College and its sister institution, Wellesley. One evening, after Corbett had screened his first tiger film and given his wildlife lecture, the pastor walked Corbett halfway home to the lake from PSC. The road is all downhill and the two chatted. After some talk, the priest came to the point and asked him what made a hunter a photographer. Here is the story of the 'conversion', as the priest called it:

> He [Corbett] had always been fond of shikar in the ordinary sense of the term, going out for hunting or shooting with not much thought of anything else but the fun and sport of it. He was known as a skilled jungle man and was often asked to lead parties out for a good shoot. One day he was out with three military officers in one of the lake and river areas of North India. They came upon a large batch of waterfowl, literally thousands of them. The officers began shooting; they went on and on, following and killing their game till they had killed over 300. They could not possibly carry them away for any use; it was simply unrestrained slaughter for the crude pleasure of it. Said Jim: 'That sickened me and opened my eyes to what ordinary uninhibited hunting and shooting meant. I resolved from that time that I would use my jungle lore for a different

kind of shooting, and in that way I began to take photographs of wild animals and jungle life. It requires much more of my skill and gives me an even greater thrill to get good pictures of my animals than when I used to hunt just to kill.

With the arrival of Hailey as Governor of UP in 1931 (this was his second term), Corbett the naturalist became a bit more articulate. An Association for the Preservation of Game in UP was formed with the Governor as its patron and Corbett and Hasan Abid Jafry as honorary secretaries. The association also brought out a journal, *Indian Wild Life,* to which the late E.P. Gee, the planter–naturalist, was an early contributor. Its aim was to 'awaken public interest in the preservation of wildlife and to take such steps as might be possible to save it from extinction'. The first issue of the monthly, priced 8 annas (half a rupee), appeared in July 1936. *The Bombay Natural History Society Journal,* Vol. 39, carried a full-page advertisement of this official organ of the Association, and also of the All-India Conference for the Preservation of Wild Life, of which the patron was Sir Harry Haig. The Association became ineffective when World War II broke out, but Corbett was a regular visitor to Government House when Hailey was in Naini Tal. He had avoided the hustle of Government Houses so far, but Hailey was a friend, and Sir Maurice Hallett, Hailey' s successor, also befriended Corbett.

Jafry, a barrister, was the political secretary of the Raja of Mahmudabad. The Association for the Preservation of Game in UP was affiliated to the All-India Sportsmen's Brotherhood, of which Jafry was also honorary secretary. The Association secured land on lease in Muhalla Purwa Ali Mirza at Lucknow on nominal payment, and the lease deed was confirmed on 8 January 1946. There was a proposal to have a building with a hall named after Hallett.

Addressing the Association at Lucknow on 5 August 1934, Hailey remarked: 'I have been much struck by the reception given to the lectures which Major Corbett has delivered; he has the advantage that he bears a reputation not only beyond compare and above approach as a sportsman, but as one who has a real love and respect for animal life.'

Catch them young, they say. Corbett now started lecturing on wildlife at Naini Tal. At Wellesley, the girls looked forward to the yearly visit of Corbett and Maggie. He usually wore shorts and shirt and a pullover. Maggie was installed on the dais and Corbett would lecture on his favourite subject, the jungle telegraph. A tiger is coming, he would announce, and then mimic a series of bird calls—the jungle babbler, drongo, peafowl—and then the animal calls—the langur, barking deer, chital and sambar. These produced a solo effect perforce, for the human soundbox has its limitations mimicking a jungle cacophony or the racket raised by frightened langur. The warning call of the deer tribe is solitary. The muntjak barks, the chital pooks or bells and the sambar ponks. The calls would change with the tiger's activities. Now, it is in cover, now moving, now stalking, and now sleeping, Corbett would announce, varying the intonations.

He would build up the climax with a tiger's growl, first subdued and then full-throated. The atmosphere for this last bit of performance was built up by putting out the lights. Before it happened, he would dramatically announce: 'Anyone with a weak heart may go now.' No one ever left, and squeals and giggles followed. All the Naini Tal boys' schools also had the benefit of this annual performance. On Hailey's suggestion he also lectured at the schools and colleges of Lucknow.

At Kaladhungi, Corbett carried on a one-man war against poachers with blandishments and threats. Old residents swear that he knocked down the gun of a poacher as he carried it on his shoulder parallel to the ground with a well-aimed rifle shot on the butt. Perhaps it was a prod from the Association that made Corbett sit down to write as well.

When did Corbett take to writing? The *Hoghunters Annual,* Vol. IV, of 1931, published the story of the Pipal Pani Tiger, now included in Corbett's *Man-Eaters of Kumaon.* The *Annual,* devoted to the 'noble art' of pigsticking and the activities of the tent clubs all over the world, was edited by Capt. H. Nugent Smith and Capt. J. Scott Cockburn from Naini Tal and published by the *Tunes of India* Press in Bombay. Corbett was not a pigsticker, though he looked after the assets of the Meerut Tent Club when it wound up at the beginning of World War II. I have it from Cockburn that Corbett was not approached to write. It could have been Ibbotson who brought the story for publication. Ibbotson, also a pigsticker, certainly competed in 1934 and 1936 for the Kadir Cup, the Hoghunter's most coveted trophy.

His second dated writing on wildlife was in the *Review of the Week* of 31 August 1932. The *Review* was published by the Assistant Superintendent-in-Charge, Government Branch Press, Naini Tal, and his article, taking up the whole of the back page, was headed: 'Wild Life in the Village: An Appeal.'

I am reproducing the article in full for three reasons: first, it shows the growth of the hunter into a conservationist; second, doubts have been raised as to Corbett's competence as a writer; and third, it is the swansong of the wildlife of the foothills. The article reads:

It was a small village of some 16 ploughs differing in no respect from hundreds of similar villages, scattered throughout the length of the tract along the Bhabar. Originally the village had been surrounded by tree jungle intercepted with grass and in this virgin jungle lived all the numerous denizens of the wild. To protect their crops the villagers erected thorn fences round their fields. As an additional safeguard a member of the depressed class was encouraged to settle in the village whose duty it was to watch the crops at night and see they were not damaged by stray cattle and wild animals. Owing to the abundance of game, tigers did not interfere with the village cattle and I cannot remember a single case of cow or bullock having been killed by a tiger. In the course of time, a great change took place not only in the villagers themselves but also in the jungle surrounding the village. Hindus, who formerly looked upon the taking of life as against their religious principles, were now clamoring for gun licenses and were competing with each other in the indiscriminate slaughter of game. As profits from the sale of game increased, field work was neglected and land began to go out of cultivation. Simultaneously, *lantana,* introduced into Haldwani as a pot plant, started to kill out the grass and *basonta* until the village was surrounded with a dense growth of this obnoxious weed. Government now stepped in and at great expense built a pucca wall all round the village. The building of this wall freed the villagers from the necessity of erecting fences and watching their crops and gave them more time to devote to the killing of the game. *This heavy and unrestricted shooting of deer had the inevitable consequence of disturbing the balance in nature with the result that tigers and leopards, that had hitherto lived on game, were now forced to live on the village cattle.* One morning in May of the present year [1932], I arrived in the village and pitched my tent in a little clearing just outside the cultivated land. News of my arrival soon spread through the village and in a short time a dozen men were squatting in front of my tent. One and all had some tale to tell. *A tiger had taken up its quarters in the lantana and in the course of two years had killed 150 head of cattle, and unless it was destroyed the village would have to be abandoned.* While the men were pouring out their tale of woe, I observed a pair of vultures circling low over a narrow stretch of lantana running between the village men and the public road. The two vultures were soon pointed at by others; so picking up a rifle, I set off to investigate. Progress through the lantana was difficult but with the aid of a good hunting knife a way was eventually cut and the remains of a horse killed the previous day found. There were plenty of pug marks round the kill, little of which remained, and it was easy to locate the tiger from his low continuous growling but impossible to see him in the dense cover. Returning to the road, which was only 40 yards round the kill and little used at this time of year, I concealed myself behind a bush in the hope that the tiger would follow me to see if I had left the locality, quite a natural thing for it to do. *Half an hour later the tiger walked out on to the road and gave me an easy shot as he stood facing me.* That evening—after I had skinned the tiger—he was *a very old animal and I took four old bullets and nine pellets of buck-shots out of him—I called the villagers together and made an appeal to them on behalf of the few remaining deer in the jungle.*

On the opposite side of the village from my camp, *irrigation water had been allowed to flow into the jungle. Over this water machans had been built in the trees and in these machans men sat through the heat of the day, and all night long on moonlit nights, and shot down animals that came to drink.*

There was no other water within miles and *if a thirst-maddened animal avoided one machan, it fell a victim to the man in the next.* I told the villagers that *God had given water free for all* and that *I was a shameful thing for men to sit over the water God had provided and shoot His creatures when they came to drink.* To do this was to 'lower themselves below a corpse-eating hyena, for even he, the lowest of all creation, did not lie in wait to kill defenceless animals while they were drinking. The men listened to me in silence and 'when I had done, said they had not looked at the matter in this light, and they promised that they would take down the *machans* they had erected and in future would not molest the animals that came to the vicinity of the village to drink. I stayed in the locality several-weeks, taking bird and animal pictures, and am glad to say the men kept their promise. I believe that much of the slaughter of deer that is daily taking place throughout the length and breadth of the Bhabar and Tarai would cease if an appeal was made to the better feelings of men. *1 do not exaggerate the damage that is being done to our fauna by shooting over water.* Let me give you but one instance. An acquaintance of mine living in a village in the Bhabar adjoining mine, in one hot season, over one small pool of water shot, with a single-barrel muzzle-loading gun, 60 heads of cheetal and sambhar which he sold in a nearby bazaar at the rate of Rs 5 per cheetal and Rs 10 per *sambhar*. It is no exaggeration to say that the *banks of every little stream and every pool of water in the vicinity of Bhabar villages are soaked with the blood of animals that never took toll of a single blade of the villagers' crops. I assert without fear of contradiction that for every shot fired on cultivated land from guns provided for crop protection, a hundred shots are fired in the jungle over water.*

Pigs and *neelgai* are the only wild animals that damage the crops in the Bhabar to any extent, and to keep them out of cultivated land Government has expended lakhs of rupees in building pucca walls. It *is asserted that in recent years tigers have increased. With this assertion I do not agree.* It is a fact that *more cattle are being killed every year; this is not due to tigers having increased but due to the balance in nature having been disturbed by the unrestricted slaughter of game,* and also to some extent to tigers having been driven out of their natural haunts where they were seldom or never seen by man, by the activities of the Forest Department. *A country's fauna is a sacred trust, and I appeal to you not to betray your trust.* Shooting over water, shooting over salt licks, natural and artificial, shooting birds in the close season and when roosting at night, encouraging permit-holders to shoot hinds, fencing off large areas of forest and the extermination by the Forest Department of all game within these areas, making unnecessary motor tracks through the forest and shooting down from motor cars, the absence of sanctuaries and the burning of forests by the Forest Department and by villagers at a time when the forests are full of young life are all combining to one end—the extermination of our fauna. If we do not bestir ourselves now, it will be to our discredit that the fauna of our province was exterminated in our generation and under our very eyes, while we looked on and never raised a finger to prevent it. (Major Corbett)

The article has been reproduced without any change, and the italics are Corbett's own for emphasis. And shall we call him a Jeremiah?

While lecturing boys and girls in Naini Tal's schools in the 1930s and 1940s, little did Corbett realize that the knowledge of the jungle he was imparting to the youngsters would be put to adult use and that he would don an army uniform again one day to instill confidence into young officers of the Allied nations to face the rigours of life in the forest. These were what were later known as the Chindits, under training for the Burma campaign.

Corbett was 64 when World War II broke out. Again he offered his services to the army, and again the army would have nothing to do with an old man. Undaunted, the Major in the Indian Army Reserve of Officers found a way to contribute his mite to the war effort. Welfare was close to his heart and he agreed to become vice-president of the District Soldiers' Board that looked after the needs of the families of serving men. This involved a lot of travelling and miscellaneous work like writing letters, tracing missing personnel, and rendering help to servicemen's families in distress.

Simultaneously, he started recruitment for the Civil Pioneer Corps. The term lasted from 1940 to 1942. In 1942, the job was interrupted by serious sickness, an attack of tick typhus. Corbett was in hospital for three months. He was released with a warning that he should now take things easy and that the strenuous life he had been leading would be dangerous to his health. But the old warrior made a quick recovery and was soon clamouring for another opportunity. This came in the form of an assignment very close to his heart: would he train troops in junglecraft?

In February 1944 Corbett was commissioned a Lieutenant-Colonel and appointed senior instructor in jungle lore at a centre the army had opened at Chhindwara, in Madhya Pradesh (formerly Central Provinces). Leading the strenuous life of a training camp, he could not keep up the pace and was laid up in 18 months with a severe attack of malaria. He went back to Kaladhungi in September 1945 and was nursed through sickness by Maggie, who herself was stricken with the same disease. But before the war ended his first book was out and he had plans to write more.

What did he teach in Chhindwara? The Defence Ministry Historical Section could not produce a thing on this and we have perforce to go back to *Jungle Lore* to dig up facts about the instructor's activities. For a month he devoted himself to a study of the flora and fauna of Burma and then reported at Chhindwara ready to teach.

The British generals had realized after the rout in Burma that the Japanese could only be defeated if their jungle war techniques could be matched with something better. Gen. Orde Charles Wingate, basing himself on his experience in Ethiopia, laid down the first principles of jungle war. In this, the orthodox principles of war are considerably diluted and detachments living off the country or fed from the air make deep forays into enemy territory as fighting patrols.

One psychologically inhibiting factor was the city upbringing *of* the Allied officer class who found themselves completely lost in the jungle. Corbett had to prepare them for the task ahead and convince them that the jungle was a living thing, and once the rudiments of junglecraft were acquired it was more hospitable than the desert which is the city. In the jungle one 'could live at peace with all wildlife' (Corbett 1999).

Corbett was in his element now. One morning at Chhindwara he gave his first performance before a draft, which luckily contained some keen bird-watchers. He demonstrated his mastery of junglecraft by calling a screaming serpent eagle low for closer scrutiny by fishing out a reed whistle from his bush-shirt pocket and blowing at it. That was an imitation of the cry of a fawn in distress. His only regret was that the eagle could not be brought closer for a photograph as well. He used the reed whistle for a signalling system, as it was a natural sound that would not attract the enemy in the forest.

The main accent was, however, on survival. These lessons were on direction-finding, identification of edible flowers, fruits and tubers, the pin-pointing of sounds, and training the eye to a full field of 180 degrees of vision. An applied junglecraft lesson sought to teach soldiers how to assess the human tracks on a jungle path, the number of men in a party, if loaded or unloaded, and if hurrying—there are then more toe marks and fewer heel marks. The apothecary's son also taught about the medicine available in the jungle for wounds, fevers, sore throat and stomach disorders. Further, he gave instructions on how to kill game without a firearm and prepare tea without a metal pot. In an aside in *Jungle Lore,* Corbett sought the forgiveness of those he trained if 'he worked them hard'. It had to be so, for 'time was short'.

## Notes

1. Indian Army Reserve of Officers (IARO); VD, a decoration for voluntary officers; Officer of the Order of the British Empire (OBE); Companion of the Indian Empire (CIE).
2. *Extrait de Mammalia*, Tome 32, No. 1.
3. *The Lantern*, XVIII(3), December 1951.
4. Courtesy British Museum (Natural History).
5. Ruby Beyt's notes.
6. Ibid., the accident is undated.
7. He is better known for his translation into English of Tulsi Das's *Ram Charit Manas*, a Hindustan Times publication. His wife, Lois Rockey, was a close friend and girl guide companion of Maggie.

## References

Corbett, Jim, 1997a, *Man-Eaters of Kumaon*, Reprint, Original edition published in 1943, OUP, New Delhi.

———, 1997b, *My India*, OUP, New Delhi.

———, 1998, *Tree Tops*, OUP, New Delhi.

———, 1999, *Jungle Lore*, OUP, New Delhi.

# Section III

# Independent India's Environment

*Photo credit Pallava Bagla*

# Introduction

In January 1968, the special guest who presided over the Convocation of Bangalore University expressed deep concern about the impact of unplanned, haphazard development on the human environment: 'In the name of modernity, we are doing violence to our landscape... We are recklessly cutting down trees and destroying wildlife. In building a project, if you have to pull down a hundred trees, you must instinctively deem it your responsibility to plant two hundred,' she went on, 'and this is also a practical need. We have today drought in parts of our country and I have no doubt it is due to the fact that we have cut down many forests.' Four years later, the same leader as Prime Minister of India would deliver another major address, this time at the first UN Conference on the Environment at Stockholm in Sweden (Gandhi 1982: 15, 17).

But concerns about the environment were by no means limited to the highest level of government. The issues of the environment in post-1947 India showed signs of both change and continuity from earlier days. As Indira Gandhi indicated, modern development such as dams and mines entailed denudation of green cover, loss of top soil, water pollution and displacement of people. Conversely, the absence of development could also pose problems, especially for the vast underclass. Here, the lack of access to clean potable water or sanitation, adequate supplies of energy or poor housing could be factors in a deteriorating environment (see Agarwal *et al.* 1982).

The issues of ecology were and often are bound up with choices of development. In a country dependent on food imports till around the time Indira Gandhi spoke at Bangalore, the priority was to increase output of cereals. The primary way to do this was to use high-yielding varieties of wheat and rice which gave more output per unit of land. This entailed greater use of chemical fertilizers, pesticides and water than earlier, which in turn generated new, often unintended and unforeseen problems of soil quality.

M.S. Swaminathan, the leading agronomist and a prime architect of the Green Revolution reflects on its context. A leading scholar of the twentieth-century history of the planet, John McNeill, reflects on its ecological consequences and costs in this section (see Swaminathan's paper in this volume, also see Vohra 1973, for a different view).

Beyond such zones of intensive agriculture, there were larger issues of land degradation, of the erosion of its productive capability due to unwise use. Growing crops is only one of the myriad ways of living off the land: grazing and gathering biomass, hunting and trapping, tending itinerant herds of goats, sheep or camels, sedentary animal husbandry, and shifting cultivation are but a few of other, often allied occupations. The expansion of population numbers to 1.02 billion by 2001 and the expansion of markets and industry generated both opportunities and new pressures. N.C. Saxena and N.S. Jodha, among the best scholars of land and common property issues, offer fresh insights into issues that affect the life and work of nearly four of five rural Indians who live partly or wholly off the land. The loss of topsoil due to erosion or the breakdown of community level systems to renew grass and fodder resources are matters of concern.

Water pollution was a serious issue by the time of Indira Gandhi's first tenure as Prime Minister (1966–77). Along with the Wildlife (Protection) Act of 1972, a Water Pollution Act was also promulgated two years later. Yet, in 1982, studies estimated that as much as 70 per cent of the surface water was polluted by human waste or industrial effluent, with major negative implications for human and animal health (Roy 1982: 16–30). Thakkar (see Chapter 15), examines the working of anti-pollution norms while Sharma (see Chapter 16) draws attention to underclass groups who live in the riverine tracts of Bihar. Known as the *diara,* these are a very different landscape in comparison with the epicentres of the Green Revolution or of urban India (see Sharma 2001; on a fresh water lake, see Shanker 1992).

The lack of privilege and sheer poverty could be accentuated by policies and programmes that did not redress the problems of the poor living in these regions. Or worse, still, there were and are choices for growth that enclose resources that are vital for wage workers or marginal farmers, or cut off sources of livelihood. In this regard, pastoralists, often seen as a hindrance to forestry or as unproductive mobile peoples, turned out to be rather different on closer examination (see Chapter 17; also see Agarwal 1999;

Saberwal 1999; Narain 2005). Similarly, systems of shifting cultivation in the Lushai Hills of the North-East can be far less destructive of forest cover, given a reasonable time of fallow between crop growing cycles (see Chapter 18; also see Sethi 2005). Shepherds and other herders or tribal peoples can, given adequate policy support and assistance, thereby actually help to restore and keep green cover and be partners in ecologically sound growth.

A very important dimension brought out by movements as much as by sensitive scholarship is that of gender. Women in India are unequal partners by virtue of their social condition (such as absence of property rights or equal educational opportunity). The environment is often critical to their livelihoods, with tasks like water and fuel collection imposing a heavy burden in physical and economic terms. The transformation of gender ties with women being equal partners cannot be accomplished in the absence of gender sensitive views of the environment (Agarwal 1985).

Finally, this section touches on a different set of concerns, of vanishing biodiversity. Urban and industrial pressure on the one side and the basic livelihoods of the vast underclass on the other, have over the years increased the pressure on habitats, species and ecosystems. In 1973, these concerns found focus in the launch of Project Tiger to save the national animal from extinction. Despite successes, poaching and, more than that, threats to habitat and prey continue to be critical issues. Further, such ecosystems constitute a living laboratory of how nature's cycles of repair and renewal work. Krishnan (Chapter 20) offers below a lucid rationale for preserving stretches of land and waterscape and Karanth (Chapter 21) details the biological logic for protection. The essays are food for thought about issues of equity at another scale: beyond species and generations (see Karanth 2005; for a contrasting picture from Arunachal Pradesh, see Datta 2005).

As Indira Gandhi noted in 1968, there are no easy choices in independent India. Markets, an ever larger force in public life, offer incentives and create opportunity but unchecked they can play havoc with resources, with extraction outpacing regeneration. Similarly, governmental regulation is a must, but bears down on those with little or no land, less purchasing power or low social status. The story of the last several decades is one of attempts to grapple with these dilemmas. Failures have often outnumbered successes. But at least there is no dearth of positive example and thoughtful initiative.

## References

Agarwal, Anil, Ravi Chopra and Kalpana Sharma (eds), 1982, *The State of India's Environment,The First Citizen's Report*, Centre for Science and Environment, Delhi.

Agarwal, Anil, 1985, 'Beyond pretty trees and tigers, The Vikram Sarabhai Memorial Lecture', New Delhi, 1985, reproduced as 'Politics of Environment: I', in Anil Agarwal and Sunita Narain (ed.), 1985, *The State of India's Environment: The Second Citizen's Report, 1984-85*, Centre for Science and Environment, Delhi, pp. 362–81.

Agrawal, Arun, 1999, *Greener Pastures,* OUP, Delhi.

Datta, Aparajita, 2005, 'Fading Fauna, Forgotten People', *Down to Earth*, 15 September, pp. 46–49.

Gandhi, Indira, 1982, 'Tradition and Change, Address to the Convocation of Bangalore University, 8 January 1967', in Indira Gandhi, 1982, revised 1983, *On Peoples and Problems*, Hodder and Stoughton, London, pp. 13–18.

Karanth, Ullas K., 2005, *A View from the Machan, How Science can Save the Fragile Predator*, Permanent Black, Delhi.

Narain, Sunita, 2005, 'The Other 'Food' Crisis', *Down to Earth*, 15 December, p. 5.

Roy, Dunu, 1982, 'Water,' in *Citizen's Report, 1982*, pp. 16–30.

Saberwal, Vasant K., 1999, *Pastoral Politics,* OUP, Delhi.

Sethi, Nitin, 2005, ' Mautak will flower', *Down to Earth*, 31 October, pp. 27–38.

Shanker, Uday, 1992, 'Chilka: A Lake in Limbo', *Down to Earth,* 31 August, pp. 1–7.

Sharma, Mukul, 2001, *Landscapes and Lives, Environmental Reportage from Rural India,* OUP, Delhi .

Vohra, BB., 1973, 'A Charter for the Land', *Economic and Political Weekly*, 31 March.

II

# Agriculture on Spaceship Earth

M.S. Swaminathan

The expression 'Spaceship Earth' was first used by Buckminster Fuller many years ago to dramatize the fact that the resources of Earth are finite. Today this viewpoint has become necessary for man. If one looks at it this way, one can see Earth as an extra large, very efficient, Nature-designed spaceship, carrying its full load of three billion passengers through space at 18.5 miles per second, equipped with its own energy, refueling, recycling and life-support systems. A thin envelope of air, soil and water around the Earth together with energy from the sun provide the major ingredients for the evolution and survival of the large number of living things, including man, on Earth, and all are closely interrelated. This complex and delicately balanced web of relationships together constitutes the biosphere within which man has his place. It is hence evident that man can ultimately make only a finite claim on the limited resources of Earth.

The utilization of Earth's resources as though they were infinite, particularly by the rich nations, has led to the present awareness of an impending environmental and energy crisis, if economic activity is continued on the same path and pace. Brown (1972), for example, has pointed out that in 1950 when the gross world product (GWP) was just over a trillion dollars, eco-catastrophes were few. In 1970, when the GWP became nearly three trillion dollars, rapid eutrophication of lakes, oil spills, and fish kills, devastating floods arising from deforestation and similar disasters became more frequent.

---

* First presented as Coromandel Lecture, February 26, 1973. First published in *Coromandel Lectures: Platform for a Common Present and Future for Humankind*, Secunderabad, Andhra Pradesh, India: Coromandel Fertilisers Limited, 2001, pp. 37–66. Printed here from Anwar Dil, 2004, *Life and Work of M.S. Swaminathan: Toward a Hunger-Free World*, East West Books, Chennai, pp. 130–47.

Growth in the consumption of natural resources, both renewable and non-renewable, is proceeding at an alarming rate. Petroleum consumption is expanding at 4 per cent yearly, fresh water at 3 per cent, marine protein at 4 per cent and the use of chemical fertilizers at 7 per cent. A 3 per cent annual growth rate means a doubling in every 24 years, while a 7 per cent growth rate leads to doubling every 10 years. In their study *Limits to Growth,* Meadows (1972) and co-workers, therefore, stressed that the exponential growth in consumption of finite resources, however large the reserves may be, cannot continue indefinitely. Many of their assumptions may be open to question.

Taking agriculture, Meadows and co-workers have calculated that at the present level of productivity about 0.4 hectares of arable land will be needed to feed a person. The total world availability of arable land is about 3.2 billion hectares. As the population grows, more land is not only needed for producing more food, but also for urban–industrial use. Calculations show that in 2000 CE, the land available for agriculture will fall short of the amount needed to produce the requisite quantity of food for the population at that time, at current productivity levels. This 'doomsday' year will come 25 years later, if the productivity can be doubled and 50 years later, if the productivity can be quadrupled. There could be a difference of opinion about these dates and calculations but the basic thesis is clear—we just cannot continue to fondly hope that the Earth's resources can be indefinitely exploited on the assumption that some new discovery would always help to open up some hitherto unknown pathway of production.

*The Ecologist* (January 1972) in a much discussed article entitled 'A Blueprint for Survival' stated the choice available to man as follows:

> The principal defect of the industrial way of life with its ethos of expansion is that it is not sustainable. Its termination within the lifetime of someone born today is inevitable—unless it continues to be sustained for a while longer by an entrenched minority at the cost of imposing great suffering on the rest of mankind.

In the consumption-oriented world of today, the rich nations consume most of the world's products, while within developing nations the richer sections of society, which may constitute only a fraction of the population, consume a large proportion of the goods produced. It is probable that the garbage cans of the richer nations may contain more protein than the total quantity of protein consumed by the poor of the developing world. One third of the world's people who live in developed lands consume nearly eight times more energy per capita than do citizens in poorer countries. The consumption revolution of the developed world obviously cannot go on at the current pace, particularly since the developing countries also wish to consume more energy now. This realization is leading more and more thinkers to begin to view the problems of the world as a whole, and not merely consider each part and aspect as an isolated problem. *Only One Earth* by Barbara Ward and Rene Dubos (1972) is an example of this new awareness.

We live in a world of sad scientific ironies and economic enigmas. The regions where man first settled down to cultivate plants and thereby initiated what we now call 'agriculture' are also the regions which contain the greatest number of hungry people today. The regions which scientists designate as the centre of origin of a crop plant are, strangely enough, the areas where the same plant is today giving one of the poorest average yields. The Jeypore tract of Orissa, which is believed to be an ancient centre of origin of rice, is a good example. The regions where technical skills were of a high order in ancient days are today characterized by relatively poor quality of output and workmanship. One need only recall in this connection the past glory of India and China in the production of textiles, pottery, porcelain and metal goods.

This trend needs to be arrested, if the developing countries are to accomplish even partially their socio-economic goals. This is, however, more easily said than done. The compound interest law of growth rate operates without abatement. We are now in the second UN Development Decade but by most economic yardsticks, the income and consumption gap between the rich and poor worlds is widening. Between 1967 and 1971, developed market economics increased their agricultural exports on an average by 11 per cent per year, while the developing market economies lost ground by one per cent (FAO statistics). Similarly in 1971, agricultural output went up by 9 per cent in North America and 5 per cent in Western Europe, while the increase in the developing countries ranged from 1 to 2 per cent. On the positive side is the fact that we have an opportunity to take to a pathway of development, which does not lead to the degradation of both man and his environment. The 'One Earth' concept is vital in any system of planning owing to the many tangible and intangible international linkages, which influence national endeavour. However, we must accept the reality of the existence of sovereign nation states and the differences of interests and priorities between the two sharply divided groups in the world—the rich and the poor nations. Hence, within the overview provided by the concept of 'One Earth' it will still be necessary for some time to come to think in terms of planning the rational and best use of resources for each country or region. I propose to confine myself, therefore, in this paper to

**Table 11.1: Population Projections (Millions)—India**

| *Year* | *NCAER** | *Planning Commission* | *Registrar General* | *Dr. Thamarajakshi* |
|---|---|---|---|---|
| 1968 | 521 | 527 | – | – |
| 1971 | – | – | 548 | – |
| 1973–74 | – | 596 | – | – |
| 1981 | 689 | 690 | 657 | – |
| 1985 | 755 | 740 | – | 703 |
| 1991 | 827 | – | – | – |
| 2000 | – | – | – | 860 |

**Source:** National Council for Applied Economic Research

**Table 11.2: Food Demand and Supply Projections for India**

| | *Gross Food-grain Demand (Million tons)* | | | | *Food-Grain Production (Million tons)* | | |
|---|---|---|---|---|---|---|---|
| | *Demand* | | | | *Supplly* | | |
| *Year* | *NCAER* | *Planning Commission* | *NCA* | *ESD* | *NCAER* | *Planning Commisson* | *ESD* |
| 1968–69 | 99.8 | 94.4 | – | – | 96.2 | 98.0 | – |
| 1969–70 | – | – | – | 102.2 | – | – | – |
| 1973–74 | – | 126.9 | – | 116.1 | – | – | 115.6 |
| 1975–76 | 120.3 | – | 121.9 | – | 121.5 | – | – |
| 1978–79 | – | – | – | 145.4 | – | – | 134.8 |
| 1980–81 | 135.7 | 163.0 | 146.3 | – | 137.7 | 165.5 | – |
| 1985–86 | 150.3 | 190.0 | 173.6 | – | 151.8 | 192.3 | – |
| 1990–91 | 167.1 | – | – | – | 168.7 | – | – |
| 2000 | – | – | 228.5 | – | – | – | – |

**Note:** Gross Demand: Net Demand (for human consumption) + (Allowance for seed, feed and wastage).

a discussion of the problems and possibilities for agricultural development in that sector of Spaceship Earth, which is India.

## Population Growth and Food Supply

I have discussed elsewhere the various projections of population growth and food needs up to 2000 CE and also the possibilities for producing the needed quantity of food. The population and food requirement projections made in different studies are summarized in Tables 11.1 and 11.2. It should be possible for us to achieve the production targets, if we systematically adapt and extend the available scientific know-how to the different agro-ecological areas. But what we need from the agricultural sector is not only more food but also more income and more avenues for remunerative employment. Therefore, we have to develop for each ecological area a land and water use strategy which can, through a careful blend of agriculture, forestry, animal husbandry and fisheries, help to diversify the sources of employment and maximized income.

## Potential for Further Improvement in Yield

I would like to take crop and fish production to illustrate current trends.

(1) Crops: It has been estimated by crop physiologists that $12.2 \times 10^5$ cal/ha/day, which is equivalent to 124 tons/ha/year of carbohydrate, can be expected to be the maximum attainable biological yield in the tropics. Assuming a harvest index of 50 per cent in crops like wheat and rice, we should expect a yield of $6.1 \times 10^5$ cal/ha/day or approximately 62 tons/ha/year of carbohydrate in the form of grains. There are reports that yield up to 80 q/ha and 100 q/ha have been obtained in wheat and rice respectively in 130 days under intensive agriculture. Due to cloudiness during the growing season, it may be difficult to expect yields exceeding $4.88 \times 10^5$cal/ha/day in rice. In All India Coordinated trials, energy conservation to the extent of $3.05 \times 10^5$ cal/ha/day has been reported in rice, which represents about 60 per cent of the potential.

In wheat, the highest grain yield has reached nearly one-third of the potential yield. In oilseeds, the single important factor responsible for poor yield is the low hardest index. Even with one irrigation and 50 kg. N/ha, it is possible to obtain 80 per cent of the highest yield. Hence, striking improvement in the productivity of these crops is possible in areas with limited water availability.

Improvements in yield in most economic plants have taken place since the beginning of agriculture through small improvements and changes. Occasionally, jumps of larger quanta accelerate the pace of progress. Usually, such jumps are associated either with an important landmark in the genetic improvement of crop plants or in advances in water and pest management. Some significant landmarks in plant breeding during this century with reference to both yield and quality are indicated in Table 11.3. It should be

**Table 11.3: Some Significant Landmarks in the Twentieth century**

| | *Plant Breeding* | | |
|---|---|---|---|
| *Development* | *Years of widespread* | *Country where originally used* | *Genetic Principle(s) involved* |
| 1. Hybrid maize | 1933 | USA | Hybrid vigor |
| 2. Hybird sorghum | 1957 | USA | Hybrid vigor |
| 3. Hybrid pearl millet | 1959 | USA | Hybrid vigor |
| 4. Dwarf wheat | 1961 | USA | Dwarfing genes from Japanese 'Norin' wheats |
| 5. Dwarf wheat | 1963 | Mexico and later India, Pakistan and Middle East | 'Norin' dwarfing genes & relative photo-insensitivity |
| 6. Dwarf indica rice | 1955 | Taiwan | Genes for Dwarfing and relative photo-insensitivity, stiff and erect leaves |
| 7. Dwarf indica rice | 1955 | Philippines, India and other countries in Asia | Genes for dwarfing and relative photo-insensitivity, stiff and erect leaves |
| 8. 'Opaque-2' maize | 1965 | USA | High lysine content |
| 9. Hybird barley | 1969 | USA | Hybird vigour |
| 10. 'Hiproly' barley | 1969 | Sweden | High protein and high lysine |
| 11. Hybird upland cotton | 1970 | India | Hybird vigour |

TABLE 11.4: YIELDS OBTAINED DURING THE 1972 KHARIF SEASON IN TRIALS UNDER THE ALL-INDIA COORDINATED RESEARCH PROJECT FOR DRYLAND AGRICULTURE

| *Rainfall Patern and Centre* | *Yield in Q/ha* | | | | |
|---|---|---|---|---|---|
| | *Jowar* | *Bajra* | *Setaria* | *Arhar* | *Soybean* |
| *Seasonal rainfall defiicent (–20 per cent to –60 per cent from normal)* | | | | | |
| Hyderabad | 17.8 | 11.8 | 16.3 | 14.1 | – |
| Indore | 39.5 | – | – | – | 29.8 |
| Dehra Dun | – | – | – | – | 26.1 |
| Rajkot | 3.3 | 9.0 | – | 8.9 | 5.4 |
| Varanasi | – | – | 12.4 | 14.3 | 7.5 |
| *Seasonal rainfall scanty (–60 per cent and below)* | | | | | |
| Agra | 6.0 | 19.9 | – | 19.9 | – |

TABLE 11.5: AVERAGE ANNUAL YIELD OF DIFFERENT ANIMAL PRODUCTION SYSTEMS

| *Area* | *Product* | *Yield (kg/ha)* |
|---|---|---|
| Pasture land | Cattle | 6–308 |
| Inland Ponds (India) | Fish | 3000 |
| Continental shelf | Ground fish | 25–715 |
| Humboldt current | Anchovies | 375 |
| Mariculture (Japan) | Oyster | 57,500 |
| Mariculture (Spain) | Mussels | 300,000 |

emphasized that the role of the new seeds in improving yield has been primarily that of a catalyst and that the real yield advance has been brought about by associated changes in agronomy, plant protection and engineering.

Advances have also taken place in improving the yield of plants and animals in semi-arid and arid areas. The yields obtained in trials conducted under the All India Coordinated Research Project for Dryland Agriculture during Kharif 1972, when widespread drought occurred, indicate that there is much scope for improving production in these areas through an integrated plan for moisture conservation, fertilizer application, pest control and crop and varietal substitution (Table 11.4). Some imaginative and elegant experiments are now underway in achieving crosses between distantly related species through somatic cell hybridization, in using pollen cultures for getting haploids, in introducing the genetic capacity for symbiotic nitrogen fixation in non-legumes and in enhancing the 'sink' capacity, or capacity for economic yield through the genetic manipulation of production physiologic processes. It is premature to guess what the ultimate impact of these researches would be on agricultural development. Meanwhile, we have enough work to do in converting the scientific breakthrough, which has already taken place in the major crops into production advance.

(2) Fish: Among farm animals, fish is the most efficient converter of plant food. World fish production rose from 21.1 million tons in 1950 to over 69 million tons in 1970. Let me illustrate the position both in marine and inland fisheries in our country

## Marine Fish Production

The realistic estimate of the minimum potential catch per year (both demersal and pelagic) is of the order of 11–12 million tons as against the present annual production of about 2.8 million tons. From the western part of the Indian Ocean, the total estimate is approximately 6 million tons consisting of 3.9 million tons of demersal fishes, 1.9 million tons of pelagic fishes and 0.2 million tons of crustaceans. From the eastern Indian Ocean, 3.2 million tons is the estimated potential catch, which includes 1.4 million tons of demersal fishes and 1.8 million tons of other fishes. The Indian Ocean also has a potential of about one million tons of tuna and tuna like fishes of which less than 0.2 million tons are being caught, mainly using the long lines. There is, therefore, considerable scope for the development of fishery for tunas, particularly for the skipjack, smaller yellow fin tunas, frigate mackerel and the little tunny.

*Continental shelf around India (the region of the sea up to 200m depth):* The minimum estimated potential catch is about 2.4 million tons as against the present exploitation of about 1.2 million tons. Hence, the scope for increase is of the order of 1.2 million tons inclusive of both pelagic and demersal resources. The break-up for the west coast and the east coast is given below:

*West coast of India:* The estimated minimum potential catch is 1.6 million tons consisting of 1.02 million tons of pelagic catch and 0.58 million tons of demersal fishes. The present catch on the west coast is about 0.9 million tons and hence the new potential available is of the order of 0.7 million tons.

*East coast of India:* The east coast can produce at least 0.8 million tons (0.67 million tons of pelagic fishes and 0.14 million tons of demersal fishes). The present catch is of the order of 0.3 million tons.

*Categories of New Resources Available in the Seas around India:* The new resources available have neither been categorized nor quantified in detail, although

**TABLE 11.6: LAND USE STATISTICS (1968–69) (THOUSAND HECTARES)**

| | |
|---|---|
| Geographical area | 328,048 |
| Reporting area | 305,814 |
| | *Per Cent Area* |
| Forests | 20.5 |
| Not available for cultivation, pastures, trees, culturable waste. and fallow | 34.5 |
| Net area sown | 45.0 |

**Source:** Fertilizer Statistics, 1970–71.

**TABLE 11.7: AREAS UNDER DIFFERENT SOILS, 1969–70**

| *Soil Group* | *Area (Million hectares)* | *Mean Annual Precipitation (mm)* | *Mean Annual Temp (°C)* |
|---|---|---|---|
| 1. Red soils Red Sandy and Red Loamy | 7.2 | Extensive area 600–1200 Small areas in West and North east 300 and 400 respectively | 25–28 |
| 2. Black soils (shallow, medium and deep) | 64.4 | 500–1200 | 25–28 |
| 3. Alluvial soils (costal, deltatic and Plains) | 63.3 | Extensive area 500–1200 Eastern Plains 4000 | 20–28 |
| 4. Red and Yellow soils | 33.5 | Extensive area 600–1000 Eastern Part 3000–4000 | 20–28 |
| 5. Desert Soils | 21.9 | 100–400 | 25–28 & > 28 |
| 6. Submontane soils | 20.2 | 1200–3000 | 10–20 |
| 7. Mixed Red & Black soils | 14.8 | 600–800 | 25–28 |
| 8. Laterite soils | 11.7 | Extensive areas 2000–3000 small areas–1200 | 20–25 |
| 9. Grey Brown soils | 8.9 | 400–500 | 25–28 |
| 10. Saline and Alkali soils | 70.0 | 200–1200 | 25–28 |
| 11. Mountain Meadow | 6.6 | 200–1200 | 10 |

**TABLE 11.8: NITROGEN BALANCE: 1969 (THOUSAND METRIC TONS)**

| | *USA* | *Australia* | *India* |
|---|---|---|---|
| I. N in Yields Non-leguminous Crops | 4,518 | 365 | 2,841 |
| Leguminous Crops | 1,996 | 6 | 460 |
| Forage and Pasture | 5,180 | 10,700 | 300 |
| II. N Input | | | |
| Legume Crops and forage | 8,610 | 12,840 | 912 |
| Free-living N-fixation | 1,400 | 1,000 | 650 |
| Fertilizer N | 6,313 | 172 | 1,222 |
| Difference (II-I) | +4,629 | +2,941 | –816 |

some general indications are available. It is now known that the catch from traditional fisheries can be increased provided the market conditions remain favourable for their utilization.

During the time when the catches are high the prices go down to uneconomic levels, thus forcing the fishermen to exercise a voluntary restraint in landing more fish. The new resources of the Indian coast can be classified as follows:

(1) Pelagic Fisheries: (*a*) Mixed pelagic species: Recent studies have shown that about 200,000 tons of mixed pelagic species consisting mainly of anchovies, lesser sardines, ribbon fishes and silver bellies are available in the month of April along the south-west coast in the region extending up to 40 meter depth. These are at present almost totally unexploited. The magnitude of total anchovies resource is probably much greater; (*b*) Tunas: In Laccadives, there is traditional fishery for tunas. However, good concentrations of surface shoals have been noted beyond the traditional fishing range in these islands and also in the region between the islands and the south-west coast. It is also known that the central Bay of Bengal, especially the area near Andaman and Nicobar Islands, supports large population of tunas; (*c*) Clupeoids off the east coast and off Goa and Maharashtra: There are resources of sardines and other clupeoids available off the east coast and the coasts of Goa and Maharashtra. These have largely remained unexploited; (*d*) Cephalopods (squids and cuttle-fishes): Recent work shows that the cephalopods are abundant in the Bay of Bengal and Gulf of Kutch. These are very poorly fished at present. However, the economics of their exploitation and utilization have not been worked out.

(2) Demersal Fisheries: (*a*) Prawns: Recent studies show that the Gulf of Kutch, south-east coast of India, Gulf of Mannar and Palk Bay and the

coastal waters of Tamil Nadu and Andhra Pradesh have very rich grounds of prawns. The fishery in the grounds off the south-west coast has developed considerably, but the other areas are not fished intensively; (*b*) Deep sea crustaceans and fishes: Off the south west coast in the depth range 75–450 metres between the latitudes 8° and 15° North (area 43,000 sq km), potential resources of the order of 100,000 tons are available. These resources have been discovered only during the last five years. Of these, at least 23,000 tons are crustaceans (deep-sea prawns, deep-sea lobsters and crabs) and the rest are fishes. The crustaceans are particularly abundant in the depth range 180–450 metres (the upper continental slope). Rich grounds of deep-sea lobsters have also been found along the paper continental slope of the Gulf of Mannar; (*c*) Perch resources: Very good grounds of perches, fishable by hook and line have been found off the south-west coast between 75 and 100 meter depths. The quantification of these resources has not yet been made, but the catch rates are expected to range between 200–300 kg per hundred hooks per hour, which is considered to be high; (*d*) Other resources: Good grounds for sharks, catfishes, sciaenids, carangids, silver-bellies, pomfrets, and perches fishable by trawls have been bound off the Indian coast, up to at least forty metres depth. These are only partly fished at present. Various forms of mariculture offer great opportunities for increasing food production (Table 11.5). In Japan, seaweed culture is also being exploited well.

## Fresh Water Fish Production

In terms of quantity, the world catch of fresh water fish is only one-tenth of the total fish production, i.e., about 6.9 million tons. Nevertheless, freshwater fish include several varieties of high value, which are in great preference in several countries. The inland fisheries have been grouped under capture and culture fisheries. Development of capture fisheries is somewhat expensive and has only a limited scope. As against this, culture fisheries can contribute significantly to food production. Freshwater fish culture has its longest history in Asia and Far East particularly in China. Stocking of streams, rivers, lakes and reservoirs, has been widely practiced for a multiplicity of reasons. Due to lack of adequate data, it is difficult to estimate the inland fish production, especially the quantum of fish produced through culture operations. According to a recent FAO estimate based on data from 42 countries, the total world production by aquaculture is over five million tons. This estimate, however, does not include production of sport, bait and ornamental fishes. In the overall spectrum of inland fishery development, aquaculture plays an important role in the economy of many countries. In countries like China, India and Indonesia, nearly 40, 38 and 28 per cent respectively of the total production of fish is accounted for by aquaculture. In land-locked countries, fish culture is developing into one of the important sources of animal protein. Although reliable information on the area utilizable for aquaculture is not available, it is known that in addition to conventional pond culture, there are extensive swamps, estuaries and low-lying areas with assured source of water supply suitable for fish culture. Data so far collected show that only a small part of this is being utilized for aquaculture.

With the development of a variety of techniques of fish culture such as closed water ponds, running water ponds, cage culture, paddy-cum-fish culture, hanging culture, culture in lakes and reservoirs etc., a manifold

**TABLE 11.9: ULTIMATE IRRIGATION POTENTIAL IN DIFFERENT STATES (DATA FROM THE REPORT OF THE SECOND IRRIGATION COMMISSION)**

| | *Area (million hectares)* | |
|---|---|---|
| *State* | *Ultimate Potential* | *Area Irrigated (1966–67)* |
| Andhra Pradesh | 10.3 | 3.9 |
| Assam | 2.2 | 0.6 |
| Bihar | 8.0 | 2.0 |
| Gujarat | 3.8 | 1.1 |
| Haryana* | 0.85 | 1.8 |
| Jammu & Kashmir | 0.6 | 0.3 |
| Kerala | 1.5 | 0.5 |
| Madhya Pradesh | 8.0 | 1.1 |
| Maharashtra | 4.5 | 1.4 |
| Mysore | 3.4 | 1.2 |
| Orissa | 4.0 | 1.1 |
| Punjab | 6.1 | 3.4 |
| Rajasthan | 5.0 | 2.1 |
| Tamil Nadu | 3.6 | 3.4 |
| Uttar Pradesh | 15.1 | 7.2 |
| West Bengal | 4.3 | 1.5 |
| All India | 81.7 | 32.8 |

**Note:** * Area under major and Medium projects included under Punjab expansion of aquaculture in most parts of the world is possible. In many countries, the possibility of substantial increase in yield by the adoption of the technique of composite fish culture has been established. Similarly, it has been proved that integrated culture practices employing a variety of compatible organisms such as prawns, frogs, ducks and even plants like *Euryale ferox* and *Trapa bispinosa* along with fish are advantageous in making full use of the three-dimensional aquatic environment. Further, rotation of crops involving fish, soybeans and rice are also being practiced with considerable success by some farmers. The labour-intensive nature of the present aquaculture operations also generates additional employment potential in the rural areas.

Water pollution and waterweeds pose some of the greatest threats to fish culture. A concerted attack on both these problems is vital to realize the potential fish offers to improve food supply and promote protein intake.

## Our Agricultural Assets

Sunlight, land, water, plant, animal and human resources form the principal foundations of any agricultural system. The extent of advance made in enhancing both agricultural productivity and stability would depend upon the progress achieved in an integrated domestication of plants, soil and water based on the principles of synergy and symbiosis. In India, domestication of economic plants and animals evoked much interest since ancient days but equal interest in the domestication of soil and water was lacking. We see as a result large areas of eroded land and problems like salinity and alkalinity. Inadequate attention to the replenishment of soil fertility has also led to an impoverished nutrient status in most soils. One of our most urgent tasks therefore is to pay greater attention to the care and build-up of soil health and fertility and to the scientific utilization of water and sunlight. I would like to refer briefly to the opportunities we have for better land, water and sunlight utilization.

### Land Use

The pattern of use of land for agriculture, forestry and other uses is indicated in Table 11.6. The soil types are of varied kinds, with red, black and alluvial soils occurring in almost equal proportions (Table 11.7). Forestry occupies a little less than half the area under agriculture but contributes only 1.4 per cent of the gross domestic product. During 1969–70, the contributions of agriculture, animal husbandry and forestry to the gross domestic product were 42.4, 4.1 and 1.6 per cent respectively. The poor contribution of forestry to GDP and the neglected state of this science have been highlighted in an interim report of the National Commission on Agriculture (NCA). NCA has pleaded that we must move away from the concept of mere conservation of natural forests to one of man-made forests. We need a dynamic programme of research in forest tree improvement including imaginative attempts to shorten the duration by accelerating the rate of growth of trees which can provide fuel and raw material for the paper industry. Our ability to promote a better ecological balance and a desirable organic recycling process would depend to a considerable extent upon what happens in the field of forestry research and development in the coming years.

The authors of *Limits to Growth* have attempted to quantify the extent of fall in land availability for agricultural use that we will have to be prepared for with population expansion. They have assumed that each additional person would need 0.08 hectares of land for purposes like housing, roads, waste disposal, power lines and other uses and 0.4 hectares for producing the food he needs. If we use this formula in our country, over five million hectares of land will be needed every year to meet the requirements of the newborn. Therefore, we have to aim at developing integrated land use systems in agriculture where the productivity of the soil can be continuously improved, without any impairment of its long-term production potential. To achieve this we have to learn both from the successes and failures of the developed countries in the matter of soil care and use. While there are several aspects to this problem, I would like to refer to two urgent ones relating to nutrient supply and soil structure.

## Soil Care and Fertility

The recent advances in the genetic enhancement of the yield potential of several economic plants have stemmed from successes in improving their capacity to utilize more nutrients and convert them into the economic product. Thus, for getting a higher output, a greater input, particularly of nutrients, becomes essential. The first question that arises in this context is whether the continuous application of fertilizer is detrimental in the long-term. Excessive use of nitrogen and phosphorus has led to difficulties in some countries. Kanwar (1972) has discussed this question with reference to our country. He has shown that our problem is not akin to those of the developed countries for a variety of reasons, among which the following are some of the more important.

We are at present using about 6.15, 1.45 and 0.92 kg. N, $P_2O_5$ and $K_2O$ respectively per hectare (according to 1969–70 statistics), which is about 1/47th of the fertilizer consumption per hectare in Japan. On a per capita basis, the amount is about 2.6 kg. Even if a fraction of this fertilizer is lost through leaching or other biological activities, the effect on the environment will be hardly measurable. However, in the monsoon season the loss of N can be high in rice. Drainage studies in heavily manured paddy fields at the Central Rice Research Institute, Cuttack, have revealed that during the Kharif season, 42 kg N, 29 kg phosphorus, 38 kg iron and 17 kg manganese were lost per hectare. No separate record is available on the nitrate content of the water. There are very few studies about the quantitative estimation of N losses in the gaseous form from submerged condition in paddy cultivation. Datta and associates at IARI using $N^{15}$ technique concluded that gaseous N loss was 24.1, 30.2 and 53.7 per cent in case of tagged ammonium sulphate, nitro phosphate (high water solubility) and nitro phosphate (low water solubility) respectively.

The loss of nutrients in gaseous forms as well as in drainage from heavily fertilized rice fields requires critical studies. Also, more studies are needed of the nitrate content of underground water in areas of intensive fertilizer application such as the grape gardens of Hyderabad.

From the available data, fertilizers do not seem to be significantly involved in raising the P levels in surface and ground waters. On the other hand, by promoting the growth of vegetation they help in reducing erosion and thus reduce the loss of P to waters. The contribution by human, animal and industrial waste and detergents to P content of streams, lakes and marine and also inland waters may be substantial as studies by the Central Inland Fisheries Research Institute, Calcutta, have revealed. Thus, organic sources of phosphorus may be a source of pollution in some areas.

Kanwar has stressed that the most important cause of pollution from the agricultural lands is soil erosion. This process not only leads to the loss of millions of tons of nutrients every year, but it is threatening our irrigation projects and cities. It is estimated that we are losing annually about 6,000 million tons of soil from 91 million ha, which are reported to be suffering from problems of erosion. Assuming that from a soil of 0.5 per cent slope, the loss of nutrients is of the order of 5.8 kg. N, 10.7 kg. $P_2O_5$, and 42.8 kg. $K_2O$/ha, the annual total loss from Indian soils could be as high as 2.5, 3.8 and 2.6 million tons of N, $P_2O_5$ and $K_2O$ respectively. This quantity is much greater than the present total consumption of NPK from fertilizer as well as organic manure sources. Moreover, there are millions of hectares which lose many times more nutrients per hectare because of greater slope and credibility. Obviously, such losses cannot be totally avoided, but we can certainly minimize them. Fortunately, a part of these leached nutrients must be going back to soil through water.

We have to assess the quantity of nutrients we need to produce about 228 million tons of food-grains by 2000 CE. We need nutrients not only for food crops but for all domesticated plants and for fish production. Research carried out under the All India Coordinated Project on the correlation between soil test values and response of crops to fertilizer has shown how we can maximize the return from fertilizer application. Based on this data, Dr. B. Ramamoorthy and his colleagues have calculated that to produce the food and other products we need by 2000 CE, we may need a minimum of 4.60, 3.94 and 1.92 million tons of the nutrients N, P and K respectively. These calculations are based on the assumption that the efficiency of fertilizer application will be very high. The present installed capacity for fertilizer production is of the order of 1.46 and 0.50 million tons for N and P respectively.

What pathways of nutrient supply should we take? The plant cannot give more yields from hungry soils. It is obvious that we should apply fertilizers judiciously and economically making full use of the research data relating to the potential for synergistic interactions. It is equally obvious that we should not use synthetic fertilizers as the sole means of nutrient supply. We have to evolve an integrated nutrient supply strategy for each cropping system, involving an appropriate combination of organic and inorganic manures and ensure that legumes find a place in the crop rotation.

1973 is a particularly suitable year to initiate the development and adoption of such an integrated procedure, since the availability of fertilizers is less than required.

In its research bulletin on 'Towards Self-reliance in Agriculture', IARI has advocated the tapping of the following sources of P to the maximum possible extent: (*a*) Basic slag, a by-product of the steel industry wasted at present; (*b*) Rock phosphate deposits in different parts of the country; (*c*) Phosphorus-rich sands of the Arabian Sea Islands.

The following are some of the major sources of organic manure, which deserve widespread conservation and use: (*a*) Urban solid wastes like garbage, night soil, sludge and slaughterhouse wastes; (*b*) Urban liquid wastes like sewage, sullage and urine; (*c*) Industrial wastes; (*d*) All farm and crop wastes; (*e*) Weeds, particularly water weeds like water hyacinth; (*f*) Cattle-shed sweepings and all animal wastes; (*g*) Inedible oil cakes.

If we are to conserve cowdung and bones for manurial purpose, we should popularize the widespread use of cowdung gas plants and bone digesters. If quick-growing fuel trees can be developed, the rural population need not have to rely so heavily on cowdung for fuel. In addition to the use of all such wastes, the cultivation of pulse and green manure crops both as a pure and as an inter-crop in rotations in irrigated areas will be very helpful both to get more pulses and to fix more nitrogen biologically.

There is a superabundance of nitrogen in nature, with about 87,000 tons in the atmosphere above each hectare. The contribution of legumes to nitrogen fixation in the soil is highest in Australia and least in India (Table 11.8). Of more than 2,000 million hectares of meadows, pastures and rough grazing in Australia, much of it arid, about 100 million hectares have been improved by introducing legumes like *Trifolium subteraneum,* inoculated with bacteria and usually sown with super phosphate, and sometimes with lime and minor elements. In the United States also, legumes provide the largest amount of nitrogen for primary production, soybeans alone contributing approximately 1.4 million tons of nitrogen. We will do well to emulate these examples.

In contrast to our country, where interest in organic recycling of all wastes has been generally poor in spite of Mahatma Gandhi's efforts to focus attention on its importance, much attention has been paid to the conservation and use of all wastes including human excreta in China. The following extract from a book by Han Suyin (1965) emphasizes the importance given to waste recycling in China.

> Tobacco requires much manure, spraying with manure being done every five days in April and May when the plants grow. The good earth of Szechuan, moist and replete and rich, glistening as with oil is nourished with human and pig excrement, treated for some days in pits. It is only now after the Revolution, that chemical fertilizer plants are being built, but today man and pig are still the main providers of manure. And therefore in the city of Chengtu in the nineteenth century and up to 1949, some of the wealthiest families were those owning the public cloacae and selling excrement to the countryside. I remember Third Uncle on this subject, which rendered him prolix, for he was always aware of the earth, the sowing and planting and reaping of it; 'Three things are necessary, water, earth and manure. Water from the sky, soil of the earth, and manure of man and animal; the excremental waste confirms the harmonious cycle of life, since out of waste comes renewal.' In an agricultural society dung has its function and natural dignity; and even today, I can become engrossed in the relative merits of 'green lush' or 'brown gold.'

In Europe and North America also human waste recycling systems are now under development and alternative methods which do not involve water to flush excreta, are rapidly being standardized. It is high time we also start seriously working on the development of efficient cycling and nutrient supply system which will lead to the proper maintenance of soil structure and fertility.

## Pollution from Pesticides

As in the case of fertilizer, the average consumption of pesticides is so low that there does not appear to exist a situation akin to those of some of the developed countries. However, a good proportion of the pesticides used by us have the property of persistence in the plant material. Hence a more extensive surveillance and monitoring system is necessary. Also the right combination of pesticides is necessary for each region, as otherwise the pest which is not sensitive to the particular chemical used, would tend to survive preferentially and cause much damage. Entomologists now advocate the use of integrated pest control methods involving an appropriate admixture of genetic, chemical, biological, agronomic and other techniques. It is planned to get such an approach tested at the field level shortly in rice. So far we have not gathered operational experience in this field.

## Water Requirements

On an average, we receive an annual rainfall of about 112 cm for 370 million-hectare metres. It has been estimated that about 120 million-hectare metres seep into the soil and about 170 million hectare metres remain in the top layers and contribute to soil moisture which support crop growth. The Irrigation Commission has made detailed projection of both water availability and scope for planned exploitation. It appears that the total ground water recharge available for utilization may be of the order of 26.75 million hectare-metres per annum, in contrast to the current utilization of about 10 million hectare metres. Major, medium and minor irrigation works have gradually expanded the irrigated area. The states, which still have less than 15 per cent of the net cultivated area with irrigation facilities, are Madhya Pradesh, Maharashtra, Mysore, Gujarat and Rajasthan. The Irrigation Commission has made a broad assessment of the irrigation potential in the different States (Table 9).

The total water requirements of crops, animals and man and of industry up to 2000 CE have been calculated by Prof. A.M. Michael and his associates of the Water Technology Centre, Indian Agricultural Research Institute, using the projections for domestic and industrial use made by Kathpalia and Varadan (1971) and by Planning Commission. These studies show that at present there is a big gap between the net available water supply and the amount required for intensive cropping. The demands on water resources, possible supply and progressive exploitation during the period 1968–69 to 2000–01 are shown in Figure 2. It appears that if the available resources can be tapped and husbanded properly, particularly through intensive efforts to increase the efficiency of water conveyance and use, we may be able to manage up to 2000 CE without having to 'in for new sources of water such as desalination of seawater. The projections of Figure 2 have been worked by Prof. A.M. Michael of IARI. However, such as overall analysis of water budgeting hides the fact that even today there are considerable locational and regional gaps in the water demand-supply position. Rapid technological developments are taking place in areas like desalination of brackish water and weather modification and obviously in the choice of areas for conducting feasibility studies in the use of the new sources of water supply, the 'regions where the water budget shows greater deficit thereby causing much human hardship, should receive priority. There is, however, immediate scope in tall parts of our country to capture more water through better watershed management and by introducing water saving concepts such as design of houses which collect rain water, development and use of waste disposal system which do not involve the use of much water and adoption of improved methods of water conveyance and distribution (IARI 1970).

## Utilization of Sunlight

The need for using productively sunlight, which is one of our greatest endowments, has been stressed repeatedly by many research workers but little has so far been done in tapping for human use more of this valuable energy. Multiple cropping and inter-cropping provide an effective agricultural method of using solar energy (IARI, 1972). Green Plants can be made to harness sunlight throughout the year, wherever there is water to sustain crop growth. There is a vast potential for using solar energy in and semi-arid areas for household purposes, thereby saving the valuable organic matter for the soil. Garg and Krishnan (2001) have shown that a flat plate collector of two sq. metres at Jodhpur and three sq. metres at Delhi can help to heat 140 liters of water up to 55°C in the afternoon of January. This can provide sufficient hot water for a family of five persons. Such techniques should not remain only in the research laboratory.

## Global Ecological Balance

I have deliberately refrained from discussing the problems of our spaceship in a global context, since there are many excellent books giving such an overview. It, however, hardly needs mention that technological changes particularly those which affect the planet balance of energy could trigger off catastrophic changes. For example, there is evidence to suggest that over the last decade, the release of $CO_2$ into the atmosphere, as a result of man's burning fossil fuels and of deforestation has been increasing by 0.2 per cent a year. If this trend continues, the earth's temperature could rise by 0.5°C by 2000 CE. Industrial emission of dust, soot and gas as well as the development of supersonic transport may all influence earth's climate. Barbara Ward and Rene Dubos have hence pleaded for concerted world action in this field. According to them,

These concerns with global air pollution lie beyond the effective protection of individual governments. It is no use one nation checking its energy use to keep the ice caps in place if no other government joins in. It is no use the developed nations suggesting lower energy use just at the moment when the developing nations see increased use as their only exit from the trap of poverty. Man's global interdependence begins to require, in these fields, a new capacity for global decision-making and global care.

## Dos and Don'ts

The environmental policy advocated in the richer nations is designed to protect the high standard of living resulting from the unprecedented growth in the exploitation of natural resources during the last century from serious damage by the very processes of such growth. It is of necessity a policy based on a series of Don'ts. This is inevitable since the aim is to undo some of the damage already done or to prevent further damage along the same lines. Thus, there is an outcry against some of the potent agents of improved agricultural productivity such as fertilizer and pesticide.

The poorer nations, however, are faced with the desire and need to produce more food from hungry soils, more clothing and more housing. They are aware that historically a rising standard of living has depended on the ability of agriculture to release manpower to other more industrial pursuits. They hence naturally wish to develop more industries and to find productive and remunerative employment for their growing population. For them, conditions of poverty and inadequate arrangements for human and other waste disposal may be greater cause of water and air pollution than the effluent from factories or fertilizer from the fields. Since the causes of pollution are by and large different, the solutions will have to be different too and it would be a grave mistake to attempt to copy the policies now being propagated in the developed world.

We are fortunately in a position to build a positive policy of economic-ecology based on a series of Dos rather than Don'ts.

The first Coromandel Lecturer, Sir Joseph Hutchinson spoke about the need for breeding soils for sustained high productivity and has in a recent book (1972) explained 'the part that farming as a way of life has played in the preservation of good rural environments and the extent to which the future depends on this precious feature of agriculture.' To the farmer, he says, the farm and family are one integrated unit. He remains poorer than his urban cousins who not only want him to produce cheap food but accuse him of destroying the amenity of the environment, if with purely economic motive he removes his hedges or uproots a tree. Too often his critics want it both ways. To Prof. Hutchinson the only hope for the future is that there are still many people who are prepared to live with less of the world's goods than they could acquire in an urban occupation.

The Second Coromandel Lecturer, Dr. Norman Borlaug, earned this theme further and reminded us that well-fed and over-fed human beings have no right to impose conditions, which will preserve hunger and starvation for the majority. There is 'no time to relax' in our efforts to eradicate malnutrition and poverty and as an illustration of choice of priorities and options in action programmes, he said that if he were faced with the choice of having to watch the death of a few birds or the slow death of starving millions, he would 'vote for the survival and feeding of man.'

I have tried in this lecture to analyse how, within the concept of a careful husbandry of finite resources, we can move forward to produce more food and better rural living. I deliberately chose crops and fish for mention, since it is in these two areas that there is the greatest scope for rapid advance. In my view, it is only integrated land use involving a relevant combination of agriculture, forestry, fishery and animal husbandry, integrated strategies of nutrient supply and pest control, greater attention to organic recycling and water conservation and extensive tapping of sunlight through the agency of green plants and simple heat collectors that can help us to improve rural economy without ecological harm. We have a unique opportunity to evolve a policy for natural resource utilization based on the principles of economics and human welfare. To do this we have to learn how to deploy two of the most potent tools employed by Nature—synergy and symbiosis. Can we do this?

We can, but for this there are three pre-requisites. First, we have to shed single-discipline-oriented and one-sided thinking and develop a problem-solving and integrated approach. Integration is needed not merely in the planning process, but more urgently at the action level. Second, we must launch a techniracy (technical literacy) programme aimed at importing technical skills to illiterate peasantry. Thirdly, we must shun the 'band-wagon' and aping approach and generate a 'do' rather than a 'don't' atmosphere. We must help our farmers to apply fertilizer, pesticide, water and other inputs wisely and should not try to create confusion about the basic truth that inputs are needed for output. A 'swadeshi' approach is even more relevant today than when Gandhiji first advocated it.

## References

Brown, L.R., 1972, *World without Borders,* Random House, New York.

Garg, H.P., and A. Krishnan, 2001, *Central Arid Zone Research Institute.*

Hutchinson, J., 1972, *Farming and Food Supply,* Cambridge University Press, Cambridge, UK.

IARI, 1970, *Handbooks of Farm Irrigation Structures. IARI Research* 3:82. 1972. Recent research on multiple cropping. *IARI Research Bulletin* 8: 148.

Kanwar, J.S., 1972, 'Are Fertilisers Pollutants?', *Fertiliser News* 17: 17–22.

Kathpalia, G.N., and N.S. Varadan, 1971, 'Need for National Water Policy', A symposium on Integrated Development of Surface and Sub-surface Water Resources, New Delhi.

Meadows, D.H., D.L. Meadows, J. Randers, and W.H. Behrens, 1972, *Limits to Growth,* Universe Books, New York.

Suyin, Han, 1965, *The Crippled Tree,* Jonathan Cape, London.

Swaminathan, M.S., 1973, 'Population and Food Supply', *Yojana* 17: 7–15.

Ward, Barbara, and Rene Dubois, 1972, *Only One Earth,* Penguin Books, Harmondsworth, UK.

12

# The Green Revolution

J.R. McNeill

Mechanization dovetailed with the Green Revolution, was a crucial departure in agriculture that depended centrally on plant breeding. The Green Revolution was a technical and managerial package exported from the First World to the Third, beginning in the 1940s but making its major impact in the 1960s and 1970s. It featured new high-yielding strains of staple crops, mainly wheat, maize and rice. Plant geneticists selected these strains for their responsiveness to chemical fertilizer and irrigation water, for their resistance to pests, and eventually for their compatibility with mechanized harvesting. Success required new inputs, new management regimes, and often new machines. The great triumph came with dwarf wheat and rice, which could hold up a heavy, grain-packed head without bending or breaking the plant stalk. Like the great political revolutions of the twentieth century, the Green Revolution drew intellectually mainly from the Western world, changed its forms when it spread elsewhere, and led to unexpected consequences.

Dwarf wheat, although finally created in Mexico with American money and expertise, had deep roots anchored in the work of an Austrian monk and Japanese agronomists. The mathematical genetics of Gregor Mendel (1822–1884), lost for decades after he published his work, was rediscovered in 1900. Scientists around the world took note, including those at work on rice and wheat breeding at the agricultural research stations sponsored by Meiji Japan. There the Ministry of Agriculture and Forestry had organized crop breeding research in the 1880s. At a time when land hunger turned many peasants into emigrants and politicians into imperialists, the ministry

---

* Extracted from J.R. McNeill, 2000, *Something New Under the Sun: An Environmental History of the Twentieth-Century World,* Allen Lane, The Penguin Press, London, pp. 219–27.

sought rice and wheat breeds suitable for Japanese circumstances: scarce land, plenty of night soil. In 1925 its research succeeded, producing after many crossbreedings of Japanese and American wheat, a semi dwarf wheat known as Norin 10. (Norin was the ministry's acronym.) The ministry distributed it to farmers in 1935, but World War II came before it had much impact on Japans food supply. In 1946 an agronomist with the U.S. Army noticed Norin 10 and imported it into the United States, where it was further crossbred with wheats in the state of Washington. Norin 10 did not solve pre-War Japan's food problems, but it eventually changed the world (see Hayami 1975; Hayami and Yamada 1991).[1]

American farmers and plant breeders were hard at work on hybrid maize, hoping to concoct higher-yielding and disease-resistant strains. The great Charles Darwin had been among the first to dabble in crossbreeding maize, publishing his results in 1876. American disciples furthered the work and by 1918 it had developed the double-cross,[2] the basis for all subsequent hybrid maize. In 1930 only 1 per cent of U.S. maize acreage was sown to hybrids, but the USDA converted to the new gospel in the 1930s. By 1939 a sixth of U.S. corn was hybrid, by 1950 three-quarters, and by 1970 over 99 per cent. U.S. corn yields rose to three or four times the levels of the 1920s (Mangelsdorf 1974: 211–14; also see Fitzgerald 1990).

The first farmer to make hybrid corn a commercial success was Henry Wallace in the 1920s.[3] Wallace, later Franklin Roosevelt's Secretary of Agriculture and Vice President, took a special interest in Latin America. As a successful farmer who liked to be called the 'father of industrialized agriculture,' he saw great opportunity for applying the fruits of modern genetics to the venerable techniques of farming in Mexico and points south. He helped persuade the Rockefeller Foundation to finance a wheat and maize research centre in Mexico (1941–1943), which soon hired another son of the Iowa soil, Norman Borlaug (born 1914).[4] Borlaug had a fresh Ph.D. in plant pathology when he arrived in Mexico in 1944. By 1953 he had set to work combining Norin 10 with Mexican and American varieties. Over the years he and his associates created new wheat strains that proved enormously responsive to heavy dosages of nitrogen and timely ones of water, and which in some cases were highly resistant, at least initially, to pests and plant diseases.

Borlaug was the father of Mexico's Green Revolution. He won a Peace Prize in 1970 for his plant breeding, and has a street bearing his name in Hermosillo in northwestern Mexico, the heartland of high-yield farming in Mexico. With the help of Ford and Rockefeller money, the United Nations Food and Agriculture Organization (FAO), U.S. Agency for International Development (AID), and other organizations, the revolution spread from Mexico. Its greatest progress came in southwestern Asia's wheat belts, from the Indian Punjab to Turkey, beginning in 1963 when Borlaug sent dwarf wheats to Indian crop-breeding stations. By 1968, 18 countries sowed dwarf wheat (Lupton 1987: 68–9).

Similar developments took place in the Philippines after the 1960 birth of the International Rice Research Institute (IRRI), also sponsored by the Rockefeller Foundation (Anderson 1991). Using dwarf rice strains first selected by Japanese breeders in the 1920s in Taiwan (then a Japanese colony); rice geneticists created high-yield rice varieties that combined the best features of tropical (*Indica*) and temperate (*Japonica*) rice. By the late 1960s, new breeds from IRRI carried the Green Revolution to the world's great rice basket, the broad arc from Bengal to Java to Korea. China developed its own high-yield rice, beginning in 1959. It came too late to mitigate the great famine of 1959 to 1961, which killed some 25 million to 30 million people—a political more than an ecological event in any case—but it did arrive in time to help Chinese agriculture weather the storm of the Cultural Revolution (1966–1976) (Dalrymple 1974:10–15, 73–75).

Borlaug saw the Green Revolution as humankind's best hope to feed rapidly growing populations, as perhaps it was. But its geographic spread suggests it had other attractions as well. The Green Revolution had little impact on sub-Saharan Africa, except for higher maize yields, notably in Zimbabwe.[5] It received its greatest support outside of Mexico on the frontier of the communist world from Turkey to Korea, and recommended itself as a means to blunt the appeal of socialist revolution, at its height in the 1960s. The rice programme in particular originated from American anxieties about the possible spread of Chinese communism after 1949. Meanwhile, socialist societies—China, Vietnam, and Cuba at least—embraced the idea of scientifically improved crops with equal vigor. In several of its manifestations, then, the Green Revolution was a child of the Cold War.

The Green Revolution also appealed keenly to most of the influential segments of society within Asian and Latin American countries. It promised to augment the incomes of landed elites and, where this was an issue, make land reform less urgent. To state bureaucracies it seemed to show a way to urban industrial society, and hence to wealth and power, without the risks of alternative paths. A more efficient agriculture, particularly an export-oriented one, could build up capital needed for industrialization and at the same time get labour off the land and into factories. Achieving this without the brutal costs of Soviet methods, or the noose of huge foreign debt, made excellent sense to the influential ministries of Mexico or Indonesia. Furthermore, the Green Revolution promised independence from American food aid, which recipients normally suspected as a political tool.[6] For all these reasons, both in the United States and in Latin America and Asia, the Green Revolution in the 1960s and 1970s was a technology package whose time had come.

Its impact was sudden, substantial, and like most revolutions, not quite what its instigators had in mind. Dozens of countries managed to keep food production ahead of population growth, mainly thanks to high-yield wheat and rice. By 1970 about 10 to 15 per cent of the Third World's wheat and rice area was under the new varieties. By 1983 the proportion was over half, and by 1991 three-quarters (Tolba and El-Kholy 1992: 296; WRI 1996: 226). In China, high-yield strains accounted for 95 per cent of rice and maize by 1990.

This dissemination of new breeds amounted to the largest and fastest set of crop transfers in world history. Bountiful yields became routine between 1960 and 1990, with effects more sudden and sizable than those of history's previous agricultural turning points. From the dawn of agriculture until the seventeenth century, yields (in Europe at least) increased only about 60 to 90 per cent. The first 'agricultural revolution' beginning in England around 1680, doubled English yields within 70 to 90 years. Most other European societies followed suit. Then between 1860 and 1910, yields doubled again—much less in the United States, more in Europe. Meanwhile, outside of Europe and the USA, yields and labour productivity stagnated or declined between 1800 and 1950, contributing to the prevailing inequalities in wealth and power (Bairoch 1989). With its basis in crop breeding and the transfer of successful strains, the Green Revolution merits comparison with the great historical crop introductions, such as the arrival of American food crops (maize, potato, cassava) in Eurasia and Africa after 1492, the importation of Southeast Asian plantains into tropical Africa, and the Arab introduction of citrus and sugarcane to the Mediterranean

**Table 12.1: Yield History in 93 Developing Coutrise, Yield (kg/ha) 1961–1992**

| *Crop* | *1961–* | *1969–* | *1979–* | *1990–* | *Increase Factor* |
|---|---|---|---|---|---|
| *1963* | *1971* | *1981* | *1992* | *1962–1992* | |
| Wheat | 868 | 1.153 | 1.637 | 2,364 | 2–7 |
| Rice | 1,818 | 2,218 | 2.653 | 3.459 | 1–9 |
| Maize | 1,157 | 1,456 | 1.958 | 2.531 | 2.2 |

**Source:** Adapted from WHI 1996: 226.

world after CE 900. Table 12.1 gives an impression of the phenomenal changes in crop yields since 1960 in the Third World.

As these yield figures show, the Green Revolution created the promised fields of plenty. It did much else besides. Ecologically it combined with mechanization to promote monoculture. Since farmers now had to purchase seed rather than use their own, and because they needed fertilizers and pesticides specific to a single crop, they saved money on inputs by buying in bulk for one crop. Monoculture, as explained earlier, invites pest problems. Often even the initially pest-resistant crops eventually proved vulnerable lo one or another infestation. Hence farmers turned to heavier and heavier doses of pesticides. This efficiently selected for resistant pests—as antibiotics did for bacteria.[7] Meanwhile, most of the pesticides missed their targets and ended up elsewhere, sometimes in water supplies, human tissues, and other awkward places. The WHO estimated in 1990 that pesticide poisoning killed about 20,000 people per year, mostly in cotton fields. Roughly a million people (as of 1985) suffered acute poisoning, two-thirds of them agricultural workers (ibid.: 296; see also Pimentel and Lehman 1993). The vast fertilizer requirements of the Green Revolution led to eutrophication of lakes and rivers. The necessary irrigation helped drive the huge dam-building programmes of China, India, Mexico, and elsewhere. The Green Revolution also altered the species and genetic diversity of agriculture: it extended the sway of rice, wheat, and maize, reducing the use of lesser crops not so responsive to nitrogen-and water-rich diets: and it vastly reduced the varieties of rice, wheat, and maize in wide use. Before the Green Revolution, farmers raised thousands of strains of wheat around the world. After it, they increasingly used only a few. In this respect, the Green Revolution was a gamble that scientific agriculture could protect a few high-yield strains from pests and diseases. By and large it has, by staying one step ahead of the evolution of pests, as antibiotics did with pathogens.[8]

It was also a gamble that oil and water would remain cheap enough to satisfy the energy gluttony and bottomless thirst of the new agriculture. So far, this too has worked, within limits.[9]

The social consequences of the Green Revolution brought more surprises. In many settings it did not defuse agrarian tensions. In Mexico and the Indian Punjab, for example, the Green Revolution strongly favoured farmers with reliable access to credit and water. Some of the less fortunate drifted to the cities, some went to work for the more successful, others went to the United States or the Persian Gulf to work for wages—in cases to accumulate enough capital to become prosperous peasants—Green Revolution kulaks—themselves. As a rule, though not without exceptions, the Green Revolution promoted income inequality among farmers. Where alternative employment for the losers was hard to find, as in Punjab or highland Ethiopia, social frictions intensified into overt class and ethnic or religious conflict. A sampling of the literature suggest that the social effects proved more favourable in land raisings rice than those with wheat.[10]

The Green Revolution, like farm mechanization, selected winners internationally as well as ecologically and socially. South Korea, China, India, and to a lesser extent Mexico improved their agricultural balance of payments, reduced or eliminated food dependence, and, whatever the ecological or social costs, improved their international economic and political position. Countries that could not create favourable conditions for the Green Revolution—those with too little water or underdeveloped credit markets—suffered in comparison. Broadly speaking, this meant sub-Saharan Africa sank in the scales against Asia and Latin America as a Cold War tool of the West—which in part it was—the Green Revolution served its purpose, although high-yield rice strengthened communist China as much as it did Asia's island fringe, which America relied upon to contain China.

The Green Revolution did something—but not much—to empower Latin America and tropical Asia vis-a-vis the West and Japan. It helped in the industrialization drives of Taiwan, South Korea, Indonesia and the other 'Asian tigers'. It made India a food exporter. But while the Green Revolution made Third-World agriculture more land-and labour-efficient, it could not match the productivity increases ongoing in the West and Japan. In 1950, agriculture in the West was seven times more labour-efficient than in the Third World; in 1985, 36 times more—and about 36 times more prosperous. The Green Revolution did not engineer income redistribution toward

Third World farmers. Nor did it achieve food independence except for a few countries. Until, 1981 the Third World had long been a net exporter of food; after 1981 it was a net importer (Bairoch 1989: 346).

These facts derive from the continuing agricultural revolution in the West and Japan from which the Green Revolution was an offshoot; and from agricultural and trade policy. Take Britain, for example, a major food importer since the repeal of the Corn Laws in 1846. British agricultural yields stagnated between 1890 and 1940. Subsidies and protection began in the 1930s. Meanwhile, the biological, chemical, and mechanical transformations of modern agriculture took hold, and yields rose beginning in 1942. The difficult food situation in Britain during World War II and the immediate post-War years predisposed all governments, Labour and Conservative, to favour agricultural subsidies, firmly buttressed in 1947 and to pursue higher yields through scientific agriculture. Yields doubled or tripled by the 1980s. To the dismay of devotees of the principle of comparative advantage, Britain, which grew 30 per cent of its grain in 1936, managed self-sufficiency in 1986.[11] Similar miracles occurred in the fields of much of Europe, Japan, Australia, New Zealand, and North America after 1945.[12] The Soviet Union partly missed out because modern genetics rankled the socialist sensibilities of Stalin and Khrushchev, delaying progress in plant breeding until the mid-1960s. Before 1960 the USSR partook of mechanization and irrigation, but only after 1965 did it undertake conversion to the doctrine of genetic manipulation and heavy use of nitrogen. Thus the total geopolitical effect of modern changes in agriculture improved the relative position of the West and Japan slightly, that of China, the Asian tigers (South Korea, Taiwan, Malaysia), and Latin America even more slightly, while contributing to the relative decline of Soviet status, and to the weakness of Africa.[13]

## Conclusion

The general transformation of farming after 1940, of which mechanization and the Green Revolution were parts, both shaped the twentieth century and reflected its dominant trends. It was energy- and knowledge-intensive. It replaced simpler systems with more complex ones, involving distant inputs and multiple social and economic linkages. It reduced family and regional autonomy, enmeshing farmers in a world of banks, seed banks, plant genetics, fertilizer manufacturers, extension agents, and water bureaucrats. It transplanted what worked in the West and Japan to other societies. It sought to harness nature tightly, to make it perform to the utmost, to make it maximally subservient to humankind or at least some subset thereof. And it sharply increased output, making us dependent upon its perpetuation. As of 1996, to feed ourselves without these changes, we would have needed to find additional prime farmland equal in area to North America.[14]

Lacking such a spare continent, the human race ended the twentieth century in a rigid and uneasy bond with modern agriculture. Our recast agro-ecosystems depended on social and international stability to safeguard required flows of inputs. Our social and political systems required the perpetuation of these agroecosystems.

The modem agricultural revolution was nearly as important as the new regime in human-microbial relations in shaping the twentieth century. Both fundamentally affected the well-being, health, and security of life for billions of people. Both helped govern the ongoing redistributions of power and wealth among classes and nations. Both represented a drift toward ever-greater complexity—and potential vulnerability to disruption—in the systems that underpin modern life.

The fact that we are not more often food for microbes depends on the precarious balances of modern public health; that we in turn have as much to eat as we do (questions of distribution aside), depends on the no less precarious balances of modern agriculture. Though you drive out nature with a pitchfork, yet it will always return,' So thought the Roman poet Horace.'[15] Is his wisdom now out of date?

## Notes

1. Italian crop breeders also enjoyed early success, using Japanese varieties from 1912: by 1932 a quarter of Italian wheat—Mostly in the north— was early-ripening breeds derived from crosses with Japanese wheat. These Italian wheats by the 1970s were in wide use in the Mediterranean world. (Dalrymple 1974: 10–11)
2. In a double-cross, four varieties are combined in two 'generations,' selecting for the most desirable characteristics.

3. Wallace (1888–1965) was a farmer, publisher of an agricultural journal and son of Warren Harding's Secretary of Agriculture. Wallace became FERS Vice President in 1941, but was replaced by Truman four years later. He fell out with Truman, partly over foreign policy, and subsequently ran for the Presidency several times as leader of the Progressive Party.
4. Jennings 1988 deals with the politics of the Rockefeller initiative. Apparently Henry Wallace and the top managers of the foundation hoped to support the president elected in 1940, Avila Camacho, from whom they hoped for some compromise with respect to American property (including that of Standard Oil) nationalized by the previous president, Lizard Cardenas. Wheat was then a minor crop in Mexico. See also Fitzgerald (1986).
5. Hybrid maize research began in Southern Rhodesia (now Zimbabwe) in 1930. A variety (SR-52) released in 1947 proved successful on the commercial farms (mainly white-owned) after 1950. SB-52 raised yields in southern Africa by about 51 per cent but the expensive inputs limited its use among poorer peasants. No other Green revolution crop had significant impact in sub-Saharan Africa, see Jahnke *et al*. (1987), and also Low (1985).
6. This occurred especially after the passage in 1964 of U.S. Public which 480 which connected food aid to improving attitudes toward the United States.
7. Resistant pest species increased about four-fold between 1955 and 1988 (estimated from Tolba and El-Kholy 1992:295).
8. A partial exception camp in 1970 when a dingus especially well adapted to U.S. corn destroyed 15 per cent of the harvest, prompting farmers to vary the genetic profile of their seed in 1971(Mangelsdorf 1974:213).
9. See Freeman (1993) for the limits as seen by a booster (a former U.S. Secretary of Agriculture). Pimentel and Heichel (1991) show that in energy terms high-yield crops are about 25 per cent as efficient as hand-and-hoe agriculture and half as efficient as ox-and-plow farming. Indeed, modern U.S. farming burns far more calories than it produces when one factors in the energy requirements of making fertilizer.
10. Critiques of the social implications of the Green Revolution are numerous. See Shiva 1991; Thandi 1994 (on Punjab); Hazell and Ramasamy 1991 (on Tamil Nadu, where things went well); Allaudin and Tisdell 1991 (on Bangladesh, where they did not); Simonian 1995:170–72 and Sonnenfield 1992 (on Mexico). On income distribution specifically, see Sharma and Poleman 1993 and David and Otsuka 1994. Bezuneh and Mabbs-Zeno 1984 write on Ethopia's Green Revolution and its impetus to unrest and political revolution in the 1970s.
11. Blaxter and Robertson (1995). The hallowed principle of comparative advantage states that countries maximize their income by producing only

what they produce most efficiently and trading for all other goods. This Britain resolutely did not preferring food security to income maximization.

12. See Cochrame (1993); Hayami and Yamada (1991). In the century 1880–1985, Japanese agriculture increased labour productivity 16-fold despite the requirements of rice. Land productivity (yields) improved by a factor of 4.4 (Hayami and Yamada 1991:253–54).
13. In addition to the works cited above, I consulted Burmeister (1990), and Freeman (1993).
14. This figure comes from Dennis Avery, cited in the *Economist*, 16 July 1996: 23. Other solutions include, of course, more equitable distribution of food and an alternative technical means of making food production more land efficient.
15. From Epistles 10:24: 'Naturam expellas furca tamen usque recurret.'

## References

Allaudin, Mohammad and Clement Tisdell, 1991, *The Green Revolution and Economic Development: The Process and Its Impact in Bangladesh*, St. Martin's Press, New York.

Anderson, Robert S., 1991, 'The Origins of the International Rice Research Institute,' *Minerva* 29: 61–89.

Bairoch, Paul, 1989, 'Les trois révolutions agricoles du monde développé: rendements et productivité de 1800 à 1985,' *Annales: Economies, sociétés, civilisations* 44(2): 317–53.

Benzuneh, Mesfin, and Carl C. Mabbs-Zeno, 1984, 'The Contribution of the Green Revolution to Social Change in Ethiopia,' *Northeast African Studies* 6(3): 9–17.

Blaxter, Kenneth and Noel Robertson, 1995, *From Dearth to Plenty: The Modern Revolution in Food Production*, Cambridge University Press, Cambridge.

Burmeister, Larry, 1990, 'State, Industrialization and Agricultural Policy in Korea,' *Development and Change* 21: 197–220.

Cochrane, Willard, 1993, *The Development of American Agriculture: An Historical Perspective*, University of Minnesota Press, Minneapolis.

Dalrymple, Dana G., 1974, *Development and Spread of High-Yielding Varieties of Wheat and Rice in the Less Developed Nations*, USDA, Washington.

David, Christina, and Keijiro Otsuka (ed.), 1994, *Modern Rice Technology and Income Distribution in Asia* (Boulder: Lynne Rienner).

Fitzgerald, Deborah, 1986, 'Exporting American Agriculture: The Rockefeller Foundation in Mexico', *Social Studies of Science* 16: 457–83.

Fitzgerald, Deborah, 1990, *The Business of Breeding: Hybrid Corn in Illinois,* 1890–1940, Cornell University Press, Ithaca.

Freeman, O., 1993, 'Perspectives and Prospects', in Douglas Helms and Douglas Bowers (eds), *The History of Agriculture and the Environment*, 'Washington: Agricultural History Society', pp. 3–11.

Hayami, Yujiro, 1975, *A Century of Agricultural Growth in Japan*, University of Tokyo Press, Tokyo.

Hayami, Yujiro, and Saburo Yamada, 1991, *The Agricultural Development of Japan,* University of Tokyo Press: Tokyo.

Hazell, Peter B.R., and C. Ramasamy, 1991, *The Green Revolution Reconsidered: The Impact of High-yielding Rice Varieties in South India*, Johns Hopkins University Press, Baltimore.

Jahnke, Hans, Dieter Kirschke, and Johannes Lagemann, 1987, *The Impact of Agricultural Research on Tropical Africa*, World Bank, Washington.

Low, Allan, 1985, *Agricultural Development in Southern Africa*, James Currey, London).

Lupton, F.G.H., 1987, 'History of Wheat Breeding', in F.G.H. Lupton (ed.), *Wheat Breeding*, Chapman & Hall, London, pp. 51–70.

Mangelsdorf, Paul C., 1974, *Corn: Its Origin, Evolution, and Improvement*, Belknap Press, Cambridge.

Pimentel, David, and Hugh Lehman (eds.), 1993, *The Pesticide Question: Environment, Economics, and Ethics*, Chapman & Hall, New York.

Sharma, Rita, and Thomas T. Poleman (eds) 1993, *The New Economics of India's Green Revolution: Income and Employment Diffusion in Uttar Pradesh*, Cornell University Press, Ithaca.

Shiva, Vandana, 1991, *The Violence of the Green Revolution*, Zed Books, London.

Simonian, Lane, 1995, *Defending the Jaguar: A History of Conservation in Mexico*, University of Texas Press, Austin.

Sonnenfeld, David A., 1992, 'Mexico's Green Revolution' 1940–1980: Toward an Environmental History', *Environmental History Review* 16(4): 28–52.

Thandi, Shinder Singh, 1994, 'Strengthening Capitalist Agriculture: The Impact of Overseas Remittances in Rural Central Punjab in the 1970s', *International Journal of Punjab Studies* 1: 239–70.

Tolba, Mostafa K., and Osama A. El-Kholy (eds), 1992, *The World Environment,* 1972–1992, Chapman & Hall, London.

WRI (World Resources Institute), 1996, *World Resources* 1996–1997, OUP, New York.

# Rehabilitating Degraded Lands

N.C. Saxena

More than two-thirds of India's population is dependant on land. It is significant that this percentage has remained almost unchanged in the last 80 years; the reason is not far to seek. Employment generation in the secondary and tertiary sectors of the economy is unable to absorb even the additional urban labour force. Therefore, scope for the rural work force finding meaningful employment in the urban sector is severely limited. With an increasing capital output ratio, the income elasticity of employment in the non-agricultural sector keeps declining, making it more difficult for the urban sector to absorb the extra manpower that is intended to be released from land.

During, the last three decades three programmes have been tried on a large scale to help alleviate rural poverty—land reforms, the Integrated Rural Development Programme (focusing on self employment in the non-agricultural sector) and wage employment programmes. These were not generally targeted to improve productivity of marginal lands in rainfed areas and, therefore, did not show a sustained increase in rural incomes. On the other hand, due to soil and water run-off, the health of the two most important resources in rural India—land and water—has fast declined. This paper traces the background of the present approach to waste-land development, and suggests some measures to improve the sustainability of the current programme.

According to the nine-fold land classification, out of 304 million hectares of land in India for which records are available, roughly 40 million hectares arc considered totally unfit for vegetation. It is either urban and under

* Reprinted with permission from *Seminar* (499), March 2001. The views expressed are personal and do not bind the organization for which the author works or has worked in the past.

there non-agricultural uses such as roads and rivers, or is under permanent now, rocks and deserts. The break-up of the remaining 264 million hectares of land that is fit for vegetation is as follows:

| *Land type* | *Area in million hectares* |
|---|---|
| Cultivated land | 142 |
| Forest land | 67 |
| Fallows/culturable wastes/ pastures/groves | 55 |
| Total area of culturable lands | 264 |

The above land use classification, however, does not say anything about the extent of land degradation or loss in (productivity overtime due to various natural and man-made causes. Not only are culturable wastes and pastures considered highly degraded—that is, producing biomass much below their potential—even a substantial part of cultivated and forest lands have lost their productivity due to inappropriate land use and over exploitation

The National Commission on Agriculture estimated that 175 million ha of land was under some form of degradation and was in need of attention. According to the commission, all rainfed paddy lands in the country were subject to water and wind erosion and thus in the 175 million ha the commission included 85 mha of cultivated land too; of the remaining 90 mha there would be (35 mha) of degraded and barren forest lands and the rest 55 mha would constitute common and revenue lands.

Avital piece of information about on-forest wastelands pertains to its ownership. There are three obvious categories: private, community and government. In addition to uncultivated lands which were historically part of the farmers' holding, especially in ryotwari semi-arid areas, many poor families have been allotted some 6 mha of wastelands under various programmes over the last 20 years. Thus, substantial culturable waste area has been privatised as a conscious policy outcome, although such lands may still be lying uncultivated. Besides, there are encroachments, mostly unrecorded.

A large part of this land may be suitable for growing grasses, shrubs or trees, but not crops. The same is true of government wastes that are owned by the government but used by the community, and grazing lands which are generally vested in village bodies. There is little *de facto* distinction between the two categories, as both are used for grazing and are generally quite degraded. These are also referred to as common or village lands. There is much regional variation as well as variation between neighbouring villages. In villages of intensive cultivation, common lands are of marginal importance, but in hilly and unirrigated villages common lands still offer livelihood possibilities for the poor.

Village lands have generally been a victim of the 'tragedy of the commons' phenomenon where they are exploited by all, without anyone considering himself responsible for maintenance. The government with respect to these lands followed a *laissez faire* policy. Neither were any funds allocated for them, nor was any specific government department made responsible for grasses and pasture development. As these lands could not meet peoples' needs, the biotic pressure on forests increased, leading to degradation of forests too. By the mid-seventies it became evident that if peoples' demands for fuel wood and fodder were not met, it would be impossible to even save productive forests. This was then sought to be achieved through a programme of social forestry on village and private lands. The philosophy around afforestation of common lands in the social forestry programme was that these lands were to be brought under the forest department's control, in the hope that the trees would be later handed over to the village communities for protection and management.

However, the forest department (FD) could not look after small patches scattered over hundreds of villages in a district, creating enormous problems of protection leading to high mortality. The continued involvement of FD in the initial years discouraged local bodies from taking over, as a result extending FD management. Besides, government failed to define, establish and publicise the rights of the people to the trees and the procedures for marketing and allocating benefits. The shares that would accrue to the individuals, village, panchayat and FD were not clearly laid down. Insecurity about benefits led to indifference on behalf of the people.

Thus, in perspective it appears that the two policies on wastelands—the land distribution policy of the 1970s and the social forestry of the 1980s—seem to have been influenced by Hardin's ideas that there are only two sustainable solutions: either the commons should be privatized, or they should be brought under the control of a coercive slate authority. A third model, of community control over commons and over the programme for their regeneration, was not given any thought in these models. As discussed later in this paper, the experience of the last 10 years shows that community control is the only option that generates sustained benefits, both social and private.

In 1985, the Government of India set up a National Wastelands Development Board to promote peoples' participation in afforestation programmes and to regenerate the health of India's wastelands. A target of afforestation of five million hectares a year was set up for the board. Later, a Department of Wastelands Development was created in 1992. New structures, however, did not mean new policies. The same old approach of bringing lands under government domain continued.

The apartment of Wastelands Development, while issuing sanctions, insisted that the government agency involved with plantations should have complete control over lands where trees were planted. Even for farmlands, the guidelines stipulated that the executing agency produce a certificate from land owners that they had authorised the agency to execute works on their lands. Thus the people, (whether on public wastelands or on their own lands, were supposed to be mere spectators; they had no role in planning or execution of the programme.

Second, the problem of degeneration of village wastelands was seen as arising from peoples' demand for fuel wood and fodder resulting in lack of tree cover. Therefore, the programme concentrated on production of fuel wood. Actually the loss of soil from such lands was linked more to lack of control of run-off of rainwater. It was not so much a demand problem as that of land management, especially common lands. Without controlling run-off it was not possible to stabilize water regimes even for crop lands, and it was a mistake to look at the degradation of common lands in isolation to the other issues of low productivity of crop lands in rainfed areas. Thus, it was difficult to rehabilitate degraded lands without introducing moisture conservation and water harvesting measures. Such measures are needed for all rainfed areas put to biomass production.

The main thrust of the programmes should have been on activities relating to soil conservation, land shaping and development, pasture development, and water resources conservation for the entire watershed, not merely afforestation on wastelands. Even when such schemes were undertaken by the agriculture department, these remained departmental in approach with no involvement of the people. Maintenance of the created assets suffered, as beneficiaries were not motivated to assume responsibility for maintenance. Besides, these were implemented in an isolated and segmented manner and watershed as a unit of area development was totally lost sight of.

It is estimated that up to the end of the Eighth Plan about 16.5 mha rainfed/degraded land had been treated/developed. However, these achievements are not reflected in the data for net sown area, which has remained almost stagnant at around 142 mha for the last 30 years. This indicates that either the treated lands were already under cultivation or an equal area is getting degraded or diverted for non agriculture purposes. The possibility of bogus reporting also cannot be ruled out.

The 1999 Report of the Comptroller and Auditor General states that despite spending Rs 2,195 crore on the Drought Prone Area Programme (DPAP) between 1973 and 1995 the drought prone area increased from 55.3 mha to 74.6 mha in the same period. There was insufficient evaluation of the quality of works. And even run-offs not measured, with the result that survival rate of plantations was very low, and structures to promote water and soil conservation were not maintained.

The evaluation reports have shown that watershed development programmes cannot succeed without full participation of project beneficiaries and careful attention to issues of social organization. This is because success depends on consensus among a large number of users. The costs and benefits of watershed interventions are location-specific and unevenly distributed among the people affected. Moreover, collective capability is required for management of commons and for new structures created during the project.

These shortcomings were taken care of, at least on paper, with the acceptance of new guidelines on watershed development based on the Hanumantha Rao Committee report with effect from April 1995. It provides for the development of an entire compact micro-watershed rather than pieces of wastelands scattered at different places. The strength of the guidelines lies in the decentralization of decision-making process by involving local panchayati raj institutions and people at grassroot level. It aims at establishing a system under which village people can actually involve themselves in the planning, implementation and monitoring of watershed development programme.

In the preparation of the watershed development plan, user groups and other people depending directly on the watershed have to be actively involved. Another strength of these guidelines lies on the flexible approach followed in the method of release of funds, the area to be covered in each watershed, as well as choice of components. Besides this, the new guidelines also attempt at making these projects sustainable by establishing a watershed development fund and involving people in deciding equity issues and usufruct sharing mechanisms.

Technical assistance was to be provided by a multi-disciplinary team comprising experts from various departments and sectors to the programme implementing agencies (PIAs), which could be a line department, a NGO, a panchayat or a user group. A well-designed training programme was given top priority in order to create awareness among functionaries.

Progressive guidelines, however, do not always lead to desired action. The field staff is not trained to solicit participation, which continues to be 'an ideology without methodology', without a road map in sight. Disregard of peoples' participation is a legacy of the 'PWD culture' governing civil works. It also represents a continuation of the colonial approach to conservation, where community stakes in resources, their knowledge and management systems do not count. The state's appropriation of community resources and dismantling of traditional management systems for CPRs and the promotion of conservation-technique-dominated, top down, government-supported initiatives are other factors which marginalised the communities and their stake in the resources.

Local collective action in the last 30 years has also been undermined by a number of political and economic processes. Village societies have become heterogeneous, and market forces have commercialized the erstwhile subsistence economies integrating them with urban and national economies. Possibilities for migration and mobility tend to work against cooperation. Moreover, anti-poverty programmes such as the Indira Awaas Yojana increase the dependence of the poor on the village elite and petty bureaucracy and at the same time make them compete within themselves for limited favours from government. This adversely affects the sustainability of people coming together for a common cause.

Cooperation works best in small groups with similarity of needs and clear boundaries, and shared norms and patterns of reciprocity. Competitive politics erodes the traditional authority structures, and modernization improves the options of both exit and voice for the common people. As old authority structures crumble, appeals to government for conflict resolution and arbitration become more common, and dependence on government for local resource management increases. Many rural communities in developing countries are now in this difficult transition period, with traditional institutions on the decline while new self-governing institutions are yet to be born.

Although five years is too short a time to evaluate the new guidelines, yet limited evidence suggests that most projects have failed to generate sustainability. A major study was done for the ICAR in 1998 (reported in J. Farrington edited *Participatory Watershed Development*, OUP, 1999) of 70 villages in Maharashtra and A.P., covering several watershed programmes. The survey revealed that increase in agricultural production did not last for more than two years. Structures were abandoned because of lack of maintenance and there was no mechanism for looking after common lands. Farmers were not convened about the need to contribute, which would happen only when they make the decisions about what kind of measures are introduced on their plots.

Often, in government projects, farmers do not have this choice and technological norms are too inflexible. The very fact that farmers are unwilling to contribute towards the cost of works shows that they have little faith in the effectiveness of the programme. Except for the villages where NGOs were active, all other programmes scored poorly. In many cases performance in control villages, which had no investment, was better. Most government watershed development investments have yielded disappointing results given the vast resources allocated to date.

For watershed projects to be sustainable, community managed systems are needed and they can succeed only with farmers' contribution and their commitment of time and resources. Although the common guidelines prescribe that those benefiting from works on private land should make a contribution of 10 per cent of the total cost and 5 per cent for works on common lands, in practise it has been difficult to either collect these charges in full or recovery is often shown on paper but adjusted by the contractor in his bill.

For sustainability it is important that the contribution from farmers be a necessary commitment before the start of a project, for that would ensure a sense of ownership leading to better quality and transparency and assured money for maintenance. It may be pointed out here that Myrada, an NGO working in southern India, insists that all works on private lands should be fully financed by the individuals concerned.

Watershed development projects are being implemented by several departments of the Government of India, often with different guidelines. This causes confusion and brings a bad name to the GOI because state governments get the impression that there is no effective coordination between different central departments. Even when the approach or guidelines are common, funds are sanctioned by different departments, and each does its own separate monitoring.

Other reasons for poor implementation of watershed projects are:

- Insecurity about availability of funding at the grassroot level, as there is no guarantee that funds would be released in time by the GOI or other funding agencies. Pressure to spend available funds by a fixed deadline so that more funds could be demanded.
- Limited time permitted for preparatory and group formation activities. A strict orientation to achieving physical targets discourages field staff from taking time to promote social organisation. Field staff has no incentive to make the effort to pursue participatory approaches.
- There is little impact assessment or evaluation of physical progress after the scheme has run for a couple of years. It is taken for granted that once money has been spent, physical progress automatically results. It is quite likely that soil conservation structures may not last for more than a few years, or plantations may not survive.
- Unclear criteria for selecting areas and villages or for withdrawing from a village.
- Limited human resource capabilities. Government staff have some times subcontracted all work related to participatory processes to NGOs without developing any internal capacity.
- As funds from the Ministry of Rural Development go straight to the districts, there is little involvement of 22 senior state government officials and line agencies. Watershed development programmes require a comprehensive and integrated approach involving several line departments and, therefore, the full involvement of state governments is essential.
- In the present form, schemes are left to be planned and executed by district level officers. The capability of district level officers to plan and write a good project is extremely limited. Moreover, once it is

realized that senior officers from the divisional and state capitals take no interest in such schemes or their interest is limited to monitoring of financial expenditure alone, the quality of project preparation and of implementation suffers a great deal.

- Horizontal linkages between various line agencies at the district level are very weak. Thus, although water shed development may require integration of soil conservation techniques with plantation, there is little likelihood of effective coordination between the district soil conservation officer and the district forest officer. The tendency in Indian administration is to look up to seniors rather than establish linkages with officers in other departments at an equal level.
- There is no arrangement for handing over of structures and maintenance of plantation after a project is completed. Therefore, sustainability of projects gets impaired.

It should be stressed here that watershed development programmes should only indirectly address the problem of poverty or unemployment. It should be aimed at increasing or stabilising the carrying capacity of land and water resources in rainfed areas. As poverty is both a cause and effect of over-exploitation of natural resources, successful implementation of watershed development programmes would result in sustainable reduction in poverty. On the other hand if production is not emphasised as the goal, one may end up by achieving neither reduction in poverty nor employment.

At the same time, given that development of agriculture on a watershed basis would be biased in favour of those who own land, it is important to ensure that the landless do not suffer in contrast. Access to land through *pattas* on common property resources, equitable distribution of grazing opportunities, usufructory rights on forest produce and development of non-farm employment would have to be ensured in order that equity considerations are not lost sight. Furthermore, the interface of the watershed associations/committees with the panchayati raj institutions, particularly the gram panchayats, is tenuous. Of course, members of the watershed association are also members of the gram sabha, but they would have to be made accountable to both the gram sabha and the gram panchayat.

A new Department of Land Resources has recently been created in April 1999 by merging the schemes of area development, such as DPAP, DDP (Desert Development Programme) and watershed development/soil conservation/social forestry part of the Employment Assurance Scheme with the present Department of Wastelands Development. In order to ensure that past mistakes are not repeated, the new department would have to lay greater emphasis on performance. Capacity building of grassroot organization in planning, monitoring, implementation and marketing should be the future strategy. Other features of the new approach should be:

- There should be constant monitoring, evaluation, impact assessment by external experts. NABARD, MANAGE, NIRD, among others, should be involved in this exercise.
- Funds in the first stage should be given for those states that set up organizational structures to properly appraise and evaluate such projects.
- High priority should be given to rejuvenation of village ponds and tanks, and recharge of groundwater.
- There should be integration of all area development with the felt needs of the people, such as drinking water and credit.
- Panchayats should be involved and their involvement should include transfer of funds to such village level bodies, including user groups, who would handle funds and the job of government agencies should be to facilitate and train rather than control funds. Where panchayats represent several villages, single village organizations, as sub-units of panchayats should be created, so that land in question is appurtenants to one village only to avoid conflict.
- The ownership and control over revenue wastelands should be transferred to PRIs and village organizations to ensure certainty of tenure. In ex-ryotwari states, transfer of revenue lands to panchayats has still not been done, which makes it problematic for the village panchayats to 'own up' efforts on such lands.
- Since cultivable wastelands in many states have already been settled with the poor, special projects should be undertaken to make such lands productive. Private ownership will help in sustained increase in land productivity.

In short, the objective of all land based intervention should be, 'to enable rural people in rainfed regions to prevent, arrest and re verse degradation of life support systems, particularly land and water, so as to produce biomass in a sustainable and equitable manner.' It is important to look at forest lands, non-forest wastelands and crop lands in an integrated manner. This is often not done, as upstream treatment to reduce soil movement does not benefit large farmers who are downstream. They see no advantage and are indifferent or opposed to this strategy. They would prefer to conserve and harvest water in the drainage line so that it can be used directly for irrigation or to replenish groundwater. However, lands in the upper catchment should be rehabilitated first for at least three reasons. To benefit the landless and the poor who depend on the upper slopes; so that groundwater recharge begins at the earliest; and by the time the lower catchment is treated any debris and erosion running down from the upper catchment has been minimized. Upper slopes are typically under the control of the forest department, which does not permit other departments to operate on its lands. The Ministry of Rural Development has recently permitted its funds to be used in watershed schemes by the FD, but a similar initiative is needed from the Ministry of Agriculture too. One of the least understood but most useful concepts is the issue of complementarity between forests and agriculture. If it is strengthened, the local community develops a stake in the preservation of forests, which can deter individual attempts at encroachments or degradation. Also, at present, the three life support systems, i.e. land, water and forests, remain unintegrated administratively and management-wise. Therefore, the government should strive towards an integrated planning approach at the village level through peoples' participation. It is only by linking the future of forests and uncultivated lands with crop lands and groundwater recharge that will ensure the sustainability of government efforts.

# 14

# Common Property Resources and the Rural Poor

N.S. Jodha

## Introduction

An important factor completely disregarded by development policies and programmes in India is the role of common property resources (CPRs) in the economy of rural people, particularly of the rural poor. CPRs, broadly speaking are the resources accessible to the whole community of a village and to which no individual has exclusive property right. In the dry regions of India, they include village pastures, community forests, waste lands, common threshing grounds, waste dumping places, watershed drainages, village ponds, tanks, rivers/rivulets, and riverbeds, etc.[1] Despite their significant contributions to the economy of rural people, these resources have seldom received enough attention from planners. The important reasons are: (i) Rural development planning has largely emphasized private property resource (PPR) centred activities, be it the promotion of high-yielding crop varieties, or distribution of cross-bred cattle, or supply of electricity for ground-water lifting devices;[2] (ii) Inadequate understanding of the survival mechanisms used by the poor as well as the complementarities between CPRs and PPR-based activities in rural areas; (iii) The contributions of CPRs are not only numerous and varied but they are often available as a matter of routine. Consequently, they often go unnoticed by rural researchers as well as by development planners. This paper, based on household and village data from 21 dry tropical districts in seven states in India, attempts to quantify some of the contributions of CPRs with special reference to the rural poor whose dependence on CPRs (excluding water-based CPRs used

* Reprinted from N.S. Jodha, 'Common Property Resources and Rural Poor in Dry Regions of India', *Economic and Political Weekly*, Vol. XXI, No. 27, 5 July 1986.

through irrigation) is greater than that of the well-to-do villagers. The paper also examines the changing status of CPRs, especially its impact on the rural poor.

After commenting on the methods used for selection of areas and collection of field data in Section II, the paper briefly describes in Section III general benefits accruing to the rural people from CPRs. Section IV presents evidence on rural households' dependence on CPRs, as well as on employment and income generated by CPRs. Section V deals with the changes in the status of CPRs, particularly the decline in the areas of selected CPRs, and how it has affected the rural poor.

## II

## Method and Material

This paper is a part of a larger study on CPRs initiated in 1982 to understand their role in the farming systems in the dry tropical regions of India. That investigation, besides indicating strong complementarities between CPRs and the PPR-based farming systems, revealed a number of important issues that are relevant to both anti-poverty programmes and strategies for development and management of waste lands in the dry regions. The present paper, however, is addressed to the specific theme of CPRs and the rural poor.

The study covers 21 districts in dry tropical regions spread over seven states. Table 14.1 provides additional information about the coverage of the study. The areas were selected purposively. Two considerations that had to be satisfied were: (i) representation of zones with different soils, agro-climatic features, and population densities; and (ii) availability of local co-operators to help in the field work. The latter was a logistic requirement imposed by the nature of the study. The study of CPRs, unlike routine agro-economic surveys, required greater flexibility and use of unconventional methods for gathering information. This involved greater reliance on investigators who have lived for a long time in the study areas and had knowledge of the local language as well as oral history of the villages. Pre-existing close contacts with the villagers and an ability to record participant observations along with other numerical data were other attributes of the potential collaborators in the field work. The identification of relevant co-operators (investigators) from different agroclimatic zones therefore preceded the purposive selection of study areas.[3]

Information on physical features and population pressure and on the extent of the CPR lands in the study villages is given in Table 14.1[4] In the study villages, the proportion of CPR land to total village area ranged from 9 to 28 per cent (Table 14.1). The extent of CPRs is affected by several factors including physical conditions, institutional environment, and demographic pressures. Though there is not a one-to-one correspondence between lower population pressure on land and higher extent of CPR lands, Table 14.1 indicates that in the villages of 10 districts where proportion of CPRs to village area was 15 per cent or higher, the population densities were also lower, ranging from 73 persons per sq km to 115 persons per sq km.

Respondent households for the study were selected in the following way. First, the households were stratified into three groups: (i) landless labourers and small farmers (<2 ha of dryland equivalent); (ii) large farmers (i.e., top 25 per cent of the households on the basis of land they owned in each village); and (iii) medium farmers (i.e., other than those covered by [i] and [ii] above). Final selection of the sample household was made randomly from these categories. The respondents from category (i) are described as 'poor' and those from category; (ii) as 'others' throughout the study unless mentioned otherwise.[5] On some specific aspects such as monitoring of grazing on CPRs, and acquisition or disposal of privatized CPR lands, the selection of households was dictated by other factors that will be described while discussing these aspects. For detailed case histories, households were selected purposively. The procedures used for collecting specific data will also be described while discussing the results.

Information collected for CPR studies was supplemented by more detailed data from ICRISAT's village level studies (VLS) that have been conducted in 10 villages in four states. Under the VLS, resident investigators collected information on various aspects of traditional farming systems from panel households regularly with an interval of 20 to 30 days (Binswanger and Jodha 1977; Jodha *et al.* 1977; Singh and Singh 1982; Singh *et al.* 1984).

## III

## CPRs and Their Contributions to Rural Economy

To understand the significant role played by CPRs in the life and economy of rural people, we briefly highlight the general benefits offered by CPRs.

The way in which these benefits and contributions emanate from different CPRs, is elaborated by me (1985a) and are briefly indicated in Table 14.2.

(a) Through supply of fodder and grazing space some CPRs help individuals in saving their land for crops. Put differently, these CPRs help in sustaining a number of animals for draft and livestock production which would not have been permitted on land owned by individuals, especially for small farmers.

(b) CPRs like dry beds of rivers/tanks used for off-season cropping and rivulets, tanks used to collect irrigation water, play an important resource-augmenting role in the PPR-based farming system.

(c) The traditional farming systems in dry areas partly derive their stability and viability through an integrated production strategy involving crops, livestock, trees/bushes, the latter being less sensitive to temporal variability of rains. Village forests, grazing lands, rivulets and watershed drainages play a significant role in this strategy.

(d) CPRs, as sources of both physical supplies as well as employment and income, cushion dryland farmers' welfare during crisis periods, e.g., droughts.

(e) Rural inequalities generated by the PPR-based farming system are partly reduced by CPRs as the resource-poor households (unlike the rich), significantly supplement their income from CPRs. CPRs also greatly contribute to the poor man's nutrition by facilitating his food gathering from forests, ponds, and other sources and strengthening his self-provisioning system.

(f) Management of CPRs should be an integral component of resource-centred prospective technologies and development strategies directed at conservation and better use of waste lands and the environment in the dry regions.

## IV

## CPRs and Rural Poor

Before initiating the discussion, a few clarifications are in order. First, ground-water as a CPR is not covered by this study at any stage. Second, as far as the data regarding dependence on employment and income from

**TABLE 14.1: THE LOCATION AND AGRO-CLIMATE FEATURES OF THE AREAS COVERED BY THE CPR STUDY[1]**

| State/District | Taluka/Tehsil | No of Villages | Rainfall[2] (mm) | Soil Type[3] | Proportion of CPRs to Total Area of Villages (Per cent) | Density of Rural Population[4]: Persons/sq km of Total Area in Taluka | Density of Rural Population[4]: Persons/sq km of Total Area in Village | Density of Rural Population[4]: Person/10ha of CPRs Area in the Village |
|---|---|---|---|---|---|---|---|---|
| Andhra Pradesh | | | | | | | | |
| Anantapur | Anantapur | 2 | 563 | A-L,G | 15 | 90 | 106 | 71 |
| Mahbubnagar | Athmakur, Kalvakurthi, Mahabubnagar | 5 | 721 | A-D, A-L | 9 | 124 | 162 | 186 |
| Medak | Medak | 3 | 834 | A-D, V-D | 11 | 117 | 158 | 145 |
| Gujarat | | | | | | | | |
| Banaskantha | Kankrej, Vadgam | 5 | 655 | S | 10 | 210 | 201 | 205 |
| Mehsana | Sidhpur, Vijapur | 5 | 633 | S | 11 | 340 | 332 | 301 |
| Sabarkantha | Prantij | 5 | 739 | S, V-M | 12 | 238 | 253 | 208 |
| Karnataka | | | | | | | | |
| Bidar | Bhalki, Bidar, Humnabad | 3 | 907 | A-M, A-D | 12 | 162 | 137 | 113 |
| Dharwar | Kalghatgi | 3 | 691 | A-M | 10 | 134 | 156 | 154 |
| Gulbarga | Gulbarga | 3 | 702 | V-M, A-D | 9 | 106 | 129 | 146 |
| Mysore | Gundulupet | 3 | 680 | A-M | 18 | 106 | 103 | 55 |

(continued to next page)

| State/District | Taluka/Tehsil | No of Villages | Rainfall[2] mm | Soil Type[3] | Density of Rural Population[4]: Proportion of CPRs to Total Area of Villages (Per cent) | Persons/$Km^2$ of Total Area in | | Person/10ha of CPRs Area in the Village |
|---|---|---|---|---|---|---|---|---|
| | | | | | | Taluka | Village | |
| Madhya Pradesh | | | | | | | | |
| Mandsaur | Mandsaur | 4 | 847 | V-M | 22 | 142 | 116 | 51 |
| Raisen | Gairatgunj | 6 | 1181 | V-M | 23 | 82 | 91 | 41 |
| Vidisha | Vidisha | 4 | 1134 | V-M, V-L | 28 | 91 | 98 | 38 |
| Mahrashtra | | | | | | | | |
| Akola | Mangrulpir, Murtizapur | 5 | 840 | V-M | 11 | 116 | 145 | 130 |
| Aurangabad | Aurangabad | 4 | 727 | V-M, G | 15 | 128 | 114 | 76 |
| Sholapur | Akkalkot, Mohol, Sholapur | 4 | 667 | V-D, V-M | 19 | 122 | 111 | 59 |
| Rajasthan | | | | | | | | |
| Jalore | Sanchor, Bhinmal | 5 | 421 | S | 18 | 67 | 98 | 54 |
| Jodhpur | Jodhpur | 3 | 319 | S | 16 | 68 | 81 | 50 |
| Nagaur | Jayal | 3 | 389 | S, G | 15 | 71 | 73 | 48 |
| Tamil Nadu | | | | | | | | |
| Coimbatore | Coimbatore, Palladam | 4 | 718 | A-L, S | 9 | 250 | 361 | 402 |
| Dharmapuri | Dharmapuri, Pennagaram | 3 | 844 | A-L, S | 12 | 286 | 210 | 169 |

**Notes:** 1. Based on district, taluka and village records and field work in the villages during 1982–85.

2. Average annual rainfall of the nearest rain gauge stations of the study villages.

3. Soil types: S = Sandy and/or sandy loam, G = Gravelly, A = Alfisol (red soils), V = Vertisol (black soils), D = Deep, M = Medium deep, L = Shallow.

4. Population and area data relate to 1981.

**Table 14.2: Benefits from Village Common Property Resources**[1]

| *Benefits* | *Contributions to Farming Systems*[2] | *CPRs*[3] | | | | | | |
|---|---|---|---|---|---|---|---|---|
| | | *A* | *B* | *C* | *D* | *E* | *F* | *G* |
| Physical Products: | | | | | | | | |
| Food | e,c | ✓ | | ✓ | | | | |
| Fodder | a, c, d, e | ✓ | ✓ | | | ✓ | | |
| Fuel/timber, etc | a.e. | ✓ | ✓ | ✓ | | ✓ | ✓ | |
| Water | b, c, d | | | | ✓ | ✓ | | ✓ |
| Manure/silt/space | b, c | ✓ | ✓ | ✓ | | | ✓ | |
| Supplement income/ employment: | | | | | | | | |
| Off-season activities | e, d | ✓ | | | | | ✓ | ✓ |
| Drought period sustenance | d | ✓ | ✓ | | | | ✓ | |
| Additional | b,d | | | ✓ | ✓ | | ✓ | ✓ |
| Petty trading/handicrafts | d,c | ✓ | | | | | ✓ | |
| Social gains: | | | | | | | | |
| Resource conversion | f | ✓ | ✓ | | | | | |
| Drainage/recharge of Groundwater | f, d, b | | | ✓ | ✓ | ✓ | | |
| Sustenance of poor | e | | | | | | | |
| Stability of farming sysems | – | ✓ | ✓ | ✓ | | | | ✓ |
| Renewable resource supply | a, b | ✓ | ✓ | ✓ | | | | ✓ |
| Better micro-climate/ environment | d | ✓ | ✓ | | ✓ | | | |

**Notes:** 1. Table adopted from Jodha (1985a).

2. The benefits can be regarded as contributions to PPR-based farming systems and can be classified under categories (a) Resource saving for PPR-based farming systems; (b) Resource augmentation for farming systems; (c) Fuller use of environment; (d) Seasonal buffer and stability of farming systems; (e) Rural equities and nutrition; and (f) Importance in resource-centred technology.

3. CPRs: (A) Community forest; (B) Pasture/waste land; (C) Pond/tank; (D) river/rivulet; (E) Watershed drainage/river banks; (F) River/tank beds; (G) Groundwater.

CPRs are concerned all CPRs (except ground-water) are considered. However, while discussing the status of CPR's (e.g, decline in their area, extent of privatization, etc) only village forests, pastures, waste lands, catchment as well as filling areas of ponds, river banks, and watershed drainages are covered. The 'rural poor' for this study, as indicated in the last section, includes landless labourers and small farm (< 2 ha of dryland equivalent) households. The dependence of the rural poor on CPRs is greater than others. To highlight this fact, the rural poor's situation has been compared with that of the large farm households, indicated by 'others' under each Table. In the dry regions the large farm households too may not be very rich, but they are certainly better off than the 'poor'. The information is presented by district along with the number of villages as well as number of households wherever necessary.

## Rural Poor's Dependence on CPRs

Dependence of a community on CPRs can be indicated in several ways. They include the proportion of households making use of CPRs, per household average number and type of benefits derived from CPRs, or actual extent of benefits in cash or kind received from CPRs.

Table 14.3 summarizes some details that indicate the dependence of rural households on CPRs in the study villages.[6] Between 84 and 100 per cent of the poor households gathered food, fuel, fodder, and fibre items from CPRs. In contrast, just 10–28 per cent of large farmers depended on CPRs for these items. However, a greater proportion of large farmers collected items like silt from CPRs, e.g., ponds to enrich soils of their fields, and timber (mature trees from the village forest that are not legally allowed to be cut) for their private use. Though not indicated in the Table, the households who freely used water from rivulets for irrigating their land did not include 'poor' households.

The number of CPR-based activities as well as the number of CPR products collected[7] by respondents, as reported and physically verified in most cases, were also higher for poor households compared to the rich households (Table 14.3).

The greater dependence of the poor on CPRs compared to the rich, i.e, large farm households, as captured in Table 14.3, implies that poor households derive more of the benefits indicated in Table 14.2 from CPRs. Both negative and positive features of poor households' resource endowment compel or induce them to make greater use of CPRs.

First, the poor households have high man-to-land ratios and in most cases high animal-to-land ratios, and CPRs indirectly provide an important means for adjusting factor proportions in farming systems. CPRs, through grazing space and cut fodder/fuel, help the poor households to devote all of their land area to food or cash crops rather than sparing part of it for fodder/fuel supplies. Similarly, the CPRs' role as a cushion during the crisis situation, non-crop season or drought period, is greater for the poor households, as unlike the rich, they do not have many other adjustment mechanisms (Jodha 1985a).

Second, and more important, the dependence of poor households on CPRs is reinforced by some degree of match between the characteristics of CPR-based activities and features of the labour endowment of the poor.

For instance, the extraction cost or cost of using CPR's, except the lifting of water from rivulets or wells, is quite low and the main input required in the process is human labour. This matches very well with the labour surplus resource position of the poor. Furthermore, because of overuse and degradation, the returns from CPRs to individual users, in most cases, are not very high and are unattractive for the rich. However, poor households, with surplus labour and low opportunity cost, readily accept the low pay-off activities possible through CPRs. Additionally, most of the CPR-based activities, except collection of seasonal fruits and spices, are neither time bound nor fulltime jobs and workers can engage in them without sacrificing alternative employment opportunities. Specific skills for harvesting CPRs are not required, except in the case of collection of honey, fish from ponds, and game from forests, and even unemployable members of the family such as young children and old people can help. Thus, a self-selection process of CPR users, governed by low pay off from CPR use and still lower opportunity cost of labour of the users, initiates and perpetuates the poor's dependence on these resources.

The oral histories revealed that in the past, until CPRs became very 'unproductive', large farmers' dependence on CPRs was also greater. Even today in areas like Madhya Pradesh, where CPRs continue to be more productive, proportionately more of large farmers depend on CPRs than in many other regions.

Table 14.3 also reveals considerable interregional differences in households' dependence on CPRs in terms of the number of CPR items collected

TABLE 14.3: INDICATORS OF RURAL HOUSEHOLDS' DEPENDENCE ON CPRS IN THE STUDY VILLAGES[1]

| Details | *States (Study Districts, Villages)* | | | | | | | | | | | | | |
|---|---|---|---|---|---|---|---|---|---|---|---|---|---|---|
| | *Andhra Pradesh (2, 5)* | | *Gujarat (3, 6)* | | *Karnatak (3, 7)* | | *Madhya Pradesh (3, 7)* | | *Maharashtra (3, 6)* | | *Rajasthan (3, 7)* | | *Tamil Nadu (2, 4)* | |
| Category of households | Poor[2] | Others[2] | Poor | Others | Poor | Others | Poor | Others | Poor | Others | Poor | Others | Poor | Others |
| Number of households | 65 | 41 | 84 | 62 | 64 | 33 | 98 | 72 | 102 | 64 | 72 | 64 | 48 | 23 |
| Per Cent households collecting CPR products: Food items, | 95 | 10 | 96 | 16 | 84 | 14 | 100 | 18 | 98 | 13 | 100 | 23 | 93 | 12 |
| fuel, fodder, fiber | 99 | 15 | 100 | 19 | 100 | 18 | 100 | 11 | 100 | 16 | 100 | 28 | 100 | 17 |
| timber, silt, etc. | 37 | 59 | 29 | 83 | 41 | 78 | 21 | 84 | 19 | 90 | 31 | 89 | 92 | 42 |
| Per household average numberof: CPR based activities[3] | 4 | 2 | 5 | 2 | 5 | 3 | 6 | 3 | 3 | 2 | 5 | 2 | 4 | 3 |
| CPR items Collected[4] | 7 | 4 | 8 | 3 | 7 | 4 | 12 | 5 | 7 | 3 | 10 | 5 | 6 | 3 |

**Notes:** 1. Based on field surveys during 1983–85.

2. 'Poor' households include agricultural labourers and small farms (< 2 ha of dry land equivalent) households. 'Others' include large farm households only.

3. Activities included product collection, grazing, processing and handicrafts based on CPR products and marketing of CPR products.

4. Items included fuel, fodder, various wild fruits and flower, roots, leaves and skin of plants and trees, honey, gum, fish, small game, silt etc. For more details (including prices of these items) see Jodha (1985a).

and CPR-based activities followed. Without information on volume or value of these entities, they are at best crude indicators of relative dependence of households on CPRs in different regions. Yet, they do indicate considerable interregional variation, e.g., Madhya Pradesh or Rajasthan compared with Maharashtra or Tamil Nadu. Several factors individually or jointly help explain the above differences. Factors operating on the supply side of CPR products include:

(1) Overall greater size of CPRs (as in the case of Madhya Pradesh and Rajasthan);
(2) Higher productivity of CPRs, in terms of type and quantity of produce, as in the case of Madhya Pradesh where village forests and waste lands offer good supplies;
(3) Availability of high value products like honey, gum, spices, soapnuts and fish as in some villages of Madhya Pradesh, Maharashtra, and Karnataka; and
(4) Coinciding of CPR-product availability with the labour surplus period, e.g., in Rajasthan when a number of wild fruits like *ker* and *sangari* become available in the summer.

The factors operating on the demand side of CPR products include the general inadequacy of the household's own resources to sustain the family, as seen in many poor families of Madhya Pradesh, Tamil Nadu and Andhra Pradesh in particular; greater profitability of PPR-based activities that are supported by CPRs, e.g., poor man's dairying in Gujarat and sheep/goat rearing in Rajasthan; easy marketability of CPR products, like minor forest products, fuel, fodder, etc., in Tamil Nadu, Karnataka and Maharashtra; distress caused by drought or crop failure which compelled people to depend on CPRs to a greater extent as in some villages of Rajasthan, Tamil Nadu and Karnataka.[8]

## CPRs as Source of Fuel

Table 14.3 gave broad indications of poor people's significant dependence on CPRs for collection of fuel and other supplies. Information about fuel requirements met from CPRs and other sources is presented in Table 14.4. Actual fuel consumption (kg of dry weight) by a limited number of households was monitored for one week during each of the three seasons in VLS villages. Though family size was more or less the same, the average quantity of fuel consumption by large farm households was substantially higher, ranging from 185 to 219 kg for three weeks compared to poor households, ranging from 104 to 185 kg for three weeks.[9]

Figure for large farmers ranged between 8 and 13 per cent except in Raisen in Madhya Pradesh, where large farmers also gathered and used considerable amount of fuel from CPRs. The poor households collected fuel material (including dung) from CPRs in small quantities almost throughout the year. Large farmers cut and collected bushes from CPRs at the end of the monsoon, initially for purpose of fencing and using them as fuel subsequently. Of course, unlike the rural poor the large farm households have a relatively self-sustaining fuel supply system where crop by-products and animal dung play an important role (Table 14.4).

## Dependence for Animal Grazing

After fuel, fodder is the major item for which practically every poor household depends on CPRs. However, the extent of dependence on CPRs for fodder will be far greater if animal grazing, along with the collection of fodder, is also considered. Accurate recording of information on the extent of grazing is a complex and difficult task. Nevertheless, monitoring of animal grazing for a small number of households was attempted in the selected villages for one week in each season. That information is summarized in Table 14.5. The procedure used for monitoring animal grazing days and their conversion into animal unit grazing days is described in footnote 3 of Table 14.5.

Except in Jalore and Nagaur districts of Rajasthan, where livestock raising was a major occupation, poor households had fewer animal unit grazing days than the large farm households. However, the proportion of animal unit grazing days provided through CPRs for poor households was two to three times higher compared to the large farmers in all districts except Raisen. Jalore, Nagaur and Raisen are the only areas where CPRs were used to meet more than 35 per cent of the grazing requirements of large farmers. Availability of better grazing material or larger grazing space besides the

**Table 14.4: Some Details about Fuel Consumption by Households in the Study Villages, 1984–85[1]**

| District (State) | Household Categories | Per Household Average Consumption of Fuel During One Week in Each Season (Kg) | | | | Proportion (Per Cent) of Fuel From: | | | |
|---|---|---|---|---|---|---|---|---|---|
| | | | | | | CPRs | Own Sources | | |
| | | Monsoon | Winter | Summer | Total | | Crop Byproduct[2] | Dung | Fuelwood and Others |
| Mahbubnagar | Poor (13)[3] | 42 | 36 | 41 | 119 | 84 | 8 | 9 | – |
| (Andhra Pradesh) | Others (7) | 74 | 67 | 49 | 190 | 13 | 20 | 41 | 26 |
| Akola | Poor (13) | 38 | 34 | 32 | 104 | 79 | 18 | 3 | – |
| (Maharashtra) | Others (7) | 72 | 65 | 48 | 185 | 13 | 43 | 24 | 20 |
| Sholapur | Poor (7) | 36 | 37 | 43 | 119 | 72 | 14 | 12 | 2 |
| (Maharashtra) | Others (7) | 77 | 63 | 65 | 205 | 10 | 38 | 34 | 18 |
| Sabarkantha | Poor (20) | 62 | 65 | 57 | 184 | 66 | 9 | 25 | – |
| (Gujrat) | Others (10) | 70 | 74 | 69 | 213 | 8 | 46 | 28 | 18 |
| Raisen | Poor (20) | 66 | 61 | 58 | 185 | 74 | 6 | 11 | 9 |
| (Madhya Pradesh) | Others (10) | 79 | 74 | 66 | 219 | 32 | 15 | 29 | 24 |

**Notes:** 1. Based on data (including through selectively weighing the items) from the panel households covered by ICRISAT's Village Level Studies (VLS). Average size of household was 4–5 persons.

2. They include stalks of cotton, pigeon pea (tur), castor and sun hemp, groundnut shells, and empty maize cobs mainly.

3. Figures in parentheses indicate number of households. For definition of 'Poor' and 'Others' see note 2, under Table 14.3.

**TABLE 14.5: THE EXTENT OF DEPENDENCE OF HOUSEHOLDS ON COMMON PROPERTY RESOURCES FOR ANIMAL GRAZING IN THE STUDY VILLAGES[1]**

| *District and No of Village* | *Type and No of Households* | *Per Household Average Number of Animal Unit Grazing Days and the Proportion of CPR G razing There dueing 3 Weeks Year[2]* | | |
|---|---|---|---|---|
| | | *Animal Unit Grazing Days[3]* | | |
| | | *Total (No)* | *On CPRs* | *(Per Cent) On CPRS* |
| Aurangabad (2) | Poor[4](11) | 63 | 43 | 69 |
| | Others[4](10) | 125 | 34 | 27 |
| Mehsana (2) | Poor (12) | 89 | 70 | 78 |
| | Others (12) | 118 | 18 | 15 |
| Sabarkantha (2) | Poor (16) | 68 | 57 | 84 |
| | Others (10) | 102 | 12 | 11 |
| Mandsaur (2) | Poor (12) | 117 | 94 | 80 |
| | Others (11) | 198 | 55 | 28 |
| Raisen (2) | Poor (10) | 91 | 70 | 77 |
| | Others (10) | 168 | 71 | 42 |
| Jalorer (2) | Poor (16) | 231 | 180 | 78 |
| | Others (13) | 189 | 71 | 38 |
| Nagaur (2) | Poor (16) | 240 | 213 | 89 |
| | Others (12) | 205 | 77 | 37 |
| Mysore (1) | Poor (12) | 68 | 56 | 83 |
| | Others (9) | 112 | 33 | 29 |

**Notes:** 1. Based on household interviews and participant observation during the field work during 1982–85.

2. Data relate to one week each in three seasons during 1983–84.

3. A grazing day implies that the animals remained in the grazing land/field at least for 6 hours in a day and no stall feeding (except concentrates) was given. The number of grazing days using the above procedure were first recorded for each animal owned by the respondent household during the reference weeks. Animal grazing days (no of animals × number of grazing days) were changed into animal unit grazing days by converting all animals into animal units on the following basis: 1 animal unit = 1 bullock/cow/he or she buffalo/ Camel/ hourse/2 calves or buffalo/5 sheep/goat.

4. For definition of 'Poor' and 'Others' households see note 2 under Table 14.3.

limited extent of stall-feeding practised in the areas, determined these inter-regional differences in grazing on CPRs by large farmers (Jodha 1985b). Mehsana and Sabarkantha were at the other extreme, where large farmers' dependence on CPRs for grazing only reached 15 per cent. In these areas, dairying was well developed and livestock management, involving higher extent of stall-feeding, was also improved. Consequently, dependence on degraded, common grazing lands had significantly declined.

Moreover, in Gujarat, as in some other areas, privatized CPR lands received by large farmers were used for grazing only.

The poor households' greater dependence on CPRs for grazing is mainly because they do not own grazing land. Furthermore, they owned a larger proportion of currently unproductive animals, including those through the salvage system[10] such as young stock, dry milch stock, which were sustained through free grazing on degraded pastures. Moreover, poor people also owned more sheep and goats, which were able to survive on degrded pastures, on which cattle could barely subsist.

## Employment Generation

CPRs are not only a source of physical supplies for the rural people. CPRs also greatly contribute to employment and income of the rural people. Depending on the type of worker, the CPR based activity, and the season, employment through CPRs may be both a continuous and an irregular activity. To highlight the contribution of CPRs towards employment of rural people we use the following four sets of data:

(1) Time spent by different household members on collection of CPR products in different seasons.

(2) Number of CPR employment days of individual workers during the whole year; these data have been collected under the Village Level Studies (VLS).

(3) Share of CPR-based activities in the household workers' total labour time allocation. These data were recorded through participant observations and interviews under the VLS.

(4) CPR-activities during days of involuntary unemployment. That information was available for the VLS-panel households.

## CPR Product Collection

The time spent on collecting CPR products varies between members of households. We prepared an inventory of total CPR products collected during the week preceding the interview day in each season. Next, we recorded amount of time spent by family members, separately for adults and children, on collecting CPR products. Two hours of child labour were set as equal to one hour of adult labour.[11] We thus obtained total adult labour hours spent on CPR product collection by a household. For the poor, the total employment generated by CPRs, during the 3 weeks, ranged between 120 and 200 hours per household (Table 14.6). Large farm households spent less time on CPR product collection than the poor. We used 8 hours to convert the above information into employment days. CPR product collection during the 3 weeks created 15–25 days of work per household for the poor. The corresponding employment for large farm households ranged between 4 and 10 days.

Owing to factors like weather, changing availability of CPR products changes in the availability of other work and hence the opportunity cost of labour, etc, the pattern of employment revealed during the 3 reference weeks may not represent CPR employment in all the weeks in the year. Yet, by assuming that during the rest of the weeks of the year employment created by these activities is at least 50 per cent of the average reported for the 3 reference weeks, one can hazard a guess about employment during a year. Under the above assumption, employment per household may range between 128 and 204 days per year or 36–64 days of work per worker per year in the poor households. An important limitation of these estimates is that they also include pooled observations on small and irregular time intervals devoted to the collection of CPR products and that they do not reflect sustained employment for whole days.

The time spent on CPR product collection indicated very minor seasonal differences (Table 14.6), though the type, value and quantity of CPR products collected during different seasons do differ.

### Employment Days Through All CPR-based Activities

Employment information about each adult worker in the VLS-panel households was regularly recorded with an interval of 20–30 days for several

**Table 14.6: Employment Provided by CPR-Product Collection Activities in the Study Villages[1]**

| *District and No of Villages* | *Per Household Average Number of Hours Spent on CPR-Product Collecting in One Week during:* | | | | | | | | *Estimated Total Employment[3] (Mandays) in* | | | |
|---|---|---|---|---|---|---|---|---|---|---|---|---|
| | *Monsoon* | | *Winter* | | *Summer* | | *Total* | | *Reference Weeks[3]* | | *Whole Year[4]* | |
| | *Poor[2]* | *Others[2]* | *Poor* | *Others* | *Poor* | *Others* | *Poor* | *Others* | *Poor* | *Others* | *Poor* | |
| Aurangabad (2) | 39 | 9 | 35 | 18 | 36 | 13 | 120 | 40 | 15 | 5 | 128 | (36)5 |
| Mehsana (2) | 58 | 21 | 64 | 26 | 52 | 25 | 174 | 72 | 22 | 9 | 187 | (46) |
| Sabarkantha (2) | 62 | 25 | 71 | 32 | 58 | 21 | 191 | 78 | 24 | 10 | 204 | (51) |
| Mandsaur (2) | 36 | 7 | 50 | 17 | 43 | 10 | 129 | 34 | 48 | 4 | 153 | (40) |
| Raisen (2) | 70 | 20 | 78 | 28 | 52 | 17 | 200 | 65 | 25 | 8 | 212 | (64) |
| Jalore(2) | 44 | 16 | 51 | 17 | 46 | 18 | 141 | 51 | 18 | 6 | 153 | (45) |
| Nagpur (2) | 54 | 28 | 60 | 20 | 49 | 17 | 163 | 65 | 20 | 8 | 176 | (54) |

**Notes:** 1. Based on field work during 1982–85. Information relates to the week preceding the interview day during different seasons.

2. For definition of 'Poor' and 'Others' see note 2, under Table 14.3.

3. Man days of employment arrived at by dividing number of hours by 8.

4. Extent of employment during the whole year is worked out by using 50 per cent of the employment rate of the reference weeks as a rate of employment on CPR-product collection per week in the year.

5. Figures in parentheses in the last column indicate the number of employment days per worker.

years. Collecting CPR products is only one component of CPR employment. Other important components are processing of CPR products and herding[12] on CPRs. Inclusion of these components on CPR employment (Table 14.7), does not change the relative position of poor and rich households as revealed by Table 14.6.

For the poor, CPRs provided exclusive employment for 43–89 days per household or 18–31 days per adult worker during the reference year. This was marginally higher than their employment on their own farms. Large farm households, in all districts except in Raisen, were occupied with CPR activities only to a small extent. Furthermore, even if only 10 per cent of

**TABLE 14.7: EMPLOYMENT IN CPR-BASED ACTIVITIES IN THE STUDY VILLAGES[1]**

| District and No of villages | | *Per Households Average Number of Employment Days in a Year (1982–83) Created by:* | | | |
|---|---|---|---|---|---|
| | | *CPR-Based Activities:*[2] | | *Own Farm Work* | |
| | | *Poor*[3] | *Others:*[3] | *Poor* | *Others* |
| Mahbubnagar | (1) | 53<br>(112)[4] | 5<br>(40)[4] | 50 | 209 |
| Akola | (1) | 48<br>(93) | 4<br>(34) | 42 | 375 |
| Sholapur | (1) | 89<br>(187) | 12<br>(51) | 86 | 289 |
| Sabarkantha | (2) | 43<br>(155) | 7<br>(69) | 44 | 301 |
| Raisen | (2) | 56<br>(124) | 25<br>(66) | 47 | 237 |

**Notes:** 1. Based on employment data collected under the VLS project of ICRISAT.
2. CPR–based activities include product collection and marketing, handicrafts based on CPR-products and animal grazing on CPRs. Data relate to household members and exclude hired grazers. Household members having regular paid jobs are not covered by the Table.
3. For definition of 'Poor' and 'Others' households see note 2, under Table 3. The number of 'Poor' and 'Others' households respectively, in the study areas were as follows: Mahbubnagar, 18, 8; Akola 16, 9; Sholapur 17, 8; Sabarkantha 30, 17; and raisen 33, 18.
4. Figures in parentheses under columns 3 and 4, indicate number of employment days when CPR-based activities were undertaken casually while doing other jobs.

**Table 14.8: Share of CPR-based Activities in Daily Time-Allocation of Households in the Study Villages[1]**

| Year | Proportion (Per Cent) of CPR-based Activities in the Total Time of Activities Covered by Daily Time-Allocation by Household Members in | | | | | |
|---|---|---|---|---|---|---|
| | *Mahbubnagar* | | *Sholapur* | | *Akola* | |
| | *Poor[2]* *(40)* | *Others[2]* *(20)* | *Poor* *(40)* | *Others* *(20)* | *Poor* *(40)* | *Others* *(20)* |
| 1957/76 | 17 | 1 | 16 | 6 | 13 | 1 |
| 1976/77 | 12 | 4 | 12 | 7 | 8 | 1 |
| 1977/78 | 15 | 9 | 15 | 3 | 12 | 2 |
| 1978/79 | 22 | 9 | 11 | 6 | 11 | 2 |

**Notes:** 1. Based on interviews and participant observation on time-allocation by each working member of the household for 1 day (i. E, the day preceding the day of interview) with an interval of 20–30 days during the year. Such information collected under the VLS project of ICRISAT, covers time-allocation details for 12–15 days in a year for each worker.

2. For definition of 'Poor' and 'Other' households see note 2, under Table 14.3. Figures in parentheses indicate number of households under each category.

the time of the partial employment-days (in parentheses in Table 14.7) is assumed to have been spent on CPR-based activities, another 9–18 days of employment would be derived from CPRs.

## Time Allocation

The data on time allocation collected under the VLS, indicated that CPR activities accounted for around 10–20 per cent of total daily time of the poor in all villages (Table 14.8). Large farmers allocated only little time to CPR activities. In some cases, particularly among the poor households, there was significant inter-year variability in the proportion of time allocated to CPR activities. These changes can be explained by weather and emerging employment and income opportunities, which modify people's relative dependence on CPRs.

The data in Table 14.8 provide information for only 12–15 days per year for each adult worker. But they are likely to be more precise compared to other recall-based data because the former were recorded through participant observation or a recall of barely 10–20 hours. Their real significance lies in confirming the broad patterns about the rural poors' employment through CPRs, already revealed by the preceding Tables.

## Follow-up Activities During Involuntary Unemployment

Special investigations of activities of workers who could not get wage employment in the village labour market were conducted under the VLS. A significant proportion of such workers resorted to CPR activities. Workers of the large farm households were not involuntarily unemployed to a significant extent. Adults of poor households were involuntarily unemployed for 28–69 days during the reference year (Table 14.9). CPR employment in most districts accounted for 23–30 per cent of the total days, for which adults of poor households otherwise would have been involuntarily unemployed (item 'e', Table14.9). If the CPR–based activities taken up along with other activities are also considered, the employment contribution of CPRs during the involuntary unemployment period would increase further (item 'f', Table 14.9).

CPRs are an important source of employment opportunity for the rural poor that can be exploited at times when no alternative employment is available. This is highlighted by greater self-employment in CPR activities during the off-season, especially after the crops are harvested, during poor crop years, and during the days of involuntary unemployment in the labour market. In some cases, self-employment of poor households in CPR generated employment was often higher than the employment created by a number of anti-poverty rural development programmes.[13]

## Income Generation by CPRs

It is quite difficult to translate all the supplies and services offered by CPRs into income flows. Furthermore, since most of the CPR-based activities have a low pay-off, their contribution to monetary income may not be high. However, the importance of income received from the use of CPRs

**TABLE 14.9: DETAILS INDICATING THE ROLE OF CPR-BASED ACTIVITIES DURING THE DAYS OF INVOLUNTARY UNEMPLOYMENT IN THE STUDY VILLAGES, 1982–83[1]**

| Details | *Involuntary Unemployment and Follow-up Activites in the Case of Different Groups of Households in* | | | | | | | | | |
|---|---|---|---|---|---|---|---|---|---|---|
| | *Mahbunagar* | | *Sabarkantha* | | *Raisen* | | *Akola* | | *Sholapur* | |
| | *Poor*[2] | *Others*[2] | *Poor* | *Others* | *Poor* | *Others* | *Poor* | *Others* | *Poor* | *Others* |
| Total household (number) | 21 | 1 | 31 | 7 | 27 | – | 31 | 2 | 33 | 8 |
| Total Workers (number) | 49 | 1 | 58 | 13 | 38 | – | 68 | 5 | 57 | 14 |
| Per worker average number of days of: | | | | | | | | | | |
| a. Involuntary unemployment[3] (number) | 43 | 35 | 63 | 32 | 28 | – | 34 | 26 | 69 | 50 |
| b. Employment through follow up activities during (a) (number) | 32 | 35 | 43 | 8 | 22 | – | 31 | 23 | 55 | 38 |
| c. Share of exclusively CPR-based employment in (b) (per cent) | 35 | 83 | 33 | 0 | 38 | – | 9 | 0 | 34 | 18 |
| d. Share of partly CRR-based employment in (b) (per cent) | 50 | 17 | 30 | 0 | 13 | – | 26 | 0 | 20 | 22 |
| e. (c) as proportion of (a) (per cent) | 26 | 83 | 23 | 09 | 30 | – | 8 | 0 | 25 | 14 |
| f. (d) as proportion of (a) (per cent) | 37 | 17 | 17 | 0 | 10 | – | 24 | 0 | 15 | 17 |

**Notes:** 1. Based on data on involuntary unemployment and follow-up activities of panel households covered by the VLS for 1982–83.

2. For definition of 'Poor' and 'Others' households see note 2, under Table 14.3.

3. Involuntary unemployment is defined as an event when the worker tried to seek employment in and around the village and failed to get it.

should be evaluated more in terms of its temporal and situational contexts, rather than in terms of its magnitude alone. Only two components of CPR-based income are discussed here.

They are income from CPR product collection[14] and proportion of income from animal husbandry that is attributable to grazing on CPRs.[15] CPR's contribution to household income and its stability, through their support to PPR-based crop farming, and processing and marketing of CPR products, are not considered. The income data in Table 14.10, therefore, considerably underestimate the total income contribution of CPRs.

Average annual income of poor households from CPRs ranged from Rs 445–Rs 830 during the reference period. CPR income of large farmers did not exceed Rs 300, except in the Rajasthan and Madhya Pradesh villages. A substantial share of this income originated from CPR product collection in most regions. The exceptions were large farm households in Nagaur, Jalore, and Mandsaur district where livestock production is important. Except in the Akola and the Aurangabad villages, the CPR-income accounted for 15–23 per cent of total income from all other sources, for the poor households. The corresponding proportions for large farm households were 1–3 per cent only.

In six of the 12 study areas, average annual income from CPRs per poor household exceeded Rs 700. CPR income constituted more than a fifth of poor households' income from all other sources in these districts (except Mehsana). In the case of two districts at the bottom, this figure was less than Rs 550. The regional differences in the income from CPRs can be explained in terms of the extent and productivity of CPRs and major activities supported by them. For instance, high CPR income areas like Raisen, Jalore and, to an extent, Nagaur are endowed with larger extent of CPRs (Table 14.1). Similarly, availability of high-value CPR products such as gum, honey, fruits and game, contributed to higher CPR income in Raisen and Dharmapuri. Poor man's animal husbandry, sustained mainly through CPRs, was responsible for higher CPR income of the poor in households of Nagaur, Mehsana, Sabarkantha and Jalore. Limited area of CPRs and limited number of CPR products on the one hand and greater availability of agricultural employment opportunities due to cotton and paddy dominated cropping patterns on the other, may explain the lower CPR incomes of the poor in the Akola and the Mahbubnagar villages.[16]

**Table 14.10: Average Annual Household Income from CPRs in the Study Villages[1]**

| District[2] | Per Households Annual Average Income in the Case of[3] | | | | | | | |
|---|---|---|---|---|---|---|---|---|
| | Poor | | | | Others | | | |
| | Households (Number) | Value of CPR-Products Collected (Rupees) | CPR-Share in Livestock Income (Rupees) | Total (Rupees) | Households (Number) | Value of CPR-Products Collected (Rupees) | CPR-Share in Livestock Income (Rupees) | Total (Rupees) |
| Mahbubnagar | 15 | 382 | 152 | 534<br>(17)[4] | 10 | 109 | 62 | 171<br>(1) |
| Mehsana | 26 | 421 | 309 | 730<br>(16) | 24 | 88 | 74 | 162<br>(1) |
| Sabarkantha | 35 | 432 | 336 | 818<br>(21) | 19 | 111 | 97 | 208<br>(1) |
| Mysore | 26 | 534 | 115 | 649<br>(20) | 11 | 112 | 58 | 170<br>(3) |
| Mandsaur | 23 | 400 | 285 | 685<br>(18) | 18 | 113 | 190 | 303<br>(1) |
| Raisen | 37 | 568 | 212 | 780<br>(26) | 15 | 283 | 185 | 468<br>(4) |
| Akola | 16 | 342 | 105 | 447<br>(9) | 9 | 85 | 49 | 134<br>(1) |

(continued to next page)

| District[2] | Per Households Annual Average Income in the Case of[3] | | | | | | | |
|---|---|---|---|---|---|---|---|---|
| | Poor | | | | Others | | | |
| | Households (Number) | Value of CPR-Products Collected (Rupees) | CPR-Share in Livestock Income (Rupees) | Total (Rupees) | Households (Number) | Value of CPR-Products Collected (Rupees) | CPR-Share in Livestock Income (Rupees) | Total (Rupees) |
| Aurangabad | 22 | 405 | 179 | 584 (13) | 21 | 110 | 53 | 163 (1) |
| Sholapur | 24 | 443 | 198 | 641 (20) | 9 | 143 | 92 | 235 (2) |
| Jalore | 24 | 447 | 262 | 709 (21) | 27 | 170 | 217 | 387 (2) |
| Nagaur | 32 | 473 | 358 | 831 (23) | 25 | 143 | 295 | 438 (3) |
| Dharmapuri | 30 | 530 | 208 | 738 (22) | 11 | 112 | 54 | 164 (2) |

**Notes:** 1. Based on field work during 1982–85.

2. Number of villages covered in Mahbubnagar, Akola and Sholapur was one each. In other districts it was two.

3. CPR income was estimated by valuation of CPR-products collected by the households on the basis of village level prices of those products. Livestock income attributed to CPRs was estimated on the basis of contribution of CPR grazing to total livestock maintenance. The procedures used, briefly mentioned in the text, tended to underestimate the contribution of CPR grazing to livestock income. In the case of villages in Mahbubnagar, Akola, Sholapur, Sabarkantha and Raisen. all incomes were estimated on the basis of cost accounting method, while in others they were based on recalls during three seasonal rounds of field work.

4. Figures in the parentheses indicated total CPR-based income as per cent to total income from other sources.

**TABLE 14.11: IMPACT OF CPRS ON INEQUALITY OF HOUSEHOLD INCOMES IN THE STUDY VILLAGES[1]**

| *District and Number of Village* | | *Mean and Gini Coefficient of Households Income from* | | | |
|---|---|---|---|---|---|
| | | *All Sources* | | *All Sources Excluding CPRs* | |
| | | *Mean (Rupees) (Per Cent)* | *Gini Coefficient* | *Mean (Rupees) (Per Cent)* | *Gini Coefficient* |
| Mahbubnagar | (1) | 7443 | 0.41 | 6777 | 0.50 |
| Akola | (1) | 9549 | 0.40 | 8928 | 0.48 |
| Sholapur | (1) | 6989 | 0.32 | 6008 | 0.37 |
| Sabarkantha | (2) | 9249 | 0.33 | 8205 | 0.45 |
| Raisen | (2) | 5186 | 0.34 | 4150 | 0.44 |

**Notes:** 1. Based on data (for 1983–84) from panel household covered by the VLS. The panel of 401 households from each village included 10 households from each category namely, landless labourers, small farmers, medium farmers, and larger farmers.

## Reduction in Income Inequalities

The pattern of inter-class differences in CPR income suggests that CPRs help reduce the rural inequalities. The inequalities of the VLS-household incomes, measured with the Gini coefficient, are reduced when CPR income are included in household income (Table 14.11). The values of the Gini coefficient varied from 0.37 to 0.50 in different areas when income from all sources excluding the CPRs was considered.

However, once CPR-based income was included in household income, the value of the Gini coefficient declined and ranged between 0.32 and 0.41 in different areas.

In view of the fact that CPRs contribute more than a fifth of the incomes from other sources for the poor, they should be recognized as an integral part of programmes directed towards reduction in rural poverty and inequality. In fact, the CPR-based income works out to Rs 112 per worker in an area like Akola and Rs 243 per worker in Nagaur, even when not all components of CPR-related income have been quantified here. If the unaccounted income components are added, CPR-based income could be much higher in many areas than the per household income generated by a number of antipoverty programmes.[17]

## V

## Changing Status of CPRs

Investigation into the village-level records and oral accounts of the recent past in the study areas indicated that the contribution of CPRs towards the income of the rural poor as well as rural equity was probably much higher in the past than it is at present. Both the area and physical productivity of CPRs, in terms of quantity and quality of products, have declined. We tried to document this change by comparing total CPR output in the early 1980s with the early 1950s, when following independence, land reforms were introduced.[18] For both periods physical productivity of CPRs and their area were recorded. The CPRs covered are village pastures, village forests, waste lands,[19] unoccupied catchments of ponds, tanks, river banks, watershed drainages and community threshing grounds.

### Decline of CPRs

Decline of CPRs can take three forms:

(i) Physical loss of resources, such as the submersion of grazing land in a newly constructed irrigation dam or area of CPR covered by roads and buildings.

(ii) Deterioration of physical productivity of resources, as revealed by degradation of pastures or forest lands.

(iii) Re-assignment of usage and property rights as indicated by transfer of CPR lands to private ownership.

As mentioned earlier, the following discussion is confined to the last category of the decline of CPRs.

In the dry areas we studied, CPRs mostly occupy sub-marginal land with undependable water supply. With increased resource scarcity, it can however be expected that common property rights will be infringed on or substituted by *de jure* or *de facto* private property rights. In the areas covered by this study, the above substitution of property rights has taken place on a large scale. For instance, during 1982–84, the extent of CPR lands in the study villages ranged from 9–28 per cent of total area (Table 14.12). These proportions ranged between 15 and 42 per cent in 1950–52. The area of CPRs during the last three decades has thus declined by 26–52 per cent.

Details of technological, institutional, and demographic circumstances causing decline in the area of CPRs, and also their physical productivity, have been discussed elsewhere (Jodha 1985a; 1985b). Here we will confine our discussion to those aspects of the decline of CPRs that affect particularly the rural poor. The main form of decline is the privatization of CPRs.[20] Under various welfare programmes CPR lands had been distributed to people for private use. CPR lands had also been illegally appropriated often with subsequent legalization. The stated intention of privatization of CPRs was to give land to the poor who were landless or who had very little land.

Hence, from the days of land reforms in early 1950s to the present day, programmes such as the 20-Point Programme, allotment of land titles to the poor of CPR lands, already used collectively by the poor, was an important measure. The consequent decline in the area of CPRs, and the resultant overcrowding and the degradation of CPRs have led to a considerable reduction in the overall quantity of CPR benefits for the poor. However, it is quite possible that the losses the poor suffered collectively might have been compensated by the individual gains from possessing more private land. This balance could be examined by comparing the magnitude of the decline in CPR-based income of the past with the returns from privatized CPR lands owned by the poor. A few studies investigating the losses and gains have been initiated and await completion.

Alternatively, one may analyse the changes in the land resources base of the poor following the privatization of CPRs.

## Privatization and Distribution of CPR Lands[21]

It is evident from Table 14.13 that, although the privatization of CPRs was promoted in the name of helping the poor, very little land was received by them. For 10 of the 19 areas covered, the proportion of poor households in the recipients of CPR lands was higher than the proportion of other

TABLE 14.12: EXTENT AND DECLINE OF CPR LAND IN THE STUDY VILLAGES[1]

| *State and District* | *Number of Village* | *Area of CPR-Lands (ha)*[2] | *CPRs Proportion of Total Village Area (per cent)* | | *Decline in the Area of CPRs since 1950–52 (per cent)* |
|---|---|---|---|---|---|
| | | | *1882–84* | *1950–52* | |
| Andhra Pradesh | | | | | |
| Anantapur | 2 | 221 | 15 | 24 | 36 |
| Mahbubnagar | 5 | 408 | 9 | 16 | 43 |
| Medak | 3 | 198 | 11 | 20 | 45 |
| Gujarat | | | | | |
| Banaskantha | 5 | 167 | 9 | 19 | 49 |
| Mehsana | 5 | 224 | 11 | 17 | 37 |
| Sabarkantha | 5 | 198 | 12 | 22 | 46 |
| Karnataka | | | | | |
| Bidar | 3 | 297 | 12 | 20 | 41 |
| Dharwad | 3 | 242 | 10 | 18 | 44 |
| Gulbarga | 3 | 291 | 9 | 15 | 43 |
| Mysore | 3 | 335 | 18 | 27 | 32 |
| Madhya Pradesh | | | | | |
| Mandsaur | 4 | 327 | 22 | 34 | 34 |
| Raisen | 6 | 770 | 23[3] | 42 | 47 |
| Vidisha | 4 | 338 | 28[3] | 38 | 32 |
| Maharashtra | | | | | |
| Akola | 5 | 192 | 11 | 19 | 42 |
| Aurangabad | 4 | 304 | 15 | 21 | 30 |
| Sholapur | 4 | 433 | 19 | 25 | 26 |
| Rajasthan | | | | | |
| Jalore | 5 | 639 | 18 | 29 | 37 |
| Jodhpur | 3 | 591 | 16 | 38 | 58 |
| Nagapur | 3 | 619 | 15 | 41 | 63 |
| Tamil Nadu | | | | | |
| Coimbatore | 4 | 187 | 9 | 17 | 47 |
| Dharmapuri | 3 | 225 | 12 | 26 | 52 |

**Notes:** 1. Based on village-level records and field work during 1982–5.

2. CPRs included are village pastures, village forests, waste lands, unoccupied catchments of ponds/tanks, river/rivulet banks, watershed drainages, and community threshing grounds, etc, which are used for free animal grazing and collection of fuel, fodder, and other items.

3. In Raisen and Vidisha villages large area of CPRs initially taken over by forest department but now available for use by the villagers is included with CPRs.

farmers (Table 14.13, col 6 and 7).[22] However, in 16 of the 19 regions, the share of the poor in the privatized lands was lower than the share of all other farmers (Table 14.13, col 4 and 5). The poor received between 0.8 and 1.6 ha per household, whereas other farmers received 1.5–4.9 ha per household. The comparison of landholding size before and after the privatization of CPR lands indicates that those who had relatively more land also got more land.

Thus, transfer of CPR land had not helped the poor to improve their resource position in relation to the better-endowed farmers.

In seven of 19 regions more than 40 per cent of CPR lands received the poor was no longer owned by them at the time of field work.[23] Land was sold, mortgaged, or leased on long-term basis as a first step towards eventual sale (Table 14.14).[24] In 9 districts the rural poor had lost control over 20–40 per cent of the CPR lands they had received. In more than half the districts 10–20 per cent of the CPRs land received by rural poor were fallowed. Thus the actual extent of privatized CPR lands cultivated by the rural poor was much less than the area indicated by Table 14.13. For other recipients, i.e., large and medium farmers, the extent of privatized CPR land being given up was negligible. A substantial part of the new land acquired by large farmers was fallowed.

## Disentitlement of the Poor

The complete process of privatization of CPRs as it affected the rural poor involved three stages: (i) they were deprived of their right to collectively use the CPRs; (ii) they were given individual title to small parts of privatized CPRs; and (iii) the circumstances disentitled them of the newly received lands.

According to Table 14.15, 63–91 per cent of the land distributed to the poor under welfare programmes subsequently changed owners. In contrast, the proportion of land that was later transacted away was only 9–37 per cent for land that was illegally appropriated initially. The appropriation of land in a way reflected the effective and felt need for acquiring land and hence after acquiring it, people kept it at any cost. The CPR land distributed under welfare programmes may have been regarded as a dole by the poor. The latter parted with these lands to satisfy their other needs and priorities.

**TABLE 14.13: DISTRIBUTION OF PRIVATIZED CPRS IN THE STUDY VILLAGES[1]**

| State, District and Number of Village | Total Land Given (ha) | Total Households (Numbers) | Proportion of Land to (per cent) | | Proportion in the Recipients (per cent) | | Per Household Land Area Received by | | Per Household Land Area Owned by | | | |
|---|---|---|---|---|---|---|---|---|---|---|---|---|
| | | | | | | | | | Poor | | Others | |
| | | | Poor[2] | Others[2] | Poor | Others | Poor | Others | Before | After | Before | After |
| 1 | 2 | 3 | 4 | 5 | 6 | 7 | 8 | 9 | 10 | 11 | 12 | 13 |
| Andhra Pradesh | | | | | | | | | | | | |
| Mahbubanagar (2) | 418 | 343 | 50 | 50 | 76 | 24 | 0.8 | 2.6 | 0.3 | 0.9 | 3.0 | 5.1 |
| Medak (3) | 75 | 58 | 52 | 49 | 59 | 41 | 1.1 | 1.5 | 1.0 | 2.2 | 3.1 | 4.6 |
| Gujarat | | | | | | | | | | | | |
| Banaskantha (3) | 75 | 29 | 18 | 82 | 38 | 62 | 1.3 | 3.4 | 0.8 | 2.0 | 5.4 | 8.8 |
| Mehsana (3) | 85 | 63 | 20 | 80 | 36 | 64 | 0.7 | 1.7 | 1.0 | 1.7 | 8.0 | 9.8 |
| Sabarkantha (3) | 127 | 74 | 28 | 77 | 55 | 45 | 0.9 | 2.8 | 0.5 | 1.1 | 7.0 | 9.8 |
| Karnataka | | | | | | | | | | | | |
| Bidar (3) | 89 | 55 | 39 | 61 | 64 | 36 | 1.0 | 2.8 | 1.0 | 2.0 | 6.4 | 9.2 |
| Gulbarga (3) | 112 | 50 | 43 | 57 | 60 | 40 | 1.6 | 3.2 | 0.8 | 2.4 | 4.5 | 7.7 |
| Mysore (3) | 161 | 98 | 44 | 56 | 67 | 33 | 1.2 | 2.9 | 0.9 | 1.9 | 4.1 | 11.6 |
| Madya Pradesh | | | | | | | | | | | | |
| Mandsaur (2) | 120 | 55 | 45 | 55 | 75 | 35 | 1.3 | 4.7 | 1.2 | 2.5 | 7.7 | 12.4 |
| Raisen (4) | 115 | 72 | 42 | 58 | 68 | 32 | 1.0 | 2.9 | 1.3 | 2.2 | 6.2 | 9.0 |
| Vidisha (4) | 123 | 77 | 38 | 62 | 48 | 52 | 1.3 | 1.9 | 1.3 | 2.5 | 4.9 | 6.8 |

*(continued to next page)*

| *State, District and Number of Village* | | *Total Land Given (ha)* | *Total Households (Numbers)* | *Proportion of Land to (per cent)* | | *Proportion in the Recipients (per cent)* | | *Per Household Land Area Received by* | | *Per Household Land Area Owned by* | | | |
|---|---|---|---|---|---|---|---|---|---|---|---|---|---|
| | | | | | | | | | | *Poor* | | *Others* | |
| | | | | *Poor*[2] | *Others*[2] | *Poor* | *Others* | *Poor* | *Others* | *Before* | *After* | *Before* | *After* |
| 1 | | 2 | 3 | 4 | 5 | 6 | 7 | 8 | 9 | 10 | 11 | 12 | 13 |
| Maharashtra | | | | | | | | | | | | | |
| Akola | (3) | 101 | 100 | 39 | 61 | 58 | 42 | 0.7 | 1.5 | 1.0 | 1.6 | 3.1 | 4.6 |
| Aurangabad | (2) | 83 | 55 | 30 | 70 | 42 | 58 | 1.1 | 1.8 | 1.1 | 2.2 | 6.4 | 8.3 |
| Sholapur | (3) | 132 | 72 | 42 | 58 | 53 | 47 | 1.5 | 2.3 | 0.7 | 2.2 | 3.4 | 5.6 |
| Rajasthan | | | | | | | | | | | | | |
| Jalore | (2) | 83 | 27 | 14 | 86 | 37 | 63 | 1.4 | 4.9 | 0.3 | 1.7 | 7.2 | 12.5 |
| Jodhpur | (3) | 405 | 318 | 24 | 76 | 35 | 65 | 0.9 | 1.5 | 0.4 | 1.3 | 2.3 | 3.8 |
| Nagapur | (3) | 147 | 81 | 21 | 79 | 41 | 59 | 1.2 | 3.1 | 1.3 | 2.5 | 2.4 | 5.2 |
| Tamil Nadu | | | | | | | | | | | | | |
| Combatore | (4) | 206 | 145 | 50 | 50 | 75 | 25 | 1.1 | 2.9 | 0.8 | 2.5 | 3.8 | 5.8 |
| Dharmapuri | (3) | 241 | 127 | 49 | 51 | 55 | 45 | 0.9 | 2.1 | 1.0 | 1.9 | 4.6 | 7.5 |

**Notes:** 1. Based on field work during 1982–85.

2. For definition of 'Poor' and 'Other' households, see note 2 under Table 14.3.

Furthermore, in several areas, unlike the appropriated land, the land distributed under welfare programmes was often poor and sub-marginal. The low quality of the land may also explain why the poor parted with it so easily.

**TABLE 14.14: CURRENT STATUS OF THE CPR LANDS GIVEN TO THE RURAL POOR IN THE STUDY VILLAGES, 1982–85[1]**

| *State and District and Number of Villages* | | *Total Land Given (ha)* | *Poor Household's (Number)*[2] | *Proportion (Per Cent) of the Received Land* | | |
|---|---|---|---|---|---|---|
| | | | | *Sold/ Mortgaged/ Leased*[3] | *Fallowed* | *Cultivated* |
| Andhra Pradesh | | | | | | |
| Mahbubnagar | (3) | 209 | 262 | 50 | 3 | 47 |
| Medak | (2) | 38 | 34 | 10 | 9 | 74 |
| Gujarat | | | | | | |
| Banaskantha | (3) | 14 | 11 | 26 | – | 74 |
| Mehasana | (2) | 17 | 23 | 59 | 10 | 31 |
| Sabarkantha | (3) | 35 | 41 | 45 | 12 | 43 |
| Karnataka | | | | | | |
| Bidar | (3) | 34 | 35 | 21 | 14 | 64 |
| Gulbarga | (3) | 48 | 30 | 25 | 8 | 67 |
| Mysore | (4) | 70 | 66 | 41 | 13 | 46 |
| Madhya Pradesh | | | | | | |
| Mandsaur | (2) | 54 | 41 | 30 | 20 | 50 |
| Raisen | (4) | 48 | 49 | 18 | 10 | 72 |
| Vidisha | (4) | 47 | 37 | 21 | 15 | 64 |
| Maharashtra | | | | | | |
| Akola | (3) | 39 | 58 | 21 | – | 79 |
| Aurangabad | (2) | 25 | 23 | 54 | 15 | 36 |
| Sholapur | (3) | 55 | 38 | 48 | 9 | 43 |
| Rajasthan | | | | | | |
| Jalore | (2) | 14 | 10 | 44 | – | 56 |
| Jodhpur | (2) | 97 | 112 | 33 | 15 | 52 |
| Nagaur | (3) | 39 | 33 | 30 | 3 | 67 |
| Tamil Nadu | | | | | | |
| Coimbatore | (4) | 103 | 109 | 21 | 6 | 73 |
| Dharmapuri | (3) | 120 | 70 | 13 | 3 | 85 |

**Notes:** 1. Based on field work during 1982–85.

2. For definition of 'Poor' and 'Others' households see note 2, under Table 14.3.

3. Lands leased on long-term basis (similar to mortaged lands) are included here.

This reasoning is supported by the high proportion (37–47 per cent) of sub-marginal lands in the land given up (Table 14.14). Thus the quality of land and the manner of obtaining it seemed to have complemented each other.

Some more factors that directly or indirectly impinge on effective desire for land can be examined in terms of resource endowments and capacities of the recipients of the CPR lands. Resource endowments are differentiated in terms of possession of land, bullocks and other complementary resources including experience in farming.

Only a small proportion of the households who gave up the newly acquired land owned complementary resources (Table 14.15). Mere distribution of land, particularly of sub-marginal land, without the provision of necessary complementary resources was not sufficient to develop and cultivate the land. The situation differed dramatically for those households that retained the newly acquired lands (Table 14.15). More importantly, family labour capacity as indicated by a household's dependence on labour income was the major asset of 62–82 per cent of the households who gave up their new lands. Rather than sticking to a small piece of land that could not be developed and used for want of other resources, these households preferred to sell or mortagage the land and concentrate on wage earnings.

Inability to develop and use the land was the reported reason for giving up the land in 48–100 per cent of the cases in different areas. Immediate cash needs of the family compelled 9–52 per cent of the households to sell out, mortgage, or lease out their land. Poor quality of land was reported as a reason for not retaining the land by more than 50 per cent of the households.

To sum up, our analysis shows that privatization of CPRs as a strategy to help the rural poor yielded a negative result. The collective loss of the poor from a decline of CPRs has not been compensated by acquisition and retention by the poor of privatized CPRs. Alternative strategies to help the poor are needed. One strategy could be to develop CPRs and improve their use through technological and institutional interventions. However, it may be pointed out that growth in the CPR productivity alone may prove counterproductive. At present the process of self-selection of CPR users, tend to induce mainly the poor to depend on these resources. CPR activities are low pay-off options. The poor chooses them as the opportunity cost of their labour is lower than the returns from CPR activities. An increase in CPR productivity will induce greater demand on CPRs through (i) increased

**Table 14.15: Some Indicators of Factors Causing Dispossession of CPR Lands Given to the Rural Poor**[1]

| Details | *Some Attributes of CPR Lands (A) Retained and (B) Given Up by Households in the Study Village of* | | | | | | | | | | | | | |
|---|---|---|---|---|---|---|---|---|---|---|---|---|---|---|
| | *Andhra Pradesh* | | *Gujarat* | | *Karnataka* | | *Madhya Pradesh* | | *Maharashtra* | | *Rajasthan* | | *Tamil Nadu* | |
| | *A*[3] | *B*[4] | *A* | *B* | *A* | *B* | *A* | *B* | *A* | *B* | *A* | *B* | *A* | *B* |
| Total land area (ha) | 139 | 108 | 39 | 27 | 83 | 69 | 114 | 35 | 71 | 48 | 99 | 51 | 189 | 34 |
| Total households involved (number)[5] | 208 | 88 | 53 | 22 | 75 | 56 | 103 | 24 | 76 | 43 | 113 | 42 | 147 | 32 |
| Proportion of lands (under A and B) through: | 12 | 68 | 38 | 63 | 22 | 72 | 43 | 91 | 24 | 83 | 35 | 78 | 19 | 70 |
| Illegal grabbing (per cent) | 88 | 32 | 62 | 37 | 78 | 28 | 57 | 9 | 76 | 17 | 65 | 22 | 81 | 30 |
| Proportion of sub-marginal lands to lands (under A and B)[7] (per cent) | 9 | 45 | 0 | 37 | 6 | 45 | 0 | 37 | 9 | 25 | 35 | 39 | 20 | 47 |
| Bullocks (per cent) | 27 | 0 | 66 | 0 | 43 | 7 | 64 | 8 | 39 | 0 | 62 | 7 | 27 | 6 |
| Heavy dependence on labour income (per cent) | 24 | 70 | 13 | 82 | 48 | 71 | 10 | 62 | 24 | 81 | 25 | 76 | 28 | 78 |
| Proportion of households giving up land due to: | | | | | | | | | | | | | | |
| Inability to develop/use the land (per cent) | – | 48 | – | 77 | – | 71 | – | 88 | – | 74 | – | 91 | – | 100 |

*(continued to next page)*

| *Details* | *Some Attributes of CPR Lands (A) Retained and (B) Given Up by Households in the Study Village of* | | | | | | | | | | | | | |
|---|---|---|---|---|---|---|---|---|---|---|---|---|---|---|
| | *Andhra Pradesh* | | *Gujarat* | | *Karnataka* | | *Madhya Pradesh* | | *Maharashtra* | | *Rajasthan* | | *Tamil Nadu* | |
| | *A*[3] | *B*[4] | *A* | *B* | *A* | *B* | *A* | *B* | *A* | *B* | *A* | *B* | *A* | *B* |
| Cash needs of the family (per cent) | – | 52 | – | 23 | – | 29 | – | 12 | – | 26 | – | 9 | – | – |
| Poor quality of land (per cent) | – | 50 | – | 58 | – | 57 | – | 61 | – | 58 | – | 63 | – | 52 |

**Notes:** 1. Based on field work during 1982–85.

2. Details are presented by pooling data for different districts for each state presented in Table 14.
3. A = CPR lands received and retained by the recipients.
4. B = CPR lands received and subsequently given up by the recipients.
5. The households included under category (B) include all those who had partly or fully given up their newly acquired lands.
6. Under the programmes ranging from introduction of land reforms in the early 1950s to the new 20–point programme in the early 1980s, land was distributed to the rural poor as a part of the welfare programme.
7. Lands were illegally encroached initially, the possession was regularized subsequently by revenue authorities.
8. Rocky/gravelly/steep and completely eroded lands, and the lands with problem soils (eg, saline soils) etc, are included under this category.

number of users; (ii) pressure for privatization including through illegal appropriation; and (iii) encroachment by government in the form of handing over of more paying CPRs to contractors as has happened in the past. The consequence in all cases would be reduced benefits to the rural poor.

Hence, the management or regulated use of CPRs is as important as any measure directed to raise their productivity. Identification and evaluation of intervention measures, to meet these objectives, form a part of our ongoing work on the subject.

## Notes

1. Ground-water is also often regarded as a CPR, although it may not (unlike other CPRs considered by us) belong to any identifiable community.
2. The neglect of CPRs in the process of rural development has been so great that even when some programmes are directed to these resources, they are adopted more as physical measures and have little concern for their CPR context. For instance, efforts in terms of research and development activities relating to forests and grasslands or measures to harness runoff and ground water, are seldom adopted in the framework of CPRs. Even in the programmes connoting group orientation, e.g., rural co-operatives or community development projects, the community' is usually by passed due to the PPR orientation of development activities (Jodha 1973).
3. They included local school teachers or other village-level functionaries, elderly villagers, persons from voluntary agencies, educated youth in the villages and workers of some research organizations engaged in the field investigations in the areas covered by our study.
4. Information on the extent of CPRs at taluka or district level was difficult to assemble. The secondary data or official records on land utilization while reporting the extent of forest, wasteland, etc, do not provide any indication about their ownership status (e.g., whether owned by government, private individuals or a village community, etc). Hence, it was impossible to estimate the extent of CPRs in the total area reported at taluka level.
5. Information from medium farm households (i.e., category (iii) was collected only from some and not from all the villages. Hence it is not reported in this paper.
6. Details about rural households, dependence on CPRs, and their employment and income gains from CPRs for Raisen and Nagaur districts as reported in this paper may slightly differ from the preliminary results about the same districts reported earlier (Jodha 1985a). This is due to the availability of additional

information and exclusion of medium farm households from this paper. This also applies to employment and income details reported in the paper.

7. The highest number of CPR items recorded from a small farm household in a village in Raisen district of Madhya Pradesh was 22. For a similar account of product collection and dependence of the rural poor on CPRs in the irrigated areas of Uttar Pradesh and tribal areas of Andhra Pradesh, see, Dasgupta (1985) and AERC (1985) respectively.
8. In view of the situation-specific relative role of these factors in different villages, it is difficult to explain the regional differences in the dependence of people on CPRs with reference to specific factors. This limitation will also apply to the regional differences in quantified benefits reported in the following discussion.
9. Large farm households used fuel for multiple purposes, e.g, cooking food and animal feed, making tea quite frequently, heating water for bath, etc. They consumed small quantities of kerosene as well. The poor used fuel mainly for cooking food and often only once a day. In most cases the fuel requirement was higher during the rainy season, partly because of the reduced fuel use efficiency due to humidity.
10. Though varying in its extent, the practise of 'salvaging' unproductive animals was observed in practically all the study areas. Large farmers gave their unproductive animals to the poor for maintenance, as it was costly to maintain them, at their own farm, using the feeding standards followed by the large farmers. When such animals became productive they were returned to the large farmer and net additions to the value of such animals {after becoming productive) was shared by the two parties. Depending on the type of animal, necessity of seasonal migration of animals due to fodder/water scarcity, and marketability of animals, the terms and conditions governing this practise differed from region to region. In areas like Gujarat and Rajasthan such herding was an important source of income for the rural poor.
11. We did it quite reluctantly as the productivity of child labour in terms of collection of a number of CPR products was often equal to or higher than that of adults.
12. The grazing employment days may involve some under reporting, as only those cases were included where grazing on CPRs (as against private lands) were clearly detectable from the worksheets of investigators.
13. George (1984) reported per household 125 mandays of additional employment, generated by IRDP in his study villages in Rajashtan. A study by NABARD (1984) reported additional employment of 109–163 mandays per beneficiary family due to various schemes (except tailoring) covered by IRDP in eight states. PEO (1985) reports 13–69 per cent increase in employment due to

IRDP schemes in different areas. However, absolute number of mandays of additional employment were not reported.

14. Income from CPR products was estimated by first making an inventory of the total products collected by respondent households during different seasons and then calculating their value on the basis of the village-level prices of the products. These prices were much lower than the corresponding prices even at the taluka town level.
15. Estimation of livestock income attributable to CPR grazing is based on data on animal-husbandry practises (e.g, stall-feeding, grazing including on CPRs) and on composition of livestock income. Apportioning of animal income to CPR grazing was simple for sheep, goats, and currently unproductive animals that were almost totally maintained through CPR grazing especially by the poor. In the case of animals involving both stall-feeding and open grazing only dung income was apportioned between CPR-grazing, private resource grazing and stall-feeding on the basis of animal-unit days involved. This, however, resulted in considerably underestimating animal income attributable to CPR grazing.
16. Levels and the regional pattern of CPR incomes (Table 14.10) may change if income from CPR product processing and marketing is considered. This was indicated by rough calculations based on limited cases of processing, marketing, and handicrafts based on CPR-products in different areas. They included soapnut in Aurangabad; forest products for basket, *pattal,* and mattress making in Mysore, Dharmapuri, Mandsaur and Raisen areas; sisal and other products used for making rope and other fibre in Akola, Sholapur and Mahbubnagar areas; wild fruits for making pickles and medicine in Nagaur and Raisen. These activities could add about 9–21 per cent to the already reported CPR-income in different areas.
17. A study by NABARD (1984) reported that for 10 per cent of the beneficiary households per household incremental income from IRDP in different states was upto Rs 500. For another 13 per cent it ranged between Rs 501–1,500. According to a PEO (1985) study, the per household incremental income from IRDP schemes ranged from Rs 1,069–Rs 2,770. However, in several areas the incremental income from different schemes was less than or around Rs 600 per family. Kutty Krishnan (1984) reported that 39 per cent of the participants in IRDP schemes in his study village in Kerala did not have any change in their income. Another 29 per cent of the families got incremental income of less than Rs 400 per family.
18. Land reforms served as an important event to facilitate the recall of the past by the people. Moreover, the status of CPRs was greatly affected by various by-

laws and side effects of land reforms (Jodha 1985a, 1985b). Hence it was quite convenient to gather details from oral history and records before this change.

19. All the village waste lands unlike other CPRs listed here do not fall under the jurisdiction of village or village panchayat. They belong to State Revenue Departments. However, since they are common access resources, for the concerned village community they are village CPRs for all practical purposes. Historically, these commons were categoriser as wastelands not on the basis of their extent of biological or economic productivity. Being uncultivable (or uncultivated in some cases), they were non-revenue generating lands. Hence, the colonial government classified them as 'wastelands'. For some implications of this categorization, see Shiva (1986).
20. In some cases, area of CPRs declined due to transfer of land to institutions (e.g., milk cooperative, village school, public works department, forest department, etc.). For a brief account of causes and consequence of transfer of forests to state control, see Singh (1986).
21. The details on this aspect were collected in the following manner. First, the land parcels (currently privately owned lands that had CPR status in the past) were identified. Then the households, which received ownership of the land under various welfare programmes or through grabbing were identified. Information about the resource position of these households was collected. Current ownership and use status of these lands were also recorded. For understandable reasons it was not possible to cover all privatized CPR land. The extent of the land left out broadly ranged between 18–32 per cent in different villages.
22. 'Others' in Tables 14.13–14.15 do not necessarily include only large farmers unlike the data reported by rest of the Tables. In the latter, 'large farmers' indicated by 'Others' were deliberately selected. In the case of privatized CPR lands, the primary step was selection of land parcel and then an enquiry into who got it. The recipients of CPR-land parcel included people other than large farmers also.
23. This suggests that poor people's approach to land is not different from other assets given to them under various anti-poverty programmes. Based on more than a dozen evaluation studies of IRDP, Rath (1985) summarizes the extent of current non-possession (22–72 per cent) of the assets initially distributed to the rural poor under the integrated Rural Development Programme in different states.
24. Since in most of the areas there were legal restrictions on purchase of land from economically and socially backward groups, leasing of land was used as a cover for acquiring such lands.

## REFERENCES

AERC, 1985, *Minor Forest Produce in the Tribal Economy in Andhra Pradesh,* India: Agro-Economic Research Centre, Andhra University (Limited distribution) Waltair, Andhra Pradesh.

Binswanger, H.P., and Jodha, N.S., 1978, *Manual of Instructions for Economic Investigators in ICRISAT's Village Level Studies*India: Economics Programme, ICRJSAT (Limited distribution), Patancheru, Andhra Pradesh.

Das Gupta, Monica, 1985, *Informal Security Systems and Population Retention in Rural India,* NCAER Working Paper No. 2, National Council of Applied Economic Research (Limited distribution), New Delhi.

George, K.M., 1984, 'Rural Development Programme: Its Strength and Weaknesses', *Indian Journal of Agricultural Economics,* 39(3): 434–43.

Jodha, N.S., 1973, 'Special Programmes for Rural Poor: The Constraining Framework', *Economic and Political Weekly,* 8(13); 633–39.

———, Ashokan, M., and Ryan, J.G., 1977, *Village Study Methodology and Resource Endowment of Selected Villages*India: Occasional Paper 16, Economics Programme, ICRJSAT (Limited distribution), Patancheru, Andhra Pradesh.

———, 1985a, 'Market Forces and Erosion of Common Property Resources', in *Agricultural Markets in the Semi-Arid Tropics,* Proceedings of the International Workshop, October 24–28, 1983, 1CRISAT Centre, India. ICRISAT, Patancheru, Andhra Pradesh, pp. 263–77.

———, 1985b, 'Population Growth and the Decline of Common Property Resources in Rajasthan, India', *Population and Development Review* 11(2), pp. 247–64.

Kutty Krishnan, A.C, 1984, 'A Case Study of Integrated Rural Development Programme in a Kerala Village', *Indian Journal of Agricultural Economics,* 39(3): 419–26.

NABARD, 1984, *Study of Implementation of Integrated Rural Development Programme,* NABARD (Limited distribution), Bombay, India.

PEO, 1985, *Evaluation Report on Integrated Rural Development Programme,* New Delhi, India: Plan Evaluation Organisation, Planning Commission, Government of India.

Rath, Nilkantha, 1985, 'Garibi Hatao': Can IRDP Do It? *Economic and Political Weekly,* 20(6): 238–46.

Shiva, Vandana, 1986, 'Coming Tragedy of the Commons', *Economic and Political Weekly,* 21(15): 613–14.

Singh, R.P., and S.B. Singh, 1982, *Features of Traditional Farming Systems in Two Villages of Gujarat, Patancheru*, AP, India: Progress Report 35, Economics Programme, ICRISAT (Limited distribution).

Singh, R.P., B.L. Jain and V.B. Rao, 1984, *Features of Traditional Farming Systems in Selected Villages of Madhya Pradesh,* Progress Report 61, Economics Programme, ICRISAT, Patancheru, Andhra Pradesh.

Singh, Chhatrapati, 1986, *Common Property and Common Poverty: India's Forests, Forest Dwellers and the Law,* OUP, Delhi.

15

# WATER POLLUTION CONTROL: ROLE OF COMMUNITY AND PUBLIC ACTION

**Himanshu Thakkar**

## INTRODUCTION

Rising, and increasingly widespread, incidence and levels of pollution of surface and groundwater sources in different parts of the country is a matter of grave public concern. Pollution of the Ganga, Yamuna, Sabarmati, Lower Bhavani, and Palai have attracted nation-wide attention. But the problem is far more widespread. According to the National Environmental Engineering Research Institute (NEERI), a staggering 70 per cent of available water in India is polluted. 'From the Dal lake in the North to the Periyar and Chaliyar rivers in the South, from the Damodar and Hooghly in the East to the Thana creek in the West, the picture of water pollution is uniformity gloomy. Even our large perennial rivers like the Ganga are today heavily polluted' (CSE 1982; 1984–85).

Groundwater pollution, which is more dangerous because it is more difficult to reverse, is also becoming a major problem. Its seriousness is illustrated by the prevalence of arsenic poisoning in Bengal, the growing salinity problem in Gujarat, and the prevalence of high levels of pesticide residues in groundwater in several areas.

1. Scientists from Jadavpur University estimate that 38 million people in eight districts of the state of West Bengal are exposed to health

---

* Reprinted with permission from Himanshu Thakkar, 2004, 'Water Pollution Control: Role of Community and Public Action' in Vaidyanathan and H.M. Oudshoom (eds), *Managing Water Scarcity: Experiences and Prospect*, Manohar, Delhi. This paper was written in mid-1997 and last updated in late 1997. The main conclusions of the paper remain valid.

risks from the high arsenic content of groundwater, an important source of drinking water. The increase in arsenic content is attributed to excessive pumping of groundwater for irrigation following the advent of high-yielding paddy varieties.

2. In Gujarat, the area affected by salinization in 19 *taluks* around the Gulf of Khambat is estimated to have increased from about 7–8 per cent in 1960 to 64 per cent in 1993. About a fifth of the state's lands are currently afflicted by this problem and, in the absence of effective corrective action, is projected to reach 30 per cent by 2000.
3. In Delhi, one-third of the groundwater samples examined were found to have nitrate concentrations above the permissible limit; the incidence of DDT in the Yamuna waters at Delhi and that of T-HCH and DDT in the Cauvery was found to be very high.

These are but illustrative examples. The available data and institutional arrangements for measurement and monitoring are inadequate to get a comprehensive picture of water quality. Nevertheless, the fact that the main sources of pollutants—namely agricultural chemicals, untreated industrial effluents, and urban sewage—are growing rapidly, with little being done to contain the process, strongly points to a progressive aggravation of the problem. The consequences to the health of the people and to the sustainable use of water resources over the long run are very serious indeed.

## The Government's Response

The government has declared its good intentions about pollution control through numerous policy statements and formal legislations.

- The National Water Policy (Ministry of Water Resources 1987) declared: 'Economic development and activities including agricultural, industrial and urban development, should be planned with due regard to the constraints imposed by the configuration of water availability.'
- The National Conservation Strategy and the Policy Statement on Environment and Development was adopted by the Government of India in June 1992 to provide the basis for the integration and internalization of environmental considerations in the various policies and programmes of the government.

- In 1994, a far-reaching notification by the Ministry of Environment and Forests (MEF) made it obligatory for almost all developmental and industrial concerns to conduct Environment Impact Assessment (EIA) studies.
- In a decision intended to make industry more accountable to the people, the Cabinet Secretariat, in January 1997, made public hearings mandatory before giving environmental clearance to any industry project or activity, whether private or governmental.
- In April 1997, by an extraordinary gazette notification, MEF established the National Environmental Appellate Authority (NEAA), in exercise of its powers under a 1997 Act. The NEAA is to deal with complaints against developmental projects that fall under the EIA of 1994. 'The aim is to enable people directly affected by developmental and industrial projects, as well as NGOs, to present their point of view and ensure transparency in the process of EIA.'

But these steps do not seem to have made much of an impact. Pollution continues to spread and increase in intensity in spite of these pronouncements. There are several reasons for this. For one, when the government acts both as a promoter and a regulator of economic activity, often there is a conflict between these two roles. In practise, the government has acted more as a promoter than as a regulator. It seems in the name of development everything is expendable, including the health and future of the people. Worse, when responses of the government are looked at in the courts, one finds that the government is only trying to defend the interests of the polluters. For effective pollution control, the pollution control bodies will have to be independent of the government.

Pollution control agencies being part of the government do not have the independence and autonomy to initiate action or enforce the provisions of the law. Monitoring mechanisms are weak, and often absent; where they exist, the information they collect is not available to the public; and there is little coordination between the concerned departments. Nor do the elected representatives and legislatures show any interest in pollution-related issues, much less in exerting pressures for remedial action. On the contrary, they act more often in ways which favour, if not actively support, the interests of the polluters.

Since the affected communities have a direct stake in controlling pollution for the health and well-being of their members, it is essential that they have access to information regarding the state of their water sources and the right not only to voice their concerns but also to seek effective redress for damage already done and to prevent potential damage in the future from development projects. The laws do not provide for consultation, not to speak of the involvement, of the affected communities in government processes concerning pollution monitoring and control. Attempts by the affected communities to seek redress through petitions and appeals to the government rarely yield a positive response.

Frustrations with the ways in which the official regulatory framework functions have led communities severely affected by water and other pollution to take recourse to public protests, mass agitations, and the courts. The responsive attitude of the judiciary to public interest litigation has many times facilitated the raising of pollution complaints in the courts. There has been a significant growth of such community–NGO initiatives. The purpose of this Paper is to focus on the experiences in certain specific cases drawn from various states in which affected communities or citizens' groups have played an active role. The strategies they followed and the role and response of other actors (namely, polluters, the government, and the courts) is also examined briefly in some cases.

These are not meant to be exhaustive accounts, but essentially to give a feel of the kind of issues involved, the attitudes of the concerned parties, and the difficulties of securing full and effective redress, to indicate the potentials as well as the limitations of such public activism by communities and NGOs. We argue that the active involvement and participation of local communities is essential to bring the problems onto public platforms, to promote informed discussions of solutions, and to force the concerned public agencies (pollution control authorities, courts, and the government) to take appropriate remedial action.

## Case Studies of Public Action

### Gujarat: Campaign to Save the Narmada River

In Gujarat, one of the most industrialized states in the country (also earning the sobriquet 'most polluted state', as noted by the Gujarat Ecology Commission (May 2001), vying to be the most industrialized, more and more instances of severe water pollution are coming to light, particularly in what is rather proudly called the golden corridor between Valsad and Mehsana. A number of cases were filed in the Gujarat High Court against polluting industries in and around 1995. These were mostly at the instance of concerned citizens; the affected communities and NGOs were not involved. The court did act and passed a number of orders for closing down hundreds of industrial units. These orders made headlines in newspapers in 1995. However, possibly due to the economic clout of the industries and the political support of the government, these units restarted operations without much change within months. The weaknesses of the cases, filed without the necessary homework, were also evident (Chakraborty 1995).

The courts' intervention resulted in some remedial action in terms of the installation of treatment facilities. But these were limited in both scale and impact. Effective pollution control remains a distant goal. Even today the industries are merrily operating without much change in the situation. However, the cases helped to generate greater awareness among the public about the polluting nature of industries and stimulated greater community and NGO involvement in these issues.

One such initiative was taken by the people of the villages around Mehgam in the Amod *taluka* of Bharuch district in central Gujarat. In 1993 they approached the Manviya Technology Forum (MTF), a voluntary organization active in the area, to say that the Gujarat Industrial Development Corporation (GIDC) was constructing a pipeline to release effluents from the proposed Vagra GIDC industrial estate into the Narmada river and was destroying their land and standing crops in the process. Soon an organization called the Narmada Pradushan Nivaran Samity was formed, and letters were written to the relevant officials asking if any environment impact study of the proposed Vagra industrial estate had been done. According to the GIDC officials only an engineering study was done and no environment impact study was done regarding the release of effluents. The Samity also discovered that the proposal did not have the sanction of the Gujarat Pollution Control Board (GPCB).

The people had no option but to stop the work of the pipeline. Soon the fishing community of the villages on the banks of the Narmada, fearing the destruction of the rich estuarine fisheries if the effluents were released into the river, also joined the struggle. Their point was that if the upcoming Indian Petrochemicals Corporation Limited complex in the same area was required to release its effluents into the sea and not into the river or the estuary, then why was the GIDC or any other agency being allowed to release the effluents into the river? The Bharuch Fisheries Directorate also opposed the release of any effluents into the river. In October 1995, the collector told the GIDC to give up its plan to release effluents into the river or the estuary and hand over the study to the NEERI to find out where the effluents could be released into the sea. The GIDC has yet to take a decision on this issue. But in the meanwhile, both the work of the pipeline and the setting up of the industrial estate remained stalled. For the people this was a victory. This was to be the first step in the struggle of the Samity to control the pollution of the Narmada river.

## More Struggles Against the Pollution of the Narmada

As the news of this breakthrough spread, the people of Piramal village in the Bharuch laluka approached the Samity, saying that they were facing a similar problem. The GIDC was constructing a channel from its upcoming Jhagadia industrial estate to release effluents into the Narmada at Haripura village, which is just upstream of the point where Amlakhadi had been releasing untreated pollutants from the Ankleshwar and Panoli GIDC estates for over two decades now. People on the banks of this 28 km-long Amlakhadi were living with this poisonous channel which had killed the fish in the rivulet, killed or maimed the domestic animals who drank water from it, flooded their lands, and destroyed their groundwater. For them, the proposed new channel was like rubbing salt in their wounds.

When the activists mobilized the people and stopped work on the channel in an attempt to pressurize the authorities, the response from the various quarters was rather quick. The GIDC, the industrialists and the administration immediately called a meeting. Fresh from their victory at Vagra, the people put up the same grounds for not allowing the release of any effluents into the river. It was soon discovered that no EIA of this new industrial estate had been done. Strangely, the GPCB had given its consent for releasing the Jhagadia effluents into the river, contrary to the statement of the GPCB member secretary that 'We are not giving consent to anyone to release effluents into a freshwater body, including the river.' The district collector, who called the meeting to resolve the issue, could not find fault with the agitating people.

But in this case the GIDC was desperate. Four industries at the Jhagadia estate were ready to start production, and if more industries were to be attracted, the problem could not be allowed to linger. In a meeting called by the collector on 18 January 1996, the GIDC agreed that a study to find out where in the sea and through what kind of channel the effluents of the Jhagadia, Ankleshwar, and Panoli estates should be released would be given to the NEERI; that till the effluent channel was ready, the existing four industries would not be allowed to start operations at the Jhagadia industrial estate; and that the last 8 km. of Amlakhadi would be cleaned up before the 1996 monsoon at the estimated expense of Rs 65 lakh to be shared by the industries, the GIDC, and the Members of Parliament from the area.

This success was a major boost for the people. The channel was desilted before the monsoon. Though only a temporary measure to reduce the flooding of the area, for the people of the village on the banks of the channel, it was a great boon indeed. However, the other clauses of the agreement of January 1996 are proving to be much more difficult to implement.

First, even as the rapid survey report by NEERI was awaited, officials are raising the finances for the effluent channel to the sea, estimated to cost over Rs 350 crore. The industries which have for decades released untreated effluents into the river for free would not be willing to spend that kind of money. When last heard, the World Bank was considering funding this project.

Second, the GIDC is already wavering about not allowing any more than the existing four industries at its Jhagadia estate to start operations. In an interview with me, the deputy chief engineer of GIDC asked:

> what is the logic in such an understanding? We are not allowing any industry to come up in Jhagadia without effluent treatment. And you know that the effluents of both the Ankleshwar and Panoli estates have been flowing into the Narmada without any treatment for decades. Which is more polluting? Anyway, that understanding was reached only by our local officer

implying thereby that they were unlikely to honour the agreement reached with the protesting people in January 1996.

While this struggle was going on, Rajshree Polyfils Ltd., a plant of the Birlas, was constructing an effluent channel through Umarva–Sir–Bamalla and other villages to release effluents into the Narmada at village Navra.

Strangely, the GPCB had given its consent for this release at Navra village just upstream of the point from where the Gujarat Water Supply and Sewage Board (GWSSB), GIDC, IPCL, Bharuch city, and a number of other units were lifting water from the river for domestic and industrial uses. When I asked GWSSB and GIDC officials if they knew of such a move, they were shocked to 'know' that this was indeed happening.

People, of course, stopped the construction of the effluent pipeline in March 1996. The GPCB said in June 1996 that they were reviewing the consent for the release of the effluents of the Birla unit. This review led to the cancellation of the consent. When it was pointed out that in a similar way the industries in the Valia *taluka* of Bharuch district were operating in a landlocked area where the effluents from the industries were entering the Amaravati river—a tributary of the Narmada—from where GWSSB was lifting water for villages of Valia *taluka* and Ankleshwar town, the GPCB member secretary said that the consent of these Valia industries too was being reviewed. While this was another round of victories for the people, it also showed the utter lack of consistent policy for pollution control in the Gujarat administration.

## Bichri Village in Rajasthan

Bichri in Rajasthan is another case where in spite of several scientific and official reports showing that an industrial establishment, Silver Chemicals, was responsible for making the groundwater—the only source of water for the villages around—toxic and non-potable, no action was taken. The case was taken to the Supreme Court. Strong protests by the local people backed by a number of NGOs led to the closure of the factory in 1988, but the guilty could not be punished. Not surprisingly, the victims were never participants in the decision-making process after 1988 when the factory stopped working.

The Supreme Court ruled in favour of the petitioners and upheld the principle of absolute liability of the polluter, that is, when an enterprise is engaged in a hazardous or inherently dangerous industry posing a potential threat to the health and safety of persons either working in or residing in the surrounding areas of the factory, it owes an absolute and non-delegatable duty to the community to ensure that no harm results to anyone as a result of this activity. If any harm does occur, the enterprise must be held liable to compensate for it.

This was indeed a landmark judgement. However, some people argue that exemplary damages should have been awarded to the people of Bichri, and that the Rajasthan government and the MEF should have been held guilty for allowing the polluting industries to come up in the first place. Till date, not a paisa has been paid as compensation. Mannarma, a Bichri farmer, pointed out that the nation's three main acts concerning the environment are defective as they put the onus on the ill-functioning PCBs, which lack adequate and popular representation (Shankar 1993; Sharma 1996).

Some seven years after the closure of the main errant units, in February 1996, the Supreme Court ordered that five units of a particular company producing toxic chemicals in Bichri village in Rajasthan's Udaipur district be closed down. In the meantime, the same company has reportedly shifted base, and is merrily doing an encore in Vapi, Gujarat, arrogantly confident that no one can touch it. Events over time have vindicated this confidence.

## Madhya Pradesh: Pollution of the Betwa River

Over the years, the Betwa river, lifeline of the obscure and sleepy town of Vidisha to the east of Bhopal, had degenerated into a sewer because of the effluents from a distillery. A movement to restore this lifeline took shape gradually, cutting across social strata and political ideologies, and bloomed into an un-ignorable protest against the pollution of the river. The incidence of dead fish in the Betwa in 1993 did lead to some turmoil, and the distillery was then ordered to shut down, but the distillers obtained a stay order from the Gwalior High Court. Subsequently, the government was not able to vacate the stay.

The local administration, however, played a crucial role of keeping open the information channel to its citizens. All political parties decided to set aside their differences to work together on this issue. The entire business community also threw in their lot. The deputy chief minister of the state agreed that the studies carried out by the Madhya Pradesh PCB had proved that Som distillery, Sehatganj in Raisen district, was responsible for the extreme pollution loads in the river. And yet the system failed to act. In fact, he was quoted in the press as saying that the state government and the official machinery had sold out to the liquor mafia, referring obviously to the infamous Som distillery. That led to an uproar in the assembly, forcing the chief minister to call a meeting and directing the MPPCB to order the closure of Som distillery. But even then the chief minister said that his government was not responsible for the mess and that only the MPPCB was responsible. To gather mass support for the movement, for three days in November 1996 the people of the town were contacted on the street corners and requested to sign letters of pledge. In the process, they were individually educated on the issue, and their future support pledged. Over twenty thousand letters were signed. Paryavaran Vahini, the citizens' forum for action on the local environment led the movement. Madanlal Sharma, affectionately called *Betwa ke baba*, has been trying to educate the local people about the necessity of keeping the river clean for the last 22 years.

The mass death of fish in the upstream *anicut* in July 1996 after such incidents had earlier led to the final ignition of the movement against pollution. Ultimately, in December 1996, a distillery owned by a powerful industrial house, identified as the major source of the pollution, was closed down. The quiet exit of an environment minister who had staunchly defended the polluters and, on the face of it, a 'concerned' government valiantly rushing to the rescue of the harassed people *after all* the fish had died were some other notable gains of this struggle (Banerji 1997).

## Goa: The Nylon Plant Struggle

The struggle against the setting up of a nylon plant in Goa is a good example of how a movement of people can prevent a potentially polluting industry from being established. The government formed by the Progressive Democratic Front was forced to appoint a Legislative Assembly House Committee, which took the unusual and unprecedented step of going in for a public hearing to decide on the matter. The Committee recommended that the project should be shelved on account of the likely threat of pollution: '... it would be an ill-advised move to allow large chemical industries discharging even their treated effluents into our eco-rich and virgin rivers. . . . We have to safeguard our rivers against any conceivable environmental and pollution threat....' The report also indicted the government and government agencies for bypassing the well-laid down procedures for grant of licenses, selection and grant of sites, and other related matters. Among the lapses noticed was the failure to get an EIA for the project, which is mandatory before any clearance is given. The Goa State Pollution Control Board was pulled up by the Committee for its failure to uphold the mandate given to it under the provisions of the Water (Prevention and Control of Pollution) Act, 1974 and the Air Act of 1981 as the documents pertaining to the no objection certificate were found to be very superficial. Despite such a scathing report, the project went ahead under a new government that followed the PDF-ruled government.

The decade-long people's agitation in Goa led by the Anti-Nylon Citizens' Action Committee also launched an awareness campaign about the harmful effects of the project, about the outdated technologies being used, and the track record of the company. The issue made such an impact that it figured in the panchayat elections held in 1991 and those who opposed the project were elected. The local panchayat refused to give permission for construction activities at the project site but the government authorities invoked some deemed provisions in the law to give the go-ahead to the project's promoters (Mohan 1995). It was only after a prolonged agitation by the people in 1994–95, resulting in police repression including arrests, firings, and deaths, that the project was finally shelved and was forced to go to another state, Tamil Nadu.

## The Zuari Agro Chemicals Factory

In another case, Zuari Agro Chemicals, a Birla concern supported by US Steel, the International Finance Corporation, and the Bank of America, started operations in 1973. Within three months fish began to die in the water bodies around the factory; within a year cattle were also dying in the area; and in another year, coconut trees began to wither away. The factory was shut down by an order under Section 133 Cr.P.C. in 1974, six months after it had formally opened. This was the beginning of a legal battle whose eventual outcome shows how even dogged perseverance may not be rewarded by adequate judicial relief.

By 1975 protests began. A Citizens Anti-Pollution Committee was formed. The Goa Government gave an ultimatum to the factory in April

1975, but without effect. The district magistrate ordered the immediate closure of the factory in the same month. At different stages of protests, different political parties supported the campaign of the villagers. Ultimately, the company had to pay compensation, provide drinking water to the affected villages, and install pollution control equipment under the control of the then Central Pollution Board.

In 1983 a case was filed for a permanent injunction against the ZAC in the district and sessions court of South Goa. This was rejected on technical grounds. After the enactment of the Environment Protection Act (EPA) in 1986, it was decided to file a case under the EPA. However, in order to prosecute under the EPA, it is necessary that air and water samples be taken and analysed by statutory agencies. Even after a notice was given of the intention to prosecute the ZAC under Section 19(b) of the EPA, neither of the authorized agencies, i.e., the state PCB or the Inspectorate of Factories, would take the required samples. Without the analysis of the samples by one of these agencies, a criminal suit cannot be entertained under the EPA or under the Air and Water Acts.

A petition was filed by the Goa Foundation and citizens praying that it was incumbent upon the two authorities appointed under Section 11(1) of the EPA to take samples of air and water within a specified period on the requests from citizens and also to disclose to the parties the results of the analysis and the inquiries carried out as a result of the complaints filed under the EPA. The Inspectorate of Factories refused to take the samples since the alleged pollution was occurring outside the factory premises and the Inspectorate of Factories had no jurisdiction outside the factory. The state PCB stated that it was not bound to take any samples under any provision of the statute! The petition drags on. The villagers of Sancoale, where the offending unit is located, have had an ongoing battle with the ZAC for over two decades now (Alvares 1994; CSE 1982).

## Maharashtra: Sterlite Industries

The story of why and how Sterlite Industries decided to cut its losses and leave Ratnagiri in Maharashtra is one of the more significant events that shows what a well-informed movement against a potentially polluting industry can achieve. The people of Ratnagiri, a coastal district in southern

Maharashtra, successfully stopped a copper smelter unit from being put up in their area' in 1994 as they feared that the unit would lead to the pollution of their land, air, and water. The 60,000-tonnes-per-annum capacity copper smelter plant costing Rs 700 crore was to be put up by Sterlite Industries. A report by an expert committee instituted by the state government vindicated the opposition of the people, saying that such an industry would endanger the region's fragile ecosystem. Sterlite nevertheless went ahead.

Unfortunately for Sterlite, the *sarpanch* of the Zadgaon gram panchayat, within which the factory promises fell, happened to be a former civil servant. He understood the law and knew that the company should have sought the permission of the gram panchayat before beginning construction. When he sent a *gram sevak* to serve notice on the factory, the man was thrown out by the general manager.

This led the gram panchayat to send a demolition notice. In May 1993, the *sarpanch* called a press conference to declare that the gram panchayat planned to rent bulldozers and demolish the plant. The response of the state government, quite significantly, was to promulgate a new ordinance saying that the approval of the Maharashtra Industrial Development Corporation was mandatory for such construction (Sharma 1995; Shrivastava 1994).

It took years of strident and sustained protest by the local people to stop this unit. The struggle included direct action by some 35,000 people who marched to the factory site and demolished the sheds of the factory. A local newspaper, *Ratnagiri Times*, played a very active role in the campaign. It set about finding out from independent environmental experts the consequences of setting up such a large copper smelter plant within a short distance of a heavily populated urban area. Once it was known from environment experts that the plant could not be located so close to the city, a variety of lay people and professionals came together and formed the Ratnagiri Bachav Saugharsh Samiti. A social boycott was declared on all the employees of the company. The employees were asked to vacate the houses by their landlords, and even autorickshaw drivers would not transport people employed by Sterlite.

As a result, Sterlite was forced to abandon its Zadgaon site. It is now putting up the plant in Tamil Nadu, where too it is facing some agitation, though not as effective as the campaign in Ratnagiri. Two copper smelter plants are also being set up in Bharuch, Gujarat despite some local resistance.

## Delhi: Pollution of the Yamuna

In early 1997, after some months of study, the Centre for Science and Environment, an NGO, came out with a report on pesticide pollution of the Yamuna river. The report was deliberately released at a public meeting by a former prime minister. The media gave the event prominent coverage as the report focused on the presence of pesticides in the drinking water supplied to Delhi.

A debate was raised in the Delhi Legislative Assembly following the release of the report. The Delhi Water Supply and Sewarage Undertaking (DWSSU) came out with blanket statements asserting that these were myth-spreading reports and that there were no pesticides even in the raw water. With time these statements were changed to the effect that the pesticide levels were within the prescribed limits, and then further that efforts were being made to meet the standards. A citizen picked up the report and filed a public interest petition in the Delhi High Court, alleging that the right to life of the citizens had been violated. The CPCB agreed with the CSE's report in the court. The Haryana PCB, however, which was made party in the case, as Haryana was the pollutant state in this case, blamed Delhi for the pollution of the Yamuna river upstream of Delhi! Dozens of reports have been filed in the newspapers. The debate and the court case go on.

Some significant gains have been achieved, some awareness has been generated about the importance of the quality of drinking water. The DWSSU has agreed that they had not been monitoring the pesticide levels in the Yamuna waters. They have now decided to buy equipment to constantly monitor the same. They have also started talking about acquiring the appropriate technology to treat water to remove the pesticides. However, as the experience of other countries shows, technological solutions are, if not impossible to implement, prohibitively costly. The authorities will have to seek management solutions to see that pesticides do not enter the water. For once pesticides enter the water it is difficult to remove them (Agarwal 1996a).

According to a study by some scientists of the Nuclear Research Laboratory of the Indian Agricultural Research Institute, 33 per cent of the groundwater shows nitrate contents exceeding the general acceptable limits (20 ppm), and 15 per cent crossed the maximum permissible limit (45 ppm). Health effects of long-term exposure to nitrates is potentially serious.

A recent report by the Industrial Toxicology Research Centre, Lucknow, says that wherever surface waters were polluted, practically all pesticides were found in higher concentrations in groundwater compared to surface water, some of them higher by over 100 per cent. Once the groundwater is contaminated, it is practically impossible to remove the pesticides and more people depend on groundwater thinking it is safer than the surface waters. The other significant observation of the report is that the range of pesticide residues in drinking water in rural areas was comparable to the concentrations in the major urban centres like Delhi, Kanpur and Kolkata. Thus, rural areas are as vulnerable to the adverse impact of such non-point sources of pollution, perhaps even more because, unlike urban areas, people in rural areas generally drink untreated water (Anon 1996; CPCB 1996; Kumar 1995).

## Karnataka: The Mangalore Refinery

In Karnataka, strong pressure from affected constituents forced the Mangalore Refinery and Petrochemicals Limited to install a Rs 200 crore effluent treatment plant and set up measures to recycle the treated effluents. In December 1995, thousands of fisherfolk assembled to stage a *rasta roko* on the national highway. The Karnataka State PCB had given a clean chit to the project on the water pollution aspect. But the data and analysis done by the National Institute of Oceanography did not correspond with the survey conducted by the state, the centre, or the MRPL. The government's stand since the beginning had been skewed in favour of the company. Finally, in December 1995, the government had to issue an order whereby the MRPL was directed to 'install suitable equipment and facilities for the recycling of waste and removal/disposal of sludge, if any; and maximum extent of recycling will be achieved to conserve water and at the same time to minimise discharge of effluents into the sea' (Sitaraman 1996).

However, the refinery started partial production without the effluent discharge pipeline in place. Within three months, by July 1996, a large amount of oil-contaminated water leaked out of MRPL, damaged crops on over 200 ha. land, and polluted practically all the sources of water. Activists of the movement immediately identified the culprits. Faced with a furore from the opposition in the legislative assembly, the state government, otherwise an apologist for the unit, had to warn the company that if it refused to pay compensation to the farmers and install a recycling plant, the company would not be allowed to restart production. The state PCB, invoking Section 32(1)(c) of the Water Act, prohibited the company from any discharge outside the plant's premises. None of these actors would have acted the way they did in this instance but for the expression of strong protest by the people. And the people could muster their courage and strength to protest only because they had knowledge and experience (Krishnaswamy 1996).

The people of Bidar in northern Karnataka launched an agitation against some drug companies because they seemed to have brought in more misery in the form of widespread pollution than prosperity to the region. In their enthusiasm to develop Bidar, the official machinery totally ignored the need to regulate the disposal of toxic substances by these factories. Soon the region's air, water, and land were in a state of deterioration. For over three years, the Karnataka Vimochana Ranga has been organizing the villagers against the polluters. The government has been compelled to constitute a committee, which came out with a very unfavourable report with respect to the industrialists. Arguing that the very purpose of developing industries is being defeated as it is becoming dangerous to human life, the report recommended urgent measures and ordered the temporary closure of five polluting factories until remedial measures were undertaken.

But some factories procured a stay order from the Karnataka High Court and continued to pollute the area. In June 1995, an expert committee was constituted whose report, known as the Kudakavi report, reiterated the need for a treatment plant. When neither the government nor the Karnataka PCB took any action to bring this into effect, an intensified agitation was launched. When even after the vacation of the stay order by the High Court, on 19 September 1995, the KPCB did not take action to close down the factory, on 11 October 1995, the people decided to march to the factories to lock the four polluting drug factories in the Kolar industrial area. The action invited violent encounters with the Bidar police over the next few Jays, Finally, live days later, unable to control the mounting pressure from people from all over the state, the chairperson of the KPCB was compelled to invoke Section 33A of the Water Act, and order the shutdown of the four units (Sitaraman 1996).

## Andhra Pradesh: Patancheru

An industrial estate in Patancheru, just off the Hyderabad–Pune highway in Andhra Pradesh, lets out untreated industrial effluents from the estate into the Nakkavagu drain. As a result, the groundwater was poisoned, affecting hundreds of acres of agricultural land and also the drinking water of about 15 hamlets.

According to the Status of Environment in Andhra Pradesh, Citizen's Report, 1990, published by the Academy of Gandhian Studies, Hyderabad,

> Since its inception, the work of the Andhra Pradesh Pollution Control Board (APPCB) has been more on the lines of pollution-promotion rather than controlling it. Most of the time, it (the APPCB) pursues loosely structured, sham legal cases.

The Campaign Against Pollution was launched by some NGOs and citizens, which led to a petition being filed in the Supreme Court, among other outcomes.

Following the filing of a PIL in 1990, the Supreme Court directed the district judge of Medak to inspect the location and submit a report. This report, submitted in November 1995, concluded that over 100 industries were responsible for causing groundwater Pollution in the region. In May 1996, the Supreme Court directed that the guilty industries reimburse Rs 1.5 crore in compensation to the people, set up a proper effluent treatment plant, and the government supply treated municipality water to the villages. The long struggle had led to some improvement in the situation, which would not have come about it the industry and the government were left to themselves. But the industries continue to operate and release untreated effluents as before. The struggle continues.

## Tannery Pollution in Tamil Nadu

The North Arcot district of Tamil Nadu is a good example of how an industry can ruin the environment. Nearly 70 per cent of India's total exports of leather and leather products come from 700 odd tanneries in Tamil Nadu. Ambur is one of the worst affected towns. Untreated effluents from about

250 tanneries, most of which are located in and around the towns of Ambur and Ranipet, have almost turned the land around these towns into wasteland. Crop production is down by more that 70 per cent. The water is undrinkable. Respiratory and skin diseases are widespread.

IIT Madras and the King Institute in Madras found in the early 1980s that the water supply in Ambur was unfit for human consumption. The factories were then given a year's time to install effluent treatment plants. But nothing was done.

Over half a million people have been affected following water pollution by the tanneries. Besides laying waste over one lakh hectares of farm land, they have contaminated ground and surface water, making it unusable. An estimated 40 million litres of untreated effluents are discharged daily into the rivers, irrigation ponds, and wells. The worst hit is North Arcot-Ambedkar, which, with the largest number of tanning units, accounts for over 80 per cent of the foreign exchange earned from leather processing in the state. Over 35,000 ha. is now unfit for cultivation and the area under paddy has been reduced by half.

The problem, as the industry sources say, began in the mid-1970s with the Centre bringing pressure on this sector to raise exports by fixing targets for the states. This led to a scramble to produce more in the shortest possible time, and the tanners switched to the faster chromium process route from the eco-friendly vegetable process that uses barks and plant extracts. The new process reduced the time and space required for tanning, but it meant the entry of highly toxic waste, which was—for want of ETPs or other pollution control measures—let out into the open. The TNPCB, set up in 1982, let the problem fester by not doing anything beyond issuing notices to the tanneries (in one case, a single unit was warned 25 times) to set up ETPs (Krishna Kumar 1996).

An NGO, Community Action for Development, has been active in mobilizing people against the pollution. The secretary of the Vellore Citizens Welfare Forum, on whose public interest petition the case was taken up by the Supreme Court, explains: 'Besides seeking the restoration of the ecological balance, this time we have also insisted on providing adequate compensation to the affected people.' Following their petition, 54 tanneries were closed in the North Arcot-Anibedkar district in May 1995 by the Supreme Court.

In another landmark judgment, the Supreme Court, in April 1996, ordered that more than 200 tanneries in Tamil Nadu be shut down for failing to install effluent treatment facilities as required by law. The century-old industry in the state has played havoc with the lives and livelihood of the people, irreversibly fouling up fields, crops, flora and fauna, and destroying living spaces by releasing toxic wastes. The managements of the tanneries were prompt in appealing to the Supreme Court to reconsider its order to close down the units. The court rejected this plea, and when the tannery owners said that the installation of pollution-control systems would cost over Rs 100 crore, the judges said: 'These units must know that they cannot operate like this.'

In arriving at its decision, the Supreme Court was aided by the National Environment Engineering Research Institute's February 1996 report on the status of CETPs. According to this, of the 30 ETPs identified in the tannery cluster of the five districts of North Arcot-Ambedkar, Erode-Periyar, Dindigul-Anna, Tiruchi, and Chengai-MGR, only seven were operational. In fact, none of these operational ETPs met the standards laid down by the Union Ministry of Environment and Forests or the TNPCB.

In August 1996, on the basis of a petition from the Vellore Citizens Welfare Forum, the Supreme Court imposed fines of Rs 10,000 on each of the 700-odd tanneries in Tamil Nadu, and asked them to pay compensation for polluting the environment and also to install pollution treatment plants in four months (Banerji 1997; CSE 1984–85; 1996–97).

## Some Lessons

### State Policy

The aforementioned cases clearly show that determined and sustained action by the affected communities, public action groups, and non-government organizations, helped by a sympathetic judiciary, have in several cases succeeded in placing the pollution problems prominently on the public agenda, getting their complaints heard and getting some redress.

But they also bring out the lukewarm attitude of the government in monitoring pollution and the reluctance even to entertain complaints under the relevant law, gather and make available relevant information. Reluctance to invoke legal sanctions and enforce court judgements is also evident in several cases. There is also reluctance to take action sanctioned by the law. Many of the court judgements ordering the closure of plants or compelling them to install treatment facilities have not been implemented or have been implemented partially and only after protracted delays.

The efficacy of such treatment plants as have been set up is also in some doubt. For example, a study by Greenpeace on toxic contamination in India found that combined effluent treatment plants have been of little use in reducing the extent of pollutants present in water. It feels that CEPTs are non-functional, and that the money being spent on such plants by the World Bank would be better spent instead to control and stop pollution at the source. This also applies to the governments and the courts, which tend to see CEPTs as the panacea for all pollution ills.

The problem lies partly in the lacunae in the law. The laws are many, scattered, and sometimes conflicting in goals and standards. They are often not based on the ground realities. They are not effectively communicated to the stakeholders. They are written in a language and in a style which an average Indian would find difficult to understand. There are also large gaps, like the absence of any legislation regarding the treatment of municipal waste and pollution control in the small industry sector and in agriculture. Standard fixing is inadequate and largely incomplete with reference to many toxic materials.

Pollution control agencies being part of the government do not have the independence and autonomy to initiate action or enforce the provisions of the law. Monitoring mechanisms are weak, and often absent. Where they exist, the information they collect is not available to the public. There is little coordination between the concerned departments. Nor do the elected representatives and legislatures show any interest in pollution-related issues, much less in exerting pressures for remedial action. On the contrary, they act more often in ways which favour, if not actively support, the interests of polluters. The laws do not provide for consultation, not to speak of the involvement, of the affected communities in government processes concerning pollution monitoring and control. Attempts of the affected communities to seek redress through petitions and appeals to the government rarely yield a positive response.

The implementation of laws is difficult in a context where several departments and agencies with different, and often inconsistent, mandates are involved. The department of industries, for instance, being concerned with attracting new industries, considers anti-pollution laws a hindrance. The agencies for monitoring and enforcing pollution laws are typically understaffed, ill-equipped, too weak relative to the development departments, and are liable to be opposed by vested interests. In some states, for example in Maharashtra, the government has authorized private consultants and laboratories to issue no-objection certificates, giving rise to manipulated results and grant of no-objection certificates even to polluting units. These difficulties are compounded because the government acts both as a promoter and regulator of economic activity and because pollution control boards lack the requisite authority and autonomy to effectively act as regulators.

A deeper problem is the perceived conflict between economic growth and the need to safeguard the environment. Insistence on meeting emission and effluent standards involves additional (often large) investments and increased operating costs and lower returns to investment. This is resisted on the ground that it would make new industrial investment less attractive to entrepreneurs and would result in a dampening of investment and growth of output and employment. Since strict enforcement of pollution standards on the existing plants also raises costs and reduces profits, they are reluctant to incur the extra costs. Instead, they successfully pressure the government, on threat of closure, to bear the costs directly or through subsidies. That closure of the plants may throw workers out of their jobs is a contingency which governments are unwilling to face.

The government currently relies almost entirely on anti-pollution laws, which lay down standards and threaten violators with civil and criminal consequences. Self-control on the part of polluters in the face of laws and even as part of business ethics and social responsibility can help but these steps are hardly effective. Strong economic disincentives are one possible measure that may help. But currently state policies work in the contrary direction by grossly under-pricing water and not charging industries for the release of polluted effluents. Moreover, there is no reason why the total expense of the government for pollution-control activities, including the staff and administrative expenses, should not be borne by the industry and other polluting actors.

Costs for pollution abatement can be reduced by more economical use of water. Among industrial units using similar technology and product mix, the specific water consumption varies over a wide range. For example, in India the paper and pulp factories consumed, during 1987–88, between

230 cu.m. to 450 cu.m. per tonne of output; the fertilizer industry used between 9 to 40 cu.m. per tonne of urea produced (during 1988–89). Overall, water use per unit of output of Indian industries is considerably more than that of their counterparts in the developed countries. The fact that even the lowest water consumption figures of the Indian plants are significantly higher than the international levels in most cases highlight the scope for saving costs by reducing water consumption.

Since water charges in India are very low, industries are not charged anything for the release of effluents and even nominal penalties for violations are not enforced. Factories and manufacturing centres throughout the country seem to have a *carte blanche* approach to water—they consume it, waste it, pollute it, all without worrying about the costs. In the pollution legislation, most of the penal provisions relate to source-specific standards. And these source standards are not really married with ambient standards. There is a need to link source-specific pollution through the total volume of pollution-causing activity. That has not been done in India. Nor is there an agency to monitor overall water quality.

In sharp contrast, many industrialized countries charge heavy levies for both the use of freshwater and for the release of effluents, thereby creating a strong incentive for industry to enhance water use efficiency. Policies that made water consumption, wastage and pollution more expensive provided a boost to water recycling and conservation. Faced with levies of this kind, industry does look for enhanced water use efficiency. In 1965, Japan's industrial output was worth US $21 per cu. m. of water. In 25 years this value had increased to US $77 in real terms. In the former West Germany, water use stabilized between 1975–90 even as industrial output increased by 44 per cent. In the US, industrial water consumption in 2000 is expected to be only one-third of what it was in 1977. All this was made possible by policies that provided a boost to water recycling and conservation and that made water consumption, wastage, and pollution more expensive (Thakur 1995).

In Sao Paulo, Brazil, water use in manufacturing dropped between 42 and 62 per cent, depending on the industry, from 1980 to 1982, after effluent charges were introduced. Consumption of water in the GDR dropped from 400 to 120 litres per capita per day, or 70 per cent, between 1989 and 1994, primarily as a result of cost-based pricing (Serageldin 1995).

## The Courts

Among the roles of various other actors, that of the courts has been the most significant in recent times and hence it has been in the limelight in the arena of pollution control activities. However, the courts can play a role only when an individual or organization or community approaches it; there are but few instances where the courts have taken *suo moto* action based on a letter or a media report. By allowing pollution-related complaints under the fundamental rights violations, the courts have greatly increased the chances of complaints getting heard and decided speedily. However, the effectiveness of the actions to the courts is directly dependent on the quality of the petitions and the commitment and dedication of the lawyers. Hence, even as the courts have acted in a number of cases, they have not been able to set examples or trends for long-term sustained control of pollution.

For example, the courts, even when taking note of the destructive nature of the impacts of water pollution, have generally failed to either quantify the impacts, or penalize the polluters to compensate the losses. This failure as the concrete cases discussed earlier in this chapter show—have not been as much of a deterrent to the polluters as one would like and expect. The complexity of environmental problems and the rationale behind laws and regulations, taken together with the quality of the evidence, difficulty of proof, and calculation of damages pose formidable challenges to the courts. This naturally affects the quality of the judgements, leading to a situation where in a number of cases the judgements are challenged or remain unimplemented.

Also, it is only the High Courts or the Supreme Court that have been active in this sphere. They have not taken cognizance of the fact that when an industry violates water pollution norms, the PCBs file cases in the lower courts which are notorious for inaction and delay. Also, even when the failure of the pollution control is evident, the courts have failed to pin-point the responsibility and penalize the guilty. Thus, the actions of the higher courts have also failed to achieve effective pollution control. It is clear that access to the courts is an option exercised by citizens because the executive has abdicated its responsibility. There is little attempt to identify the key problems and reverse the trend.

One other notable problem with judicial action has been that while handing down orders for the closure or shifting of industries from certain areas, they have not ensured that the workers do not suffer due to these actions. Moreover, they also have not ensured that at the new places, the industries follow the minimum pollution norms.

## The Media

In almost every instance of successful struggle by the communities to control or remove polluting activities, the media have played a significant role. More effective action has come about when the communities, awakened by some voluntary organization, individuals, or the media, have risen to take action. It is also seen that such cases, publicized by the media, have the potential of inspiring other communities to also act to curb the destructive impacts of pollution. The role of the media in both informing the people about the pollution and in publicizing the communities' and courts' actions has been generally exemplary, though one thing lacking in this is that the media have not taken the issue to such levels where the government has been forced to take institutional steps to control the pollution effectively. Also, the media have many times failed to effectively analyse the situation and to follow up on court or other actions to show how useful or harmful such actions have been. (In the late 1990s, the role of the media in this regard was less encouraging.)

## Communication and NGOs

There is an urgent need to learn lessons from such instances and to institutionalize the role of communities in pollution control, which is non-existent today. This institutionalization can be achieved, for example, by stipulating that every large industry or group of industries should have a citizens' committee from the surrounding area, with the right to inspect the plants and records of any company, whenever necessary. Second, each industry should be asked to make periodic reports to the community about its pollution-control activities and records. Third, the local pollution control officers must be in some way accountable and answerable to the local communities on a regular basis.

The resources, energy, and expertise which have to be mobilized for pollution control are often beyond the means of the communities and even the NGOs. They have to contend with a paucity of relevant information and struggle to remove the veil of secrecy which shrouds such data as are available. Some of the needed information can be collected by training local people and by utilizing the resources in local colleges and research institutions. But organizing it takes effort and far more responsive and active institution in civil society than is currently in evidence. In any case, a great deal of data analysis requires facilities and skills, which only the government and higher-level research laboratories have. The pollution-control agencies must obviously be strengthened and made transparent and accountable. Greater involvement, on a sustained basis, of universities and of research institutions outside of government would help increase the credibility as well as the accessibility of the data. In either case, full and prompt publication of the information and a legal right for interested members of the public to gain free access to such information is imperative.

## Academia and Research Organizations

As we go through the cases described in this chapter, we see that academic and research institutions have a very important role to play when communities are agitated and mobilized over water pollution. Their role is not only to help the communities and NGOs assess the levels and extent of pollution and its constituents, but also to play a more proactive role of monitoring the state of affairs. They also need to make their reports public in such a way that the reports reach the people who are affected most. While there have been instances where someone or the other agency has helped the affected communities to increase the general level of awareness and understanding of pollution problems and their solutions, one cannot escape the conclusion that the scientific community has largely failed to conserve scarce water resources or come to the aid of the affected communities.

What is important after coming out with reports which have a significant bearing on national resources is to at least see to it that the conclusions and recommendations get widespread publicity, raise awareness, and build up some pressure so that the authorities are made to change faulty policies and programmes. However, unfortunately, most research institutions have singularly failed to even publicize their findings of a serious nature, leave aside the question of lobbying for policy changes. It is generally left to the media, NGOs and concerned citizens to do that (Chaudhuri 1996; GEC 1997; Kumar 1995).

Moreover, learned professionals and institutes of higher learning do not seem to have got into the act despite their better access to information and their greater clout in decision making. If local communities are to be informed, it will be necessary to cultivate the social responsibilities of the professions in building civil society and in organizing good governance.

## CONCLUSION

There is a need to institutionalize the right to information of the communities regarding the activities that affect their lives. Various actors including the government, industry, and the aid agencies, must be made to make public reports about their activities. The communities must also have a right to inspect and a right to participate in the decision-making process about all activities that may affect their lives. The pollution-control bodies have to be independent of the government in every sense of the term. Third, every user of water must be made to pay the full costs that have been incurred in providing the water to the user. This is true even for community resources like groundwater. Similarly, pollutants must be made to pay for the release of effluents even when if they are treated. There is no reason why the total expenditure incurred by the government for pollution-control activities, including the staff and administrative expenses, should not be borne by the industry and other polluting actors.

The research and academic institutions have got to play a much more significant role in identifying relevant issues and in informing all the concerned parties about the status of pollution and its impacts. They must also push for changes in policies that they have found to be necessary for reducing and checking the pollution of resources. They need to work hard to increase the water resource literacy and the availability of the necessary knowledge about the state of pollution of our water resources.

## REFERENCES

Academy of Gandhian Studies, 1990, *Status of Environment in Andhra Pradesh,* Citizens Report, Hyderabad.

Agarwal, Anil, 1996a, *Homicide by Pesticides: What pollution does to our bodies,* CSE, New Delhi.

———, 1996b, 'Pay-oils to Progress', *Down to Earth,* 15 October.

Alvares, Norma, 1994, 'Pollution Battles: Limited Legal Redress', *The Hindu Survey of the Environment.*

Anon., 1996, 'Water Alarm', *Down to Earth,* 30 November.

Banerji, Rajat, 1997, 'Dead Fish Tell Tales', *Down to Earth,* 31 January.

Banerji, Rajat and Max Martin, 1997, 'Courting Green', *Down to Earth,* 15 May.

Central Pollution Control Board, 1996, *Groundwater Quality in Flood Affected Areas of* Delhi, CPCB, March.

CSE (Centre for Science and Environment), 1982, *The State of India's Environment: The First Citizen's Report,* New Delhi.

———, 1984–5, *The State of India's Environment: The Second Citizens' Report,* New Delhi.

———, 1996–7, *Water Newsletter* (various issues).

Chakraborty, Sujit, 1995, 'Green justice: Up in smoke?' *Down to Earth,* 30 September.

Gujarat Ecology Commission, 1997, *Ecological Degradation around the Gulf of Khambhat, Gujarat: A Status Report,* February.

———, 2001, *State of the Environment: Gujarat,* May.

Krishna Kumar, Asha, 1996, 'Facing the music', *Frontline*, 31 May.

Krishnaswamy, Chetan, 1996, 'Paying for a spill', *Frontline*, 23 August.

Kumar, Sanjay *et al.,* 1995, *Organochlorine Residues in Rural Drinking Water Sources of Northern and Northeastern India,* Industrial Toxicology Centre, Delhi.

Madan, M. Mohan, 'Nylon 6, 6 Project: Advantage Goa', *The Hindu Survey of the Environment*, 1995.

Ministry of Water Resources, 1987, *National Water Policy,* Government of India, New Delhi.

Serageldin, Ismail, 1995, *Towards Sustainable Management of Water Resources*, World Bank, Washington D.C.

Shankar, Uday, 1992, 'Purifying the Ganga', *Down to Earth,* 30 September.

———, 1993, 'Groundwater: Disappearing Act', *Down to Earth,* 15 July.

Sharma, Anju *et al.*, 1996, 'The Blind Court', *Down to Earth,* 30 April.

Sharma, Kalpana, 1995, 'Konkan Smelter Project: What People's Power Means', *The Hindu Survey of Environment.*

Shrivastava, Rahul, *et al.*, 1994, 'Polluting industries: Just Shifting the Issue', *Down to Earth,* 31 May.

———, 'Copper unit comes a cropper', *Down to Earth,* 15 May 1994.

Sitaraman, Sudha, 1996, 'By hook or by crook ...', *Down to Earth*, 15 April.

——— , *et al.*, 1996, 'Beyond Backwardness', *Down to Earth,* 30 April.

Thakur, K.B., 1995, 'That Sinking Feeling', *Down to Earth*, 30 June.

The World Bank, 1993, *Water Resources Management: A World Bank Policy Paper,* Washington, D.C.

16

# Diara Diary

Mukul Sharma

Come monsoons and the cities of Bhagalpur, Munger, Kahalgaon and Sahebganj witness a sudden influx of people on the main roads, along the southern banks of the Ganga river. Villagers, displaced by annual floods, take shelter on railway platforms, bus stands and pavements. They are the diara people, foraging for food and shelter, and receiving reprimands everywhere. Hundreds of people huddle in small plastic tents. They get very little work in towns and are forced to work for very low wages, with men often pulling rickshaws and women working as housemaids.

The land between the Ganga and its tributaries—Burhi Gandak, Gandak, Kosi and Some is called the diara. The diara region faces floods and erosion every year. Using the main stream of the river as a reference, diara can be divided into three categories. First, there are the main river bed diara lands, which lie inside the river bed. These get flooded every year and are available for cultivation only during the summer months. Second, there is a slightly elevated mid-region, located on the banks in varying widths, known as the mainland diara land, which is frequently inundated during the flood season and is available for cultivation for eight or nine months a year. Third, there are the upland diara lands, lying above the mainland diara lands, which have got a higher elevation because of alluvial deposits in the past. They are even less frequently flooded.

July 1992: heavy rains in Bhagalpur, Bihar have flooded its diara region. Thatched huts, household items, bhadai crops—all have been swept away. Many animals have drowned. The people of the diara take refuge under the shelters of plastic sheets they have erected on footpath, roads, and Sendi

* This is a substantially revised and updated version of the original article that appeared in the *Navabharat Times* (in Hindi) between 12 and 20 January, 1990.

compound. They stay here during the rainy season, and return to their villages in the diara after the fury of the monsoon is spent. For the inhabitants of the diara, this is a recurring annual feature.

The diara soil is mostly sandy and light being constantly subjected to flooding and erosion. Low-lying diara areas, and those lying close to the currents of the Ganga and its tributaries, are flooded annually; while those located higher, and away from the current, are inundated sporadically. Erosion normally accompanies the receding floods. Due to continuous erosion and floods, and the changing course of the rivers, some areas emerge out of water and some others get submerged. Some diaras are like islands between two rivers, or between two rivers, or between two currents of the same river, but mostly they occur on both sides of the rivers.

A prominent marker of the Bhagalpur diara was the Bhagalpur Blinding Case, when many convicts were blinded in jails. Most of these belonged to the diara, and were locked up on charges of petty theft. Prior to this and even afterwards, the diara was a witness to the massacres of Galwa (1982), Medini Chowki (1983), Pipra (1984), Shambho (1985), Sonbarsa (1985), and Taufir (1985), in which an estimated 1500 people have been killed.

In the villages, the diara is also called *doab*, which literally means a tract between two rivers. Local villagers distinguish between two kinds of diaras: those which suffer the onslaught of floods and erosion once, or several times in a decade, and others, called *karar* diara, where erosion takes place after many decades or once in a 100 years. Among the several stories and myths connected to the diara, some can be traced back to the Puranas. According to a legend, the diara was an area under the curse of the king Bhagirath, being the place he had travelled to, to get rid of his ailment.

Diaras are spread all over Bihar, covering nearly 200–250 kilometers. They cove nearly 9 lakh hectares of the land: 2.40 lakh hectares in the Ganga diara, 2.30 lakh hectares in the Ganga diara, 2.30 lakh hectares in the Burhi Gandak, 1.40 lakh hectares in the Gandak, 1.50 lakh hectares in the Kosi, and 1.10 lakh hectares in the Sone. The inhabitants of the diara area are mainly of the middle and the backward castes, like Gangots, Mallahas, Dhanuks, Gwalas, Khairwars, Tantis, Bhumihars, and Rajputs. Herbert Risley identified the Gangots and Mallahas as the original inhabitants of the diara region. P.C. Roychoudhury's *Bihar District Gazetteers* (1962) also refers to Gangots, Mallahas, and Yadavs as the main inhabitants on the banks of the Ganga. The Bhumihars, Rajputs, Kurmis, and Koeris may not be among the original inhabitants of the diara, but they have always played a major role in the socio-economic structure of the area. Before independence, the Ganga diara land was mainly in the hands of an Agarwal family of Raj Baneili, Dilip Narain Singh, and Kamleshwari Prasad Singh of Monghyr and a trader family, Ram Gulam Sahu. Since independence, there has been little change in the socio-economic structure of the diara land. Some diaras have witnessed the rise of new landed castes like the Yadavs and the Kurmis. In addition to the land, the landowner and a few rich people of the town today control most other means of production, like fisheries, milk, and vegetable production in the diara. Even the Ganga was under zamindari control in the area until the 1980s. Till date, there has been no official survey of the diara land, nor has there been any effort to properly measure it. Diara inhabitants increasingly complain that the regularity of floods and the extent of erosion have increased over the years, and hence the environment of the diara is worsening.

It has become an annual phenomenon for the villagers in the diara region to be displaced. Life in the villages which suffer displacement and resettlement every year is full of uncertainty, insecurity, poverty, and disease. Ten to twelve feet below Bhagalpur town lies the Shankarpur diara, which was the worst affected by floods. There are right villagers in this area. Some are slowly eroding every year. Families maintain their links with the place because of their agricultural land. The houses, made of *kash* grass, bamboo and straw, lie disordered. 'Since the last fifteen years, floods have occurred every year, destroying our village. There is regular loss of life and property—houses fall, crops are swept away; animals are killed and diseases spread. Government assistance does not reach here,' said Ramprasad of Shankarpur diara.

Villagers displaced by floods face a tough time for four to five months of the year. They take shelter in Bhagalpur town, where they find it difficult to get space to settle even in the open. 'Like animals we roam the roads and the streets, looking for shelter and food—at the railway platform or the bus stand, on the roadside or on government land. Almost every year, we have to go to the town, because we cannot go elsewhere. But the Bhagalpur administration has made no efforts for our temporary relief and rehabilitation, complained another villager, Jaikishan. The police extort money at such times. Lallan Prasad said that during the days of displacement, the police come frequently and arrest them on charges of theft. They have to settle illegally, which is why the police keep harassing them.

Food and employment become a grim problem in these months. The people can neither get work on the land, nor are there any jobs in the town. Kailash of Chauvanniya village found it very difficult to survive. Their stock of grains and money was exhausted within a month. Very few were left with enough money to hire a cart, set up a pavement shop or rear a cow. The wages for labour are also very low in the rainy season. Some ply rickshaws, and women work as maids in households. Loans even have to be taken from moneylenders.

Disease and death make life all the more difficult. In 1989, after the floods, Roshni Devi of Babu tola lost her two sons to diarrhoea. Harisharan of Chauvanniya died of snakebite. Choki of Shakarpur died of high fever. Cows and buffaloes also die of diseases, or lack of fodder. With their feeble means of livelihood also taken away, people barely manage to keep alive.

Erosion also displaces people. During the erosion of 1971, hundreds of families of Shakarpur diara were permanently displaced. Since their agricultural lands were in the diara, they could not settle far away. From 1971–85, they were driven from place to place, sometimes near the Shankar Talkies, sometimes to Buranath, or the railway station, or to Nathnagar. Ultimately in 1986, they were settled away from town, near the airport, in Jhurkhuria hamlet, a desolate, uneven place, where the government gave 40 bighas of land to settle about 300 families. Because of bungling in land distribution more than half the displaced population did not get land titles. Many families whose place of residence had not been eroded, obtained land titles through bribe. Also, members of the same family managed to get separate land entitlements. Many people from the town also got in, with the result that only 52 genuinely affected families could get the land.

Jhurkhuria has no water, electricity, nor a proper road. Many of the displaced have no means of employment. Their essential means of livelihood is petty shop-keeping and wage labour. Some have arable land in the diara. They sow the crop but the fields are so far away that farming and tending the crop has become extremely difficult. Fodder shortage affects cattle rearing and, by and large, buying fodder from the town is beyond their means.

Another example of the neglect of the diara-displaced people is Company Bagh, the area of the town which touches the boundary of Bhagalpur University. A hundred displaced families of Shakarpur, Chauvanniya, Faridpur-Noordipur, and Ajmeripur came here in 1948 and their colony was declared legal in 1956. But there is nothing here which speaks of a town—no roads, drains, electricity, ration cards. A single hand pump supplies drinking water for the whole colony. For their livelihood they are still dependent on their old diara. They go there for farming, labour and cattle rearing. They are scattered here and there, without any fixed address.

In the Taufir and Anthavan diara villages in the Kahagaon block of Bhagalpur district, the majority of the villagers live in constant fear that the Ganga can consume their villages any day. Before the floods in 1990 the current of the Ganga was about 4 kilometers away, but now it is a mere half kilometer off. Some of the outer areas of the villages have already been lost to the river, and the erosion presents a frightful sight—an eroded quagmire of land, uprooted trees and seething water. The villagers know that their villages could disappear in one stroke, but they are forced to stay on, as there is no alternative.

When the villagers tried to find some land, the local administration did not approve of it.

Taufir, a village of nearly 200 houses, had 1500 bighas of agricultural land before 1971. According to the villagers, so far 1200 bighas have gone under water. Anapur, Nandgola, Rani, Ikchari and other villages have met a similar fate. Since the floods of the 1990s, many more villages are on the verge of erosion. Villagers of the area stress that in 1992 the Ganga changed its course with great force and violence. There was increased flooding and erosion, and a greater loss of life and property. Even after the flood waters receded, the main current of the Ganga did not settle, and the threat of erosion remained.

Krishan Das, a venerable old man of Taufir, says, 'I have been living in this village since my grandfather's time. There have been many vicissitudes and many a time the land was eroded. We faced it all together. But this time, the erosion has left us stranded.' Middle-aged Harihar furnished more information about how they had sent their women and children to their relatives. The village school had closed down. Some people were living in the neighbouring villages. They came to the village by day, but went back before it became dark.

According to the village headman, they were prepared to settle anywhere. Their only condition was that the land should be away from the Ganga and safe from erosion, so that displaced persons were not displaced again within a year or two. The administration had been repeatedly requested to either stop erosion or consider giving the villagers alternative land, but did nothing. Some years ago, when a similar erosion threat loomed over some prosperous diara villages of Sultanganj and Munger districts, the administration immediately arranged for huge boulders to form a *bund* on that end of the Ganga. The current stopped moving towards the village, and slowly changed course. There was extensive agricultural land in Nandgola which belonged to absentee landlords, where Taufir villagers used to work on wages. The villagers resolved in September 1990, to take over four bighas of this land and settle there. The sharecroppers also did not object to this. Some constructed their huts; most others had marked out their plots. The local administration then intervened on the complaint of the landlord, had the huts demolished, and arrested some people.

The inhabitants of the diara die a thousand deaths. Since 1938 Ganga floods and erosion have annually affected the diara villages of Vikrampur, situated 1 kilometer south of the Bihpur railway station, in the Naugachiya block of Bhagalpur district. Everyone in the village has a tale of woe of the misfortunes that have befallen them over the last 60 years and more—how their houses collapsed, how the land was lost, how farming came to an end, how they ran from pillar to post for a livelihood, how for a year or more daily food had been a problem, how educating their children, or repairing destroyed homes, or getting out of the moneylender's net was impossible and so on.

Vikrampur was originally situated about 6 kilometers south of its present location. There were about four hundred houses, spread over 30 bighas of land. After the erosion in 1964, no residential area was available to the villagers. Some distance away, there were 15 bighas of land, belonging to landowners of the neighbouring village, some of which the displaced people had bought according to their means. In Harijan tola (or quarter), about 60 houses are cramped in 2 bighas, each house getting not even one *kattha* to itself. There is no community or grazing land in the village. In the crowded village, open space of any kind is scarcely to be found.

Agricultural land in the village has eroded thrice: in 1937, 1975 and 1987. Since 1990 the village has been free from erosion, but most of the agricultural land is now under water on account of the annual floods. So much sand has spread over the land because of the floods that it cannot yield anything except *parval* (*Tricosanthes dioica*—its fruit is eaten as vegetable). Even when farming is done, it is capable of yielding only one crop, rabi.

Waterlogging is another problem. While constructing bridges and highways, proper provision was not made for water outlets. Where such provisions were made, there was inadequate maintenance. Flood waters remain for many months, harming the rabi crop. Some areas are permanently waterlogged. 'Almost one hundred acres of Vikrampur land has become permanently waterlogged. There are two reasons for this. First, there is no provision for water outlets. The second is the vested interest of powerful people and criminal gangs. Such water logged areas have a lot of fish, and fish traders do not want the area evacuated until they have killed the entire fish population,' says Devchandra Jha who teaches in a local school.

No family in the village has more than ten bighas of land, and 10 per cent of the families, belonging to the Scheduled Castes, have no land at all. Of 450 families, except for the sixty Brahmin families, all the rest—Kurmis, Koeris, Nai, Ravidas and Muslim families—have been reduced to working as labourers to earn their livelihood. Brahmin men folk have some employment in government or private services, though not of a kind that makes them secure and well-to-do.

Far from initiating any special project to stop erosion and provide water outlets, even routine government development programmes are absent in Vikrampur village. It has no electricity or drinking water supply. The school and the library virtually disappeared; there are no over fish in the waterlogged area, too weak to resist the powerful fishing interests. Caught in their poverty, insecurity and helplessness, they live merely on the hope that 'Ganga *maiya* would give back their land some day'.

The Ganga diara of Bihar has nearly three lakh hectares of arable land, which regularly gets layers of new soil and sand. The land closer to the river has a layer 30–60 cm thick, which gradually decreases in the land away from the river. Only vegetables can be grown on land with new soil left on it by floods.

The diara arable comprises upland, medium land and low land. The upland gets flooded only during very heavy floods, or during August–September. Given irrigation facilities it can be cultivated through the year. The diara has very little upland. The medium land remains submerged in floodwaters for a maximum of three months, from July to October. The low land remains flooded longer, and can be cultivated only for four to five months. The land which emerges out of water from time to time takes many years to become cultivable. Two kinds of wild grasses—*kas* and *jhauva*—with their roots help make the land firm and heavy.

Floods quite often destroy the kharif crop, grown from June–July to September–October. Even the rabi crop, grown from October–November to March–April is sometimes delayed because of floodwaters. Drought compounds the problem further. According to one estimate only 3 per cent of the diara has irrigation facilities.

Villagers estimate that only 15 per cent of diara land is cultivable. Primarily, vegetable are grown for sale in the town market. A trip to the town market entails trudging over miles of sand, through accumulated water and slush, and crossing the river by boat. The villagers, therefore, prefer to sell their produce to the local merchant.

From the Shankarpur diara, cow and buffalo milk is sold on a large scale at Bhagalpur and other adjoining towns. For the cattle rearers a major problem are floods, which carry away cattle and make the area disease-prone. In addition, obtaining fodder is difficult during and after the months of flood. Unable to buy fodder from the town, the milkmen are often forced to sell off their cattle.

Two types of milk trade are found in Bhagalpur town. Landowners, who are also prosperous cattle owners, hire local labourers and small farmers to buy milk from all around the village. Their work starts in the early morning and ends late in the evening, and they are paid daily. On a smaller scale, is the milk trade by the *daihars* independently. They do not own cattle but buy milk from the villagers, paying cash with an advance taken from the town shopkeeper. Losses associated with milk business like spilt milk can set them back considerably.

In the Ganga diara, from Sultanganj to Peripainti in Bhagalpur district, there are about 40,000 fishermen and about 15,000 fishermen in the Kahalgaon diara. The general pattern of the fish trade here is that prosperous people in villages take *pattas* for fishery rights at particular places along the diara. They generally get their work done through contractors, who in turn get in touch with local fishermen. Different agreements operate in different areas—at some places they get a monthly payment, while in some cases they work in a kind of bondage.

When agricultural land goes under water, the owner of the land is not entitled to fish in the water on his land. He may not even be allowed to come near the newly formed pond. The powerful in the village or criminal gangs grab it. Every year many lives are lost in the fights between landlords and criminal gangs over the control of these ponds. From Sultanganj to

Pirpainti in the Ganga diara, dozens of well-armed criminal gangs operate, which also have a stake in fishery. Even after making the patta payment, local fishermen have to pay extortion money to these criminal gangs. The Superintendent of Police in Bhagalpur gave an account of this for 1987: On 24 May, the Nandana Mandal gang kidnapped a fisherman, Durga Sahni. On 3 June, Rajkumar Sahni was kidnapped. In September, 11 fishermen were kidnapped and later released on payment of ransom. On 14 October, the Hulo Yadav gang looted the catch of eight fishermen. On 15 October, a fisherman, Radhey Sahni was abducted and killed and his boat stolen. On 16–17 November, at least 12 fishermen were killed near the diara ghat in Pirpainti police station. A contractor, Buddu Sau, contracted these fishermen from Kahalgaon. In Rani diara, a rival contractor Bhola Mandal arrested them. This led to a clash of two criminal gangs, leading to the deaths. On 1 December, criminals injured Ram Lal while he was fishing.

Bihar's diara is mostly unsurveyed. Even where there have been surveys, either before or after independence, new land-related disputes have cropped up because of faulty surveys. The changing course of the river after floods, the land erosion, and the appearance of new land compounds this problem.

An example is the Dhamdaha area in the western part of Purnea district, which was badly affected by the Kosi floods. Between 1901–11, the Kosi current moved westward towards Bhagalpur, thus leaving the area flood and erosion free while also throwing up new agricultural land. According to a survey of forty villages in 75 square miles, in 1923-6 (*Final Report on the Survey and Settlement Operation of the Kosi Diara Area in the District of Purnea, 1923–26,* Government of Bihar and Orissa, 1927) there were many changes in the Kosi. First, the old tenants occupied their own land. Since the land boundaries had been wiped out, the stronger tenants encroached on the land of their weaker neighbours. The Santhal and Gangot sharecroppers were removed from the land after three to four years. People from Chapra and Muzaffarpur came and occupied the newly emerged land. The Darbhanga Maharaja gave them land rights. Santhals and Musahars normally tilled this land. The Maharaja also arranged land for some co-sharers. These co-sharers paid rent to the Maharaja and got their tilling done by farmers who came from Munger, Bhagalpur and Santhal Parganas. Farmers who worked under the co-sharers were called *kolatdars.*

In 1922, the collector of the area acknowledged that the entire area was teeming with land disputes and consequent criminal activities. A survey of the area was necessary, Particularly that under the Darbhanga Maharaja (Bihar and Orissa Revenue proceedings, No 11324, 1926). In 1923, the Revenue Department started a survey settlement of the Kosi diara in Purnea. The *Final Report on the Kosi Diara* in the district of Bhagalpur and Purnea, 1926 states that during the survey, powerful people of the area got the tillers and local poor dispossessed, and got a legal stamp over their seizures. The survey officer remarked

> After a tour of the area it is difficult for any officer to believe that the poor people of the village have merely some dismal residential land in some corner of the village and the entire agricultural land of the village belongs to people of some different district.

A second survey in 1950, intended to establish the right of the landowners on their land, was faulty from the beginning, being essentially based on the 1923 survey settlement. The new survey settlement once again dispossessed the poor, the landless and the sharecroppers. With this land grabbing began a new wave of land disputes and land struggles. In the Kosi diara and other areas of Purnea district, the pattern of landownership is as follows: Kursela family 12,000 acres, Ramgulam Sahu 20,000 acres, Maulchand 10,000 acres, Jagdish Chaudhary 10,000 acres, Vaksh Chaudhary 12,000 acres, Prithvi Chand 10,000 acres, Saryu Mishra 4500 acres and Moihur Rehman 4000 acres.

The survey settlement of Bhagalpur district was done in 1902-10 under P.W. Murphy, ICS, but the diara was not included in this survey. There was a special Ganga diara area survey during 1908–15, but its aim was only to fix the boundary of the diara area. The diara area was under Raja Todarmal, and different landlords were given land rights. The tillers were poor Gangots, but the ownership of the land was with the landlords and their leaseholders. This Ganga diara area was also disturbed because of the dispossessing of the small farmers. However, no effort was made to reduce tensions or to have a fresh survey in order to resolve the land disputes.

The survey settlement of Bhagalpur district, which commenced in 1950, ignored the diara. A survey settlement of the area was, however, ordered in 1958, but only after land disputes in the diara became extremely violent during 1950–5. The survey started in 1959 and lasted until March 1965. In this survey, the first in a hundred years in the area, farmers and sharecroppers were dispossessed on a large scale, as had happened during other surveys. In many areas, owing to the influence of kisan sabhas and other organizations of farmers and sharecroppers, records were made and maps prepared in favour of the land tillers. The illegal occupations of big landlords were also brought to light. Nevertheless, as soon as the survey settlement was completed, the Bihar government scrapped the survey settlement had no 'permanent authoritative significance'. No official explanation was given for this decision nor was a fresh survey ordered. Similar is the plight of the diara in other districts of the state. The Bihar government's *Estate Manual* 1953 provides for the annual survey of the diaras, and instructions have been provided for implementation (viz., the land thrown up by the Ganga will first belong to the tenant who owned it before it went into water; if there is no previous owner, the land will rest with the government, etc.), but in the absence of authoritative and genuine land sur.vey settlements, these provisions and instructions are meaningless. The result is dispossession of small farmers and sharecroppers, the complete control by landowners, the terror of criminals and official connivance in all this. In Mahadevpur mauja in the Shankarpur diara, for example, the district administration had apparently made a map of the area in the early 1900s. Thereafter the entire area suffered erosion and the land remained submerged in the Ganga for years. When this land emerged out of the water in the 1980s, landowners Nathuni Singh of Shankund and Bhurimal Marwari of Naugachiya seized about 1000 bighas. In the fight between the old small farmers and the new land grabbers about 50 lives have been lost.

The land grabbers started making use of criminal gangs, in return allowing these gangs to cultivate hundreds of acres of illegally occupied land. Abhay Kumar Sinha captured about 500 bighas of land in Shankarpur diara in 1990 and distributed in to a criminal gang for sharecropping. Before that, he had hundreds of acres of crop destroyed and then had the sharecroppers dispossessed. Similarly, in Ubari Pat village of Chauvanniya diara, the crops and granaries of sharecroppers were looted.

Many in Bhagalpur, Munger, Naughchiya, Kahalgaon, Purnea, and Katihar say that over the years, the flood havoc has been increasing. The floods reach the diara earlier than before, and the entire *bhadai* crop, which is reaped in August, gets destroyed. The floods have accelerated, and the loss of life and property has increased. The peace and extent of erosion has also increased. Earlier, villages on higher land were safe from erosion.

It is believed that the floods bring new soil and chemical elements, which raise the productivity of such land. This also reduces the cost of fertilizers and irrigation. For this diara farming had been called the cheapest. Now villagers complain that production in this area has reduced. Ram Sevak Srivastava, an agriculture scientist in Sabour observed, 'New kinds of diseases have been plaguing crops and animals. The chemical nature of the soil and water brought by the Ganga has also changed, because of which the land is not becoming fertile, but sterile.'

The difficulties of the people in diara may be traced to the pollution of the Ganga, the felling of trees, and the wrong planning of bunds and embankments. A study by two Bhagalpur University professors, K.S. Bilgrami and J.S. Dattamunshi, showed that in the 256 kilometers between Barauni and Farakka, the Ganga is the most polluted near the Mokamah Bridge. Here, industries like the Bata Shoes Factory, McDowell Distillery, the oil refinery, the thermal power station and the chemical fertilizer factories discharge their effluents into the Ganga. Bata and McDowell discharge 250,000 litres of used and dirty water in the Ganga every day. According to Bilgrami, in experiments done to gauge the poisonous nature of this part of the river, fish introduced in the Bata area died in 48 hours, and in the McDowell area water, within five hours. The effluvium from the oil refinery is so much that once the river caught fire near Munger.

Not only is the Ganga getting polluted, but its rate of siltation has also increased. From Patna to Farakka, the strong current of the Ganga in the diara carried a great amount of eroding mud. This mud piles up in the Ganga at different places. The riverbed is rising due to industrial wastes as well.

Diara forests have also depleted on a large scale. Vikrampur had an abundance of rosewood trees, which are hardly visible now. *Kas* and *jhauva* grass is destroyed in the competition to quickly grab the land. Landowners cut this grass and tried to cultivate the land. Commercial exploitation of these has also begun.

The several bridges, embankments, pumping stations and irrigation projects erected on the Ganga adversely affect the water flow in diara. A few days of heavy rain in the diara are enough to accumulate water everywhere. Many large diara areas of Rupauli and Kadava sectors in Purnea and Katihar districts have become permanently waterlogged. In the Naugachiya area of Bhagalpur district, flood prevention embankments have been built for roads, railway and the canals without much thought given to water exit mechanisms. The result is that diara areas prone to submergence face more intense submersion. When the embankments are breached, the people of the diara lose their houses.

The region today has a disturbed existence, the constant dispersion making people more inclined to crime, murder and litigation. Caught in an increasing river wrap, the region is breaking under the weight of water. Every area shows signs of instability; every village is haunted by the fear of imminent non-existence.

In the case of the diara, distress is a daily phenomenon. Death is slow, and it also regularly taints the survivors, due to the recurring fear of floods. Not able to exercise control over their lives, the diara people feel a sense of defeat, caused not only by indiscriminate Nature, but also by the discriminate oppression of landlords criminals and state agencies. Their everyday living is accompanied by repeated humiliating experiences, which leave them with a diminished sense of their own being. Scattered here, there and everywhere, their pain and anxiety is spread over time, space and region.

# 17

## South Asian Pastoralism: The Environmental Question

**Arun Agrawal and Vasant Saberwal**

Many have argued that pastoralism is inimical to the environment.[1] At the root of such negative preconceptions are beliefs that pastoralists' animals compete with wildlife populations, that their feeding habits negatively affect the regenerative capacities of forested landscapes, and that the unpredictability of movements make it impossible to put institutional arrangements in place that would regulate environmental use.[2] Yet, we also know of other instances in which herding has been encouraged. Prior to the establishment of a profitable timber market in the Indian state of Himachal Pradesh, for example, local kings encouraged herding in an attempt to maximize their tax returns from lands grazed by herders (Singh 1998). Even today, in Himachal Pradesh and in the Indian states of Uttaranchal in the Himalaya as well as in different parts of western Uttar Pradesh, Haryana, Gujarat, and Madhya Pradesh in plains, pastoralists enter into contracts with both state agencies and cultivators where they sometimes pay for their animals to graze and at other times are welcomed for the manure of their animals.

But it is certainly the case that with the growing presence and authority of the forest department in British India, the rhetoric against herding acquired greater urgency. This was in line with a general hostility towards a variety of land use practises that interfered with the regeneration of Indian forests—including swidden and shifting cultivation and pastoralism. Borrowing from a growing international rhetoric against grazing, foresters in the early twentieth century fuelled fears of a changing hydrological cycle in areas intensely grazed by pastoralists. In dry regions of western India, they

---

* This paper is an extract from Arun Agrawal and Vasant Saberwal, 2004, 'Whither South Asian Pastoralism?: An Introduction', *Nomadic Peoples* 8(2).

claimed that unrestricted grazing by animals of pastoralists would translate into an eastward expansion of the Thar Desert. In the Himalaya, overgrazing was associated with a reduction in vegetation cover, and in consequence, expectations of a drying of mountain streams, massive soil erosion, and, eventually, intensified flooding in the Indo-Gangetic plains. Framed in the context of the collapse of the Harappan Civilization, overgrazing was explicitly linked to an anticipated demise of this civilization.[3]

It is not surprising then that fears about an impending environmental collapse, reiterated over the past century, have also produced a widespread societal acceptance of the belief that grazing, particularly grazing by nomadic pastoralists, leads to environmental deterioration. In many instances, state conservation policies have had damaging effects on the viability of pastoralists' economies and existing social organizations (McCabe *et al*. 1992). Yet, much of this rhetoric remains unsubstantiated by rigorous science (Leach and Means 1999; Saberwal 1999). More importantly, the science that has explicitly examined the links between human land-use practises on the one hand and desertification, flooding and soil erosion on the other provides a substantial basis for contesting the easy connection between grazing by animals of pastoralists and the various indicators of environmental degradation.

Much of the more discerning careful research on the subject has been carried out in sub-Saharan and Sahelian Africa. Experimental research in the region has pointed to the fact that the impact of grazing by African pastoralism is far more complicated than has typically been assumed (Hjort 1981; Ellis and Swift 1988; Turner 1993), specifically countering the notion that the Saharan desert boundary is not unidirectional (Binns 1990; Mace 1991). Rather, the boundaries move in response to localized and annual rainfall patterns. Natural phenomena appear to play a far more important role in the expansion or contraction of the desert rather than the grazing practises of ever-growing pastoralism populations. Precisely the same points can be made with regard to the Indian Thar desert as well.

But the ecological debate on the impacts of grazing on the environment is quite vigorous in Africa (Ellis and Swift 1988; Homewood and Rodgers 1991), especially so in comparison to India. This is at least in part a reflection of the absence of cutting edge ecological research in the sub-continent. Even so, many of the findings in Africa about the relative impact of human and natural factors on changes in desert boundaries, along with advances in theoretical ecology that point to arid ecosystems being in a constant state of flux, and highly responsive to variations in rainfall, hold relevance for the drier parts of western India.

Existing research in the Himalayas has also questioned standard assumptions regarding the ecological consequences of human land use practises (Gooch 1998; Sharma 1997). Recent research provides compelling evidence that counters the forest department's and many environmentalist claims that in the absence of greater controls over grazing, soil erosion in the hills and flooding in the plains would be accelerated (Gyawali 2001; Metz 1991). Instead, this research that although anthropogenic pressures likely play a minor role in soil and water conservation, soil erosion in these geologically unstable mountains and downstream flooding are primarily influenced by natural processes: tectonic movement on the hand, and rainfall on the other (Hamilton 1987; Ives and Messerli 1989).

We should emphasize that the research that we are citing does not suggest the absence of all links between grazing and vegetation. Rather, this research effectively questions the overarching rhetoric that sees the absence of grazing as necessarily good, does not examine who the absence of grazing is good for and in what ways, and provides little by way of measurements of the precise impact of grazing in field settings. This research acknowledges that grazing pressures may, under certain conditions, lead to localized degradation and lower productivity. But it also shows that few universally generalizable consequences result from grazing pressures, and also argues that the consequences of such degradation are likely to be greatest for herders themselves. Overall, the adverse impact of grazing is likely to be far more limited than it is routinely made out to be. It convincingly shows that there is an urgent need to move away from an alarmist rhetoric that calls upon herders to make sacrifices in the national interest, and a need to understand these issues more specifically in localized contexts.

A different set of issues is opened up when we talk of the relationship between grazing and biological diversity. Once again, and in parallel with the assumptions of a negative relationship between grazing and the environment, many scholars advance the argument that grazing is inimical to the conservation of biological diversity and protected areas (Prins 1992). This assumption, and more generally the assumption that all human activities are inimical to wildfire conservation, finds its way into the Indian Wildlife (Protection) Act, which requires state governments to extinguish all grazing and other rights within national parks, and to curtail these rights within sanctuaries.[4] Recent research has questioned these ideas as well, demonstrating in many instances a facilitative relationship between grazing and biological diversity—such as has been recorded in grasslands of Africa (Homewood and Rodgers 1991); Europe (Grubb 1976; During and Willems 1986); the United States (Hobbs and Huenneke 1992); and in India, in the alpine meadows of the Himalaya, in the Bharatpur Wildlife Sanctuary in Raahasthan, and elsewhere (Saberwal 1999; Naithani *et al*. 1992).

From the perspective of the politics of institutional control over resources, it is easy to see that an organization such as the forest department that claims to be responsible for environmental conservation would be more likely to exaggerate the extent of environmental degradation since the presence of degradation would require a greater role and involvement for the agency to manage the environment. The mechanisms to bring about improved management would depend on the nature of identified causal relations: if grazing is a menace, then it would be necessary to reduce grazing pressures; if pastoralists' animals are responsible, then they should be excluded, and so forth. Such assumptions naturally facilitate growing control by specific government departments, and even when institutional arrangements succeed in limiting grazing in one area, it only means the displacement of grazing pressures elsewhere.

But the environment–pastoralism equation resolves itself in many different ways, and pastoralists have been at the receiving end of an environmental rhetoric in more ways than one. To take just one example, consider the Raikas in western Rajasthan. Over the past several decades, they have faced a closure of village commons that is justified by appeals to better environmental protection by the more powerful cultivator population of the village. Such closures force herders to spend more time away or their home village, thereby reducing their ability to participate in village-level politics (Agrawal 1999). Indeed, the relationship between herders and their animals, institutional arrangements, and environmental outcomes is hardly mapped out by the standard oppositions that are set up by categories such as settled versus migrant, agricultural versus pastoral state versus local, or common versus private.[5]

A key component of state/societal descriptions of herders' rationality is the expectation that all herders are continually attempting to increase the size of their herds—for symbolic as well as monetary reasons. In all of this, there is a routine homogenization of the herder, with little recognition of the many dimensions that stratify herders themselves. Much recent research points to the fluidity within certain herder societies, and the fact that individuals move in and out of herding, in response to a wide variety of factors—market conditions that may alter the profitability of herding, the availability of alternative options, including cultivation and jobs in existing or emerging markets, and the very real problem of accessing forage owing to competing interests staking a right to land resources.

## Pastoralism and the State

State policies regarding irrigation, forests, agriculture, fodder, famine relief, and migration are only some of the instruments that contribute to the political-economic context in which pastoralists exist, and which shape the channels along which their lives flow. Of these various policies, some are aimed specifically at pastoralists. Attempts to improve fodder availability or to provide water for cattle and sheep count among such policies affect herders even if they were not the conscious targets of the policy. For example, extension of irrigation aims at improving crop yields. But it also simultaneously accomplishes a shrinking of common lands, enclosure of private lands that might be available for grazing during the fallow period and water-logging and salinity. All of these consequences simultaneously reduce the grazing area available to pastoralists' animals.[6]

But pastoralists often also encounter state efforts at development that tragically misapprehend the nature of pastoralism livelihood strategies and result in consequences that are almost inbuilt into the design of programes aimed precisely at development. Almost all effort by governments to sedentarize pastoralists fall into this category. Recall that pastoralists' mobility allows them to take opportunistic advantage of spatio-temporal variations in levels of production across ecotones. Efforts at sedentarization hobble this strategy without introducing compensating mechanisms that pastoralists can deploy in defense of their livelihood. Although many pastoralists are settling down voluntarily (Mohamed-Salih *et al*. 1995), government efforts to accomplish the same goal seldom produce attractive outcomes.[7] In part, the failure of state policies can be traced to the desire of state officials within the rhetorical of development policies that claim to attempt improvement of pastoralists lives, but the repeated and consistent failures of these policies force us to query the relationship between intentions and outcomes and indeed, to question intentions of policies as they are stated.[8]

But territorial regulation through spatial fixing and monitoring is not the only form of control available to state officials. In the case of the Raika shepherds in western Rajasthan, the state also attempts to monitor their movements through a system of passes that are issued to the leaders of each mobile camp, the presence of border guards at the boundaries of different provinces, and quotas fixed to regulate the entry of a fixed number of animals into particular forested areas for grazing. Each of these strategies of control is founded in what Sack (1986: 19) and Van der Geest and Pelusso (1995: 38) call 'territoriality'. But these strategies of monitoring and information management also move beyond the attempt to fix objects within a governed space. They constitute, instead, an adaptation to the necessity of movement of the part of pastoralists, and show that the exercise of power by the state is ultimately aimed at control over crucial information that can be systematically collected, stored and analysed.

Of course, it is not just a developmental state that pastoralists have had to deal with. Environmental strategies of states also equally often lead to perverse outcomes for herders. As pointed put earlier, cultural sensibilities have combined with an overarching concern with human impacts on the environment, to generate conservation rhetoric on the need to keep people and livestock out of protected areas. National Parks are internationally defined as areas within which there can be no consumptive use of natural resources—i.e., there can be no human settlements, no grazing, and no collection of timber or firewood, no shifting cultivation and so on. Examples of legislation aimed at keeping people out of National Parks, Wildlife Sanctuaries, Game Reserves and other, similarly designated areas, abound in every continent.

But the state is not always successful in restricting pastoralist' and others' access to forests and grasslands that are a designated part of a national protected area network. Wildlife Sanctuaries in India have fewer legal strictures for keeping people out than National Parks. Most Sanctuaries continue to be used in many ways and by many communities. More importantly, however, the bulk of India's national parks continue to be used by people in direct contravention of the legal requirement of excluding all consumptive uses within these areas. Given the potential electoral costs of forcibly denying people access to forests, state governments have rarely attempted to enforce the Indian Wildlife (Protection) Act (1972). Directed by an environmentally active Supreme Court in 1997 to curtail all consumptive activity within National Parks, many state governments have threatened to denitrify entire areas. Rather than risk alienating local communities. Alternatively, politicians have simply ordered officials of the forest department to ignore activities that are prohibited by the law (Baviskar 1998).

But it is not merely electoral pressures that keep governments from enforcing such restrictive policies. Agenda-based divisions within the bureaucracy itself have served on occasion to dilute the impacts of restrictive policies on pastoralist communities. That such over the past century and a half, over control of forest resources, as well as owing to differences in perceptions over optimal use of the landscape, have been reported from many parts of the country (Rangarajan 1996; Saberwal 1999; Sivaramakrishnan 2000). Pastoralists and others have often been able to exploit such differences to circumvent potentially oppressive policies.

Far more than the developmentalist states then, the environmentalist state is fractured, often lacking internal coherence with regard to territorial ambitions or with regard to the implementation of policies that have a negative impact on specific communities. And it is within implementation of policies that have a negative impact on specific communities. And it is within the interstices created by differing environmental agendas within the state that paternalists have most successfully managed to negotiate access to critical resources. It may be useful to contrast such a representation of herder–state relations, with less ambiguous accounts of pastoralist experiences in Africa, where the overarching theme is one of steady pastoralist impoverishment. Our account of the Indian experience may partly be coloured by the work one of us has done in Himachal Pradesh, where the Gaddi herders represent a significant and increasingly powerful vote bank. Even so it is also likely that pastoralist groups in India are better able to participate in the political process than herders in many parts of sub-Saharan or Sahelian Africa. This may partly be explained by political processes in pastoralist Africa that do not have the geographic 'reach' that the Indian system does, thereby denying many pastoralist communities much of a voice in the shaping of their own destinities.

## Notes

1. The best known of such arguments is undoubtedly the schematic, misplaced, and popular model presented by Garrett Hardin (1968).
2. Hogg (1992: 131) argues 'the tragedy of pastoralism today derives from a misunderstanding of the position and motivation of paternalists who, often under extreme circumstances, are trying to come to terms with a rabidly changing political, economic, and natural environment. The only sensible pastoral development strategy in this context is to provide a facilitating environment for pastoralists. Any more traditional, prescriptive "range management" approach is likely to fail.' His statement is as applicable to pastoralism and government policies in India.
3. See Saberwal (1999) for a discussion on this ratcheting of the alarmist discourse of the Punjab Forest Department, particularly vis-à-vis grazing induced degradation in the western Himalaya.
4. A rich literature now points to cultural preferences for bereft of human as the overarching influence on the continuing attempts to separate culture from nature, rather than demonstrable scientific evidence in support of the negative impacts of human activities on biological diversity (see, for example, Cronon 1995; Saberwal and Rangarajan 2003).
5. For a careful elucidation of some of these points in the Rajasthan context, see Robbins (1998).
6. For a discussion of unintended policy consequences see also Ferguson (1994: 20f). Peters (1994) provides related discussions of boreholes in a context. For an elaboration of how the dynamics of systems might defeat intentioned policies, see Behnke *et al*. (1993), and Holling (1973).
7. Some of the most adverse impacts of seediest policies are visible in the nutrition and health of paternalists, especially of their children as a raft of studies have demonstrated (Klepp *et al*. 1994; Nathan *et al*. 1996).
8. See Ferguson (1994) for a devastating critique of the stated intentions of state policies aimed to develop paternalists' lives.

## References

Agrawal, Arun, 1999, *Greener Pastures: Politics, Markets, and Community among a Migrant Pastoral People,* Durham, NC.

Baviskar, Amita, 1998, 'Nature at Stake', *Seminar* 466: 74–78.

Behnke, R.H. Jr., J. Scones and C. Kerven (eds), 1993, *Range Ecology at Disequilibrium: Pastoral Adaptation in African Savannas,* London.

Binns, T., 1990, 'Is Desertification A Myth?', *Geography* 75:106–13.

Cronon, W. (ed.), 1995, 'Introduction' in *Uncommon Ground: Towards Reinventing Nature,* New York.

During, J. and J.H. Williams, 1986, 'The Impoverishment of the Bryophyte and Lichen Flora of the Dutch Chalk Grasslands in the Thirty Years 1953–1983', *Biological Conservation* 36: 143–58.

Ferguson, James, 1994, *The Anti-Politics Machine: 'Development', Depoliticization, and Bureaucratic Power in Lesotho.* Minneapolis.

Gooch, P., 1998, *At The Tail of The Buffalo: Van Gujjar Pastoralists Between the Forest and the World Arena (India).* Unpublished PhD Thesis. University of Lund, Sweden.

Grubb, P.J. 1976. 'A Theoretical Background to the Conservation of Ecologically Distinct Groups of Annuals and Biennials in the Chalk Grassland Ecosystem', *Biological Conservation* 10: 53–76.

Gyawali, D., 2001, 'Water and Science: Hydrological Uncertainties, Developmental Aspirations, and Uningrained Scientific Culture', *Futures* 33(8–9): 689–708.

Hamilton, L.S., 1987, 'What are the Impacts of Himalaya Deforestation on the Ganges–Brahmaputra Lowlands and Delta? Assumptions and Facts', *Mountain Research and Development* 7: 256–63.

Hardin, G., 1968, 'The Tragedy of the Commons', *Science* 162: 1243–48.

Hjort, A., 1981, 'A Critique of "Ecological Models" of Pastoral Land Use', *Ethnos* 46 (3–4): 171–89.

Hobbs, R.J. and L. F. Huenneke, 1992, 'Disturbance, Diversity, and Invasion: Implications for Conservation', *Conservation Biology* 6: 324–37.

Hogg, R. 1992. 'Should Pastoralism Continue As A Way Of Life?', *Disasters* 16(2): 131–37.

Holling, C.S., 1973, 'Resilience and Stability of Ecological Systems', *Annual Review of Ecology and Systematics* 4: 1–23.

Homewood, K., 1995, 'Development, Demarcation, and Ecological Outcomes in Maasailand', *Africa* 65(3): 331–50.

——— and W. Rodgers, 1991, *Maasailand Ecology: Pastoralist Development and Wildlife Conservation,* Cambridge.

Ives, J.D. and B. Messerli, 1989, *The Himalayan Dilemma: Reconciling Development and Conservation,* London.

Kavoori, P., 1996, 'Pastoralism in Expansion: The Transhuming Sheep Herders of Western Rajasthan. Unpublished PhD Thesis, The Institute of Social Studies, The Hague.

Klepp, K., P. Biswolo, and A. Talle, 1994, *Young People at Risk: Fighting Aids in Northern Tanzania,* Oslo.

Leach, M. and R. Mearns (eds), 1999, *The Life of the Land: Challenging Received Wisdom on the African Environment,* London.

Mace, R., 1991, 'Conservation Biology: Overgrazing Overstated', *Nature* 349: 280–81.

McCabe, T.J., S. Perkin, and C. Schofield, 1992, 'Can Conservation and Development be Coupled among Pastoral People? An Examination of the Maasai of the Ngorongoro Conservation Area, Tanzania', *Human Organization* 51(4): 353–66.

Mohamed–Salih, M., J. Baker, and J. Baekers (eds), 1995, *The Migration Experience in Africa,* Uppsala.

Naithani, H. B., J.D.S. Negi, R.C. Thapliyal, and T.C. Pokhriyal, 1992, 'Valley of Flowers: Need for Conservation or Preservation', *Indian Forester* 117: 371–78.

Nathan, M., E. Fratkin, and E. Roth, 1996, 'Setetarism and Child Health among Rendille Pastoralists of Northern Kenya', *Social Science and Medicine* 43(4): 503–15.

Peters, P., 1994, *Dividing the Commons: Politics, Policy and Culture in Botswana,* Charlottesville.

Prins, H., 1992, 'The Pastoral Road to Extinction: Competition between Wildlife and Traditional Pastoralism in East Africa', *Environmental Conservation* 19(2): 117–23.

Rangarajan, Mahesh, 1996, *Fencing the Forest: Conservation and Ecological Change in India's Central Provinces* 1860–1914, Delhi.

Robbins, Paul, 1998, 'Authority and Environment: Institutional Landscapes in Rajasthan, India', *Annals of the Association of American Geographers* 88(3): 410–35.

Saberwal, V.K., 1999, *Pastoral Politics: Shepherds, Bureaucrats and Conservation in the Western Himalaya,* Delhi.

———, and M. Rangarajan, 2003, 'Introduction' in V.K. Saberwal and M. Rangarajan (eds), *Battles Over Nature: Science and the Politics of Conservation,* Delhi.

Sack, R.D., 1986, *Human Territoriality: Its Theory and History,* Cambridge.

Sharma, K.P., 1997, *Impact of Land Use and Climatic Changes on Hydrology of the Himalayan Basin: A Case Study of the Kosi Basin (Nepal),* Unpublished PhD Thesis, University of New Hampshire.

Singh, C., 1998, *Natural Premises,* Delhi.

Sivaramkrishan, K., 2000, *Modern Forests: State making and Environmental Change in Colonial Eastern India,* Delhi.

# 18

# The New Land Use Policy: People and Forests in Mizoram

**Daman Singh**

The system of shifting cultivation in Mizoram has a certain enduring quality which may be interpreted as resistance to change. In contrast to several other areas, here the reason for its persistence is not to be found in deeply rooted social and cultural tradition, but rather in the resilience of the production system itself. This method of agriculture performs the essential function of feeding the rural family, whose occupational options are severely restricted. Although time-consuming, strenuous and subject to uncertainty, it has the advantage of being self-contained. The shifting cultivator in a remote village needs neither road, nor market, nor government facility to get on with his work.

Even today, *jhumming* for subsistence is a viable proposition for the community and its environmental implications for society at large are fairly innocuous. The rich ecology and currently low pressure of population ensure its sustainability for at least some time to come. These positive pronouncements should not, however, camouflage the urgent need to find an alternative to shifting cultivation.

The urgency stems from past demographic changes which led the people of some rural areas to newly founded towns, while leaving others with relatively high levels of population density. It also arises from the historical trajectory of development that provided neither incentive nor opportunity for growth. Finally, it springs from a large and growing class of educated young people with high hopes and low prospects. The principal flaw in the system of shifting cultivation is that far more remunerative options exist in

---

* Reprinted with permission from Daman Singh, 1996, *The Last Frontier: People and Forests in Mizoram,* TERI, New Delhi, pp. 197–210.

the present technological and economic context. Communities and individuals who fail to grasp these options are in danger of rapid economic, social, and political marginalization.

The fact that the status of natural resources and their management have not yet approached a stage of crisis is the best possible starting point for the quest for alternatives to shifting cultivation, rather than being forced at a later stage to accept whatever few choices remain. This paper outlines past and present efforts of people and the government to introduce new farming systems in Mizoram. For this purpose it relies on numerous official reports as well as interaction with people and information collected from sample villages.

## A Review of Past Strategies

Throughout the years of British rule and well into the post-Independence period, the people of Mizoram remained cut off from their neighbours and almost entirely lacked exposure to methods of cultivation that differed from their own. Their insulation was reinforced by the general disinclination of the government to induce changes that were unlikely to take place of their own accord.

Initially, new ideas came from the tiny class of ex-servicemen and local traders, who brought home stories and samples of what they had seen on their travels. These pioneers persevered in novel ventures at their own initiative and expense. Their conviction led them to comb the region for planting material and to use their ingenuity to innovate locally suitable techniques through trial and error. Where these efforts fructified, their proponents became the founders of an incipient class of entrepreneurs. At the other end, the government sought to demonstrate new crops and varieties at its own farms and on farmers' land, but was impaired by weak farm management and casual attempts at extension. There was, however, one safe option that guaranteed returns: cultivation of paddy in valleys.

### Wet Rice Cultivation

This technique was demonstrated at Champhai in Mizoram as early as 1898. The cultivator had to acquire new skills in reclaiming the land, using bullocks for ploughing, experimenting with different varieties of paddy, raising nurseries, transplanting, irrigation and pest control. The resident Gorkhalis were the first to popularize valley cultivation, but were soon joined by other communities. In the course of time, people ventured to take land on lease in the few large valleys with running streams.

Over the years, wet rice cultivation became less subsistence oriented. Apart from interior villages, wet rice cultivation has all the ingredients of capitalist farming in and around urban areas. There is fierce competition for allotment of land, demand from residents and non-residents escalate land prices, and labour is hired for preparation of land and transplanting. Share cropping or cultivation on contract is a common practise, and a class of absentee landlords has come into being.[1]

After Independence, government assistance for land reclamation started in the First Five Year Plan period and continued under the central Grow More Food programme and schemes of the Assam State Plans, augmented by special central provisions for the Autonomous Districts. Financial allocations increased when Mizoram became a union territory and later a state. The schemes were implemented by the soil conservation department and later by the agriculture department. A subsidy of 50 per cent was provided for labour in land preparation, use of bulldozers, minor irrigation, purchase of draught animals and distribution of seed, fertilizer and plant protection equipment, and chemicals. The efforts of the government to promote wet rice cultivation intensified with its concern to achieve self-sufficiency in foodgrain production. As long as people were engaged in *jhumming,* they were able to produce enough to meet most of their needs. Once a large proportion moved out of agriculture, beginning in the 1970s, the import of foodgrain became an annual exercise. During the 1980s, *jhums* were substantially supplemented by wet rice cultivation. Their share of the total area under rice production decreased from 89 per cent in 1980–81 to 72 per cent in 1992–93. As the yield of fields in the valleys is about double that of those in the hills, wet rice cultivation accounted for 28 per cent of the total area under rice but provided 42 per cent of the total rice output in 1992–93. Table 18.1 provides comparative figures of *jhumming* and wet rice cultivation.

The recent developments in rice production have brought the average productivity in Mizoram not far below the regional average. In 1990–91 the reported rice yield was 1244 kg/ha in Mizoram and 1361 kg/ha in the

Table 18.1: The Area under Rice Production

| *Year* | *Total Area (ha)* | | *Percentage Share of Jhums* | |
|---|---|---|---|---|
| | *Jhums* | *Valleys* | *Rice area* | *Rice output* |
| 1980–81 | 55,264 | 6,923 | 89 | 63 |
| 1981–82 | 55,858 | 7,220 | 89 | 74 |
| 1982–83 | 52,484 | 7,195 | 88 | 81 |
| 1983–84 | 41,281 | 6,390 | 87 | 71 |
| 1984–85 | 43,378 | 7,486 | 84 | 75 |
| 1985–86 | 45,920 | 9,092 | 83 | 85 |
| 1986–87 | 38,893 | 9,452 | 80 | 72 |
| 1987–88 | 37,803 | 10,661 | 78 | 72 |
| 1988–89 | 36,616 | 12,772 | 74 | 62 |
| 1989–90 | 38,349 | 14,640 | 72 | 59 |
| 1990–91 | 36,716 | 14,607 | 72 | 55 |
| 1991–92 | 39,175 | 16,464 | 70 | 54 |
| 1992–92 | 43,858 | 17,439 | 72 | 58 |

**Source:** Directorate of Agriculture, GoM

north-eastern region, when the national average was 1751 kg/ha (NEC 1992). Yet the state is unable to meet its demand for rice. Assuming a per capita annual requirement of 183 kg, it is clear that Mizoram could not have survived without substantial import of rice in the past 20 years. The per capita net production in 1992–93, regarded as a good year for agriculture, was 118 kg, which still represented a shortfall of 36 per cent. It is however encouraging that the annual deficit has tended to decline, particularly since 1987–88. The annual rice production and estimated deficit is shown in Table 18.2.

For Mizoram to produce all the rice that its population requires, there is no option but to increase the land under wet rice cultivation both to meet current demand and to keep pace with the rate of population growth. The possibilities of doing so may be weighed on the basis of available statistics. In 1992–93, wet rice cultivation took place on 17,439 ha against the estimated potential of 50,021 ha. Assuming that all this land is brought under cultivation, and taking the average area under *jhum* paddy and the average rice yields of *jhums* and wet rice cultivation for the period 1990–91 to 1992–93 as constant, the resultant rice output would feed a population of 7.63 lakhs at a per capita annual consumption of 183 kg. This is only

TABLE 18.2: TOTAL RICE PRODUCTION AND PER CAPITA AVAILABILITY

| *Year* | *Area (ha)* | *Output MT* | *Per capita output (kg)* | *Deficit (percentage)* |
|---|---|---|---|---|
| 1974–75 | 47,500 | 28,490 | 76 | 58 |
| 1975–76 | 49,201 | 31,164 | 80 | 56 |
| 1976–77 | 73,000 | 38,462 | 70 | 62 |
| 1977–78 | 68,000 | 16,310 | 39 | 79 |
| 1978–79 | 64,100 | 21,000 | 48 | 74 |
| 1979–80 | 63,200 | 20,440 | 45 | 76 |
| 1980–81 | 62,187 | 59,633 | 126 | 31 |
| 1981–82 | 63,078 | 30,000 | 61 | 67 |
| 1982–83 | 59,679 | 45,953 | 90 | 51 |
| 1983–84 | 47,671 | 36,713 | 70 | 62 |
| 1984–85 | 50,864 | 40,894 | 725 | 59 |
| 1985–86 | 55,012 | 64,780 | 115 | 37 |
| 1986–87 | 48,374 | 68,745 | 118 | 36 |
| 1987–88 | 48,464 | 49,227 | 82 | 55 |
| 1988–89 | 49,388 | 53,000 | 85 | 54 |
| 1989–90 | 52,989 | 59,236 | 92 | 50 |
| 1990–91 | 51,323 | 63,794 | 96 | 48 |
| 1991–92 | 55,639 | 70,974 | 103 | 44 |
| 1992–93 | 61,297 | 83,954 | 118 | 36 |

**Source:** Area and production from Directorate of Agriculture.

**Note:** Population extrapolated from census data; per capita requirement assumed as 183 kg/year

11 per cent more than the population in 1991. Considering that the rate of population growth during 1981–91 was 40 per cent, even the maximum possible increase in the area under wet rice cultivation would not permit self-sufficiency beyond this decade.

Given its topographical constraints, Mizoram is likely to remain an importer of rice, but it does have the potential of increasing its production above the existing level. Apart from the statistics quoted above, the prospects for wet rice cultivation are illustrated by sample data. Out of 36 villages, the potential exists in 24 (67 per cent). In five villages (21 per cent) it has been fully tapped, in 15 (63 per cent) it is partly tapped, and in four (17 per cent) it is entirely untapped.

## Terracing

Constructing bench terraces on penile slopes is a normal practise in lull areas. Experiments at ICAR, Shillong, have shown it to be an effective check on run-off and erosion, and as being economically feasible. Beginning in 1972, the government assisted farmers in terracing their land at 80 per cent subsidy together with additional support for minor irrigation and purchase of draught animals, seed, fertilizer, and plant protection chemicals and equipment. The Mizoram directorate of agriculture estimates that a total area of 87,824 ha is suited to bench terracing.

Unlike the sure shot wet rice cultivation, terracing did not gain momentum. A review of the progress of soil conservation schemes in the north-eastern region in 1979 studied the performance of 10,000 families which had participated in the terracing programme in Mizoram. It was found that only 300 families actually gave up *jhumming*. These were families in the vicinity of large towns, and had been allotted large plots of more than 4 ha each. The majority grew cash crops on terraces, but continued to depend on *jhumming* for food crops. According to the study, the programme also succeeded in creating an affluent elite in society, which employed labour to cultivate the terraces, while busying itself with dairying, pig rearing, and other enterprises (ICAR 1983).

The problems are not entirely to do with the nature of government programmes, there being technical drawbacks also. In the first few years after a terrace is constructed, there is heavy soil loss. Although erosion gradually reduces with time, annual maintenance is necessary to repair the damage wrought by torrential rains. Stone, which is successfully used in many places to strengthen earthen bunds, is not locally available. The physical properties of the soil gradually decline over a period of six to eight years, often resulting in the short life of the terrace. Though runoff loss may be controlled, serious loss of moisture and nutrients due to high soil porosity is a continuing feature. There is a constant need for heavy input of inorganic fertilizer to maintain reasonable returns to the farmer. There is, besides, exaggerated weed potential under repeated cropping at the same site (Ramakrishnan 1993).

Terrace cultivation in Mizoram thus works out to be a high cost activity, requiring not only high investment but also high recurring expenses. Irrigation is not available to most terraces, so only a single crop can be grown.

Large doses of fertilizer and plant protection chemicals have to be applied to maintain the yield. Promotion of terracing has however not yet been abandoned by the government.

## Diversifying Food Crops

Climatic conditions in the hill region of the north-east are conducive to a large number of crops. For many of these, varieties have been screened and developed at the regional centre of the ICAR. The Mizoram department of agriculture has also conducted trials and demonstrations of different crops and their varieties.[2] Both indigenous and improved paddy varieties are suited to the plains and terraces. However, improved varieties of upland paddy have not yet been tried in *jhums*. Pulses are scarcely cultivated. Black and green grams are still under trial. Though lentil trials were encouraging, extension has been poor. *Arhar* trials failed. Among oilseeds, rapeseed, mustard, sesamum, and soyabean are widely grown, and improved varieties for the first two have become popular.

Chillies, ginger, turmeric, tapioca, sweet potato and colocasia are traditionally grown, both for consumption and for sale. Varietal improvement in ginger is picking up, and colocasia may follow in its footsteps. An enormous number of vegetables can be cultivated, but this is only viable in and around towns. *Chou-chou* is the only vegetable, which is exported from Mizoram, and has had dramatic success at Sihphir and Lungdai villages. What started out as individual enterprise has been strengthened by government support? In 1993, these two villages exported 3,509 MT of produce valued at Rs 4,077,128. Large quantities of all other vegetables are imported into Aizawl, Lunglei, Saiha, and other towns from Assam.

## Horticulture

The British introduced orange orchards to Mizoram and other fruits trickled in from neighbouring areas through individual endeavour. For a number of years the lack of roads and opportunities for sale limited the spread of horticulture. This changed with the construction of the arterial road from Silchar in Cachar district of Assam through Aizawl and Lunglei to Saiha, other important link roads, and the establishment of new towns. Today, in many roadside villages, horticulture is a lucrative enterprise, and orange, lemon, banana, *batkora,* papaya, and passion fruit find ready buyers in the towns. A total of 10,211 ha was reported under horticulture in 1992–93.

The 36 sample villages of this study reflect the early beginning made in raising orchards. Of the 26 villages (79 per cent) which have orchards, or gardens as they are called, the year of initiation is traced by 52 per cent to 1920–71, a period which stretches through British rule, Independence, and the time of the Mizo Hills Autonomous District Council. It is between 1972–86 (the Union Territory days) for 25 per cent villages, and post-1987 (when Mizoram state was formed) for the remaining 16 per cent. An estimated 29 per cent families own gardens, some of which are yet to reach fruit-bearing stage and fully occupy their owners. The number of such families is higher in villages which have approach roads and where there is a relatively large proportion of households with government jobs.

Unlike wet rice cultivation, horticulture depended a great deal on individual initiative, as its active promotion by the government is fairly recent. The first scheme was launched in 1978 on a small scale. Farmers who participated in the 'garden colony' scheme, were given cash assistance for clearing the land and digging pits and received free planting material. Subsequently, orchards became the principal component of more comprehensive programmes for *jhum* control, funded both by the centre and the state.

A unique feature of horticulture in Mizoram is its completely organic character. Numerous tropical fruits, such as orange, lemon, pineapple, and banana, have fared well. Among temperate fruits, passion fruit is an outstanding success, and its squash is bottled and marketed by the MIFCO (Mizoram Food Processing Corporation). The latest reports on strawberry are encouraging. Though many indigenous fruits are locally consumed and reach the urban markets, there has been no effort to promote their cultivation.

## Other Perennial Crops

Tea was unknown to Mizoram till it was popularized by the Christian Missions, to replace the forbidden rice beer. The plant was introduced as far back as the 1930s. Its early cultivators would prick out sprouts from fallen seed, before they took it to raising nurseries. Cultivation was both for consumption and sale. Later, in the 1970s, the soil conservation department demonstrated tea plantation on its farms. A recent assessment by officials of the Tea Board found the tea to be of good quality, similar in flavour to the high altitude tea of Darjeeling and the Nilgiris. The complete absence of chemical fertilizer and pesticides is an exclusive value of Mizoram tea. There is, however, need for scientific management, in pruning, regular plucking, reducing shade intensity, and increasing the bush population (Tea Board 1992).[3] At present the area under tea cultivation is negligible.

Large and small cardamom, coffee, rubber, black pepper, arecanut, and betel vine have all done well on government farms. Extension has been low key, which is why none have gained popularity, apart from arecanut which has been taken up by farmers on their own with excellent results. A recent study of the potential for citronella cultivation and oil extraction confirmed its suitability. Though it did well at a few government farms, it has not been extended and extraction has not been attempted (Mahishi n.d.). Sericulture was initiated by the Salvation Army in the 1920s, and later promoted by the government to the extent of setting up a separate sericulture directorate. The scale of cocoon rearing and silk production is still infinitesimal.

## The Old New Land Use Policy

The Mizoram government devised its New Land Use Policy (NLUP) in 1984 to do away with shifting cultivation. Its first form was implemented by the forest department, starting in 1984–85 and terminating in 1989–90. Under this programme 6086 families were assisted in implementation. Thereafter, there were no further government initiatives to offer long-gestation farm forestry as an alternative to *jhumming*.

### The Jhum Control Project

In 1987, the Mizo National Front took over from the Congress and replaced the old NLUP with the centre-sponsored Jhum Control Project. The pilot project in Aibawk block was implemented between 1987–88 and 1991–92. Work in Tuipang block started in 1990–91, but the scale was curtailed, as the National Development Council decided to discontinue the Special Area Development Programmes in eight states, including the Jhum Control

Project in Mizoram. Originally, the department of agriculture was the nodal agency for implementation, but in 1990–91, the rural development department took over.

Detailed information is available for the project in Aibawk block, in which 1936 beneficiaries from 16 villages participated. The aim was to progressively reduce and replace *jhumming* till its complete abandonment by 1993. Horticulture was the most favoured option. Where topography permitted, wet rice cultivation, bench terracing, contour bunding, and minor irrigation were to be introduced. Improved piggery, poultry and duck rearing, cattle development and upgradation of local cows were also planned. Crafts, such as carpentry, tailoring, and knitting were to be promoted. Departmental works were proposed for soil and water conservation, social forestry, roads, and other support infrastructure (GoM 1987).

Every family was a beneficiary under the programme, including those in government service. Of the 1936 participants, 47 per cent opted for horticulture, 20 per cent for animal husbandry, 10 per cent for wet rice cultivation, nine per cent for coffee plantation, five per cent for dry terrace cultivation, five per cent for sericulture, and three per cent for cottage industries. Assistance in cash and kind was to the extent of Rs 30, 000 per family (GoM 1991).

The rural development department conducted an internal evaluation in 1991–92. According to its report, 52.24 per cent of the participants reported an increase in net annual income in 1991 over the pre-project period. The average household income increased by 21.44 per cent, from Rs 8,227 to Rs 9,991. A survey of the assets revealed that they were fully intact in only 34.64 per cent cases, had partly perished in 45.4 per cent, fully perished in 9.55 per cent, were defective in 8.62 per cent, and had been sold in 1.75 per cent cases.

Wet rice cultivation was found to have fared the best. Orange orchards did well, barring the 10–40 per cent mortality due to long distance transport of seedlings. Terraces had fared satisfactorily, but credit was required to finance irrigation, and fencing was necessary. Fishery was partly successful. The prospects of sericulture were reportedly slim, as there was no market for cocoons. Coffee plants had high mortality and remained stunted for reasons not known. Animal husbandry fared poorly due to the tendency of the beneficiaries to sell the stock, as well as problems in quality, feed, and health care (GoM 1992).

The change in land use may be gauged by comparing the latest available statistics for 1993–94 with those of the pre-project year. The area under horticulture increased substantially from 449 to 1644 ha in 1993–94. Wet rice cultivation improved its coverage from 268 to 439 ha, of which 19 per cent land was dedicated to high yielding varieties. As pre-project data is not available for *jhumming,* data for 1984 is used instead; since that year the area under shifting cultivation decreased from 891 to 623 ha.

As the first comprehensive programme to create alternative employment opportunities, the Jhum Control Project was a good start. Obviously there was a need to tighten extension services, technical support, and monitoring for better results. However, it is rather more disturbing that seven years after the project began, reduction *in jhumming* has only been around 30 per cent. Hopefully, the figures should improve once greater incomes are realized from horticulture.

## Changes in the New Land Use Policy

With the Congress reverting to power in 1989, the NLUP was revived, but with major improvements over both the first attempt at farm forestry as well as the Jhum Control Project. An ambitious programme began in 1990–91 with the intention of covering the entire state by 1997. The department of rural development was revamped and strengthened with the necessary financial, technical, and supervisory powers to ensure smooth and coordinated implementation.

All possible administrative impediments to the new programme were eliminated. The restriction on allotment of land in roadside and riverine reserves was set aside. The Block Development Officer could ignore existing temporary land leases to facilitate compact areas being developed in each village. In NLUP villages, the village council could no longer recommend cases of land allotment to the revenue department. The Local Administration Department placed a ban on *jhumm*ing in villages covered by the programme.

In order to overcome the problem of long gestation, the programme introduced composite schemes rather than single schemes. Each principal trade was linked to a subsidiary one designed to yield quick returns. The design of schemes was improved in accordance with their specific technical and financial requirements. Activities were to be selectively promoted in areas with market openings. The people were given a wide range of options.

The agriculture and allied sector included wet rice cultivation, terrace cultivation, horticulture (mandarin orange, Valencia, *jamir, batkora,* Assam lemon, *kagzi* lime, mango, passion fruit, grape, banana, pineapple), sugarcane, citronella, *tung,* coffee, tea, pisciculture, and sericulture. Subsidiary trades offered were piggery and vegetable gardening.

In the animal husbandry sector the principal trades were dairy farming, rearing of hill cattle, pigs, goats, poultry, ducks, sheep or rabbits, while subsidiary trades were horticulture, vegetable garden, and piggery. The principal trades in the industry sector were bakery, rice hulling, knitting, tailoring, handloom, cane and bamboo works, blacksmithy, watch repairing, radio repairing, art painting, tinsmithy, *chow* making, motor works, shoe repair, steel fabrication, and carpentry. Subsidiary trades were poultry, piggery, and vegetable gardening.

Assistance was to be phased over two to three years, with a ceiling of Rs 30,000 per family, relaxable by Rs 5,000 in deserving cases. In contrast to the Jhum Control Project, the beneficiaries were to be restricted to those practising shifting cultivation or other people lacking permanent means of livelihood, but the Block Development Officer was given discretionary powers in the selection process. Each beneficiary was required to give a written undertaking not to dispose of the assets.

**TABLE 18.3: COVERAGE OF THE NEW LAND USE POLICY**

| *Year* | *Block* | *Number of beneficiaries* | *Beneficiaries/Total Families* (Percentage)* |
|---|---|---|---|
| 1990–91 | Thingsulthliah | 4654 | 98 |
| | West Phaileng | 5445 | 145 |
| | Reiek | 1981 | 92 |
| | Lungsen | 5079 | 105 |
| | Sub-total | 17159 | |
| 1991–92 | Thingdawl | 6419 | 80 |
| | Zawlnuam | 4108 | 74 |
| | *Sub-total* | 10527 | – |
| 1992–93 | – | – | – |
| 1993–94 | Serchhip | 4568 | 89 |
| | East Lungdar | 1657 | 30 |
| | Sub-tolal | 6225 | |
| | Total | 3391 1 | 97 |

**Note:** * 1991 census figures

The logistical quandaries in organizing activities on the scale indicated here can only be imagined. After the state cabinet has selected a block for the year, the rural development department staff launches a public awareness campaign, prepares a list of bona fide beneficiaries in each village with the assistance of the village council. The Block Development Officer exercises his powers of discretion and finalizes the trade selection of each beneficiary. Thereafter, the requirement of inputs is assessed and arrangements are made for procurement. Planting material for orange and banana orchards is bought from private individuals in Mizoram. Mango, coconut, and betelnut seedlings are procured through the Assam Seed Corporation. Select local farms (for instance at Sihphir, Durtlang) provide animals; some are brought from Meghalaya, Arunachal, and other states. A large quantity of implements and machinery is bought from Punjab.

All inputs have to reach the village on time, by vehicle if the village is on the road, otherwise by headload. Periodic cash instalments have to be disbursed, their correct use verified, and technical guidance and troubleshooting ensured. For all this, the Block Development Officer is assisted by his extension staff and one village level worker for every two or three villages.

It is too soon to comment on the performance of this massive programme, as it needs more time to make an impact. It is however useful lo keep an eye on the extent of shifting cultivation in NLUP blocks. This is not easy, as the department of rural development does not monitor shifting cultivation in its programme area. The area under *jhumming* is recorded by the agriculture department for regional units, or circles, which overlap block boundaries. However, West Phaileng circle falls entirely within the block by the same name, as does Kanghmun and Sairang (barring three villages) circles in Reiek block. Both of these blocks were taken up by the NLUP in 1990–91, and should therefore indicate a decline in *jhumming*. Figures for shifting cultivation for West Phaileng circle indeed showed a decline from 2954 ha in 1989–90 to 2100 ha in 1994–95; a rather modest drop of 29 per cent. Paradoxically, during the same period of five years, Sairang circle showed an increase of 72 per cent (690–1185 ha) and Kanghmun an increase of 106 per cent (185–382 ha). Despite this extremely damaging view of the NLUP, it would be unfair to draw a general conclusion on the basis of this limited data.

The design of the programme is a topic that may usefully be discussed. The NLUP is one of the first programmes of its kind that plans to reach every *jhumming* family in the state. As is evident from the description of its components, it is extremely comprehensive, both in the range of opportunities and the level of assistance provided to participants. It serves to fill crucial lacunae: the access to technology and inputs. Two critical issues that remain are the efficiency of the delivery system and the prospects for marketing the newly produced goods and services. The previous record of government programmes, and numerous real and practical problems, cast serious doubts on the fulfillment of both these requirements.

Government services in extension and input supply have been notoriously weak in small schemes with limited targets, let alone massive and highly centralized programmes such as the NLUP being executed on a war footing. The bureaucracy has been given overwhelming powers that override local village councils. It is difficult for the shifting cultivator to trust an administration that is yet to prove its credibility. There is also the question of management of funds in this heavily financed programme which has plenty of scope for leakages. A clear case of speculation on the use of funds exists in West Phaileng block, where the numbers of beneficiaries exceed the number of families enumerated in the 1991 census by an amazing 45 per cent.

Commodity procurement by government has prompted the creation of generously staffed mega-institutions to protect the interests of producers. Yet, these bodies are not directly involved in either procurement or marketing, since they authorize commission agents to perform both these tasks. The nodal agency for the procurement of even a single product is often changed from year to year, procurement prices are uncertain, and lifting of stocks is not assured, so farmers tend to sell their produce to the first possible buyer. Though the government has made a considerable effort to create food and fruit processing facilities, its performance so far has not been very encouraging. The units tend to run at low capacity and rarely make a profit.

The current situation of market opportunities in Mizoram is very disappointing. The most immediate option is the domestic urban market. Spatially, the urban centres are not uniformly distributed. Tlangnuam block in which the state capital and Lunglei block, where the district headquarters is located, each has a huge urban population. Of the 20 blocks in Mizoram, as many seven have no towns. These are West Phaileng, Aibawk, Lawngtlai, Sangau, Ngopa, West Bunghmun, and Chawngte. Except in blocks such as

Aibawk where all villages are on the road and within easy reach of Aizawl, villages in these blocks have a natural disadvantage when it comes to market opportunities.

The second option is inter-state trade. Assam is the only outlet for Mizoram, as the adjoining stretches of Tripura and Manipur have similar terrain, poor road connections and no major market. The hills of the state melt into the plains of Cachar district of Assam, where Silchar and Baiga are big trade centres. Meghalaya, Tripura, Mizoram, and Manipur compete in these markets. Except for chillies, passion fruit, and *chou-chou* or squash, Mizoram's market share is negligible.

Trans-border trade with Bangladesh and Burma has been a subject of recent negotiations. After Partition in 1947, Mizoram lost its market for timber and other forest produce, sesamum, cotton, and oranges, which were exported via the Karnaphuli river to Chittagong. If ultimately regularized, border trade may have the potential to stimulate the local economy.

Besides the limited market openings for goods produced in Mizoram, there are also constraints in access to them. On an average, only 17 per cent villages are connected by *pakka* road, 42 per cent by fair weather roads, and a large 41 per cent by footpaths or waterways. The railway line stops at Bairabi, situated at the extreme northern border of the state. The air route to Calcutta is operated by an 18-seater Dornier which is purely for passenger travel. This leaves the national highway as Mizoram's only lifeline. The road was metalled in the 1970s but remains vulnerable to blockage due to landslides in the monsoons. Frequent trouble with Cachar, as well as internal unrest, immediately disrupt road traffic. Caught in this stranglehold, the major casualties are the incoming essential commodities, for exports are in any case very limited. Strikes, protests, and border disputes are regular features that squeeze Mizoram's economy.

## The Alternative to Shifting Cultivation

The story of innovation in land use in Mizoram is a short one, which started late and unfolded slowly until recent times when it began to race to make up for lost time. The NLUP symbolizes Mizoram's answer to the future. It is a bold and sweeping strategy, aimed at revolutionizing the lives of the people. This is not just any other government programme, for its object is to assimilate every family that depends on shifting cultivation. Five years into the programme have revealed how susceptible it is to going off target. Unless the NLUP succeeds in overcoming the odds against it, there is tremendous cause for concern. An obvious fallout would be the serious erosion of the credibility of the government. If people continue with *jhumming* even after the immense resources poured into the villages, they are hardly likely to have faith in a second rescue operation.

There are however far more worrying matters. The NLUP operates in such a way as to dislocate the well-organized system of shifting cultivation. In the villages where *jhumming* has been banned, people continue the practise covertly in small groups that lack the discipline and foresight of the village councils. In negating the authority of the village council over issues of land use and management, the NLUP disables the very institution which has been instrumental in the effective administration of resources. Divested of its powers and superseded by the bureaucracy, the village council that was once a reactive and responsible body representing the interests of the people, is in danger of becoming a mere figurehead.

These ominous observations provoke an examination of other alternatives to shifting cultivation. As has been emphasized in the preceding chapters, the objective must be for economic rather than ecological betterment. The farming sector is an obvious starting point. The move from subsistence farming to producing for the market is more than just a question of technology or capital. At the very root of this transition must lie an understanding of how markets work. This is by no means a minor accomplishment for the small or marginal farmer, for whom the market is riddled with uncertainty. Left to their own devices, subsistence cultivators would take an age to develop market confidence. There is an acute necessity for more and better roads, communications, procurement, transport, storage, processing, as well as farmers' organizations to deal with the market.

Some parts of Mizoram are reasonably well-equipped to make a market entry, but others still have a long way to go. For such areas, till the time is favourable for the transition, it would be sensible to have a strategy of enhancing the existing system of cultivation, while at the same time preparing the people for the opportunity to change. So far, little attention has been paid to this aspect since both research and extension have aimed at doing away with shifting cultivation rather than improving it.

For people to accept changes in the method of *jhumming*, the improvement would have to be more than convincing, for it can only be in few locations that shifting cultivation has been forced down a path of diminishing returns. It would require a rigorous study of local practises to suggest feasible options. These may well relate to crop and variety selection; their positioning or placement in the field; the extent and manner of mixing species; the design of tools and implements; treatment of pests and plant diseases, and storage and processing of produce.

All these are aspects which need greater efforts at research and demonstration. So far, limited trials by the ICAR indicate that local tall varieties of paddy are more adapted to give higher yields in comparison to improved varieties like Pusa-33. On the other hand, maize can be bettered by introducing Vijay or Ganga-5 varieties. Though pulses are part of the diet, they are not included in the cropping pattern. These legumes could serve as cover crops to shield the soil from the direct impact of early rains, effectively use nitrogen through fixation, and provide an inexpensive source of protein that can be easily stored for consumption and sale.

Agro-forestry is a more radical step towards improved shifting cultivation, derived from the Naga tradition of preserving alder trees on their *jhums* and lopping them during cropping. The introduction of native trees and shrubs, particularly legumes and non-leguminous nitrogen fixers, may be designed in order to increase the gross output of the *jhum*. The species selected should necessarily be fast growing, with narrow crown to allow adequate light penetration into the earth, with boles having the ability to slough off lateral branches so that the clean bole length of the tree can be increased, and the ability to recycle nutrients through a rapid turn over of the leaf population. In addition, it has been observed that a belt of bamboo on the periphery of the *jhum* could act as a valuable wind break to prevent the blow-off of ash, which can be as much as 50 per cent of the total ash obtained on burning. This loss of ash translates into a high loss of nutrients, which could otherwise be utilized by the cultivated crops (Ramakrishnan 1993).

The route to the future development of Mizoram lies equally in the advancement of the non-farm sector. There are limits to the population supporting capacity of land in the hills, and this may all too soon become apparent. Besides, the primary sector remains stunted in the absence of forward linkages with the secondary and tertiary sector. The processing and manufacturing industries lag far behind, services are scarcely diversified, and the basic support infrastructure is feeble. Clearly there is an enormous amount of work to be done to ensure that the people of Mizoram enter the twenty-first century on a strong footing to determine their own destiny.

## Notes

1. In some villages along the northern and eastern border, hired labour comes from Cachar district in Assam and Burma.
2. The experience with some crops and varieties is summarized in Appendix 30, in Daman Singh, *The Last Frontier*.
3. A large tea nursery was set up in 1993–94 at Luangmul, near Aizwal, and a processing plant is proposed.

## References

GoM (Government of Mizoram), 1987, *Control of Shifting Cultivation Project, Aibawk Block, Mizoram*, Department of Agriculture, Government of Mizoram, Aizawl.

———, 1991, Evaluation of Aibawk Jhum Control Project: The Main Findings of he Survey during January–February 1991, State Level Monitoring Cell, Rural Development Department, Government of Mizoram, Aizawl.

———, 1992, *Statistical Handbook, Mizoram, 1992*, Department of Economics and Statistics, Aizawl.

ICAR (Indian Council of Agricultural Research), 1983, *Shifting Cultivation in North East India*, ICAR Research Complex for North Eastern Hill Region, Shillong.

Mahishi, S.R., n.d., *Project Report for Cultivation and Distillation of Citronella Grass*, prepared for the Department of Industries, Government of Mizoram, Emkay Aromatics Ltd, Madras.

NEC (North-Eastern Council), 1992, *Basic Statistics of the North-Eastern Region*, North-Eastern Council, Shillong.

Ramakrishnan, P.S., 1993, *Shifting Agriculture and Sustainable Development: An Inter-disciplinary Study from North-Eastern India*, OUP, New Delhi.

Tea Board, 1992, *Report on the Feasibility of Commercial Tea Cultivation in the State of Mizoram*, submitted to the Department of Industries, Government of Mizoram.

# 19

# The Gender and Environment Debate: Lessons from India

Bina Agarwal

What is women's relationship with the environment? Is it distinct from that of men's? The growing literature on ecofeminism in the West, and especially in the United States, conceptualizes the link between gender and the environment primarily in ideological terms. An intensifying struggle for survival in the developing world, however, highlights the material basis for this link and sets the background for an alternative formulation to ecofeminism, which I term 'feminist environmentalism'.

In this paper I will argue that women, especially those in poor rural households in India, on the one hand, are victims of environmental degradation in quite gender-specific ways. On the other hand, they have been active agents in movements of environmental protection and regeneration, often bringing to them a gender-specific perspective and one which needs to inform our view of alternatives. To contextualize the discussion, and to examine the opposing dimensions of women as victims and women as actors in concrete terms, this paper will focus on India, although the issues are clearly relevant to other parts of the Third World as well. The discussion is divided into five sections. The first section outlines the ecofeminist debate in the United States and one prominent Indian variant of it, and suggests an alternative conceptualization.

The next three sections respectively trace the nature and causes of environmental degradation in rural India, its class and gender implications, and the responses to it by the state and grassroots groups. The concluding section argues for an alternative transformative approach to development.

---

* This paper was originally published in *Feminist Studies*, Vol. 18, No. 1 (Spring 1992), and is reprinted here, by permission of the publisher, Feminist Studies Inc., c/o Women's Studies Program, University of Maryland, College Park, MD 20742, USA.

## Some Conceptual Issues

### Ecofeminism

Ecofeminism embodies within it several different strands of discourse, many of which have yet to be spelled out fully, and which reflect, among other things, different positions within the Western feminist movement (radical, liberal, socialist). As a body of thought, ecofeminism is as yet underdeveloped and still evolving, but carries a growing advocacy. My purpose is not to critique ecofeminist discourse in detail, but rather to focus on some of its major elements, especially in order to examine whether and how it might feed into the formulation of a Third World perspective on gender and the environment. Disentangling the various threads in the debate, and focusing on those more clearly articulated, provides us with the following picture of the ecofeminist argument(s):[1] (*a*) There are important connections between the domination and exploitation of nature; (*b*) In patriarchal thought, women are identified as being closer to nature and men as being closer to culture. Nature is seen as inferior to culture; hence, women are seen as inferior to men; (*c*) Because the domination of women and the domination of nature have occurred together, women have a particular stake in ending the domination of nature, in healing the alienated human and non-human nature' (King 1989: 18); and (*d*) The feminist movement and the environmental movement both stand for egalitarian, non-hierarchical systems. They thus have a good deal in common and need to work together to evolve a common perspective, theory and practise.

In the ecofeminist argument, therefore, the connection between the domination of women and that of nature is basically seen as *Ideological*, as rooted in a system of ideas and representations, values and beliefs, that places women and the non-human world hierarchically below men. And it calls upon women and men to reconceptualize themselves, and their relationships to one another and to the non-human world, in non-hierarchical ways.

We might then ask: In what is this connection between nature and women seen to be rooted? The idea that women are seen as closer to nature than men was initially introduced into contemporary feminist discourse by Sherry Ortner who argued that

> woman is being identified with—or, if you will, seems to be a symbol of—something that every culture devalues, defines as being of a lower order of existence than itself ... [That something] is 'nature' in the most generalized sense ... [Women are everywhere] being symbolically associated with nature, as opposed to men, who are identified with culture. (Ortner 1974: 72–73)

In her initial formulation, the connection between women and nature was clearly rooted in the biological processes of reproduction although, even then, Ortner did recognize that women, like men, also *mediate* between nature and culture.

Ortner has since modified her position which was also criticized by others (particularly social anthropologists) on several counts, especially because the nature–culture divide is not universal across all cultures, nor is there uniformity in the meaning attributed to 'nature', 'culture', 'male', and 'female'.[2] Still, some ecofeminists accept the emphasis on biology uncritically and in different ways reiterate it. An extreme form of this position is that taken by Ariel Kay Salleh who grounds even women's consciousness in biology and in nature. She argues:

> Women's monthly fertility cycle, the tiring symbiosis of pregnancy, the wrench of childbirth and the pleasure of suckling an infant, these things already ground women's consciousness in the knowledge of being coterminous with nature. However tacit or unconscious this identity may be for many women ... it is nevertheless 'a fact of life' (Salleh 1984: 340).

Others such as Ynestra King and Carolyn Merchant argue that the nature–culture dichotomy is a false one, a patriarchal ideological construct which is then used to maintain gender hierarchy. At the same time they accept the view that women are ideologically constructed as closer to nature because of their biology (Merchant 1980: 144).

Merchant, however, in an illuminating historical analysis, shows that in premodern Europe the conceptual connection between women and nature rested on two divergent images, coexisting simultaneously, one which constrained the destruction of nature and the other which sanctioned it. Both identified nature with the female sex. The first image, which was the dominant one, identified nature, especially the earth, with the nurturing mother, and culturally restricted 'the types of socially and morally sanctioned human actions allowable with respect to the earth. One does not readily slay a mother, dig into her entrails for gold, or mutilate her body…' (Merchant 1980: 144). The opposing image was of nature as wild and uncontrollable which could render violence, storms, drought and general chaos. This image culturally sanctioned mastery and human dominance over nature.

Between the sixteenth and seventeenth centuries, Merchant suggests, the scientific revolution and the growth of a market-oriented culture in Europe undermined the image of an organic cosmos with a living female earth at its centre. This image gave way to a mechanistic world-view in which nature was reconceived as something to be mastered and controlled by humans. The twin ideas of mechanism and of dominance over nature supported both the denudation of nature and male dominance over women. Merchant observes:

> The ancient identity of nature as a nurturing mother links women's history with the history of the environment and ecological change…. In investigating the roots of our current environmental dilemma and its connections to science, technology, and the economy, we must re-examine the formation of a world view and a science that, by reconceptualizing reality as a machine rather than a living organism, sanctioned the domination of both nature and women.

Today, Merchant proposes,

> juxtaposing the egalitarian goals of the woman's movement and the environmental movement can suggest 'new values and social structures, based not on the domination of women and nature as resources but on the full expression of both male and female talent and on the maintenance of environmental integrity. (Merchant 1980: xx–xxi, xix)

Ecofeminist discourse, therefore, highlights (*a*) some of the important conceptual links between the *symbolic* constructions of women and nature and the ways of *acting* upon them (although Merchant alone goes beyond the level of assertion to trace these links in concrete terms, historically); (*b*) the underlying commonality between the premises and goals of the women's movement and the environmental movement; and (*c*) an alternative vision of a more egalitarian and harmonious future society.

At the same time the ecofeminist argument as constructed is problematic on several counts. First, it posits 'woman' as a unitary category and fails to differentiate among women by class, race, ethnicity, and so on. It thus ignores forms of domination other than gender which also impinge critically on women's position.[3] Second, it locates the domination of women and of nature almost solely in ideology, neglecting the (inter-related) material sources of this dominance (based on economic advantage and political power). Third, even in the realm of ideological constructs, it says little (with the exception of Merchant's analysis) about the social, economic and political structures within which these constructs are produced and transformed. Nor does it address the central issue of the means by which certain dominant groups (predicated on gender, class, etc.) are able to bring about ideological shifts in their own favour and how such shifts get entrenched. Fourth, the ecofeminist argument does not take into account women's lived material relationship with nature, as opposed to what others or they themselves might conceive that trace the connection between women and nature to biology may be seen as adhering to a form of essentialism (some notion of a female 'essence' which is unchangeable and irreducible).[4] Such a formulation flies in the face of wide-ranging evidence that concepts of nature, culture, gender, and so on, are historically and socially constructed and vary across and within cultures and time periods.[5]

In other words, the debate highlights the significant effect of ideological constructs in shaping relations of gender dominance and forms of acting on the non-human world, but if these constructs are to be challenged it is necessary to go further. We need a theoretical understanding of what could be termed 'the political economy of ideological construction', that is, the interplay between conflicting discourses, and the means used to entrench views embodied in those discourses. Equally, it is critical to examine the underlying basis of women's relationship with the non-human world at levels other than ideology (such as through the work women and men do and the gender division of property and power) and to address how the material realities in which women of different classes (castes/races) are rooted might affect their responses to environmental degradation. Women in the West, for instance, have responded in specific ways to the threat of environmental destruction, such as by organizing the Greenham Commons resistance to nuclear missiles in England and by participating in the Green movement across Europe and the United States. A variety of actions have similarly been taken by women in the Third World, as discussed later. The question then is: Are there *gendered* aspects to these responses? If so, in what are these responses rooted?

Vandana Shiva's work on India takes us a step forward. Like the eco-feminists, she sees violence against nature and against women as built into the very mode of perceiving both. Like Merchant, she argues that violence against nature is intrinsic to the dominant industrial/developmental model, which she characterizes as a colonial imposition. Associated with the adoption of this developmental model, Shiva argues, was a radical conceptual shift away from the traditional Indian cosmological view of (animate and inanimate) nature as Prakriti, as 'activity *and* diversity' and as 'an expression of Shakti, the feminine and creative principle of the cosmos' which 'in conjunction with the masculine principle (Purusha) ... creates the world.' In this shift, the living, nurturing relationship between man and nature as earth mother was replaced by the notion of man as separate from and dominating over inert and passive nature. 'Viewed from the perspective of nature, or women embedded in nature,' the shift was repressive and violent. 'For women...the death of Prakriti is simultaneously a beginning of their marginalization, devaluation, displacement, and ultimate dispensability. The ecological crisis is, at its root, the death of the feminine principle....' (Shiva 1988: 39, 42)

At the same time, Shiva notes that violence against women and against nature are linked not just ideologically but also materially. For instance, Third World women are dependent on nature 'for drawing sustenance for themselves, their families, their societies'. The destruction of nature thus becomes the destruction of women's sources for 'staying alive'. Drawing upon her experience of working with women activists in the Chipko movement—the environmental movement for forest protection and regeneration in the Garhwal hills of north-west India—Shiva argues that 'Third World women' have both a special dependence on nature and a special knowledge of nature. This knowledge has been systematically marginalised under the impact of modern science:

Modern reductionist science, like development, turns out to be a patriarchal project, which has excluded women as experts, and has simultaneously excluded ecology and holistic ways of knowing which understand and respect nature's processes and interconnectedness as *science*. (Shiva 1988: 14–15)

Shiva takes us further than the western ecofeminists in exploring the links between ways of thinking about development, the processes of developmental change, and the impact of these on the environment and on the people dependent upon it for their livelihood. These links are of critical significance. Nevertheless her argument has three principal analytical problems. First, her examples relate to rural women primarily from north-west India, but her generalizations conflate all Third World women into one category. Although she distinguishes Third World women from the rest, like the ecofeminists she does not differentiate between women of different classes, castes, races, ecological zones and so on. Hence, implicitly, a form of essentialism could be read into her work, in that all Third World women, whom she sees as 'embedded in nature', *qua women* have a special relationship with the natural environment. This still begs the question: What is the basis of this relationship and how do women acquire this special understanding?

Second, she does not indicate by what concrete processes and institutions ideological constructions of gender and nature have changed in India, nor does she recognize the coexistence of several ideological strands, given India's ethnic and religious diversity. For instance, her emphasis on the feminine principle as the guiding idea in Indian philosophic discourse in fact relates to the Hindu discourse alone and cannot be seen as applicable to Indians of all religious persuasions.[6] Indeed, Hinduism itself is pluralistic, fluid, and contains several coexisting discourses with varying gender implication.[7] But perhaps most importantly, it is not clear how and in which historical period(s) the concept of the feminine principle in practise affected gender relations or relations between people and nature.

Third, Shiva attributes existing forms of destruction of nature and the oppression of women (in both symbolic and real terms) principally to the Third World's history of colonialism and to the imposition of western science and a western model of development. Undeniably, the colonial experience and the forms that modern development has taken in Third World countries have been destructive and distorting economically, institutionally and culturally. However, it cannot be ignored that this process impinged on pre-existing bases of economic and social (including gender) inequalities.

Here it is important to distinguish between the particular model of modernization that clearly has been imported/adopted from the West by many Third World countries (with or without a history of colonization) and the socio-economic base on which this model was imposed. Pre-British India, especially during the Mughal period, was considerably class/caste-stratified, although varyingly across regions.[8] This would have affected the patterns of access to and use of natural resources by different classes and social groups. Although much more research is needed on the political economy of natural resource use in the pre-colonial period, the evidence of differentiated peasant communities at that time cautions against sweeping historical generalizations about the effects of colonial rule.

By locating the 'problem' almost entirely in the Third World's experience of the West, Shiva misses out on the very real local forces of power, privilege and property relations that predate colonialism. What exists today is a complex legacy of colonial and pre-colonial interactions that defines the constraints and parameters within which and from which present thinking and action on development, resource use and social change have to proceed. In particular, a strategy for change requires an explicit analysis of the structural causes of environmental degradation, its effects and responses to it. The outline for an alternative framework, which I term feminist environmentalism, is suggested below.

## Feminist Environmentalism

I would like to suggest here that women's and men's relationship with nature needs to be understood as rooted in their material reality, in their specific forms of interaction with the environment. Hence, in so far as there is a gender and class (caste/race)-based division of labour and distribution of property and power, gender and class (caste/race) structure people's interactions with nature, and therefore, structure the effects of environmental change on people and their responses to it. And where knowledge about nature is experiential in its basis, the division of labour, property and power shaping experience also shapes the knowledge based on that experience.

For instance, poor peasant and tribal women have typically been responsible for fetching fuel and fodder and in hill and tribal communities have also often been the main cultivators. They are thus likely to be affected adversely in quite specific ways by environmental degradation. At the same time, in the course of their everyday interactions with nature, they acquire a special knowledge of species varieties and the processes of natural regeneration. (This would include knowledge passed on to them by for example, their mothers.) They could thus be seen as both victims of the destruction of nature and as repositories of knowledge about nature, in ways distinct from the men of their class. The former aspect would provide the gendered impulse for their resistance and response to environmental destruction. The latter would condition their perceptions and choices of what should be done. Indeed, on the basis of their experiential understanding and knowledge, they could provide a special perspective on the processes of environmental regeneration, one that needs to inform our view of alternative approaches to development. (By extension, women who are no longer actively using this knowledge for their daily sustenance, and are no longer in contact with the natural environment in the same way, are likely to lose this knowledge over time and with it the possibility of its transmission to others.)

In this conceptualization, therefore, the link between women and the environment can be seen as structured by a given gender and class (caste/race) organization of production, reproduction and distribution. Ideological constructions such as of gender, of nature, and of the relationship between the two, may be seen as (interactively) a part of this structuring but not the whole of it. This perspective I term 'feminist environmentalism'. In terms of action such a perspective would call for struggles over *both* resources and meanings. It would imply grappling with the dominant groups who have the property, power and privilege to control resources, and these or other groups who control ways of thinking about them, via educational, media, religious and legal institutions. On the feminist front there would be a need to challenge and transform both *notions* about gender and the *actual* division of work and resources between the genders. On the environmental front there would be need to challenge and transform not only notions about the relationship between people and nature but also the actual methods of appropriation of nature's resources by a few. Feminist environmentalism underlines the necessity of addressing these dimensions from both fronts.

To concretize the discussion, consider India's experience in the sections below. The focus throughout is on the rural environment.

## Environmental Degradation and Forms of Appropriation

In India (as in much of Asia and Africa) a wide variety of essential items are gathered by rural households from the village commons and forests for every day personal use and sale, such as food, fuel, fodder, fiber, small timber, manure, bamboo, medicinal herbs, oils, materials for house building and handicrafts, resin, gum, honey and spices.[9] Although all rural households use the village commons in some degree, for the poor they are of critical significance given the skewedness of privatized land distribution in the subcontinent.[10] Data for the early 1980s from 12 semi-arid districts in seven Indian states indicate that for poor rural households (the landless and those with less than two hectares dryland equivalent) village commons account for at least 9 per cent of total income, and in most cases 20 per cent or more, but contribute only 1 to 4 per cent of the incomes of the non-poor (Table 19.1). The dependence of the poor is especially high for fuel and fodder: village commons supply more than 91 per cent of firewood and more than 69 per cent of their grazing needs, compared with the relative self-sufficiency of the larger landed households. Access to village commons reduces income inequalities in the village between poor and non-poor households. Also there is a close link between the viability of small farmers' private property resources and their access to the commons for grazing draft as well as milch animals (see Blajkei 1985; Jodha 1986).

Similarly, forests have always been significant sources of livelihood, especially for tribal populations, and have provided the basis of swidden cultivation, hunting and the gathering of non-timber forest produce. In India, an estimated 30 million or more people in the country depend wholly or substantially on such forest produce for a livelihood (Kulkarni 1983). These sources are especially critical during lean agricultural seasons and during drought and famine (see Agarwal 1990; Pingle 1975).

The health of forests, in turn, has an impact on the health of soils (especially in the hills) and the availability of ground and surface water for irrigation and drinking. For a large percentage of rural households, the water for irrigation, drinking and various domestic uses comes directly from rivers and streams in the hills and plains. Again there are class differences in the nature of their dependency and access. The richer households are better able to tap the (relatively cleaner) ground water for drinking and irrigation by sinking more and deeper wells and tube wells, but the poor are mainly dependent on surface sources.

However, the availability of the country's natural resources to the poor is being severely eroded by two parallel, land inter-related trends—first, their growing degradation both in quantity and quality; second, their increasing statization (appropriation by the state) and privatization (appropriation by a minority of individuals), with an associated decline in what was earlier communal. These two tends, both independently and interactively, underlie many of the differential class-gender effects of environmental degradation outlined later. Independently, the former trend is reducing overall availability, and the latter is increasing inequalities in the distribution of what is available. Interactively, an altered distribution in favour of the state and some individuals and away from community control can contribute to environmental degradation is so far as community resource management systems may be more effective in environmental protection and regeneration than are the state or individuals. These two trends I call the primary factors underlying the class–gender effects of environmental change. Several intermediary factors impinge on these primary ones, the most important of which, in my view, are the following: the erosion of community resource management systems resulting from the shift in 'control rights' over natural resources away from community hands,[11] population growth and technological choices in agriculture and their associated effect on local knowledge systems. These also need to be seen in interactive terms. Consider each in turn.

## Forms of Environmental Degradation

Although there is as yet an inadequate data base to indicate the exact extent of environmental degradation in India and its cross-regional variations, available macro-information provides sufficient pointers to warrant considerable concern and possibly alarm. Degradation in India's natural resource base is manifest in disappearing forests, deteriorating soil conditions and depleting water resources. Satellite data from India reveal that in 1985–87, 19.5 per cent of the country's geo-area was forested and declining at an estimated rate of 1.3 million hectares a year (GoI 1990). Again, by official estimates, in 1980, 56.6 per cent of India's land was suffering from

**TABLE 19.1: AVERAGE ANNUAL INCOME FROM VILLAGE COMMONS IN SELECTED DISTRICTS OF INDIA (1982–85)**

| *State*[1] *and districts* | *Per Household Annual Average Income from Village Commons* | | | |
|---|---|---|---|---|
| | *Poor Households*[2] | | *Other Household*[3] | |
| | *Value (Rs)* | *Per cent of total household income* | *Value (Rs)* | *Per cent of total household income* |
| *Andhra Pradesh* | | | | |
| Mahbubnagar | 534 | 17 | 171 | 1 |
| *Gujarat* | | | | |
| Mehsana | 730 | 16 | 162 | 1 |
| Sabarkantha | 818 | 21 | 208 | 1 |
| *Karnataka* | | | | |
| Mysore | 649 | 20 | 170 | 3 |
| *Madhya Pradesh* | | | | |
| Mandsaur | 685 | 18 | 303 | 1 |
| Raisen | 780 | 26 | 468 | 4 |
| *Maharashtra* | | | | |
| Akola | 447 | 9 | 134 | 1 |
| Aurangabad | 584 | 13 | 163 | 1 |
| Sholapur | 641 | 20 | 235 | 2 |
| *Rajasthan* | | | | |
| Jalore | 709 | 21 | 387 | 2 |
| Nagaur | 831 | 23 | 438 | 3 |
| *Tamil Nadu* | | | | |
| Dharmapuri | 738 | 22 | 164 | 2 |

**Source:** Jodha, 1986, p. 1176.

**Notes:** 1. 'State' here refers to administrative divisions within India and is not used in the political economy sense of the word as used in the text.

2. Landless households and those owning < 2 hectares (ha.) dryland equivalent.

3. Those owning > 2 ha. Dryland equivalent (1 ha. = 2.47 acres).

environmental problems, especially water and wind erosion. Unofficial estimates are even higher. In some canal projects, one-half the area that could have been irrigated and cultivated has been lost due to waterlogging (Joshi and Agnihotri 1984), creating what the local people aptly call 'wet deserts'. The area under periodic floods doubled between 1971 and 1981, and soil fertility is declining due to the excessive use of chemical fertilizers. Similarly, the availability of both ground and surface water is falling. Groundwater levels have fallen permanently in several regions, including in northern India with its high water tables, due to the indiscriminate sinking of tubewells—the leading input in the Green Revolution technology (see, for instance, Bandhyopadhyay 1986; Dhawan 1982). As a result, many drinking water wells have dried up or otherwise been rendered unusable. In addition, fertilizer and pesticide run-offs into natural water sources have destroyed fish life and polluted water for human use in several areas (CSE 1986).

## The Process of Statization

In India, both under colonial rule and continuing in the post-colonial period, state control over forests and village commons has grown, with selective access being granted to a favoured few. To begin with, several aspects of British colonial policy have had long-lasting effects (Guha 1983). First, the British established state monopoly over forests, reserving large tracts for timber extraction. Second, associated with this was a severe curtailment in the customary rights of local population to these resources, right of access being granted only under highly restricted conditions, with a total prohibition on the barter or sale of forest produce by such right-holders. At the same time, the forest settlement officer could give considerable concessions to those he chose to so privilege. Third, the colonial state promoted the notion of 'scientific' forest management, which essentially cloaked the practise of encouraging commercially profitable species, often at the cost of species used by the local population. Fourth, there was virtually indiscriminate forest exploitation by European and Indian private contractors, especially for building railways ships and bridges. Tree clearing was also encouraged for establishing tea and coffee plantations and expanding the area under agriculture to increase the government's land revenue base. In effect these policies (*a*) severely eroded local systems of forest management; (*b*) legally cut off an important source of sustenance for people, even though illegal entries continued; (*c*) created a continuing source of tension between the forestry officials and the local people; and (*d*) oriented forest management to commercial needs.

Post-independence policies show little shift from the colonial view of forests as primarily a source of commercial use and gain. State monopoly over forests has persisted, with all the attendant tensions, as has the practise of scientific forestry in the interests of commercial profit. Restrictions on local people's access to non-timber forest produce have actually increased, and the harassment and exploitation of forest-dwellers by the government's forest guards is widespread (Chand and Bezboruah 1980; Swaminathan 1982).

## The Process of Privatization

A growing privatization of community resources in individual (essentially male) hands has paralleled the process of statization. Customarily, large parts of village common lands, especially in north-west India, were what could be termed 'community-private', that is, they were private in so far as use rights to them were usually limited to members of the community and therefore exclusionary; at the same time, they were communal in that such rights were often administered by a group rather than by an individual.[12] Table 19.2 reveals a decline in village commons ranging between 26 and 63 percentage points across different regions, between 1950 and 1984. This is attributable mainly to state policy acting to benefit selected groups over others, including illegal encroachments by farmers, made legal over time; the auctioning of parts of commons by the government to private contractors for commercial exploitation; and government distribution of common land to individuals under various schemes, which were, in theory, initiated for benefiting the poor but in practise benefited the well-off farmers.[13] For 16 of the 19 districts covered, the share of the poor was less than that of the non-poor (Table 19.2). Hence the poor lost out collectively while gaining little individually.

**Table 19.2: Distribution of Privatized Village Commons in Selected Districts of India**

| *States and Districts* | *VCs as Per Cent of Village Area (1982–4)* | *Per Cent Decline in VC Area (1950–84)* | *Per Cent of Land to:* | | *Per Cent Recipients among* | | *Per Household Area Owned (ha)* | | | |
|---|---|---|---|---|---|---|---|---|---|---|
| | | | *Poor* | *Other* | *Poor* | *Other* | *Poor* | | *Other* | |
| | | | | | | | *Before*[1] | *After*[2] | *Before* | *After* |
| *Andhra Pradesh* | | | | | | | | | | |
| Mahbubnagar | 9 | 43 | 50 | 50 | 76 | 24 | 0.3 | 0.9 | 3.0 | 5.1 |
| Medak | 11 | 45 | 51 | 49 | 59 | 41 | 1.0 | 2.2 | 3.1 | 4.6 |
| *Gujarat* | | | | | | | | | | |
| Banaskantha | 9 | 49 | 18 | 82 | 38 | 62 | 0.8 | 2.0 | 5.4 | 8.8 |
| Mehsana | 11 | 37 | 20 | 80 | 36 | 64 | 1.0 | 1.7 | 8.0 | 9.8 |
| Sabarkantha | 12 | 46 | 28 | 72 | 55 | 45 | 0.5 | 1.1 | 7.0 | 9.8 |
| *Karnataka* | | | | | | | | | | |
| Bihar | 12 | 41 | 39 | 61 | 64 | 36 | 1.0 | 2.0 | 6.4 | 9.2 |
| Gulbarga | 9 | 43 | 43 | 57 | 60 | 40 | 0.8 | 2.4 | 4.5 | 7.7 |
| Mysore | 18 | 32 | 44 | 56 | 67 | 33 | 0.9 | 1.9 | 4.1 | 11.6 |
| *Madhya Pradesh* | | | | | | | | | | |
| Mandsaur | 22 | 34 | 45 | 55 | 75 | 25 | 1.2 | 2.5 | 7.7 | 12.4 |
| Raisen | 23 | 47 | 42 | 58 | 68 | 32 | 1.3 | 2.2 | 6.2 | 9.0 |
| Vidisha | 28 | 32 | 38 | 62 | 48 | 52 | 1.3 | 2.5 | 4.9 | 6.8 |

*(continued to next page)*

| *States and Districts* | *VCs as Per Cent of Village Area (1982–4)* | *Per Cent Decline in VC Area (1950–84)* | *Per Cent of Land to:* | | *Per Cent Recipients among* | | *Per Household Area Owned (ha)* | | | |
|---|---|---|---|---|---|---|---|---|---|---|
| | | | | | | | *Poor* | | *Other* | |
| | | | *Poor* | *Other* | *Poor* | *Other* | *Before*[1] | *After*[2] | *Before* | *After* |
| *Maharashtra* | | | | | | | | | | |
| Akola | 11 | 42 | 39 | 61 | 58 | 42 | 1.0 | 1.6 | 3.1 | 4.6 |
| Aurangabad | 15 | 30 | 30 | 70 | 42 | 58 | 1.1 | 2.2 | 3.4 | 5.6 |
| Sholapur | 19 | 26 | 42 | 58 | 53 | 47 | 0.7 | 2.2 | 3.4 | 5.6 |
| *Rajasthan* | | | | | | | | | | |
| Jalore | 18 | 37 | 14 | 86 | 37 | 63 | 0.3 | 1.7 | 7.2 | 12.5 |
| Jodhpur | 16 | 58 | 24 | 76 | 35 | 65 | 0.4 | 1.3 | 2.3 | 3.8 |
| Nagaur | 15 | 63 | 21 | 79 | 41 | 59 | 1.3 | 2.5 | 2.4 | 5.2 |
| *Tamil Nadu* | | | | | | | | | | |
| Coimbatore | 9 | 47 | 50 | 50 | 75 | 25 | 0.8 | 2.5 | 3.8 | 5.8 |
| Dharmapuri | 12 | 52 | 49 | 51 | 55 | 45 | 1.0 | 1.9 | 4.6 | 7.5 |

**Source:** Jodha, 1986, pp. 1177–78.

**Notes:** 1. Before the distribution of VC land.

2. After the distribution of VC land.

Similarly, in the tapping of groundwater through tubewells, there are dramatic inequalities in the distribution of what is effectively an underground commons. Tubwells are concentrated in the hands of the rich and the noted associated fall in water tables has, in many areas, dried up many shallow irrigation and drinking water wells used by the poor. In some regions, they have also depleted soil moisture from land used by poor households (Bandhyopadhyay 1986).

Now consider the intermediary factors mentioned earlier: the erosion of community management systems, population growth and choice of agricultural technology and local knowledge systems.

## Erosion of Community Resource Management Systems

The statization and privatization of communal resources have, in turn, systematically undermined traditional institutional arrangements of resource use and management. The documentation on this is growing, but even existing work reveals systems of water management, methods of gathering firewood and fodder, and practises of shifting agriculture which were typically not destructive of nature.[14] Some traditional religious and folk beliefs also (as noted) contributed to the preservation of nature, especially trees or orchards deemed sacred.[15]

Of course, much more empirical documentation is needed on how regionally widespread these traditional systems of management were and the contexts in which they were successful in ensuring community cooperation. However, the basic point is that where traditional community management existed, as it did in many areas, *responsibility for resource management was linked to resource use* via local community institutions. Where control over these resources passed from the hands of the community to those of the state or of individuals, this link was effectively broken.

In turn, the shift from community control and management of common property, to state or individual ownership and control, has increased environmental degradation.[16] As Daniel W. Bromley and Michael M. Cernea note, the *appearance* of environmental management created through the establishment of government agencies, and the aura of coherent policy by issuance of decrees prohibiting entry to—and harvesting from—state property, has led to continued degradation of resources under tolerant eye of government agencies (Bromley and Cernea 1989: 25).

Property rights vested in individuals are also no guarantee for environmental regeneration. Indeed, as will be discussed at greater length later, individual farmers attempting tree planting for short-term profits have tended to plant quick-growing commercial trees such as eucalyptus, which can prove environmentally costly.

## Population Growth

Excessive population growth has often been identified as the primary culprit of environmental degradation. And undoubtedly, a rapidly growing population impinging over time on a limited land/water/forest base is likely to degrade the environment. However, political economy dimensions clearly underlie the *pace* at which this process occurs and *how the costs of it are distributed*. The continuing (legal and illegal) exploitation of forests, and the increasing appropriation of village commons and groundwater resources by a few, leave the vast majority to subsist on a shrinking natural resource base. Added to this is the noted erosion of community resource management systems which had enforced limitations on what people could and did take from communal resources, and which could perhaps have ensured their protection, despite population pressure (ibid.).

Population growth can thus be seen as exacerbating a given situation but not necessarily as its primary cause. It is questionable that interventions to control population growth can, in themselves, stem environmental degradation, although clearly, as Paul Shaw argues, they can 'buy crucial time until we figure out how to dismantle more ultimate causes' (Shaw 1989: 7).

What adds complexity to even this possibility is that in the link between environmental degradation and population growth, the causality can also turn in the opposite direction. For instance, poverty associated with environmental degradation could induce a range of fertility-increasing responses—reduced education for young girls as they devote more time to collecting fuel, fodder, and so on, leading to higher fertility in the long term, given the negative correlation between female education and fertility; higher infant mortality rates inducing higher fertility to ensure a given completed family size; and people having more children to enable the family to diversify incomes as a risk-reducing mechanism in environmentally high-risk areas (Rosenzweig and Wolpin 1985). These links are another reminder that it is critical to focus on women's status when formulating policies for environmental protection.

## Choice of Agricultural Technology and Erosion of Local Knowledge Systems

Many of the noted forms of environmental degradation are associated with the Green Revolution technology adopted to increase in crop output. Although dramatically successful in the latter objective in the short run, it has had high environmental costs, such as falling water tables due to tubewells; waterlogged and saline soils from most large irrigation schemes; declining soil fertility with excessive chemical fertilizer use; and water pollution with pesticides. Moreover, the long-term sustainability of the output increases achieved so far, itself appears doubtful. Deteriorating soil and water conditions are already being reflected in declining crop yields.[17] Genetic variety has also shrunk, and many of the indigenously developed crop varieties (long-tested and adapted to local conditions) have been replaced by improved seeds which are more susceptible to pest attacks. The long-term annual growth rate of agricultural production in India over 1968–75 was 2.6 per cent, that is, slightly *lower* than the pre-Green Revolution, 1950–5, rate of 3.08. Crop yields are also more unstable (Rao *et al.* 1988). All this raises questions about the long-term sustainability of agricultural growth, and more generally of rural production systems, under present forms of technology and resource management in India, and indeed in South Asia.

The choice of agricultural technology and production systems cannot be separated from the dominant view of what constitutes scientific agriculture. The Green Revolution embodies a technological mix which gives primacy to laboratory-based research and manufactured inputs and treats agriculture as an isolated production system. Indeed, indiscriminate agricultural expansion, with little attempt to maintain a balance between forests, fields and grazing lands, assumes that the relationship between agriculture, forests and village commons is an antagonistic, rather than a complementary, one. By contrast, organic farming systems (now rapidly being eclipsed) are dependent on maintaining just such a balance. More generally, over the years, there has been a systematic devaluation and marginalization of indigenous knowledge about species-varieties, nature's processes (how forests, soils and water are formed and sustained interrelatedly), and sustainable forms of interaction between people and nature. These trends are not confined to countries operating within the capitalist mode. Similar problems of deforestation, desertification, salination, recurrent secondary pest attacks on crops and pesticide contamination are emerging in China (Glaeser 1987).

What is at issue here is not modern science in itself but the process by which what is regarded as 'scientific knowledge' is generated and applied and how the fruits of that application are distributed. Within the hierarchy of knowledge, that acquired via traditional forms of interacting with nature tends to be deemed less valuable (see also, Marglin 1988). And the people who use this knowledge in their daily lives—farmers and forest-dwellers and especially women of these communities—tend to be excluded from the institutions which create what is seen as scientific knowledge. These boundaries are not inevitable. In Meiji Japan, the farmer's knowledge and innovative skills were incorporated in the broader body of scientific knowledge by a systematized interaction between the farmer, the village extension worker and the scientist. This enabled a two-way flow of information from the farmer to the scientist and vice-versa: 'Intimate knowledge of the best of traditional farming methods was thus the starting point for agricultural research and extension activities.' (Johnston 1969: 61)

Such attempts contrast sharply with the more typical top-down flow of information from those deemed experts (the scientists/professionals) to those deemed ignorant (the village users). The problem here is only partly of class differences. Underlying the divide between the scientists/professionals (usually urban-based) and the rural users of innovations (including user-innovators) whose knowledge comes more from field experience than from formal education, are also usually the divides between intellectual and physical labour, between city and countryside, and between women and men.

## Class–Gender Effects

We come then to the class–gender effects of the processes of degradation, statization and privatization of nature's resources, and the erosion of traditional systems of knowledge and resource management. These processes have had particularly adverse effects on poor households because of the noted greater dependency of such households on communal resources. However, focusing on the class significance of communal resource provides only a partial picture—there is also a critical gender dimension, for women and female children are the ones most adversely affected by environmental degradation. The reasons for this are primarily three-fold. First, there is a pre-existing gender division of labour. It is women in poor peasant and tribal households who do much of the gathering and fetching from the forests, village commons, rivers and wells. In addition, women of such households are burdened with a significant responsibility for family subsistence and they are often the primary, and in many female-headed households the sols, economic providers.

Second, there are systematic gender differences in the distribution of subsistence resources (including food and health care) within rural households, as revealed by a range of indicators: anthropometric indices, morbidity and mortality rates, hospital admissions data, and the sex ratio (which is 93 females per 100 males for all-India).[18] These differences, especially in health care, are widespread in India (and indeed in South Asia).[19]

Third, there are significant inequalities in women's and men's access to the most critical productive resources in rural economy, agricultural land and associated production technology.[20] Women also have a systematically disadvantaged position in the labour market. They have fewer employment opportunities, less occupational mobility, lower levels of training, and lower payments for the same or similar work (Agarwal 1984, 1986; Bardhan 1977). Due to the greater task specificity of their work, they also face much greater seasonal fluctuations in employment and earnings than do men, with sharper peaks and longer slack periods in many regions and less chance of finding employment in the slack seasons (Agarwal 1984; Ryan and Ghodake 1980).

Given their limited rights in private property resources such as agricultural land, rights to communal resources such as the village commons have always provided rural women and children (especially those of tribal, landless, or marginal peasant households) a source of subsistence, *unmediated by dependency relationships* on adult males. For instance, access to village commons is usually linked to membership in the village community and therefore women are not excluded in the way they may be in a system of individualized private land rights. This acquires additional importance in regions with strong norms of female seclusion (as in north-west India) where women's access to the cash economy, to markets, and to the marketplace itself is constrained and dependent on the mediation of male relatives (Agarwal 1989; Sharma 1980).

It is against this analytical backdrop that we need to examine what I term the 'class–gender effects' (the gender effects mediated by class) of the processes of environmental degradation, statization and privatization. These effects relate to at least six critical aspects: time, income, nutrition, health, social survival networks and indigenous knowledge. Each of these effects is important across rural India. However, their intensity and interlinkages would differ cross-regionally, with variations in ecology, agricultural technology, and distribution and social structures, associated with which are variations in the gender division of labour, social relations, livelihood possibilities and kinship systems.[21] Although a systematic regional decomposition of effects is not attempted here, all the illustrative examples are regionally contextualized.

## On Time

Because women are the main gatherers of fuel, fodder and water, it is primarily their working day (already averaging 10 to 12 hours) that is lengthened with the depletion of and reduced access to forests, waters and soils. Firewood, for instance, is the single most important source of domestic energy in India (providing more than 65 per cent of domestic energy in the hills and deserts of the north). Much of this is gathered and not purchased, especially by the poor. In recent years, there has been a several fold increase in firewood collection time (see Table 19.3). In some villages of Gujarat, in western India, even a 4-to-5-hour search yields little apart from shrubs, weeds and tree roots which do not provide adequate heat.

Similarly, fodder collection takes longer with a decline in the village commons. As a woman in the hills of Uttar Pradesh (north-west India) puts it:

> When we were young, we used to go the forest early in the morning without eating anything. There we would eat plenty of berries and wild fruits...drink the cold sweet (water) of the *Banj* (oak) roots....In a short while we would gather all the fodder and firewood we needed, rest under the shade of some huge tree and then go home. Now, with the going of the trees, everything else has gone too. (Bahuguna 1984: 132)

The shortage of drinking water has exacerbated the burden of time and energy on women and young girls. Where low caste women often have

**TABLE 19.3: TIME TAKEN AND DISTANCE TRAVELLED FOR FIREWOOD COLLECTION**

| *Country/region* | *Year of Data* | *Firewood Collection** | | *Data Source* |
|---|---|---|---|---|
| | | *Time taken* | *Distance travelled* | |
| *India* | | | | |
| Chamoli (hills) | | | | |
| (a) Dwing | 1982 | 5 hr/day | over 5 km | Swaminathan |
| (b) Pakhi | 4 hr/day | (1984) | | |
| Gujarat (plains) | | | | |
| (a) Forested | | once every 4 days | n.a. | |
| (b) Depleted | 1980 | once every 2 days | 4-5 km | Nagbrahman and Sambrani (1980) |
| (c) Severely depleted | 4-5 hr/day | n.a. | | |
| Madhya Pradesh (plain) | 1980 | 1-2 times/ week | 5 km | Chand and Bezboruah (1980) |
| Kumaon (hills) | 1952 | 3 days/ week | Folger and 5-7 km | Dewan (1983) |
| Karnataka (plains) | n.a. | 1 hr/day | 5.4 km/trip | Batliwala (1983) |
| Garhawal (hills) | n.a. | 5 hr/day | 10 km | A. Agarwal (1983) |
| Bihar (plains) | c. 1972 | n.a. | 1-2 km/day | Bhaduri and Surin (1980) |
| Rajasthan (plains) | 1988 | 5 hr/day (winter) | 4 km | personal observation |
| *Nepal* | | | | |
| Tinan (hills) | 1978 | 3 hr/day | n.a. | Stone (1982) |
| Pangua (hills) | late 1970s | 4-5 hr/ bundle | n.a. | Bajracharya (1983) |
| WDA (lowlands) | | | | |
| (a) low deforestation | 1982–83 | 1.5 hr/day | n.a. | Kumar and Hutchkiss (1988) |
| (b) high deforestation | 3 hr/day | n.a. | | |

**Notes:** *Firewood collected mainly by women and children.
@Average computed from information given in the study.
n.a. Information not available.

access to only one well, its drying up could mean an endless wait for their vessels to be filled by upper-caste women, as was noted to have happened in Orissa.[22] A similar problem arises when drinking water wells go saline near irrigation works (Agarwal 1981).

In Uttar Pradesh, according to a woman grassroots activist, the growing hardship of young women's lives with ecological degradation has led to an increased number of suicides among them in recent years. Their inability to obtain adequate quantities of water, fodder and fuel causes tensions with their mothers-in-law (in whose youth forests were plentiful), and soil erosion has compounded the difficulty of producing enough grain for subsistence in a region of high male outmigration (Bahuguna 1984).

## On Income

The decline in gathered items from forests and village commons has reduced incomes directly. In addition, the extra time needed for gathering reduces time available to women for crop production and can adversely affect crop incomes, especially in hill communities where women are the primary cultivators due to high male outmigration. For instance, a recent study in Nepal found that the substantial increase in firewood collection time due to deforestation has significantly reduced women's crop cultivation time, leading to an associated fall in the production of maize, wheat and mustard which are primarily dependent on female labour in the region. These are all crops grown in the dry season when there is increased need for collecting fuel and other items (Kumar and Hotchkiss 1988). The same is likely to be happening in the hills of India.

Similar implications for women's income arise with the decline in common grazing land and associated fodder shortage. Many landless widows I spoke to in Rajasthan (north-west India) in 1988 said they could not venture to apply for a loan to purchase a buffalo under the government's anti-poverty programme as they had nowhere to graze the animal and no cash to buy fodder.

As other sources of livelihood are eroded, selling firewood is becoming increasingly common, especially in eastern and central India. Most 'headloaders', as they are called, are women, earning a meager 5.50 rupees a day for 20 kilogrammes of wood (Bhadhuri and Surin 1980). Deforestation directly impinges on this source of livelihood as well.

## On Nutrition

As the area and productivity of village commons and forests fall, so does the contribution of gathered food in the diets of poor households. The declining availability of fuelwood has additional nutritional effects. Efforts to economize induce people to shift to less nutritious foods which need less fuel to cook or which can be eaten raw, or force them to eat partially cooked food which could be toxic, or eat leftovers that could rot in a tropical climate, or to miss meals altogether. Although as yet there are no systematic studies on India, some studies on rural Bangladesh are strongly indicative and show that the total number of meals eaten daily as well as the number of cooked meals eaten in poor households are already declining (Howes and Jabbar 1986). The fact that malnutrition can be caused as much by shortages of fuel as of food has long been part of the conventional wisdom of rural women who observe: 'It's not what's in the pot that worries you, but what's under it.' A trade-off between the time spent in fuel gathering versus cooking can also adversely affect the meal's nutritional quality.

Although these adverse nutritional effects impinge on the whole household, women and female children bear an additional burden because of the noted gender biases in intra-family distribution of food and health care. There is also little likelihood of poor women being able to afford the extra calories for the additional energy expended in fuel collection.

## On Health

Apart from the health consequences of nutritional inadequacies, poor rural women are also more directly exposed than are men to waterborne diseases and to the pollution of rivers and ponds with fertilizer and pesticide run-offs, because of the nature of the tasks they perform, such as fetching water for various domestic uses and animal care, and washing clothes near ponds, canals and streams (Agarwal 1981).

The burden of family ill-health associated with water pollution also falls largely on women who take care of the sick. An additional source of vulnerability is the agricultural tasks women perform. For instance, rice transplanting, which is usually a woman's task in most parts of Asia, is associated with a range of diseases, including arthritis and gynecological ailments

(Mencher and Saradamoni 1982; UNDP 1979). Cotton-picking and other tasks done mainly by women in cotton cultivation expose them to pesticides which are widely used for this crop. In China, several times the acceptable levels of DDT and BHC residues have been found in the milk of nursing mothers, among women agricultural workers (Wagner 1987). In India, pesticides are associated with limb and visual disabilities (Mohan 1987).

## On Social Support Networks

The considerable displacement of people that results form the submersion of villages in the building of major irrigation and hydroelectric works, or from large-scale deforestation in itself, has another (little recognized) class and gender implication—the disruption of social support networks. Social relationships with kin, and with villagers outside the kin network, provide economic and social support that is important to all rural households but especially to poor households and to the women.[23] This includes reciprocal labour-sharing arrangements during peak agricultural seasons; loans taken in cash or kind during severe crises such as droughts; and the borrowing of small amounts of foodstuffs, fuel, fodder, and so on, even in normal times. Women typically depend a great deal on such informal support networks, which they also help to build through daily social interaction, marriage alliances that they are frequently instrumental in arranging, and complex gift exchanges (see Sharma 1980; Vatuk 1981). Also the social and economic support this represents for women in terms of strengthening their bargaining power within families needs to be recognized, even if it is not easy to quantify.[24] These networks, spread over a range of nearby villages, cannot be reconstituted easily, an aspect ignored by rehabilitation planners.

Moreover for forest-dwellers, the relationship with forests is not just functional or economic but also symbolic, suffused with cultural meaning and nuances, and woven into their songs and legends of origin. Large-scale deforestation, whether or not due to irrigation schemes, has eroded a whole way of living and thinking. Two close observers of life among the tribal people of Orissa in eastern India note that 'the earlier sense of sharing has disappeared.... Earlier women would rely on their neighbours in times of need. Today this has been replaced with a sense of alienation and helplessness... the trend is to leave each family to its own fate' (Fernandes and Menon 1987: 115). Widows and the aged are the most neglected.

## On Women's Indigenous Knowledge

The gathering of food alone demands an elaborate knowledge of the nutritional and medicinal properties of plants, roots and trees, including a wide reserve knowledge of edible plants not normally used but critical for coping with prolonged shortages during climatic disasters. An examination of household-coping mechanisms during drought and famine reveals a significant dependence on famine foods gathered mainly by women and children for survival. Also, among hill communities it is usually women who do the seed selection work and have the most detailed knowledge about crop varieties.[25] This knowledge about nature and agriculture, acquired by poor rural women in the process of their everyday contact with and dependence on nature's resources, has a class and gender specificity and is linked to the class specificity and gendering of the division of labour.

The impact of existing forms of development on this knowledge has been two-fold. First, the process of devaluation and marginalization of indigenous knowledge and skills, discussed earlier, impinges especially on the knowledge that poor peasant and tribal women usually possess. Existing development strategies have made little attempt to tap or enhance this knowledge and understanding. At the same time, women have been excluded from the institutions through which modern scientific knowledge is created and transmitted. Second, the degradation of natural resources and their appropriation by a minority results in the destruction of the material basis on which women's knowledge of natural resources and processes is founded and kept alive, leading to its gradual eclipse.

## Responses: State and Grassroots

Both the state and the people most immediately affected by environmental degradation have responded to these processes, but in different ways. The state's recognition that environmental degradation may be acquiring crisis proportions is recent land as yet partial; and, as we have seen, state developmental policies are themselves a significant cause of the crisis. Not surprisingly, therefore, the state's response has been piecemeal rather than comprehensive. For instance, the problem of deforestation and fuelwood shortage has been addressed mainly by encouraging village communities and individual farmers to do so.

However, most state ventures[26] in the form of direct planting have had high failure rates in terms of both tree-planting and survival, attributable to several causes—a preoccupation with monocultural plantations principally for commercial use, which at times have been replaced mixed forests; the takeover of land down implementation. Hence, in many cases, far from benefiting the poor these schemes have taken away even existing rights and resources, leading to widespread local resistance. Also, women either do not feature at all in such schemes or, at best, tend to be allotted the role of caretakers in tree nurseries, with little say in the choice of species or in any other aspect of projects. Community forestry schemes, on the other hand, are often obstructed by economic inequalities in the village community and the associated mistrust among the poor of system that cannot ensure equitable access to the products of the trees planted.

Ironically, the real 'success' stories, with plantings far exceeding targets, relate to the better-off farmers who, in many regions, have sought to reap quick profits by allotting fertile cropland to commercial trees. As a result, employment, crop output and crop residues for fuel have declined, often dramatically, and the trees planted, such as eucalyptus, provide no fodder and poor fuel.[27] The recent government policy in West Bengal (eastern India) of leasing sections of degraded forest land to local communities for collectively planting, managing and monitoring tree plantations for local use, holds promise. But in several other parts of country large tracts of such land have also been given to paper manufacturers for planting commercial species.

As some environmentalists have rightly argued, this predominantly commercial approach to forestry, promoted as 'scientific forestry', is reductionist—it is nature seen as individual parts rather than as an interconnected system of vegetation, soil and water; the forest is reduced to trees, the trees to biomass. For instance, Shiva notes that in the reductionist world-view only those properties of a resource system are taken into account which generate profit, whereas those that stabilize ecological processes, but are commercially non-exploitable, are ignored and eventually destroyed (Shiva 1987).

Indeed, the noted effects of development policies on the environment—be they policies relating to agriculture or more directly to forests and water use—point to a strategy, which has been extractive/destructive of nature rather than conserving/regenerative. The strategy does not explicitly take account of the long-term complementarities between agriculture and natural resource preservation and therefore raises serious questions about the ability of the system both to sustain long-term increases in agricultural productivity and to provide sustenance for the people.

But should we see people in general and women in particular solely as victims of environmental degradation and of ill-conceived top-down state policies? The emergence of grassroots ecology movements across the subcontinent (and especially India) suggests otherwise. These movements indicate that although poor peasant and tribal communities in general, and women among them in particular, are being severely affected by environmental degradation and appropriation, they are today also critical agents of change. Further, embodied in their traditional interaction with the environment are practises and perspectives which can prove important for defining alternatives.

The past decade, in particular, has seen an increasing resistance to ecological destruction in India, whether caused by direct deforestation (which is being resisted through non-violent movements such as Chipko in the Himalayan foothills and Appiko in Karnataka) or by large irrigation and hydroelectric works, such as the Narmada Valley project covering three regions in central India, the Koel-Karo in Bihar, the Silent Valley project in Kerala (which was shelved through central government intervention and local protests in 1983), the Inchampalli and Bhopalpatham dams in Andhra Pradesh (against which 5,000 tribal people, with women in the vanguard, protested in 1984), and the controversial Tehri dam in Garhwal. Women have been active participants in most of these protests.

Although fuelled by differing ideological streams which Ramachandra Guha identifies as Crusading Gandhian, Appropriate Technology, and Ecological Marxism, these resistance movements suggest that those affected can also be critical agents of change. Common to these streams is the recognition that the present model of development has not succeeded either in providing sustenance or in ensuring sustainability. However, the points for which the differing ideologies initiate this critique are widely dispersed. In particular, they differ in their attitudes to modern science and to socio-economic inequalities. As Guha puts it, under the Crusading Gandhian approach, 'modern science is seen as responsible for industrial society's worst excesses,' (Guha 1988) and socio-economic inequalities within village communities tend to get glossed over. Ecological Marxism sees modern science and the 'scientific temper' as indispensable for constructing a new social order, and there is a clear recognition of and attack on class and caste inequalities (although the position on gender is ambiguous). Appropriate Technology thinking, which falls within these two strands, is not as well worked out a philosophic and theoretical position as Gandhism and Marxism. It is pragmatic in its approach to modern science and emphasizes the need to synthesize traditional and modern technological traditions. Although problems relating to socio-economic hierarchies are recognized, there is no clear programme for tackling them. Over the past decade there has been some cross-fertilization of thinking across these different ideological streams.

However, it is important to distinguish here between the perspectives revealed by an examination of *practise* within environmental movement and the explicit *theoretical* formulation of an environmental perspective. Although dialectically interlinked, the two do not entirely overlap. The three ideological streams, as identified by Guha, relate to different ways in which groups adhering to pre-existing ideological and philosophic positions (Marxist, Gandhian) have incorporated environmental concerns in their practise. In a sense, environment has been added on to their other concerns. This does not as yet represent the formulation of a new theoretical perspective (that an environmental approach to development needs) by any of these groups.

In terms of practise within the movement, women have been a visible part of most rural grassroots ecological initiatives (as they apparent in the Chipko movement. However, women's participation in a movement does not in *itself* represent an explicit incorporation of a gender perspective, in either theory or practise, within that movement. Yet such a formulation is clearly needed. Feminist environmentalism as spelled out earlier in this essay is an attempt in this direction.

To restate in this context, in feminist environmentalism I have sought to provide a theoretical perspective that locates both the symbolic and material links between people and the environment in their specific forms of interaction with it, and traces gender and class differentiation in these links to a given gender and class division of labour, property and power. Unlike Gandhism and Marxism, feminist environmentalism is not a perspective that is consciously subscribed to by an identifiable set of individuals or groups. However, insofar as tribal and poor peasant women's special concern their responses to it, which have been articulated both in complementary and oppositional terms to the other ideological streams, could be seen as consistent with the feminist environmentalist framework. The Chipko movement is an interesting example in this respect. Although it emerged from the Gandhian tradition, in the course of its growth it has brought to light some of the limitations of an approach that does not explicitly take account of class and gender concerns. More generally too it is movement of considerable historical significance whose importance goes beyond locational specificity, and is a noteworthy expression of hill women's specific understanding of forest protection and environmental regeneration.[28]

The movement was sparked off in 1972–73 when the people of Chamoli district in north-west India protested the auctioning of 3,000 ash trees to a sports good manufacturer, while the local labour cooperative was refused permission by the government to cut even a few trees to make agricultural implements for the community. Since then the movement has spread not only within the region but its methods and message have also reached other parts of the country (Appiko in Karnataka is an offshoot).[29] Further, the context of local resistance has widened. Tree felling is being resisted also to prevent disasters such as landslides, and there has been protest against limestone mining in the hills for which the villagers had to face violence from contractors and their hired thugs.

Women's active involvement in the Chipko movement has several noteworthy features that need highlighting here. First, their protest against the commercial exploitation of the Himalayan forests has been not only jointly with the men of their community when they were confronting non-local contractors but also, in several subsequent instances, even in opposition to village men due to differences in priorities about resource use. Time and again, women have clear-sightedly opted for saving forests and the environment over the short-term gains of development projects with high environmental costs. In one instance, a potato-seed farm was to be established by cutting down a tract of oak forest in Dongri Paintoli village. The women supported the scheme because it would take away their only local source of fuel and fodder and add 5 kilometers to their fuel-collecting journeys, but cash in the men's hands would not necessarily benefit them or their children.[30] The protest was successful.

Second, women have been active and frequently successful in protecting the trees, stopping tree auctions, and keeping a vigil against illegal feeling. In

Gopeshwar town, a local women's group has appointed watchwomen who receive a wage in kind to guard the surrounding forest, and to regulate the extraction of forest produce by villagers. Twigs can be collected freely, but any harm to the trees is liable to punishment.

Third, replanting is a significant component of the movement. But in their choice of trees the priorities of women and men don't always coincide—women typically prefer trees that provide fuel, fodder, and daily needs, the men prefer commercially profitable ones.[31] Once again this points to the association between gendered responsibility for providing a family's subsistence needs and gendered responses to threats against the resources that fulfill those needs.

Fourth, Chipko today is more than an ecology movement and has the potential for becoming a wider movement against gender-related inequalities. For instance, there has been large-scale mobilization against male alcoholism and associated domestic violence and wasteful expenditure. There is also a shift in self-perception. I have seen women stand up in public meetings of the movement and forcefully address the gathering. Many of them are also asking: Why aren't we members of the village councils?

Fifth, implicit in the movement is a holistic understanding of the environment in general and forests in particular. The women, for instance, have constructed a poetic dialogue illustrating the difference between their own perspective and that of the foresters (quoted in Shiva 1988).

Forests: What do the forests bear?
Profits, resin and timber.
Women (Chorus): What do the forests bear?
Soil, water and pure air.
Soil, water and pure air.
Sustain the earth and all she bears.

In other words, the women recognize that forests cannot be reduced merely to trees and the trees to wood for commercial use, that vegetation, soil land water form part of a complex and interrelated ecosystem. This recognition of the interrelatedness and interdependence between the various material components of nature, and between nature and human sustenance, is critical for evolving a strategy of sustainable environmental protection land regeneration.

Although the movement draws upon, and indeed is rooted in, the region's Gandhian tradition, which predates Chipko, the women's response go beyond the framework of that tradition and come suggested by their beginning to confront gender and class issues in a number of small but significant ways. For instance, gender relations are called into question in their taking oppositional stands to the village men on several occasions, in asking to be members of violence. Similarly, there is clearly a class confrontation involved in their resistance (together with the men of their community) to the contractors holding licenses for mining and felling in the area.

At the same time, ecology movements such as Chipko need to be contextualized. Although localized resistance to the processes of natural resource appropriation and degradation in India has taken many different forms, and arisen in diverse regional contexts, resistances in which entire communities and village have participated to constitute a movement (such as Chipko, Appiko and Jharkhand) have emerged primarily in hill or tribal communities. This may be attributable particularly to two factors: the immediacy of the threat from these processes to people's survival, and these communities being marked by relatively low levels of the class and social differentiation that usually splinter village communities in South Asia. They therefore have a greater potential for wider community participation than is possible in more economically and socially stratified contexts. Further, in these communities, women's role in agricultural production has always been visibly substantial and often primary—an aspect more conductive to their public participation than in many other communities of northern India practicing female seclusion.

In emphasizing the role of poor peasant and tribal women in ecology movements, I am not arguing, as do some feminist scholars, that women posses a specifically feminine sensibility or cognitive temperament, or that women *qua women* have certain traits that predispose them to attend to particulars, to be interactive rather than individualist, and to understand the true character of complex natural processes in holistic terms (see, for a critique of these arguments, Longino 1988). Rather, I locate the perspective of poor tribal and peasant women (perspectives which are often holistic and interactive) in their material reality—in their dependence on and actual use of natural resources for survival, the knowledge of nature gained in that process, and the broader cultural parameters which define people's activities and modes of thinking in their communities. By this count, the perspectives and responses of men belonging to hill or tribal communities would also be more conducive to environmental protection and regeneration than those of men elsewhere, but not more than those of the women of such communities. This is because hill and tribal women, perhaps more than any other group, still maintain a reciprocal link with nature's resources—a link that stems from a given organization of production, reproduction and distribution, including a given gender division of labour.

At the same time, the positive aspects of this link should not serve as an argument for the continued entrenchment of women within a given division of labour. Rather, they should serve, as an argument for creating the conditions that would help universalize this link with nature, for instance, by *declassing* and *degendering* the ways in which productive and reproductive activities are organized (within and outside the home) and how property, resources, knowledge and power are distributed.

## Conclusion

The Indian experience offers several insights and lessons. First, the processes of environmental degradation and appropriation of natural resources by a few have specific class–gender as well as locational implications—it is women of poor, rural households who are most adversely affected and who have participated actively in ecology movements. 'Women' therefore cannot be posited (as the ecofeminist discourse has typically done) as a unitary category, even within a country, let alone across the Third World or globally. Second, the adverse class–gender effects of these processes as manifest in the erosion of both the livelihood systems and the knowledge systems on which poor rural women depend. Third, the nature and impact of these processes are rooted interactively, on the one hand, in ideology (in notions about development, scientific knowledge, the appropriate gender division of labour, and so on), and, on the other hand, in the economic advantage and political power predicated especially, but by no means only, on property differentials between households and between women and men. Fourth, there is a spreading grassroots resistance to such inequality and environmental destruction—to the processes, products, people, property, power and profit-orientation that underlie them. Although the voices of this resistance are yet scattered and localized, their message is a vital one, even from a purely growth and productivity concern and more so if our concern is with people's sustenance and survival.

In particular, the experiences of women's initiatives within the environmental movements suggests that women's militancy is much more closely linked to family survival issues than is men's. Implicit in these struggles is the attempt to carve out a space for an alternative existence that is based on equality, not dominance over people, and on cooperation with and not dominance over nature.

Indeed what is (implicitly) being called into question in various ways by the movements is the existing development paradigm—with its particular product and technological mix, its form of exploitation of natural and human resources, and its conceptualization of relationships among people and between people and nature. However, a mere recognition that there are deep inequalities and destructiveness inherent in present processes of development is not enough. There is a need for policy to shift away from its present relief-oriented approach toward nature's ills and people's welfare, in which the solution to nutrient-depleted soils is seen to lie entirely in externally added chemical nutrients; to depleting forests in monoculture plantations; to drought starvation in food-for-work programmes; to gender inequalities in ad hoc income-generating schemes for women, and so on. These solutions reflect an aspirin approach to development—they are neither curative nor preventive, they merely suppress the symptoms for a while.

The realistic posing of an alternative (quite apart from its implementation) is of course not easy, nor is the purpose of this paper to provide a blueprint. What is clear so far are the broad contours. An alternative approach, suggested by feminist environmentalism, needs to be *transformational* rather than welfarist—where development, redistribution and ecology link in mutually regenerative ways. This would necessitate complex and interrelated changes such as in the *composition* of what is produced, the *technologies* used to produce it, the processes by which decisions on products and technologies are arrived at, the *knowledge systems* on which such choices are based, and the class and gender *distribution* of products and tasks.

For instance, in the contest of forestry programmes, a different composition of the product may imply a shift from the currently favoured monocultural and commercial tree species to mixed species critical for local subsistence. An alternative agricultural technology may entail shifting from mainly chemical-based farming to more organic methods; from monocultural high-yielding variety seeds to mixed cropping with indigenously produced varieties; from the emphasis on large irrigation schemes to a plurality of water provisioning systems; and from preoccupation with irrigated crops to a greater focus on dryland crops. A change in decision-making processes would imply a shift from the present top-down approach to one that ensures the broad-based democratic participation of disadvantaged groups. Indeed, in so far as the success stories of reforestation today relate to localized communities taking charge of their environmental base, a viable solution would need decentralized planning and control and institutional arrangements that ensure the involvement of the rural poor, and especially women, in decisions about what trees are planted and how the associated benefits are shared. Similarly, to encourage the continued use and growth of local knowledge about plants and species in the process of environmental regeneration, we would require new forms of interaction between local people and trained scientists and a widening of the definition of 'scientific' to include plural sources of knowledge and innovations, rather than merely those generated in universities and laboratories. This last is not without precedent, as is apparent from the earlier discussion in Meiji Japan's interactive teams which allowed a flow of information not only from the agricultural scientist to the farmer but also the reverse. The most complex, difficult, and necessary to transform is of course the class and gender division of labour and resources and the associated social relations. Here it is the emergence of new social movements in India around issues of gender, environment, and democratic rights, and especially the formation of joint fronts between these movements on a number of recent occasions that point the direction for change and provide the points of hope.

Indeed, environmental and gender concerns taken together open up both the need for re-examining, and the possibility of throwing new light on, many long-standing issues relating to development, redistribution and institutional change. That these concerns preclude easy policy solutions underlines the deep entrenchment of interests (both ideological and material) in existing structures and models of development. It also underlines the critical importance of grassroots political organization of the poor and of women as a necessary condition for their voices to be heeded and for the entrenched interests to be undermined. Most of all it stresses the need for a shared alternative vision that can channel dispersed rivulets of resistance into a creative, tumultuous flow.

In short, an alternative, transformational approach to development would involve both ways of *thinking* about things and ways of *acting* on them. In the present context it would concern both how gender relations between people and the non-human world are conceptualized, and how they are concretized in terms of the distribution of property, power and knowledge, and in the formulation of development policies and programmes.

It is in its failure to explicitly confront these political economy issues that the ecofeminist analysis remains a critique without threat to the established order.

## Notes

* This is a substantially revised and abridged version of a paper presented at a conference on 'The Environment and Emerging Development Issue', at World Institute of Development Economics Research, Helsinki, 3–7 September 1990. A longer version is also available as Discussion Paper No. 8: 'Engendering the Environment Debate: Lessons from the Indian Subcontinent', CASID Distinguished Speaker Series (Michigan State University, 1991). I am grateful to several people for comments on the earlier versions: Janet Seiz, Gillian Hart, Nancy Folbre, Jean Dreze, Lourdes Beneria, Gail Hershatter, Pauline Peters, Tariq Banuri, Lmyra Buvinic, and *Feminist Studies* editors and anonymous reviewers. I also gained from some lively discussions following seminar presentations of the paper at the Centre for Population and Development Studies, Harvard University, February 1991; the Centre for Advanced Study in International Development, Michigan State University of Minnesota, April 1990; the Hubert Humphrey Institute of Public Affairs, University of Minnesota, April 1990; and the Departments of City and Regional Planning and Rural Sociology, Cornell University, May 1990.

1. See especially King (1981, 1989, 1990); Salleh (1984); Merchant, (1980); and Griffin (1978). Also see discussions and critiques by Zimmerman (1987); Warren (1987); Cheney (1987); and Longino (1981).
2. See the case studies, and especially Carol P. MacCormack's introductory essay in MacCormack and Strathern (1980:13), also see Moore (1989).
3. King in 'Feminism and the Revolt' (1981, unlike in her earlier work) does mention the necessity of such a differentiation, but does not discuss how a recognition of this difference would affect her basic analysis.

4. For an illuminating discussion of the debate on Essentialism and Constructionism within feminist theory, see Fuss (1989).
5. See case studies in MacCormack and Strathern (1980).
6. Also see the discussion by Dietrich (1989). Apart from the religion-specificity of the discourse on the feminine principle, an interesting example of the relationship between different religious traditions and the environment is that of sacred groves. These groves, dedicated to local deities and sometimes spread over 100 acres, were traditionally preserved by local Hindu and tribal communities and could be found in several parts of the country. Entry into them was severely restricted and tree-cutting usually forbidden (see Gadgil and Vartak 1975). These groves are now disappearing. Among the Khasi tribe of Northeast India, elderly non-Christian Khasis I spoke to identify the main cause of this destruction to be the large-scale conversion of Khasis to Christianity which undermined traditional beliefs in deities and so removed the main obstacle to the exploitation of these groves for personal gain.
7. For instance, the *Rig Veda,* the collection of sacred Sanskrit hymns preserved orally for over 3,000 years, which constitutes the roots of Brahminic Hinduism, is said to have been traditionally inaccessible to women and untouchable castes, both of whom were forbidden to recite the hymns on the ground that they would defile the magic power of the words (for elaboration, see Flaherty 1990). In contrast, the Bhakti movement, which began around the sixth century, sought to establish a direct relationship between God and the individual (without the mediation of Brahmin priests) irrespective of sex or caste and gave rise to numerous devotional songs and poems in the vernacular languages. Many women are associated with the movement, one of the best known being the sixteenth-century poet-saint, Mirabai. Today, the Bhakti tradition coexists with the more ritualistic and rigid Brahminic tradition. In fact, a significant dimension of the growing Hindu fundamentalism in recent years is precisely the attempt by some to give prominence to one interpretation of Hinduism over others—a visible, contemporary struggle over meanings. Similarly, several versions of the great epic, *Ramayana* have existed historically, including versions where the central female character, Sita, displays none of the subservience to her husband that is emphasized in the popular version (treated as sacred text) and which has moulded the image of the ideal Indian woman in the modern mass media. Feminist resistance to such gender construction has taken various forms including challenging popular interpretations of female characters in the epics and drawing attention to alternative interpretations (see, for instance, Chakravarty 1983; Agarwal 1985).
8. See Habib (1984) and his essay in Ray Chaudhuri and Habib (1982).

9. See especially Kerala Forestry Research Institute (1980: 235).
10. It is estimated that in 1981-2, 66.6 per cent of landowning households in rural India owned I hectare or less and accounted for only 12.2 per cent of all land owned by rural households (GoI 1987). The distribution of operational holdings is almost as skewed.
11. I prefer to use the term 'control rights' here, rather than the commonly used term 'property rights', because what appears critical in this context is less who owns the resources than who has control over them. Hence, for instance, the control of state-owned resources could effectively rest with the village community.
12. However, the degree to which the village community acted as a cohesive group and the extent of control it exercised over communal land varied across undivided India: it was much greater in the north-west than elsewhere. See Baden-Powell (1957).
13. For a detailed discussion on these causes, see Jodha (1986).
14. On traditional systems of community water management, see Sengupta (1985); Leach (1967); and Seklar (1981). On communal management of forests and village commons, see Guha (1985); Gadgil (1985); Moench (1988). On firewood gathering practises, see Agarwal (1987). firewood for domestic use in rural households was customarily collected in the form of twigs and fallen branches, which did not destroy the trees. Even today, 75 per cent of firewood used as domestic fuel in northern India (and 100 per cent in some other areas) is in this form.
15. The preservation of sacred groves described in note 6 is one such example.
16. Also see discussion in Dasgupta and Maler (1990).
17. Under some large-scale irrigation works, crop yields are *lower* than in the period immediately prior to the project (Joshi and Agnihotri 1984).
18. For a review of issues and literature on this question, see Agarwal (1986).
19. These sex ratios are particularly female-adverse in the agriculturally prosperous north-western regions of Punjab and Haryana, where these figures are respectively, 88 and 87 females per 100 males. For a discussion on the causes of this regional variation, see Agarwal (1986); Miller (1981).
20. Women in India rarely own land, land in most areas also have limited access to personal assets such as cash and jewelry.
21. For a detailed cross-regional mapping of some of these variables in the context of women's land rights in South Asia, see Agarwal (1994).
22. Personal communication, Chitra Sundaram, Danish International Development Agency (DANIDA), Delhi, 1981.
23. These are apart from the widely documented patron-client types of relationships.

24. See Sen (1990), for a discussion on the bargaining approach to conceptualizing intra-household gender relations, and Agarwal (1990), for a discussion on the factors that affect intra-household bargaining power.
25. Among the Garo tribals of north-east India in the early 1960s, Burling, 1963, found that the men always deferred on this count to the women, who knew of approximately 300 indigenously cultivated rice varieties. In Nepal, even today, it is women who do the seed selection work among virtually all communities. See Acharya and Bennett (1981).
26. For a detailed discussion on these schemes and their shortcomings, see Agarwal (1986a).
27. Chandrashekar, Murti and Ramaswamy (1987); Shiva (1988).
28. Among the many writings on the Chipko movement, see especially Bandyopadhyay and Shiva (1987, 1988); Jain (1984); Dogra (1984).
29. I understand there have also been cases of people hugging trees to protect them from loggers in the United States, although they appear to have no apparent link with Chipko.
30. There is a growing literature indicating significant gender differences in cash-spending patterns, with a considerable percentage (at times up to 40 per cent) of items they alone consume, such as liquor, tobacco and clothes, and much of what the women earn going toward the family's basic needs. See especially Mencher (1988).
31. This gender divergence has also been noted elsewhere. See Brara (1987).

## References


Acharya, Meena and Lynn Bennett, 1981, 'Women and the Subsistence Sector in Nepal', *World Bank Staff Working Paper* No. 526, World Bank, Washington DC.

Agarwal, Anil, 1983, 'The Cooking Energy Systems: Problems and Opportunities' Centre for Science and Environment, New Delhi.

Agarwal, Bina, 1981, 'Women and Water Resource Development' (photocopy), Institute of Economic Growth, Delhi.

———, 1984, 'Rural Women and the High Yielding Variety Rice Technology in India', *Economic and Political Weekly*, 31 March, A39–A52.

———, 1985, 'Sita Speak' (poem), *Indian Express,* 17 November.

———, 1986a, 'Women, Poverty and Agricultural Growth in India', *Journal of Peasant Studies,* 13 July, pp. 165–220.

Agarwal, Bina, 1986b, *Cold Hearths and Barren Slopes: The Woodfuel Crisis in the Third World*, Zed Books, London.

Agarwal, Bina, 1987, 'Under the Cooking Pot: The Political Economy of Domestic Fuel Crisis in Rural South Asia', *IDS Bulletin,* Vol. 18, No. 1, pp. 11–22.

———, 1988, 'Who Sows? Who Reaps? Women and Land Rights in India', *Journal of Peasant Studies,* 15 July, pp. 531–81.

———, 1989, 'Women, Land and Ideology in India', in Haleh Afshar and Bina Agarwal (eds), *Women, Poverty and Ideology: Contradictory Pressures, Uneasy Resolutions*, Macmillan, and London.

———, 1990, 'Social Security and the Family: Coping with Seasonality and Calamity in Rural India, *Journal of Peasant Studies,* 17 April, pp. 341–412.

———, 1994, *A Field of One's Own: Gender and Land Rights in South Asia,* Cambridge University Press, Cambridge.

Baden-Powell, B.H., 1957, *The Indian Village Community,* HRAF Press, New Haven, Connecticut.

Bahuguna, Sundarlal, 1984, 'Women's Non-violent Power in the Chipko Movement', in Madhu Kishwar and Ruth Vanita (eds), *In Search of Answers: Indian Women's Voices in 'Manashi'*, Zed Books, London.

Bajracharya, Deepak, 1983, 'Deforestation in the Food/Fuel Context: Historical and Political Perspectives from Nepal', *Mountain Research and Development,* Vol. 3, No. 3.

Bandyopadhyay, Jayanta, 1986, 'A Case Study of Environmental Degradation in Karnataka', paper presented at a Workshop on Drought and Desertification, India International Centre, New Delhi, 17–18 May.

——— and Vandana Shiva, 1987, 'Chipko', *Seminar,* No. 330, February.

Batliwala, Srilata, 1983, 'Women and Cooking Energy', *Economic and Political Weekly,* 24–31 December.

Blaikie, Piers, 1985, *The Political Economy of Soil Erosion in Developing Countries, Longman,* London and New York.

Bardhan, Kalpana, 1977, 'Rural Employment, Welfare and Status: Forces of Tradition and Change in India', *Economic and Political Weekly,* 25 June, A34–A48; 2 July, pp. 1062–74; 9 July, pp. 1101–18.

Bhaduri, T., and V. Surin, 1980, 'Community Forestry and Women Headloaders', in *Community Forestry and People's Participation—Seminar Report,* Ranchi Consortium for Community Forestry, 20–22 November.

Brara, Rita, 1987, 'Commons Policy as Process: The Case of Rajasthan, 1955–58', *Economic and Political Weekly,* 7 October, pp. 2247–54.

Bromley, Daniel W., and Michael M. Cernea, 1989, 'The Management of Common Property Natural Resources', *World Bank Discussion Paper* No. 57, World Bank, Washington DC.

Burling, Robbins, 1963, *Rensanggri: Family and Kinship in a Garo Village,* Pennsylvania University Press, Philadelphia.

Centre for Science and Environment, 1986, *The State of India's Environment: A Citizen's Report*, 1985-86, Delhi,

Chakravarty, Uma, 1983, 'The Sita Myth', *Samya Shakti,* 1 July.

Chand, Malini and Rita Bezboruah, 1980, 'Employment Opportunities for Women in Forestry', in *Community Forestry and People's Participation—Seminar Report,* Ranchi Consortium for Community Forestry, 20-22 November.

Chandrashekar, D.M., B.V. Krishna Murti and S.R. Ramaswamy, 1987, 'Social Forestry in Karnataka: An Impact Analysis', *Economic and Political Weekly,* 13 June, pp. 935-41.

Cheney, Jim, 1987, 'Ecofeminism and Deep Ecology', *Environmental Ethics,* 9, Summer, pp. 115-45.

Dasgupta, Partha and Karl-Goran Maler, 1990, 'The Environment and Emerging Development Issues', paper presented at a Conference on Environment and Development, Wider, Helsinki.

Dhawan, B.D., 1982, *Development of Tubewell Irrigation in India,* Agricole Publishing Academy, Delhi.

Dietrich, Gabriele, 1989, 'Plea for Survival: Book Review', *Economic and Political Weekly,* 18 February, pp. 353–54.

Dogra, Bharat, 1984, *Forests and People,* published by the author, Delhi,

Fernandes, Walter and Geeta Menon, 1987, *Tribal Women and Forest Economy: Deforestation, Exploitation and Status Change,* Indian Social Institute, Delhi.

Flaherty, Wendy O., 1990, *Other People's Myths,* Macmillan, New York and London.

Folger, Bonnie and Meera Dewan, 1983, 'Kumaon Hills Reclamation: End of Year Site Visit', OXFAM America, Delhi.

Fuss, Diane, 1989, *Essentially Speaking,* Routledge, New York.

Gadgil, Madhav, 1985, 'Towards an Ecological History of India', *Economic and Political Weekly*, Special Number, November, pp. 1909-38.

Gadgil, Madhav and V.D. Vartak, 1975, 'Sacred Groves of India: A Plea for Continued Conservation, *Journal of the Bombay Natural History Society,* Vol. 72, and No. 2.

Glaeser, Bernhard (ed.), 1987, *Learning from China? Development and Environment in Developing Countries,* Allen and Unwin, London, Government of India, 1987, *Thirty-seventh Round Report on Land Holdings—I, Some Aspects of Household Ownership Holdings,* National Sample Survey Organization, Department of Statistics.

———, 1990, *Forest Survey of India,* Ministry of Environment and Forests, New Delhi.

Griffin, Susan, 1978, *Women and Nature: The Roaring Within Her,* Harper and Row, New York.

Guha, Ramachandra, 1983, 'Forestry in British and Post-British India: A Historical Analysis', *Economic and Political Weekly,* 29 October, pp. 1882–96.

———, 1985, 'Scientific Forestry and Social Change in Uttarakhand', *Economic and Political Weekly,* Special Number November, pp. 1939–52.

———, 1988, 'Ideological Trends in Indian Environmentalism', *Economic and Political Weekly,* 3 December, pp. 2578–81.

Habib, Irfan, 1984, 'Peasant and Artisan Resistance in Mughal India', *McGill Studies in International Development,* No. 34, Centre for Developing Area Studies, McGill University.

Howes, Michael and M.A. Jabbar, 1986, 'Rural Fuel Shortages in Bangladesh: The Evidence from Four Villages', *Discussion Paper 213,* Institute of Development Studies, Sussex.

Jain, Shobita, 1984, 'Women and People's Ecological Movement: A Case Study of Women's Role in the Chipko Movement in Uttar Pradesh', *Economic and Political Weekly,* 13 October, pp. 1788–94.

Jodha, N.S., 1986, 'Common Property Resources and Rural Poor', *Economic and Political Weekly,* 5 July, pp. 1169–81.

Johnston, Bruce F., 1969, 'The Japanese Model of Agricultural Development: Its Relevance to Developing Nations', in Kazushi Ohkawa, Bruce F. Johnston and Hiromitsu Kaneda (eds), *Agricultural and Economic Growth—Japan's* Experience, Princeton University Press, Princeton.

Joshi, P.K. and A.K. Agnihotri, 1984, 'An Assessment of the Adverse Effects of Canal Irrigation in India', *Indian Journal of Agricultural Economics,* 39, July–September, pp. 528–36.

Kerala Forestry Research Institute, 1980, *Studies in the Changing Patterns of Man-Forest Interaction and Its Implications for Ecology and Management,* Trivandrum.

King, Ynestra, 1981, 'Feminism and the Revolt,' *Heresies,* No. 13, Special Issue on Feminism and Ecology, pp. 12–16.

———, 1989, 'The Ecology of Feminism and the feminism of Ecology', in Judith Plant (ed.), *Healing the Wounds: The Promise of Ecofeminism,* New Society report 69, International Food Policy Research Institute, Washington DC, pp. 18–28.

King, Ynestra, 1990, 'Healing the Wounds: Feminism, Ecology and the Nature/ Culture Dualism', in Irene Diamond and Gloria Orenstein (eds), *Reweaving the World: The Emergence of Ecofeminism,* pp. 98–112.

Kulkarni, Sharad, 1983, 'Towards a Social Forestry Policy', *Economic and Political Weekly,* 5 February, pp. 191–96.

Kumar, Shubh and David Htchkiss, 1988, 'Consequences of Deforestation for Women's Time Allocation, Agricultural Production and Nutrition in Hill Areas of Nepal', *Research Report,* 69, International Food Policy Research Institute, Washington DC.

Leach, Edmund R., 1967, *Pul Eliya—A Village in Ceylon: A Study of Land Tenure and Kinship, Cambridge* University Press, Cambridge University Press, Cambridge.

Longino, Helen E., 1981, E., 1981, Review of Merchant (1980) in *Environmental Ethics,* Winter, pp. 3265–69.

———, 1987, 'Can There Be a Feminist Science?', *Hypatia,* 2, Fall, pp. 51–64.

MacCormack, Carol P. and Marilyn Strathern (eds), 1980, *Nature, Culture and Gender,* Cambridge University Press, Cambridge.

Marglin, Stephen A., 1988, 'Losing Touch: The Cultural Conditions of worker Accommodation and Stephen A. Marglin (eds), *Knowledge and Power,* OUP, Oxford.

Mencher, Joan P., 'Women's Work and Poverty: Women's Contribution to Household Maintenance in Two Regions of South India', in Daisy Dwyer and Judith Bruce (eds), *A Home Divided: Women and Income in the Third World,* Stanford University Press, Stanford.

Mencher, Joan P. and K. Saradamoni, 1982, 'Muddy Feet and Dirty Hands: Rice Productions and Female Agricultural Labour', *Economic and Political Weekly,* 25 December, pp. A149–A167.

Merchant, Carolyn, 1980, *The Death of Nature: Women, Ecology and the Scientific Revolution,* Harper and Row, San Francisco,

Miller, Barbara, 1981, *The Endangered Sex: Neglect of Female Children in North-West India,* Cornell University Press, Ithaca.

Moench, M., 1988, 'Turf and Forest Management in a Garhwal Hill Village', in Louise Fortmann and John W. Bruce (eds), *Whose Trees? Proprietary Dimensions of Forestry,* Westview Press, Boulder, Colorado.

Mohan, Dinesh, 1987, 'Food vs. Limbs: Pesticides and Physical Disability in India', *Economic and Political Weekly,* 28 March, pp. A23–A29.

Moore, Henrietta L., 1989, *Feminism and Anthropology* University of Minnesota Press, Minneapolis.

Nagbrahman, D. and S. Salmbrani, 1983, 'Women's Drudgery in Firewood Collection', *Economic and Political Weekly,* 1–8 January.

National Sample Survey Organization, 1987, *Thirty-seventh Round Report on Land Holdings—I, Some Aspects of Household Ownship Holdings,* Department of Statistics, Government of India.

Orther, Sherry, 1974, 'Is Male to Female as Nature is Culture?' in Michelle Z. Rosaldo and Louise Lamphere (eds), *Women, Culture and Society,* Stanford University Press, Stanford.

Pingle, V., 1975, 'Some Studies of two Tribal Groups of Central India, Part 2: The Importance of Food Consumed in Two Different Seasons', *Plant Food for Man,* 1.

Rao, C.H. Hanumantha, S.K. Ray and K. Subbarao, 1988, *Unstable Agriculture and Drought,* Vikas, Delhi.

Ray Chaudhuri, Tapan and Irfan Habib (eds), 1982, *Cambridge Economic History of India,* Cambridge University Press, Cambridge.

Rosenzweig, Mark and Kenneth I. Wolpin, 1985, 'Specific Experience, Household Structure, and Intergenerational Transfers: Farm Family Land and Labour Arrangements in Developing Countries', *Quarterly Journal of Economics,* 100, Supp., pp. 961–87.

Ryan, James G., and R.D. Ghodake, 1980, 'Labour Market Behaviour in Economic Status', Progress Report, Economic Programme 14, International Crop Research Institute for Semi-Arid Tropics (ECRISAT), Hyderabad.

Salleh, Ariel Kay, 1984, 'Deeper than Deep Ecology: The Eco-Feminist Connection', *Environmental Ethics,* 16, Winter, pp. 339–45.

Seklar, David, 1981, 'The New Era of Irrigation Management in India', mimeo, Ford Foundation, Delhi.

Sen, Amartya, 1990, 'Gender and Cooperative-Conflict', in Irene Tinker (ed.), *Persistent Inequalities,* OUP, New York.

Sengupta, Nirmal, 1985, 'Irrigation: Traditional versus Modern', *Economic and Political Weekly,* Special Number, November, pp. 1919–38.

Sharma, Ursula, 1980, *Women, Work and Property in North-West India,* Tavistock, London.

Shaw, Paul, 1989, 'Population, Environment and Women: An Analytical Framework', paper prepared for the United Nations Fund for Population Activities (UNFPA), Inter-Agency Consultative Meeting, New York, 6 March.

Shiva, Vandana, 1987, 'Ecology Movements in India', *Alternatives,* 11, pp. 255–73.

———, 1988, *Staying Alive: Women, Ecology and Survival,* Zed Books, London.

Stone, Linda, 1982, 'Women and Natural Resources: Perspectives from Nepal', in Molly Stock, Jo Ellen Force and Dixie Ehrenreich (eds), *Women in Natural Resources: An International Perspective,* University of Idaho Press, Moscow.

Swaminathan, Madhura, 1984, 'Eight Hours a Day for Fuel Collection', *Manushi,* March–April.

Swaminathan, Srilata, 1982, 'Environment: Tree versus Man', *India International Centre Quarterly,* 9, Nos. 3 and 4.

United Nations Development Program (UNDP), 1979, 'Rural Women's Participation in Development', *Evaluation Study,* No. 3, UNDP, New York, June.

Vatuk, Sylvia, 1981, 'Sharing, Giving and Exchanging of Foods in South Asian Societies', University of Illinois at Chicago Circle, October.

Wagner, Rudolf G., 1987, 'Agriculture land Environmental Protection in China', in Bernhard Glaeser (ed.), *Learning from China?*

Warren, Karen J., 1987, 'Feminism and Ecology: Making Connections', *Environmental Ethics,* 9, Spring, pp. 3–20.

Zimmerman, Michael, E., 1987, 'Feminism, Deep Ecology and Environmental Ethics', *Environmental Ethics,* 9, Spring, pp. 21–44.

# 20

## A Great Legacy Dissipated

M. Krishnan

To have a fair idea of the present worth of our wildlife preserves (of all kinds: sanctuaries, National Parks and other protected habitats) and to know what further steps need to be taken to effectively safeguard our heritage of nature, it is necessary to take a quick, overall look at the past. Many of these preserves were set up long before Independence, in British India and in the princely states—well-known examples are Vedanthangal in Tamil Nadu, Bandipur in Karnataka and Corbett Park (our first authentic national park, set up in 1934) in Uttar Pradesh.

Soon after the *shikar* statutes that followed Independence, the Indian Board for Wild Life was constituted, a number of fresh preserves were created (some quite major ones) and conservation tightened. Illustrative examples of this are the metamorphosis of Keola Deo Ghana at Bharatpur in Rajasthan from a wild-fowlers' paradise into the most important water-bird sanctuary in Asia in 1956, the Point Calimere and Anaimalai sanctuaries of Tamil Nadu, and the intensification of conservation in the Gir Forest of Gujarat (the last home of the Asiatic lion) and in Kaziranga in Assam (the major stronghold of the Indian rhinoceros and the no less Indian wild buffalo). The trend has been continued right into the present: Sultanpur Jheel in Haryana, another favourite resort of wildfowlers, was converted into a fine sanctuary recently and the ambitious Desert National Park of Rajasthan was initiated earlier this year.

We now have almost 150 wildlife preserves, big and small, all over India. But neither the old nor the new preserves have taken note of the basic fact

* This paper first appeared in *The Illustrated Weekly of India,* 24 August 1980. It is reprinted here from M. Krishnan, 'A Great Legacy Dissipated', *Seminar* 466, June, 1998. Though written over two decades ago, Krishnan's overview of the history and prospects of wildlife conservation in India remain relevant.

that the wild vegetation is as integral and vital a part of the wildlife of any region as its fauna. In 1970, with the acceptance by the central government of the definition of 'wildlife' as the entire uncultivated flora and fauna of a tract, this profound truth gained formal official recognition—it still awaits recognition in the field. In fact, our wildlife preserves have all been set up solely for the larger animals, and a few for water-birds, and the flora has been considered purely incidental, as providing cover and fodder, and the lesser life not considered at all.

Hydel projects have been sited close by, even right inside, some of the best preserves. With a few exceptions, diverse forestry operations (all highly destructive of the natural vegetation) are carried on in our preserves, with even the supply of raw materials at subsidised rates to industries from them undertaken: cattle grazing, the collection of firewood and forest produce and other human activities are permitted, motor roads intersect the preserves, and human traffic on foot (highly unsettling to the wild animals) allowed. All our wildlife habitats (including all preserves) are solely in charge of our various governments: naturally, then the responsibility for protection is entirely theirs, and even otherwise only governments have the sanction necessary for protection. It follows, inescapably, that by their permissiveness, our governments have been the chief depredators of our wildlife.

Actual, factual evidence as proof of every generalization made so far in this overall survey of the past is on record, but the detailing of even selected examples of such proof is beyond the scope of this essay. But before going on to them, let me comment on the outraged defence of their past policy and administration that governments are likely to raise on this criticism, that they have been preoccupied with more important things such as providing for the elemental needs of our growing populations and industries.

Yes, I do realise that they have had that Sisyphean task right from the start, but then all our preserves together constitute only about 1 per cent of the total land area. Surely they could have provided for our populations and industries with the overwhelming percentage of the land not given over to wildlife at their disposal. Furthermore, conserving the wealth of nature we still have left is also an elemental national need. Much needs to be done towards more realistic conservation to rectify the apathy and wastefulness of the past, and it has to be done right now—to delay further is to be assured of not having enough worth the saving.

That is the counsel, not of despair but of hope. The depletion of the past 50 years has been staggering: places noted for their wildlife within my own recollection are now denuded and bare and their animals have declined to rarity, even to local extinction. But the wonderful, the heartening thing is this—in spite of everything, India is still second to no country in its wild fauna, and perhaps the richest in its flora. Though heavily depleted, our fauna is still there, and notably free from exotic introductions: that cannot be said of our exotics-ridden flora, but it will regain its pristine glory if conserved, that is, if left alone and allowed to regenerate.

The policy of leaving it alone and resisting the almost overpowering urge to improve on nature has been proved to be the best, actually the only sound policy in conservation. This was called trust in the balance of nature in the past, and is now termed total environmental conservation, and it will suffice to ensure the future of our wildlife. It means emancipating preserves of viable area from all forms of human exploitation, and the provision of strict protection (now sadly lacking) against human disturbance and depredation. This is all that is needed, but since natural regeneration is a slow process, patience and faith are also needed. Incidentally, it is only in the few reserves of Project Tiger that total environmental conservation is now being attempted, after a fashion.

In the past decade, governmental cognizance of responsibility for the country's ravaged opulence of nature has displayed a certain nascent percipience. The Wild Life (Protection) Act of 1972 is constructive, if only for its insistence on a specific wildlife organization in each state and it is the first realistic attempt at a national policy of conservation. There are other tokens of awakening governmental concern for our wildlife, even of awakening public concern in a minority of our people, but for the next generation at least governments must bear the entire responsibility for protection.

Today, there is no informed popular interest in our culture, in our magnificent assets of nature. The human and natural curiosity of our children in the wild things around is sternly nipped in the bud by our traditions of life and instruction instead of being informed and stimulated. In the West, with a comparatively small store of wildlife, the great therapeutic value of such an interest in relieving the stresses and frustrations of civilized life has been fully appreciated after the Second World War: wildlife recreation is widely organized and popular, and education at all levels features nature importantly. With no reliable natural history in the written or oral literatures of our languages, and no popular feeling for it, it will be no easy task to inform and stimulate this dormant interest in our people, particularly in the younger sections. But this is a vital national need and will endow future generations with a joy and sustenance in life that we have been without.

If I have only conveyed the impression so far that it is of national importance to conserve our wildlife and wildlife habitats, I have failed fundamentally in my argument. This is no matter of mere importance, but a primary patriotic duty, quite essential for the survival of the identity of this ancient country. Surely no country depends for its identity mainly on the conglomerate accretions of its cultural past or its mutable humanity—it depends overwhelming on its own peculiar physical integrity, its geomorphology and the flora and fauna that belong to it distinctively. Oddly enough, it is our poets and not our rulers or politicians who have realized this profound truth—it is they that have sung of our mountains and valleys and rivers, sounding seas and vast coastlines, great forests and lovely flowers, and of our birds and beasts.

The dissipation of India's physical integrity has now reached the stage where further indiscriminate demands on its natural resources will certainly erode its very identity. For this reason, it is imperative that adequate tracts, typical of the country's quiddity, should be freed from human exploitation and protected efficiently. Only that can ensure the continuance of India's identity. With our vast populations and growing industrialization, it will be unrealistic to ask for much territory even for this vital national purpose, but a modest 5 per cent of the total land area should suffice, in the circumstances. Naturally, this will include all existing wildlife preserves, so that no further demands need be made on our forests; for the rest it will embrace notable geomorphological features and areas not provided for so far in our wildlife effort, such as swamps, estuaries, offshore islands, mountain tops and, by no means last, adequate expanses of plains for the wildlife of the open country, now so sadly lacking sanctuary.

21

# Sacred Groves for the Twenty-First Century

K. Ullas Karanth

India has now seriously grappled with contentious problems of conserving its wildlife for over a half century with a marked intensification of the effort in the second half of this period. This effort has been made against a backdrop of a massive population increase, rise in democratic aspirations and burgeoning consumer needs and greed, all of which fuelled a rapidly developing economy. Yet, against all odds, India has managed to hang on to, however precariously, several intact faunal assemblages of highly extinction-prone large vertebrates within the confines of its protected areas that cover only 4 per cent of its landmass.

Compared to the virtual elimination or large-scale range restriction of large vertebrate fauna of North America, Europe and China, which accompanied comparable periods of demographic and economic growth, this is not a small achievement. This half-century of rich empirical experience in conserving wildlife in India is strewn with both successes and failures. India's success in holding on to species such, as the rhino, elephant and tiger is quite spectacular when compared to the record of other Asian countries (with the possible exception of Nepal), which have undergone similar social transitions. However, major failures have also marred this conservation effort. The inability to manage wildlife scientifically and limitations on account of an official mind-set too rigid to deal with issues affecting local peoples are two glaring examples.

We have to analyse the problem of 'wildlife conservation' against the above background of empirical experience and ecological evidence. I would

---

* Reprinted with permission from K. Ullas Karanth, 'Sacred Groves for the Twenty-First Century', *Seminar* 466, June, 1998, pp. 25–31. The references have been updated.

argue that, for this debate to progress beyond the usual sterile polemics of 'wildlife versus people', the terms 'wildlife' and 'conservation' have to be defined a little more rigorously. I approach this task by eliminating some of the definitions which muddy rather than clarify the essence of the problem.

A recent study (Madhusudan and Karanth, 2002) in the Kudremukh National Park area showed widespread local hunting and collection of forest products as causing severe depression of densities of large mammals and in the quality of their habitats. However, another study that focused on general biodiversity at the same site (but was thereby forced to use less rigorous methods, in our opinion) suggested that biodiversity levels were quite high and holding out well. Both studies concurred that logging in the past had made the forest structure less 'pristine'. However, because the second study did not attempt to collect any quantitative data on large mammal species more vulnerable to human disturbance, it failed to identify their serious population declines or their causal factors.

I believe the central debate here should not be about saving overall biodiversity, which includes all living creatures ranging from soil organisms to cultivated plants, and is measured by scientists through parameters such as species richness or diversity indices. Most forms of biodiversity (e.g. soil bacteria, fungi, crops crows, rats, jackals) can survive on intensively human-dominated landscapes such as cities, towns and farmlands. Several other forms of biodiversity may need more natural landscapes to survive, but can still withstand intensive disturbances or even extractive practises (several forest trees, birds, reptiles, smaller mammals, wild pigs).

I exclude such organisms from the scope of this paper because, obviously, there are no major contentious issues about adversities imposed by human society when we deal with their conservation. In fact, most such biodiversity is likely to survive without any effort on our part or even because of our single-minded pursuit of better lifestyles and profits (gene banks, cloning, genetic engineering and so on; Therefore, I will focus here only on the conservation of a subset of this broader biodiversity, which can be termed 'wildlife'. Thus defined, 'wildlife conservation' becomes largely (but not exclusively) an issue of preserving viable populations, communities and landscapes of free ranging, extinction-prone, larger vertebrates (e.g., rhino, tiger, elephant, lion-tailed monkey, great hornbill, great Indian bustard).

The term conservation as traditionally defined includes activities such as preservation (as in protected areas), sustainable harvesting (as in fisheries), and control or regulation of organisms harmful to human interests (as in the case of man-eating tigers or crop-raiding elephants). I broadly follow this definition, but with a thrust on discussing issues related to wildlife preservation. The term wildlife conservation here does not include concepts such as animal rights and *ahimsa*, which are moral and social issues beyond the scope of this paper. Having defined wildlife conservation specifically in these terms, I will readily concede a point around which there is a surfeit of polemics in India. Many forms of biodiversity can and do coexist with human exploitation of their habitats. Some forms may even be capable of surviving higher pressures of the future. These forms of biodiversity do not need inviolate sacred groves (of whatever size) to survive in. Further, I specifically discuss wildlife preservation in the Indian context of rising human and livestock population densities. I do so against a backdrop of increasing aspirations for material consumption levels among all social strata (including rural poor and tribal), and the burgeoning commercial markets for an increasing array of products from wildlife and wildlife habitats, in a growth-friendly economic climate.

What are the inherent characteristics of wildlife species or communities, which render them vulnerable to extinctions? Several biological traits, which we simply cannot alter through progressive legislation or dedicated social activism, are clearly recognizable. Body size and diet are two such traits. Large body size is linked to greater dietary needs and hence to large home ranges and landscape level movements, even for herbivorous species such as elephants, whose food may be relatively abundant. As elephants move across large landscapes, they are attracted to raid agricultural crops, even when natural forage is abundant. Where natural forage is depleted by livestock grazing, bamboo extraction or annual man-induced forest fires, the problem of the elephant's conflict with agriculture becomes even more acute.

A carnivorous diet further accentuates this need for wide ranging behaviours among large vertebrates. To survive and reproduce, a single tiger may need a prey base of about 400 deer-sized animals. A home range supporting such a prey base may extend over 15 to 500 sq km, depending on prey density. A tigress raising cubs has to kill 60–70 prey animals in a year. Even if there is wild prey, she will kill cattle if she can. Where the wild prey base is depleted, she will prey exclusively on cattle, and occasionally humans.

Such landscape creatures therefore, live at relatively low densities. Social spacing behaviours such as territoriality, dispersal and seasonal migrations further accentuate their need for landscape level movements. For ensuring the survival of reproducing and viable populations (even using the more easily attained demographic criteria rather than the more demanding genetic criteria) of such species, relatively extensive landscapes are needed. Demographic models of tiger populations (Karanth and Stith, 1999) show that a demographically viable, but small population of 24 breeding tigresses would need a protected area of 400–4000 sq km, depending on prey density. Therefore, traditional conservation areas such as sacred groves are not big enough to hold tiger populations, nor can temple ponds sustain sea turtle populations. If these creatures are to survive the twenty-first century we need larger 'sacred groves'.

When such necessary large landscape matrices are embedded with attractive sources of food and water, such as agricultural crops or livestock populations, severe conflict between wildlife and humans becomes inevitable. In India, the extirpation of the cheetah and the near extirpation of the wolf, the restricted distribution of breeding populations of tiger and lions to less than 0.5 per cent of their former distributional range, show clearly that their coexistence with humans has cost them dearly. Elimination of the elephant from extensive forested areas, which are honey-combed with agricultural enclaves, such as in Uttara Kannada district with 80 per cent forest cover, shows that, ultimately, when human-wildlife conflict becomes inevitable, human interests will clearly prevail.

Not all such extinction-prone species are large, wide-ranging or carnivorous. Many occupy narrowly defined ecological niches. The lion-tailed macaques and great hornbills of the Western Ghats have critical needs for food obtained from rainforest trees or lianas, and they shelter in old-growth tall timber. If their habitats are to be exploited intensively for timber or forest products such as fruits, nuts, roots, bark, resins, climbers or canes, they cannot survive in these habitats anymore. Such niche specialist wildlife species cannot shift their ranges elsewhere to make room for the needs of industries such as tanning, dyeing, *agarbatti*, food processing, timber or dairying which are dismantling components of their habitats, piecemeal, to cater to expanding local, regional and global markets. While exploitation of timber is at least recognized by conservationists as a factor generally inimical to wildlife conservation, the equally severe pressures resulting from livestock grazing and from the extraction of other forest products for meeting the needs of an urban industrial market, are barely recognized as threats by most conservationists or social activists.

The fact that India's rural poor or tribal are commonly hired as cheap labour to exploit forest products which end up in distant markets, is often mistaken for an ecologically sensible, traditionally wise, harmonious coexistence of wildlife and local communities. The argument that the collection of *tendu* leaves to feed the *beedi* industry of Mangalore or *sal* seeds for making hydrogenated fats in Mumbai, or cinnamon bark for the export-led growth of agarbatti industry in Mysore, are all 'traditional' practises of wise use, simply ignores any critical ecological or historical analysis.

There is no doubt that the issue of cash income needs, which drive rural poor in India to such wage labour, must be addressed squarely on its own. This is best done without confounding the wildlife conservation issues with the livelihood rights issues. An overdose of market friendly humanism of this sort is not a substitute for site-specific ecological analysis of the real impact of such human induced, market-driven disturbances on wildlife communities.

There is no evidence, *prima facie,* that the ecology of many endangered wildlife species, communities and habitats has evolved an evolutionarily adaptive relationship to such market-driven pressures. While it is true that some levels of disturbance regimes exist in nature, and sometimes human activities have formed a part of such disturbance regimes, there is no data to argue that all human induced disturbance, at all intensities, regardless of origins, is beneficial to wildlife.

Increasingly, many wildlife species themselves have become targets of direct exploitation for meat and body parts. The hunting of animals also occurs to eliminate crop raiding, cattle killing or man-killing. Hunting often results from a need for luxury consumption or supply to the local market. Very rarely is it for true subsistence in India, as it is, for example. In the case of rainforest pygmies of central Africa, there exists massive evidence that most of the hunting in tropical forests all over the world is unsustainable (Kramer *et al*. 1997: Robinson and Bennett, in press).

Even under potentially favourable factors such as low human population densities and extensive wilderness areas in Central Africa, hunting by pygmies seems to be turning unsustainable. The combination of traditional field craft with modern hunting accessories (guns, explosives, torch lights) or even simple technological advances in metallurgy (such as use of wire cables for snares or use of steel and springs for traps), impose pressures on many species which evolution has not equipped them to cope with. When external markets are linked to such situations, wildlife declines become precipitous.

In India there is not a single site where it can be shown that unregulated hunting does not lead to virtual extirpation of wildlife. This point is particularly well illustrated by the state of wildlife in the Northeast Indian hill states and elsewhere in South East Asia. In both these places wildlife has been extirpated over extensive areas of community-owned forests, despite there being in place strong traditional structures of authority and resource management. In Northeastern India, a short list of widely distributed species driven to extinction hunting entirely or over most of their ranges during the last two centuries includes: Javan, Sumatran and Indian rhinos, barasingha and brow-antlered deer, wild-water buffalo, Malay sun-bear, takin, and the Gangetic dolphin. Not surprisingly, all these are large bodied vertebrates.

Apparently, the altered human demographics, improved technologies and increasing market penetration have made traditional hunting of these species unsustainable. Therefore, at this point in time 'sustainable' hunting in itself, whether regulated by a local or distant authority, does not seem to offer much scope for wildlife conservation in India. When biologists argue in favour of such incentive driven hunting practises to encourage local communities to conserve wild-life, based on the African or American experiences, they seem to ignore Indian realities—just as social scientists tend to, when they advocate unhindered forest product collection by locals.

Therefore, in the face of substantial evidence to the contrary, one must be cautious while extrapolating the relevance of examples of coexistence of humans and wildlife from other ecologically and socially dissimilar contexts to India. We simply cannot extrapolate from examples of the low-density populations in the Amazon rainforest living off wild meat to rainforests of the Western Ghats. We know that hunting in rainforests of Asia has virtually extirpated large mammals because of higher human population densities, better technologies and greater market connectivity.

We cannot assume, because the Maasai herders in Africa coexist with lame carnivores in landscapes over several thousand sq km, such coexistence is a feasible option inside wildlife reserves of a few hundred sq km area. We cannot ignore the fact that Maasai reduce livestock losses to predators to tolerable levels by hunting to depress densities of lions and hyenas. While the poor quality land on which Maasai graze their cattle has few other uses in their context, the forests of India can have many profitable uses than pressuring either wildlife or rural grazers. We also need to note that in the African context such multiple use areas *exist outside of, and in addition to* large and strictly protected preserves with which they often have source-sink relationships.

The Maldhari herders of Gir are sometimes presumed to coexist harmoniously with lions, because they traditionally tolerated some livestock losses to lions. Yet, with the relocation of Maldhari herders from parts of Gir, the abundance of wild ungulates and their contribution to lion diet; both increased. The simple predator-prey system comprising of lions and domestic buffaloes turned into, arguably, a more 'natural' multiple prey-predator system. I would argue that the latter should be the goal of wildlife conservation, in a social and biological context where intact lion-ungulate communities occupy probably less than 0.1 per cent of their former distributional range. If the conservation goal were to merely have lions living off domestic water buffalo (a species as exotic to Gir as Eucalyptus trees), one could conceivably argue in favour of raising captive lions on buffalo meat at a fraction of the social cost of trying to preserve remnant pieces of the Gir ecosystem.

There is no doubt that conservation decisions result from a partly political process driven by social pressures. My point here is that this process should also incorporate substantial knowledge about biological needs of wildlife, and not merely about the human needs and aspirations. With the early biometric preservation having yielded ground to the user-friendly and market friendly conservation paradigm of 'sustainable use' over the last two decades, starting with the platitudinous World Conservation Strategy, we need to re-survey the ground we have covered.

To flip the other side of an argument tossed up a few years ago, we now need conservation not only as if people mattered but also as if large animals mattered. To describe such arguments in favour of incorporating of specific biological knowledge into the formulation of wildlife conservation policies, merely as manifestations of whimsical 'authoritarianism' on the part of socially insensitive wildlife scientists seem rather native. A social scientist would be horrified if a biologist were to state that because Bishnois of Rajasthan do not eat antelopes, the Inuit of the Arctic or the Dayaks of Sarawak could also live as vegetarians. Unfortunately, it is precisely such dateless, free wheeling extrapolation, which appears to underlie most articulations about the feasibility of peaceful, perpetual—and allegedly traditional—coexistence between human societies and extinction-prone wildlife species.

Given that biological knowledge is necessary for practicing wildlife conservation, a related issue is the distinction often made between the local community's wildlife knowledge versus external, 'authoritarian' scientific knowledge. While the former is supposed to envision multiple, holistic pathways to wise use and coexistence with wildlife, the latter type of knowledge is presumed to be linear and inflexibly driven by scientific arrogance. I believe this presumed dichotomy is often in the minds of the inexperienced or untrained observers who pronounce such judgments, rather than based on any reality.

It is common experience to many of us biologists who have put some time in the field that interpretation of a field observation made by a good wildlife scientist and a native expert usually converge. The problem of dichotomous perception on what causes a phenomenon observed in the field usually emerges when the knowledge of the arrogant 'outsider' is solely derived from books and/or when assertion of the wise 'native' is driven by an externally tutored response to justify a current resource use practise. How many people know that the famous speech of the Red Indian Chief, which extols the traditional harmony between Amerindians and Mother Earth, was actually drafted by a Seattle lawyer? Time and again, over the last two decades, when I tested my perception of the causes underlying negative factors affecting wildlife in field situations against those of an experienced but uninvolved local villager or tribesman, we usually agreed.

During a recently concluded study around Nagarahole (Madhusudan and Karanth in Robinson and Bennet, in press) our line transect survey data showed substantial differences in densities of large ungulates between two sites. One site was rigidly protected from local hunters while the other was less well protected. Our perception that this difference was largely due to greater access and less rigid enforcement of anti-hunting measures in the second site was soon confirmed by a questionnaire survey of several local informants including neutral villagers and even illegal hunters themselves. When questioned in a non-ideological context, unrelated to their immediate needs, both neutral observers and the poachers themselves agreed that local hunting was the most important cause of depressed animal densities. Yet most of the literature pertaining to the issue of wildlife conservation generated by social activists in India, fails to point out this widely prevalent ground reality.

While rigorous scientific studies may be needed to 'prove' such hypotheses in a peer-reviewed ecological journal, the driving forces behind wildlife declines over most parts of India are *usually* apparent to any knowledgeable, experienced observer. To perceive basic reasons for such declines, all that the observer has to do is, at least for a moment, deconstruct intellectual predilections such as local communities do not damage their environments or 'Indian foresters scientifically harvest timber'.

Let us agree for a moment that it is socially desirable to save a small subset of biodiversity, which I have defined as wildlife earlier. Let us be aware that this goal is achievable for a few wildlife species through establishment of gene-banks or zoos, with practically no attendant costs in terms of social conflicts, which we are facing now. In fact, wildlife authorities in many Southeast Asian countries, in practise, basically focus their entire energies on such *ex-situ* measures, helping species like the Sumatran rhino to head towards extinction. The power of the international captive-breeding lobby among Third World wildlife managers stems from its magical ability to project glitzy techno-wizardry of *ex-situ* solutions as a substitute for the messy and hated job of on-ground protective tasks.

However, let us say we really want to save wildlife species *in situ*, with all ecological relationships and processes which sustain their populations, biological communities and landscapes which harbour them. If we so decide, there is no escape from having protected areas, which are relatively free from commercial market pressures, and excessive human induced disturbance. The future challenge lies in setting up regulatory mechanisms, which can realistically work to reduce pressures and disturbances.

I am not talking here of fine tuned wildlife management measures, such as deliberate setting-back of forest succession, use of fire to promote grazing by ungulates or establishing connectivity through new corridors. I am talking of the more basic need to regulate removal of plant biomass and hunting, which often leads to conflicts between managing authorities and local communities. If we are to have relatively inviolate wild life protected areas in the future, with acceptable levels of social conflict around them (assuming for the moment that conflict-free society is still a distant dream), the following components may become necessary ingredients of any wild life conservation policy.

At present the designated protected areas which are relatively intact and have *de facto* protection probably comprise less than 1 per cent of India's area (it would exclude major national parks such as Namdapha, Nagarjuna Sagar and Manas from my category of *de facto* protected areas). Clearly, this minute fraction has to be maintained and, if possible, increased. Where and how this should be done—and can be done—are issues open to debate. But the process has to be based on site-specific data. It is clear that wherever such protected areas survive, they will be accessible to an equally small fraction of the local communities of India. Most local communities (urban, semi-urban and rural) in India, who even now manage to survive without such proximity to protected areas, will continue do so in the future.

Although ideally the state and its regulator arms should ultimately wither away and all their social policing functions replaced by incentive driven or voluntary changes in human behaviour, at this point in our history, the use of force to protect wildlife has to be acknowledged as a fundamental need. How best to minimize this need and how best we can combine it with the practise of participatory democracy and upholding of human rights, are still evolving issues.

Even in the few successful examples of extractive joint forest management such as those of West Bengal, there exists a local authority, which punishes those who break the law. In extreme situations such as that of Kaziranga, which is a 500 sq km natural enclave under immense threat from hardcore rhino poachers, a protective force of 200 men patrolling round the clock is needed. Whether such protective forces work better under a localized authority or under the government's control is an issue, which needs to be explored in a site-specific context.

We also need to recognize the fact that if only local needs of bio-mass not driven by market forces are considered, management of outside landscapes to cater to just these needs can occur under state sponsorship (social forestry) or NGO sponsored technocratic interventions (Ranthambhore Foundation in India, WWF–USA in Nepal). On a larger scale, such landscape management, which is complementary to wildlife conservation goals, can occur only under the force of economics. On the eastern boundaries of Nagarahole National Park, annual crops are grown. There is heavy pressure on the forests to graze livestock, which produce draft power and dung, as well as for collecting firewood, small timber and bamboo for local use and sale in markets. Regulatory measures here come into conflict with a relatively large fraction of the local population engaged in wood cutting and grazing. Regulation turns into a 'law and order' problem, a common scene around many protected areas of India.

On the western boundaries of Nagarahole, where the land is under coffee cultivation, the reduced dependence on livestock, production of woody biomass in the plantations and almost guaranteed conditions of full employment at relatively high wages have combined to virtually eliminate such biomass pressures on forests. The regulatory application of force in this case becomes easier, and locally acceptable, because it is specifically targeted only at a small fraction of the local population involved in illegal hunting or smuggling of valuable timber or non-timber forest products. This is somewhat similar to the situation, which prevails around Kaziranga where most genuine local needs of substance biomass and employment are met from productive agriculture and tea plantations. The protective force therefore concentrates almost exclusively on dealing with illegal hunters.

Conservation options are, in the final analysis, political options. Scientific, moral, aesthetic and pragmatic arguments have been advanced to show that sacrificing the remaining 3 per cent or so area under wildlife reserves is unlikely to make any dent on human problems, which we have been unable to solve by using and abusing the remaining 97 per cent of the land area. Do we just assume that political representatives of local communities do not understand this?

Conservationists need to use all their persuasive powers and arguments to convince them. These are powerful arguments ranging all the way from traditional tolerance of rights of other creatures to the twenty-first century needs of biotechnology for human welfare. Coupled with proper economic incentives and deployment of an adequate deterrent force, sometimes we may even succeed.

Endangered wildlife species and their habitats are coming under increasing pressure from a variety of social forces: consumer, urban and industrial interests who benefit from 'mining' the wildlife habitats for minerals, water, power, forest products, and even recreational use; Local communities whose hunger for the land, employment and forest products keeps increasing; An articulate, humanitarian, urban middle class constantly fuelling its own numerical growth in both the government and NGO sectors. The path of least resistance against all these pressures has converged on wildlife protected areas in the form of a conservation model which advocates market driven resource exploitation linked to increased local people's access into wildlife protected areas. This model is envisaged to work under the oversight of a massive developmental bureaucracy which administers the protected areas and surrounding landscapes, without any use of force.

Somehow we seem to hope that natural animal communities and their intricate interaction have the evolutionary adaptations and resilience to survive this onslaught of market-driven humanism. I believe there is neither empirical evidence nor ecological data justifying the acceptance of this user-friendly model of wildlife conservation. Clearly the paradigm of sustainable use everywhere has outlived its practical utility, as did the earlier one of state controlled authoritarian preservationism. I believe that the alternative concept of 'sustainable landscapes' advocated by Robinson (1993), in combination with the ideas of the emerging discipline of ecological economics, may provide useful tools for protecting the sacred groves in which our wildlife has to survive into the twenty-first century.

## References


Madhusudan, M.D., and Ullas K. Karanth, 2002, 'Local Hunting and the Conservation of Large Mammals in India', *Ambio*, 31: 48–54.

Karanth, K. Ullas, and B.M. Stith, 1999, 'Prey Depletion as a Critical Determinant of Tiger Population Viability', in J. Seidensticker, S. Christie and P. Jackson (eds), *Riding the Tiger: Tiger Conservation in Human Dominated Landscapes,* Cambridge University Press, Cambridge, UK.

Kramer, R.C. van Shaik and J. Johnson, 1998, *Last Stand: Protected Areas and the Defense of Tropical Biodiversity*, OUP, Oxford, UK.

Robinson, J.G., 1993, 'Limits to caring: Sustainable living and the loss of biodiversity', *Conservation Biology* 7:20–28.

———., and E.L. Bennett (eds), 1998, *Hunting for Sustainability in the Tropical Forests,* Columbia University Press, New York, USA.

# Section IV

# Movements and Alternatives

*Photo credit Pallava Bagla*

# Introduction

Over a decade ago, two leading scholars of environmental movements in India stepped back to suggest a vision that was more holistic in terms of alternatives. A movement, they argued, needed, 'a vision of what it should be fighting against; and what is perhaps more important still, with a vision of what it should be fighting for. A truer picture then of the India that is and the India that should be' (Gadgil and Guha 1995: 6).

In July 2005, the great coastal city of Mumbai (formerly Bombay) not only experienced unprecedented high rainfall, it also witnessed the havoc of floods for days. The transformation of flood plain and riverside into a built environment exposed many to the threat of floods. Revi's account of the Mithi river and its environs holds clues to something very fundamental about unwise land use along river banks (see Chapter 26). While the focus in urban India is often on large metropolises, small towns are no exception to the deterioration in air and water quality.[1]

A range of citizen initiatives are labeled under the omnibus category of 'environmental movements'. Some gained fame due to their sheer determination and their ability to capture the popular imagination. Chief among these was the Chipko Andolan, literally Hug the Trees Movement, spearheaded by village women in the Uttarakhand Himalaya. Resistance to contractors felling timber trees soon broadened into a wider quest for local control over the forest resource (Guha 2004: 24–33). Along the coastline, trawler fishing, promoted with generous subsidies by government soon became a menace to fish spawn and marine ecologies, while threatening the livelihood of fisher folk. As with the forest-based protests, these coastal struggles defy any easy label. Ajantha Subramanian shows a complex tapestry of ecological concerns and changing ideas of community on the Coromandel Coast (see Chapter 25).

Many groups combine protest with positive action. In Maharashtra (western India), water conservation and harvesting took place through mobilizing the landless poor and the cultivators (Chapter 29). Such ideas gained considerable currency in the context of recurrent droughts, with water storage via civic action being a major focus for a whole cluster of initiatives at the local and regional level. In contrast to large engineering projects, these required little by way of capital. Lower-caste groups and women, who are among those with the least access to resources, have a central role in decision making. Environmental repair is part of the process of their empowerment (Agarwal and Narain 1997). To multiply such instances across rural and urban India will be a major challenge of our times (Agarwal and Narain 1989).

Larger issues of justice have often played a role in protest over enclosure of commons vital for sustenance, in the protests of fisher folk against trawler fishing and of usufruct right holders for access to government forests. What is significant is the articulation of larger issues of sustainable use in the long term, taking the issue beyond one of survival and dignity in the present to alternative ways of planning for the future.

This is a complex task for any society, all the more so in India where one person's idea of an ideal landscape can be another's nightmare. Conservation projects aimed to preserve a range of life forms by minimizing human intrusion have extracted heavy costs in an unequal manner from those who reside in and near such nature reserves. There is a need for regulation and for classifying resources to ensure larger natural systems remain intact. But as Savyasaachi shows in an elegant little study, *Adivasi* ideas on how to classify a forest may be markedly different from those of foresters and biologists. The honeybee may hold clues to the forest that can illumine even those who track the tiger. Without idealizing or rendering romantic such practises, it is possible to assimilate the best of them (see Chapter 27; also see Savyasachi 2005; Barucha 2005).

In a very different vein, the issues of pollution from the Mathura refinery or foundries and more recently, the threat of large-scale malls near the Taj Mahal exemplify a similar dilemma. Science-based ideas for action still have to contend with *realpolitik*. The ideal scene would incorporate a transparent bureaucracy sensitive to new insight and a political leadership aware of its immense responsibility to a heritage, whether natural or human made. Both hinge on an aware citizenry and a responsible media. In the latter respect, few have a record to match Darryl D'Monte's. His reflections on the Taj case shed light on much more than the monument that for many is a symbol of our shared past (see Chapter 23; see also, D' Monte 1985).

In recent years, displacement due to large projects has moved to the centre of controversy even more than several other issues. If in the 1970s, in the storm over the Silent Valley, Kerala, protest stalled a dam that would submerge and destroy a species-rich rain forest, whereas over the last decade and more, the dams on the Narmada have divided those who oppose from those who support them. There are differences even over how much the series of dams will deliver in terms of water and power. Even more so, who should determine how many will be displaced and how these lakhs of persons will fare has drawn the attention of people across India and over world. The valley of the Narmada is but one instance of such a controversy, of displacement induced by development. Sanjay Sangvai draws on his own extensive experience of the movements against the dams to provide a participant's view.[2] As with much else, there are deep divisions on who will gain and how much, and how losers' concerns should be addressed. Even supporters of such projects agree that India's record of resettlement leaves much to be desired.[3]

India's twenty-first century will centre to a large extent on such choices and how these are made. A deep awareness of the different points of view is essential.

One person who was at the forefront of weaving together disparate strands into a mosaic of alternative visions, Anil Agarwal died tragically young in 2002. In an interview a little earlier, he spelt out precautionary principles for all concerned with the environment in an era of rapid changes. Science and its insights ought to be absorbed quickly. Yet, he favoured 'an active distrust of large technologies'. This was to be combined with a basic principle of proactive approaches to all emergent issues of the environment. No one ought to wait for a crisis to take shape, he warned. They ought to plan and act well in advance.

Anil Agarwal's words require careful reflection. Large technical reponses were not ruled out but were best not seen as a panacea or cure-all. As is evident, this calls for a state of mind that is both sensitive and receptive. Whether it will be heeded is up to the generation that will make, or unmake our shared future.[4]

## Notes

1. On small towns, see CSE (1994–95: 24–26). Also see, Mahadevia (2001). I am grateful to Professors Neema Kudva and Mary Norman Woods of Cornell University for help with these references.
2. Sangvai (1999) can be read with a recent account of dam related issues, Khagram (2005). On experiences of the displaced, see Sainath (1996: 69–132).
3. The pro-dams case is made by veteran journalist Verghese (1994); For another view, see D'Souza (2002).
4. Anil Agarwal, 'Interview', *Seminar,* 486 (2000), pp. 73–77.

## References


(CSE) Centre for Science and Environment, 1994–95, 'Probing for change', *Director's Report,* CSE, Delhi, pp. 24–26.

Agarwal, Anil and Sunita Narain, 1989, *Towards Green Villages*, CSE, Delhi.

———, 1997, *Dying Wisdom*, CSE, Delhi.

Bharucha, Rustam, 2005, *An Oral History of Rajasthan, Conversations with Komal Kothari,* Penguin, Delhi.

D'Monte, Darryl, 1985, *Temples or Tombs? Industry versus Environment, Three Controversies*, CSE, Delhi.

D'Souza, Dilip, 2002, *The Narmada Dammed: An Inquiry into the Politics of Development,* Penguin, Delhi.

Gadgil, Madhav and Ramachandra Guha, 1995, *Ecology and Equity*, Penguin, Delhi.

Guha, Ramachandra, 2004, 'A Gandhian in Garwhal: Chandi Prasad Bhatt' in *The Last Liberal,* Permanent Black, Delhi.

Khagram, Sanjeev, 2005, *Dams and Development*, OUP, Delhi.

Mahadevia, Darshini, 2001, 'Sustainable urban development in India: an inclusive perspective,' *Development in practise*, Vol. 11, pp. 2–31.

Sainath, P., 1996, *Everybody Loves a Good Drought: Stories from India's Poorest Districts*, Penguin, Delhi.

Savyasaachi, 2005, *Between Earth and Sky: An Anthology of Forest Writings*, Penguin, Delhi.

Verghese, B.G. , 1994, *Winning the Future, From Bhakra to Narmada, Tehri, Rajasthan Canal*, Konarak, New Delhi.

# Ecological Conflicts and the Environmental Movement in India

Madhav Gadgil and Ramachandra Guha

Nature-based conflicts have increased in frequency and intensity in India. They revolve around competing claims over forests, land, water and fisheries, and have generated a new movement struggling for the rights of victims of ecological degradation. The environmental movement has added a new dimension to Indian democracy and civil society. It also poses an ideological challenge to the dominant notions of the meaning, content and patterns of development.

## The Sites of Struggle

### Introduction

As the centre of power and patronage, the Indian city of New Delhi is the venue of year round demonstrations by organizations representing different classes, castes and ethnic groups. Farmers demanding the provision of subsidized power and fertilizer, industrial workers campaigning for higher pay, and ethnic minorities fighting for a separate state all recognize the symbolic significance of a show of strength in the national capital. Assured of widespread coverage by the print media, these demonstrations are often held at the Boat Club lawns, a stone's throw from the houses of Parliament and the government secretariat.

May 1990 saw a series of events unprecedented even in New Delhi: a demonstration followed, within a week, by a counter demonstration. First,

---

* Reprinted with permission from Madhav Gadgil and Ramachandra Guha, 1995, *Ecology and Equity: The Use and Abuse of Nature in Contemporary India*, Penguin, Delhi.

villagers to be displaced by the massive Sardar Sarovar dam, being built on the Narmada river in Central India, assembled in a peaceful *dharma* (sit-down strike) on Gol Methi Chowk in the heart of New Delhi, and very close to the residence of the then Prime Minister, V.P. Singh. Consisting mostly of poor peasants and tribals, the demonstration lasted for several days, with singing, dancing and exhortative speeches by the protest leaders. Most of the demonstrators had come from Madhya Pradesh, the state containing a majority of the villages to be submerged by the dam. They dispersed only after the Prime Minister met a delegation of the protesters, and assured them that the Sardar Sarovar project would be reviewed. Immediately, politicians in Gujarat, the state that stands to benefit most from the project, set about organizing a counter demonstration. After a public meeting at the Boat Club, the Gujarat protesters themselves went to meet V.P. Singh. The Prime Minister granted them an audience immediately (he had kept the Madhya Pradesh peasants waiting for days) and told them what they wanted most to hear—that he and his government were fully committed to the implementation of the Sardar Sarovar project.

A few months later, the two opposing groups were involved in a face to face encounter hundreds of miles from New Delhi, on the Madhya Pradesh–Gujarat border. On 25 December 1990, the 'Narmada Bachao Andolan' (Save Narmada Movement), an organization working among the potential outsets of the dam, began a 250 km march from Rajghat in Madhya Pradesh to Kevadia colony, the site in Gujarat of the Sardar Sarovar dam. The marchers, several thousand in all, were stopped by the Gujarat police at the border village of Ferula, and prevented from entering the state. On the Gujarat side, a large group, including students and plainclothes policemen, had assembled to heckle the marchers. A stalemate lasting several days ensued, with the pro-dam agitations shouting slogans in favour of the dam and against the Narmada Bachao Andolan and one of its leaders, the respected Gandhian Baba Amte. For their part, the protesters insisted on their right to march peacefully to the dam site at Kevadia.

On the second day of the New Year, a group of 25 protesters, with their hands tied to emphasize the non-violent nature of their struggle, entered Gujarat, only to be stopped by the police 150 m inside the state. Two more groups, again with their hands tied, joined them the next day. On 5 January, Baba Amte and another group of 25 also entered Gujarat. After being allowed to cross the border but not proceed further, they began an indefinite *dharna* (sit-down strike) on the Gordah river bridge, barely 30 m inside

Gujarat. The next day, a group of anti-dam activists including Medha Patkar, perhaps the movement's most important leader, went on a hunger strike on the Madhya Pradesh side of the border. With the Gujarat government unrelenting, the stalemate continued for several weeks, until on 28 January, with their lives in danger, Patkar and her associates were persuaded to give up their fast (Anon. 1991).

The Narmada controversy is just one, especially charged, example of a wide spectrum of social conflicts over natural resources in contemporary India. Competing claims over water and forests, in particular, are now a visible presence on the social landscape. They arise, typically, when one group of resource users—for example, industry or commercial farmers—is seen as violating (often with the aid of the state) a prior claim of another set of resource users—for example, subsistence peasants or tribals. With the resources in question becoming increasingly scarce owing to environmental degradation, these conflicts seem certain to intensify.

Social conflicts over nature and natural resources add a third category to the two generic forms of conflict widely studied by social scientists—those over cultivated land and its produce, and those within the factory. Struggles between landlords and agricultural labourers/sharecroppers over wages and the disposal of produce, or between peasants and the state over taxes and prices, have been closely studied for decades, as have conflicts in the industrial sector, whether between capitalists and workers, or between industrial enterprises and the state. By contrast, nature-based conflicts are as yet hardly documented and very poorly understood, both within the social science community and outside. Like conflicts over land and in the workplace, conflicts over natural resources typically pit against each other two unequal antagonists. To return to the example of the demonstrations in New Delhi in May 1990, the transport and food needs of the opponents of the Sardar Sarovar dam were met by modest voluntary contributions from numerous individuals and organizations. By contrast, the defenders of the dam came all the way from Gujarat (a distance of more than 800 km) in buses owned by the Gujarat government, their transport and living expenses fully subsidized by the state and its ruling political party. When the same protagonists squared off at the Ferkuva border over the New Year, the Madhya Pradesh tribals were staying in tents at the height of winter, cooking by open fire, while the Gujarati supporters of the dam were ensconced in schools and other public buildings, again well looked after by their state government.

In contemporary India, conflicts over nature, just as much as the more conventional agrarian and industrial conflicts, raise important questions, about distributive justice and economic efficiency. The distinguishing feature of this third generic form of socio-economic conflict is that it simultaneously raises issues of environmental sustainability. In so far as the natural resources in question are also vital to the agrarian and industrial sectors, the fate of these conflicts is intimately connected to the development process as a whole.

The first part of this paper provides a broad based survey and analysis of natural resource conflicts in contemporary India: we shall demonstrate that nature-based conflicts lie at the heart of the Indian environment debate. The second part investigates the vocabularies of protest characteristic of the Indian environmental movement, and its ideological expressions. The paper ends with a brief comparison between First World and Third World environmentalism.

## Forests: For Whom and for What?

In the last decade of this century, water-based conflicts—of which the Narmada controversy is at the moment the most contentious—are likely to dominate the environment debate in India. For much of the past 20 years, however, conflicts over forest resources were more visible, and perhaps more widespread. Indeed, the origins of the Indian environmental movement can be fairly ascribed to that most celebrated of forest conflicts involving the Chipko movement of the Central Himalaya. In April 1973, the peasants of Mandal, an interior village in the Garhwal Himalaya, effectively thwarted commercial felling in a nearby forest by threatening to 'hug the trees'. This brought to the fore a simmering but widespread resentment among the hill peasantry, directed at state forest policies which had consistently favoured outside commercial interests at the expense of their own subsistence needs for fuel, fodder and small timber. Thus the 'Chipko' (Hug the Trees) movement was born. In the following decade, a wave of protests against commercial logging swept the Himalayan foothills, coordinated by Gandhian as well as left wing activists. Notably, the region had a long history of peasant protest which Chipko both drew upon and furthered. As a powerful statement against the violation of customary rights by state forestry, Chipko brought into sharp focus a wide range of issues concerning forest policy and the environment debate as a whole (Guha 1989a).

Because of its novel techniques and Gandhian associations, the Chipko movement rapidly acquired fame. Yet it was representative of a far wider spectrum of forest-based conflicts. In the tribal areas of Central India, economic dependence on the forests is possibly even more acute than in the Himalayan foothills where Chipko originated. Here the 1970s witnessed escalating conflict between villagers and the forest administration in tribal districts of the states of Bihar, Orissa, Madhya Pradesh, Maharashtra and Andhra Pradesh. In tribal India, moreover, forest conflicts often have a sharper political edge. Thus in Bihar, they have been an integral element in the popular movement for a tribal homeland, while in the four other states mentioned, the question of tribal forest rights has been actively taken up by revolutionary Maoist groups (Calman 1985; PUDR 1982; Sengupta 1982).

Academic research inspired by the forest conflicts of the 1970s also revealed their long lineage. Indeed, local opposition to commercial forestry dates from the earliest days of state intervention. Before the inception of the Indian Forest Department in 1864, there was, by and large, little state intervention in the management of forest areas, which were left in the control of local communities. The takeover of large areas of forest by the colonial state this constituted an important watershed in many ways: a *political* watershed, in that it represented an enormous expansion of the powers of the state, and a corresponding diminution of the rights of village communities; a *social* watershed, in that by curbing local access it radically altered traditional patterns of resource use; and an *ecological* watershed, in so that the emergence of timber as an important commodity was to fundamentally alter forest ecology (Gadgil and Guha 1992: ch. 5 and 6).

The imperatives of colonial forestry were largely commercial. From the point of view of this analysis, its most significant consequence was the intensification of social conflict between the state and its subjects. Almost everywhere, and for long periods of time, the takeover of the forest was bitterly resisted by local populations for whom it represented an unacceptable infringement of their traditional rights of access and use. Hunter-gatherers, shifting cultivators, peasants, pastoral nomads, artisans—for all these social groups free access to forest produce was vital for economic survival, and they protested in various ways at the imposition of state control. Apart from forest laws, new restrictions on *shikar* for local populations (while allowing freer hunting for sport by the British and the Indian elite) were another contributory factor in fuelling social conflict (Rangarajan 1992).

Throughout the colonial period, popular resistance to state forestry was remarkably widespread and sustained. In 1913, a government committee in the Madras Presidency was struck by the hostility towards the forest department, which was the most reviled government agency along with the Salt Department (likewise concerned with a commodity ostensibly low in value but of inestimable worth to every village household). Two thousand miles to the north, in the Garhwal Himalaya, a British official wrote at almost the same time that 'forest administration consists for the most part in a running fight with the villagers' (quoted in Guha 1989a: 105). Popular resistance to state forestry embraced forms of protest that minimized the element of confrontation with authority, such as covert breaches of the forest law, as well as organized rebellions that challenged the right of the state to own and manage forest areas (Gadgil and Guha 1992: see especially ch. 6 and 8).

Ironically, in the post-independence period the process only accelerated. Economic development implied more intensive resource use that, in the prevailing technological and institutional framework, led inevitably to widespread environmental degradation. In the forestry sector, the industrial orientation became more marked, exemplified by the massive monocultural plantations begun in the early 1960s, while other development projects like dams and mines exerted a largely negative influence on the forests. Not surprisingly, the conflicts between the state and its citizens have persisted, and the forest department continues to be a largely unwelcome presence in the Indian countryside. However, forest conflicts in independent India have differed in one important respect from conflicts in the colonial period. The earlier conflicts emerged out of the competing claims of state and people over a relatively abundant resource; now these conflicts are played out against the backdrop of a rapidly dwindling forest resource base. In other words, a newer *ecological* dimension has been added to the moral/political/economic dimensions of social conflicts over forests and wildlife.

Cumulatively, these processes have worked to further marginalise poor peasants and tribals, the social groups most heavily dependent on forest resources for their subsistence and survival. A long-time student of Indian tribals poignantly captured their frustration with state forestry:

> The reservation of vast tracts of forests, inevitable as it was, was ... a very serious blow to the tribesman. He was forbidden to practise his traditional methods of (Swidden) cultivation. He was ordered to remain in one village

> and not to wander from place to place. When he had cattle he was kept in a state of continual anxiety for fear they would stray over the boundary and render him liable to what were for him heavy fines. If he was a forest villager he became liable at any moment to be called to work for the Forest Department. If he lived elsewhere he was forced to obtain a license for almost every kind of forest produce. At every turn the forestry laws cut across his life, limiting, frustrating, destroying his self-confidence. During the year 1933–34 there were 27,000 forest offences registered in the Central Provinces and Berar and probably ten times as many unwhipped of justice. It is obvious that so great a number of offences would not occur unless the forest regulations ran counter to the fundamental needs and sentiments of the tribesmen. A Forest Officer once said to me: 'Our laws are of such a kind that every villager breaks one forest law every day of his life' (Elwin 1964: 115).

Popular movements in defiance of customary rights have focused on two issues central to the direction of forest management. First, they have contended that the control of woodland must revert to communal hands, with the state gradually withdrawing from ownership and management. Second, those opposing forest management have pointed to the contrast between the subsistence orientation of villagers and the commercial orientation of the state. This contrast can be illustrated by two strikingly similar incidents, separated in time by a few months and in space by some 2000 miles. The first took place in Kusnur village in the Dharwad district of the southern state of Karnataka. Protesting against the allotment by the state of village pasture land to a poly fiber industry which intended to grow eucalyptus on it, the peasants of Kusnur and surrounding villages organized a 'Pluck-and-Plant' *satyagraha* demonstration on 14 November 1987, when they symbolically plucked a hundred eucalyptus saplings and replaced them with useful local species. Less than a year later, and probably without knowledge of the Kusnur precedent, Chipko activists in the northern state of Himachal Pradesh were arrested on charges of causing 'damage to public property'. Their 'crime' had been to lead villagers in uprooting 7000 eucalyptus saplings from a forest nursery in Chamba district, planting indigenous broad-leaved species in their stead. The Dharwad and Chamba episodes vividly illustrate the continuing cleavages between village interests and the commercial bias of state forestry (Kanvalli 1991; Modi 1988).

The clash between subsistence agriculturists and industry over the usufruct of state lands is only the most visible of forest conflicts. Localized opposition has also arisen amongst village artisans facing increasing difficulty in obtaining raw material from forest areas. Typically, the state has diverted to industrial enterprises, resources previously used for generations by artisans. Thus reed workers in Kerala, bamboo workers in Karnataka, and rope makers using wild grass in the Sivalik hills of Uttar Pradesh have all resisted the Forest Department's plans to give preferential treatment to the paper industry in the supply of biomass from forests owned by the state.

In most areas, forest-dependent artisans have yet to be politically organized. That is no longer the case with millions of tribals in Central India for whom the collection and sale of 'minor' (i.e., non-wood) forest produce is vital to survival. For decades, tribals collecting non-wood forest produce have been severely exploited by merchants who control the trade. For these merchants, the most lucrative of all 'minor' forest produce is the *tendu* leaf, used in making the *bidi* or Indian cheroot. Over the last two decades, social activists have organized *tendu* leaf pickers in a bid to increase their collection wages. On the eve of the 1991 plucking season, 24 organizations working among tribals in five contiguous states of central India announced that they had fixed the price of *tendu* leaves at Rs 50 per 5000 leaves (the merchants' acquisition rates varied from Rs 9 per 5000 leaves in Bihar to Rs 25 in Maharashtra). In several areas, tribal forest labourers have been organized by left wing revolutionaries, leaving the alarmed traders to seek the protection of the state. Sadly but perhaps inevitably, violence has escalated in the tribal forest districts of Madhya Pradesh, Maharashtra and Orissa (PUCL 1985; *The Statesman,* 1991).

These varied protests against state forestry coalesced in the coordinated opposition to the Draft Forest Bill of 1982, an act that sought to greatly strengthen the punitive powers of the Forest Department. Several dozen grassroots organizations lobbied hard against the proposed legislation, which the government finally withdrew (Fernandes and Kulkarni 1983; PUDR 1982). Popular opposition has also forced some notable changes in forest policy, such as the abandonment of programmes for clear felling natural forests to replace them with plantations of industrially useful exotic species. These modest successes, along with the eventual loss of interest in any single issue that is a characteristic of democratic politics, has led to an attenuation—if not on the ground at least in the public imagination—of forest based conflicts in recent years.

## Dams and the Damned

In the Indian environment debate, the space vacated by forests has been quickly filled by major dams. A small but revealing indication of this shift is contained in the dedications of the first two citizens' reports on the state of India's environment (CSE, 1982 and 1985). While the first was dedicated to the 'Women of Chamoli' who were amongst the originators of the Chipko movement, the second was dedicated simply to the 'Dam-displaced people of India'. Through the 1980s and beyond, different river valley projects—from Tehri in the north to Silent Valley in the south, Koel Karo in the east to Sardar Sarovar in the west—have been the subject of bitter controversy. The critics of multipurpose river valley projects have operated on several flanks. From an economic perspective, they have argued that the cost-benefit ratios derived by the government to justify various dams invariably overvalue benefits and undervalue costs. Using official data, they have also shown that situation rates have usually been much higher than anticipated, thereby shortening the life of reservoirs. From an ecological perspective, the high incidence of water logging and the wholesale submergence of forests and wildlife have been presented as examples of the unacceptable costs of dam building. The construction of large dams has also been shown to seriously disrupt fish life and to assist the spread of water borne diseases (CSE 1985; Kalpavriksh 1988; Paranjpye 1989; Sharma and Sharma 1981; and, for a global survey and critique, Goldsmith and Hildyard 1984).

These economic and environmental criticisms have considerable force and yet it is the *social* implications of dam construction that have evoked a major popular response. It has been estimated that in the last three decades, more than 11.5 million people have been displaced by development projects in India without being properly rehabilitated—and it is indisputable that major dams have been the major contributor to this process of forcibly uprooting people from their traditional homes (Fernandes and Ganguly-Thukral 1988). With evidence steadily accumulating of the deprivation—cultural and psychological as well as economic—suffered by the displaced communities of past projects, new dams have been increasingly opposed by populations anticipating such dislocation. Movements representing dam-displaced people have gathered force in the last 20 years: we shall come to these contemporary protests presently, but we must first note one important, though as yet little known, precursor. Known as the *Mulshi Satyagraha,* this was the opposition to a dam being built near Bombay by the flourishing industrial house of the Tatas. This episode is virtually unknown to Indian environmentalists, but in view of the remarkable parallels between the Mulshi Satyagraha and ongoing protest against large dams, its history is worth recording at some length.

In the years following World War I, the Tatas had in fact planned an ambitious series of dams on the Sahyadri hills, chiefly to supply power to the rising industrial city of Bombay. When the first dam was built near the hill station of Lonavala, the farmers whose lands were submerged were paid no compensation whatsoever. When the Tatas came to Mulshi for the next phase of the project, however, they ran into trouble. At first, the company moved on to the farmers' lands and began their test trenches without any legal formalities. But Mulshi was very close to Pune (Poona), then an epicenter of the Indian freedom movement. So when a peasant objected to a trench being dug in his field and a British engineer threatened him with a pistol, there were strong protests in Pune. The ensuing opposition to the dam, led by a young Congressman called Senapati Bapat, succeeded in halting construction of the dam for a year. The Bombay government then promulgated an ordinance whereby the Tatas could acquire land on payment of compensation. This caused the resistance to the dam to split into two factions: while the Brahmin landlords of Pune, who owned much of the land in the Mulshi valley, were eager to accept compensation, the tenants and their leader, Senapati Bapat, were totally opposed to the dam project. With the landlords, the power company and the state all ranged against them, there was little the peasants could do, and the movement collapsed in its third year. Tragically, the compensation was pocketed by the landlords, and the actual tillers of the soil were left high and dry. None the less, the movement had at least succeeded in forcing the Tatas to provide reasonable, negotiated compensation for the submerged lands, one consequence of which was that they did not proceed with the other hydroelectric projects they had intended for the Sahyadris.

When the Mulshi Satyagraha broke out, the British District Collector had toured the area, extolling the virtues of the dam. He remarked that the electricity produced by it would light up the latrines of the Bombay *chawls,* the dwelling homes of the city's industrial workers. This drew the sharp retort that the government and the Tatas sought to extinguish wick lamps in thousands of rural homes in order to light up the latrines of Bombay (Bhuskute 1968).

This exchange, apocryphal as it might be, could just as well have taken place in 1990—in either Ferkuva or New Delhi—between proponents and opponents of the Sardar Sarovar dam. In fact, when the Narmada controversy was at its height, the *Times of India,* whether by accident or design, reproduced in its archive section a report on the Mulshi Satyagraha, dated 2 May 1921. Here the paper's correspondent had succinctly represented the main objections to the Tata project, as well as its most powerful justifications. The origins of the Mulshi satyagraha, he concluded, lay in:

1. A strong sense of wrong and deep feeling of resentment among the peasantry whose lands are affected by the project, against the Government for sanctioning the scheme more than two years ago, without taking them in its confidence, i.e., without consent, knowledge or consultation of the peasant-owners of the land...
2. Suspicion and distrust in both the Government and the Company, due chiefly to the Procedure of acquisition, as to the confides of their intentions to award full compensation, or equivalent...land somewhere else, and other facilities already enjoyed by them or necessary for fresh colonization...
3. Reluctance to part with the land on account of its extreme productivity, the natural facilities of irrigation and nominal amount of land revenue.
4. Reluctance to part with lands, ancestral homes, and traditional places of worship and see them submerged under water.
5. Natural reluctance in this class of peasantry to emigrate from one place to another...

For the other side, the main claims of the project promoters were listed:

1. One and a half lakh (150,000) electrical horse-power would be created by the Mulshi Peta dam
2. It would save 525,000 tons (of) coal every year. This quantity of coal at the present rate costs Rs 183,00,000.
3. The saving of coal means a corresponding saving of Rs 1,05,50,000 worth of fuel to the mill industry of Bombay.
4. The quantity of coal saved on account of the scheme would require 26,250 wagons for transport. These would be saved and utilized for other public purposes.

5. Water once used can be directed for agricultural purposes after electrical power is created.
6. Electricity thus created would give work to 300,000 labourers. If is utilized for cotton mils, every day 51 lakh yards would be manufactured.
7. The projected electrification of the Bombay suburban railway lines would give to Bombay city much faster and more frequent trains, thus enabling the development of housing schemes in purer air and healthier circumstances. (*The Times of India,* 2 May 1921)

Here lies an uncanny anticipation of the ideological roots of the conflicts over large dams that were to erupt half a century later. On the one side, the interests of subsistence oriented peasants, on the other, the interests of urban centres and industry. When the major push towards river valley projects took place after Indian independence, it was easy to represent the former as static and backward, the latter as dynamic, forward looking and coterminous with the national goals of progress and development. The villages to be submerged by the new projects were then expected to make way for the greater national interest, all the more so as the new schemes (unlike Mulshi) were owned and executed not by private capitalists but by the state, itself the legatee of a broad-based, popular national movement.

Of course, displaced people were not entirely unyielding. A comprehensive but somewhat euphoric survey by the political scientist Henry Hart of the first wave of large dams built in independent India, noted the resentment of villagers confronted with the prospect of displacement. Thus in 1953, the residents of the town of Narayan Deva Keri, in present day Andhra Pradesh, hoped desperately that the new reservoir on the Tungabhadra river would not fill up to capacity, thereby sparing their town. Disregarding the warnings of engineers, the townspeople stayed on till the last moment, having to be evacuated in haste when surrounded on three sides by water. Despite these signs, there was general agreement, at least amongst the devotees of darn building, that 'the suffering of the displaced people was for the good of the greatest number'; nor was there much doubt of the 'willingness of the Indian villager to make way for a nation building project, provided he is convinced that the sacrifice he is called upon to make is unavoidable' (Hart 1956).

It is true that the massive—one might, following Hart, call them heroic—river valley projects of the 1950s met with little opposition. They included the Bhakra-Nangal dam in Punjab, the Tungabhadra project in

Andhra Pradesh, the Hirakud dam in Orissa and the Rihand dam in Uttar Pradesh, each displacing tens of thousands of people. Yet, over time, the Indian villager has developed a marked unwillingness to make way for 'nation-building' projects. A major reason for this growing hostility is the actual experience of communities displaced by earlier projects. The resentment of dam evacuees has been uniform: rates of cash compensation have been very low; the promise of land for land has very rarely been fulfilled (and where it has, the new lands are invariably of much poorer quality); there are problems making a new home in unfamiliar, and often hostile, surroundings, and so on (CSE 1985; Fernandes and Ganguly-Thukral 1988; Ganguly-Thukral 1992). A significant acknowledgment of these failures has been the substitution, in recent years, of the term 'displacement' by the euphemistic 'resettlement' in public discussions of this process.

Meanwhile, organized opposition to new projects gathered force in the early 1970s, with movements emerging independently in different parts of the country. The most long-standing opposition has been to the Tehri dam, being built on the river Bhageerathi in the Garhwal Himalaya. For more than a decade, the dam's construction has been opposed by the Tehri Baandh Virodhi Sangarshan Samiti (Committee for the struggle against the Tehri Dam), a forum founded by the veteran freedom fighter, Virendra Datt Saklani. The respected Chipko leader Sunderlal Bahuguna has also been very active in the movement, undertaking several hunger fasts to pressurize the government to stop construction. The objections to the dam range from the seismic sensitivity of the fragile mountain chain (and hence the possibility of a dam burst), through the submergence of large areas of forest, agricultural land and the historic town of Tehri, to the threat to the life of the reservoir from deforestation in the river catchments (D'Monte 1981). These criticisms have gathered force since the massive earthquake in the upper Bhageerathi valley in October 1991, but the government appears resolved to go through with the dam none the less.

At the same time, the other well-known Chipko leader, Chandi Prasad Bhatt, has been leading the resistance to the building of a dam at Vishnuprayag, on the Alakananda river in eastern Garhwal. This construction is taking place very close to the famed Valley of Flowers, and fears that the ecology of the valley would be permanently disturbed are compounded by the geological features of the Vishnuprayag area, which is peculiarly prone to landslides (Bhatt 1992). At the time of writing, and due in part to such opposition, the Vishnuprayag project has been indefinitely shelved. The participation of Chipko activists in these protests is hardly accidental. Having largely lost their forests to commercial exploitation, Himalayan peasants now face further suffering owing to external pressures on the other resource in which their hills are abundant—water. As with the forests, the benefits of intensive exploitation have accrued almost exclusively to the inhabitants of the plains.

The water-rich and heavily forested tribal areas of central India have also witnessed a surge of opposition to new hydroelectric projects. Two of the more notable movements have arisen in opposition to the Koel Karo dam in Bihar, and the Bhopalpatnam-Inchampalli project on the Maharashtra–Madhya Pradesh border. In both cases, threatened tribal groups have put up spirited defenses, organizing demonstrations and work stoppages. The Koel Karo struggle has been coordinated by established left-wing political groupings such as the Jharkhand Mukti Morcha and the Communist Party of India, whilst opposition to the Bhopal-Inchampalli project has been initiated by unaffiliated voluntary organizations and inspired by the veteran Gandhian Baba Amte (CSE 1985).

Groups affected by large dams have not always been tribal, however: one successful movement was actually led by prosperous orchard owners. The Bedthi project which was under construction in the Uttara Kannada district of Karnataka had to be abandoned after it was opposed by influential spice-garden farmers, largely Brahmin, whose lands were to be submerged by the project. The Uttara Kannada farmers organized a national seminar in the project's early days, and after hectic lobbying with political leaders, forced the State Government to abandon the dam (Sharma and Sharma 1981).

Another, more striking, success was the abandonment of the Silent Valley hydroelectric project in the state of Kerala. No human community was to be displaced by this 120 KW dam, but it did involve submerging one of the last surviving patches of rain forest in peninsular India. Opposition to the project was led by the Kerala Sastra Sahitya Parishad, an organization dedicated to popular science education, which has a wide reach and influence in Kerala. This Marxist movement of school and college teachers built up an unlikely collaboration with wildlife conservationists. Each group had its own reasons for opposing the project: while the KSSP rested its case on a technoeconomic appraisal of energy generating alternatives, its allies invoked the need for plant and animal conservation. Eventually, the desire of the Prime Minister of the day, Indira Gandhi, to enhance her image among the international conservation community appears to have been critical in 'the government's decision to shelve the project' (D'Monte 1985).

There is, then, a considerable prehistory to the movement against the construction of a dam on the Narmada river. The Narmada river valley project—which the writer Claude Alvares has termed the 'world's greatest planned environmental disaster'—is a truly Utopian scheme, envisaging the construction of 30 major dams on the Narmada and its tributaries, not to speak of an additional 135 medium and 3000 minor dams (Kalpavriksh, 1988). With two of the major dams already built, the focus of popular opposition has been the Sardar Sarovar reservoir, the largest of the project's individual schemes. Sardar Sarovar is unique in the history of dam building in India, in that the command area of major beneficiaries lies in one state, Gujarat, while the major displacement (193 of the 243 villages to be submerged) will affect another state, Madhya Pradesh. According to official estimates based on the outdated 1981 census, over 100,000 people, of whom approximately 60 per cent are tribal, will be rendered homeless (V. Raina, personal communication).

As early as 1977, villagers in the Nimad region of Madhya Pradesh began protesting against the prospect of eviction due to Sardar Sarovar. Somewhat ironically, social activists like Medha Patkar (now one of the Narmada Bachao Andolan's moving spirits) first began working towards the proper rehabilitation of potential outsets: it was only after realizing that there was no land available in Madhya Pradesh/Maharashtra or Gujarat for the proclaimed 'land for land' policy that they turned to opposing the construction of the dam itself. Although more than 10 years old, the movement has really gathered momentum only since 1989. It has used a varied repertoire of protest to put forward its demands: the blockade of roads and traffic (*rasta rooks*), public meetings (including some where oustees have pledged not to leave their homes even if the dam waters rise and drown them), hunger strikes, and demonstrations, especially at state capitals. In one dramatic incident, villagers from the neighbourhood of Badwani town uprooted stone markers from the dam's submergence area, transported them several hundred miles to the state capital of Bhopal and flung them outside the Madhya Pradesh legislature (*Narmada*, 1989–90).

While localized protests have been occurring all along the Narmada valley, wider public attention has been drawn through the more spectacular events. Two of these have already been mentioned—the congregation in New Delhi and the 'Sangarsh Yatra' (struggle march) from Rajghat to Ferkuva. However, the most successful of these public events was a huge rally in the town of Harsud, held on 29 September 1989. Upwards of 60,000 volunteers, mostly of tribal and peasant background, gathered in the town, itself destined to be submerged under 50 ft. of water; representatives of citizens' groups from all over India came to demonstrate their solidarity with the Narmada movement; a large public meeting, addressed by Amte, Patkar, Bahuguna and others culminated in a collective oath to resist the pattern of 'destructive development' exemplified by the Sardar Sarovar dam (Alvares 1989).

There are several features, which help distinguish the Narmada movement from other protests against large dams. Two of the most notable are its spread—it has activist groups working in three states and many supporting organizations elsewhere—and its tenacity in the face of government repression. Although the movement itself has been, in the main, non-violent, its leaders and participants have been repeatedly harassed, and occasionally beaten and jailed. Also unlike many other movements, the Narmada Bachao Andolan has been widely, and often sympathetically, covered in the print media, while it also has well-established links with environmental groups overseas. Thus Japanese environmentalists have persuaded their government not to advance money for the Narmada Valley project, while US groups sympathetic to the movement have tried hard to convince the World Bank to do likewise.[1] A final testimony to the movement's vigour is the active counter movement it has generated in support of the dam. Political leaders and social activists in Gujarat have rallied strongly behind the state's rich farmers, who stand to gain most from the project, organizing demonstrations and press campaigns and mounting an ideological offensive which portrays the Narmada movement's leaders as 'anti-development' and 'anti-national'. The Narmada activists have even been accused in Gujarat of wanting to deny tribals the fruits of economic growth by keeping them in a perpetual state *of* nakedness, hunger and illiteracy (Anklesaria 1988; *EPW* 1991).

## Struggles in the Sea

The social base of forest and anti-dam movements has been amongst the tribals and poor peasantry. Our third category of nature-based conflicts involves artisanal fisherfolk whose dependence on a living resource has also been undermined in recent decades. Distinct endogamous groups of fisherfolk, both along the sea coast and on rivers, have long been a feature of the Indian landscape. These communities, which depend more or less exclusively on the catching and sale of fish, have recently been threatened by massive encroachments on their territory.

The problems of ocean-going fisherfolk have been well documented, particularly in the studies of the economist John Kurien. The clash between artisanal fisherfolk and modern trawlers, at its most intense in the southern state of Kerala, provides a chilling illustration of what can happen when one group's exclusive control over living resources is abruptly challenged by forces more economically and politically powerful. For centuries, the coastal fish economy was controlled by artisanal fisherfolk operating small, unmechanized craft, which supplied fish to inland markets. In the 1960s, big business began to enter the fisheries sector. The advent of large trawlers, catching fish primarily for export, led to major changes in the ecology and economy of fisheries in Kerala. A rapid increase in fish landings in the early years of trawling was followed by stagnation and relative decline. While some artisnal fishermen were able to make the transition to a more capital- and resource-intensive system, the majority faced the brunt of direct competition from the trawlers. This conflict gave rise to a widespread movement—involving strikes, processions and violent clashes with trawler owners—in which small fishermen pressed for restrictions on the operations of trawlers. The movement also called for a ban on trawling during the monsoon, the breeding season for several important fish species. A partial ban which was finally imposed in 1988 and 1989 did in fact result in an increased harvest following the monsoon months (Kurien and Achari 1990).

So far as inland fisheries are concerned, there have been intermittent reports of localized opposition by fisherfolk affected by industrial pollution (see below). In a class of its own, however, is a unique movement to 'free the Ganga', involving fisherfolk in the Bhagalpur district of Bihar. Here, in a bizarre relic of feudalism, two families asserted hereditary rights of control over a 50 mile stretch of the Ganga. Claiming that these 'panidari' (water) rights originated in Mughal times, the waterlords levy taxes on some 40,000 fisherfolk living along the river. A protracted court case has so far been unsuccessful in abolishing these rights, which by an anomaly escaped the provisions of the law abolishing landlordism (*zamindari*), which was enacted after 1947. Since the early 1980s, the fisherfolk have been organized by young socialists into the 'Ganga Mukti Andolan' (Free the Ganga) movement. With fish catches also declining due to industrial pollution, the movement has been waging battle on two fronts simultaneously—against effluents and against an anachronistic system of monopoly rights over water (Narain 1983).

## Mines and Misery

Like forest conflicts, struggles over fish stocks have arisen out of the competing claims of different groups, each coveting the same resource but for different reasons. By contrast, the conflicts, which we now highlight are a consequence of the negative externalities imposed by one kind of economic activity, open cast mining, upon another, subsistence agriculture.

The most celebrated of mining conflicts took place in the Doon valley in northwest India. Home to the Indian Military Academy as well as the country's most famous public school, this beautiful valley is a favourite watering hole of the Indian elite. Here, the intensification of limestone mining since 1947 has led to considerable environmental degradation—deforestation, drying up of water sources, and the laying waste through erosion and debris of previously cultivated fields. Opposition to limestone quarrying, which gathered force in the late 1970s and early 1980s, has come from two distinct sources. On the one side, retired officials and executives formed the 'Friends of the Doon' and the 'Save Mussoorie' committees to safeguard the habitat of the valley. They were joined by hotel owners in Mussoorie, worried about the impact of environmental degradation on the tourist inflow into this well known hill station. These groups may fairly be characterized as NIMBY (not in my backyard) environmentalists, preoccupied above all with protecting a privileged landscape from overcrowding and defacement. On the other side, villagers more directly affected by mining were organized by local activists, many of whom had cut their teeth in the Chipko movement. While the first group lobbied hard with politicians and senior bureaucrats, the latter resorted to sit-ins to stop quarrying. Finally, in the wings they collaborated in a public interest litigation that resulted in a landmark judgement of the Supreme Court, recommending the closure of all but for just six limestone mines in the Doon Valley (Bandyopadhyay 1989; Dogra *et al*. 1983).

At the height of the Chardon limestone controversy, one of the valley's NIMBY environmentalists called—with characteristic disregard for the inhabitants of those areas—for mining to be shifted to the interior hills so that Dehradun and Mussoorie would be spared (Dalal 1983). Apparently she was unaware that mining was already proceeding apace in the inner hills. As could have been expected, it has met with resistance. In Almora and Pithoragarh districts of Kumaun, soapstone and magnesite mining—by either using, or by leading to the degradation of, common forest and pasture land—has greatly reduced local access to fuel, fodder and water. With the onset of the monsoon, the debris accumulated through mining descends onto the fields of adjacent villages. Meanwhile, with mining leasees preferring to bring in outside labour to act as a buffer between management and villagers, any tangible benefits to the village economy are few—and certainly inadequate in offsetting the losses caused by declining agricultural productivity and biomass availability.

Kumaun has a long heritage of social movements (Guha 1989a; Pathak 1987) and this has been invoked in the continuing struggles against unregulated mining. Social activists have worked hard to form village level *Sangarsh Samitis* (struggle committees) in the affected areas; the Laxmi Ashram in Kausani, started by Gandhi's disciple Sarla Devi, has been quite successful in involving women in these movements. In other instances, villagers have acted independently to protest against the damage done by open cast mining, using such varied forms of struggle as sit-ins, hunger strikes, and efforts to persuade mining labourers to go on strike. In many of these protests, women—whose own domain is most adversely hit by mining—have played a leading role. Several mines have been forced to close down, whereupon villagers have turned their energies towards land reclamation through afforestation (ISST 1991; Joshi 1983a, 1983b).

Another movement with broadly similar contours has been directed against bauxite mining in the southeastern state of Orissa. In the Gandhamardan hills of the Sambalpur district of the state, the public sector Bharat Aluminium Company (BALCO) has been granted permission to mine a heavily forested area of about 900 ha. The foundation stone for the project was laid by the Chief Minister of Orissa in May 1983 and mining commenced two years later. By the end of 1986, however, BALCO operations had been forced to a halt. As in the Himalaya, bauxite extraction in Gandhamardan led quickly to deforestation, erosion and the pollution of water source. Blasting operations were perceived as a threat to the region's ancient temples, which are visited by pilgrims from long distances. Characteristically, protest originated in a series of petitions being sent to senior officials and politicians. When this had no effect, students and social activists began forming village committees. A three day strike in front of the Block Development Office in October 1985 was followed two months later by a blockade which prevented BALCO vehicles from proceeding up the Gandhamardan plateau to the mines. Private vehicles carrying materials for BALCO operations were also blocked and unloaded. In the first two months of 1986, the movement shifted to the site of BALCO's proposed railway line, close to the Orissa–Madhya Pradesh border. According to figures collected by a civil liberties group that visited the area, a total of 987 people were jailed in the course of the year-long struggle, including 479 women and 51 children (Concerned Scholars 1986; PUDR 1986).

## The Polluter Does Not Pay Principle

Open cast mining cannot help but cause environmental degradation; it is in trying to pass on the costs of such degradation to surrounding villages that miners have encountered resistance. The textbook case of such negative externalities is, of course, industrial pollution. With air and water being free goods, it makes perfect economic sense for a private entrepreneur to pollute his surroundings instead of investing in technology to properly treat and safely dispose of effluents. The state, as the repository (in theory) of the welfare of the public, then emerges as the agency most likely to pass legislation to check pollution and take punitive action against offenders. Indeed, in the industrialized world a major focus of the environmental movement has been on pressurizing the state to pass legislation and create enforcement agencies to check air and water pollution (Hays 1987).

In India, too, pollution control legislation is in the statute books, but with administrative efficiency and honesty of lamentably low standards, industrial pollution has gone largely unchecked. In its executive functions, the Indian state apparatus alternates between being 'soft' and 'predatory'; in the first incarnation, laws are not enforced, while the second allows offenders to buy official compliance. Yet, in a democratic political system, citizens' actions can act as a partial corrective even when the state abdicates its role.

Among the most notorious of industrial polluters are paper and rayon factories. Three units of the Gwalior Rayons—owned by India's largest industrial house, the Birlas—have been indicted by environmentalists for affecting the economic welfare of downstream villagers through pollution. The Gwalior Rayons factory on the Chaliyar river in Kerala was closed for seven years after a spirited movement, led by the KSSP. In the adjoining state of Karnataka, Harihar Polyfibres (owned by the same parent company) has faced concerted opposition and a long drawn out court case for discharging untreated effluents into the Tungabhadra river. Villagers have complained of new diseases, declining fish yields and the reduced availability of irrigation water (Hiremath 1987). A similar charge has been laid at the door of the massive Gwalior Rayons factory at Nagda in Madhya Pradesh, while in the same state, in the district of Shahdol, the Birla-owned Orient Paper Mills has also been criticized by social activists for its pollution of the Some river.

Two other illustrations of the conflict between private profit and the public good come from Maharashtra, a state with a highly developed industrial sector and a long tradition of social activism. In October 1987, farmers and fisherfolk from the Devananda creek area of the Raigadh district protested against the discharge of effluents from 40 units operating in an industrial area owned by the Maharashtra State Industrial Development Corporation. Accusing the MIDC of not treating effluents before discharging them into a nearby river, peasants jammed a wooden log into the discharge pipeline (*The Times of India* 1987). Some months later, villagers in Ahmednagar district of the state united to oppose the pollution of land and water by the discharge from South Asia's largest distillery. Despairing of remedial action, the villagers filed a suit in the Bombay High Court, seeking Rs 10 million in damages from the offending company, the Western Maharashtra Industrial Corporation, and the State Pollution Control Board (*Indian Post* 1988a).

One final example of citizen protest against pollution comes from the district of North Arcot in the state of Tamil Nadu. Here, effluents from a cluster of tanneries abruptly raised the chloride content of drinking water and contributed to declining crops by causing soil salinity. On World Environment Day 1984, the town of Ambur, site of several tanneries, observed a total strike or *hartal*. Many women and children from the affected villages went in a procession through the town: here, women broke pitchers containing contaminated water, demanding that the authorities protect the health of their children. A huge effigy of an 'effluent monster' was burnt on the same day (*The Hindu* 1984).

## Conflicts in Context

Conflicts over forests, water and other natural resources have been widespread across human history. In premodern times they arose typically as a consequence of competing property claims and economic interests (Gadgil and Guha 1992). In the modern world, however, these conflicts have increasingly acquired a sharp ecological edge, being played out against the backdrop of increasing resource scarcities and shortages.

In India too, although nature-based conflict was by no means unknown in the past, the proximate cause of the struggles analysed here has been the pattern of development followed since independence in 1947. The distortions in resource flows, preferential subsidies and short term horizons of capitalists and the state have all worked to sharply circumscribe the access of the poor to the gifts of nature. The ensuing conflicts have been generated both by ongoing processes, such as the history of forest mismanagement, and massive new projects such as large dams. The variety and range of nature-based conflicts notwithstanding, two particular movements stand out for their symbolic importance to the Indian environment debate. These are the Chipko movement, which has now passed into history, and the ongoing struggles against the Narmada Valley Project, whose eventual outcome is still uncertain. Both conflicts illustrate the deep inequities in access to resource use in contemporary India. As paradigms of the conflict between the low and the mighty, both have relied—quite remarkably—on non-violent forms of protest. In each case, folk knowledge and anguish have forced ecologists and economists to reconsider the efficacy of dominant forms of resource use widely justified as 'scientific'. Their contribution to scientific debates apart, both movements arc invested with a deeper cultural, almost religious, significance. Chipko originated in the watershed of the holiest river of Hinduism; while for the people of central India the Narmada is no less sacred than the Ganga. Both struggles have attracted a dedicated core of activists who, in their selflessness and courage, exemplify the best in what remains of Gandhism. Finally, the two movements have helped generate a far reaching debate on the direction of economic development in India, and on the kind of society (and ecology) most appropriate to the needs of its culturally diverse, yet sharply fragmented, population. This debate and the various strands within it are examined more closely in the second part of this essay.

## Interpreting Indian Environmentalism: Tactics and Theories

### What is the Indian Environmental Movement?

In analysing the Indian environmental movement, we may distinguish between its *material, political* and *ideological* expressions. The material context is provided by the wide-ranging struggles over natural resources, the theme of this paper's first part. Broadly speaking, these conflicts have set in opposition, on the one side, social groups who have gained disproportionately from economic development whilst being insulated from ecological degradation (in particular, industrialists, urban consumers and rich farmers), and on the other, poorer and relatively powerless groups such as small peasants, pastoral nomads, tribal and fishing communities, whose livelihoods have been seriously undermined through a combination of resource flows biased against them and a growing deterioration of the environment. Our analysis suggests that the origins of these conflicts lie in the process of development itself. While forests, water and other natural resources are diverted to produce energy and commodities for the rich, the poor are made to bear the social and environmental costs of economic development, whether in the form of the declining availability of natural resources, a more polluted environment, or—increasingly physical displacement (for a detailed analysis, see Gadgil and Guha 1994).

With these struggles as its backdrop, the *political* expression of Indian environmentalism has been the organization by social action groups of the victims of environmental degradation. Action groups have embarked upon three distinct, if interrelated, sets of initiatives. First, through a process of organization and struggle they have tried, with varying degrees of success, to prevent ecologically destructive economic practises. Second, they have promoted the environmental message through the skillful use of the media, and mire innovatively, via informal means such as walking tours and eco-development camps. Finally, these groups have also taken up programmes of environmental rehabilitation (afforestation, soil conservation, and so on), restoring degraded village ecosystems and thereby enhancing the quality of life of the inhabitants.

Although these myriad initiatives may be construed, in the broad sense, as being political in nature, they have been almost entirely undertaken by groups falling outside the sphere of formal party politics. Across the ideological spectrum of party politics in India—from the Bharatiya Janata Party on the right to the Communist Party of India (Marxist) on the left—the established parties have turned a blind eye to the continuing impoverishment of India's natural resource base, and the threat this poses to the lives and livelihoods of vulnerable populations. At the same time, all parties have supported resource wasteful, ecologically destructive and centralizing technologies such as nuclear power plants and large dams. In the circumstances, it has been left to social action groups not owing allegiance to any political party—what the political scientist Rajni Kothari (1984) has termed 'non party political formations'—to focus public attention on the linkages between ecological degradation and rural poverty.

Through the process of struggle, the spreading of consciousness, and constructive work, action groups in the environmental field have come to develop an incisive critique of the development process itself. Responding to the conflicts over natural resources that have become so sharp in recent years, environmental activists and intellectuals sympathetic to their work have raised major questions about the orientation of economic planning in India, its in-built biases in favour of the commercial-industrial sector, and its neglect of ecological considerations. More hesitantly, they have tried to outline an alternate framework for development, which they argue, would be both ecologically sustainable and socially just. Although perspectives within the movement are themselves quite varied, in its totality this fostering of a public debate on development options constitutes the *ideological* expression of the environmental movement.

By highlighting the variety and intensity of conflicts over nature, the first part of this paper provided the material context for the Indian environment debate. We now present an analysis of the political and ideological contexts of this debate. In conclusion, we briefly contrast the Indian case, as a paradigm of Third World environmentalism, with the more intensively studied phenomenon of First World (i.e., western) environmentalism.

## ORGANIZING FOR ACTION

As already noted, struggles over the uses of nature have a long history. Popular upsurges like the Kumaun forest movement of 1921, and the Mulshi Satyagraha of the same year, may justifiably be claimed as part of the prehistory of modern environmental conflicts in India. Insofar as there is a marked continuity in forms of resistance, the contemporary environmental movement is, to a considerable extent, a peasant movement draped in the cloth of environmentalism (Guha 1989a). Thus many of the methods by which communities have resisted environmental degradation and/or external control of; natural resources, fall under the overall rubric of *satyagraha* (literally 'truth force', but used more generally to denote non-violent resistance). Here there are obvious parallels, and not merely terminological, with peasant protest in the Gandhian mode; although it must not be forgotten (Spodek 1971) that Gandhi himself drew upon long standing traditions of peasant resistance.

Among the variety of protest forms used by groups resisting environmental degradation, we may single out six. First comes the *pradarshan,* a collective show of strength by communities at the receiving end of environmental degradation, be they peasants opposing commercial forestry or fisherfolk protesting the ravages of trawling. Characteristically, this will take the form of a procession, culminating in a meeting near a locus of official power—perhaps a dam project site, or the residence of the district magistrate—in which leaders make exhortative speeches and a petition may be presented to the authorities.

The *pradarshan* is intended primarily to demonstrate popular disaffection and the strength of numbers. It shades imperceptibly into a more militant form of protest, the *dharna* or sit-down strike. In contrast to an ordinary procession or protest meeting, the *dharna* often aims specifically at stopping economic activities that threaten the survival options of resource dependent communities. Examples include attempts to stop the work at a dam site or, as was undertaken with some success in the Chipko movement, a large congregation in the forest to stop tree felling. A more sharply focused variant of the *dharna* is the *gherao.* Here, a key authority figure—a senior bureaucrat or politician perhaps—is surrounded by protesters and heckled till he accedes to their demands or is rescued by the police. More militant still is the *rasta roko* (literally, road blockade). Whereas the *dharna* has a narrow target, the *rasta roko*—born out of a more general disgust with state policy—blocks channels of communication that may not even be directly linked to the object of disaffection. Exasperated by the attitude of the Madhya Pradesh and Maharashtra governments, supporters of the Narmada Bachao Andolan sat for days on the National Highway between Delhi and Bombay, blocking passenger and commodity traffic on a vital artery.

Fifth, we have the resurrection of a classic technique of Gandhian nationalism—the *jail bharo andolan* (literally, movement to fill the jails). Here, protesters deliberately court arrest by violating a law, most frequently Section 144 of the Criminal Procedure Code, used to prohibit large gatherings. At the same time, the inadequacy of Indian jails to handle large numbers of prisoners assures them a relatively swift release. Our sixth and final technique also vividly recalls Gandhi. This is the *bhook hartal* or hunger strike. Whereas the other forms of protest highlighted above are characteristically collective, the *bhook hartal* is most frequently the preserve of one charismatic figure. The fast unto death by a widely respected popular leader is a coercive technique to compel the state to yield, in fear of the consequences of the leader succumbing to the fast.

Environmental action groups in India have thus utilized a varied and flexible repertoire of protest. These distinctive forms of struggle are, of course, both overlapping and complementary: nor is our list exhaustive; new forms are being created even as we write. The Narmada movement has already witnessed a major *sangharsh yatra* (struggle march), while its participants have frequently threatened a spectacular *jal samadhi* (literally, water burial)—in other words, to immerse themselves in the rising waters of the reservoir rather than be displaced from their ancestral lands. All in all, this repertoire of protest has helped to focus public attention on specific natural resource conflicts. In a democracy, which allows dissent, but where the state tilts markedly towards the rich and powerful, these forms of protest collectively constitutes the 'weapons of the weak' (Scott 1985).

## Communication and Education

In most such conflicts, collective protest against the agencies of the state, using one or more of the tactics described above, has been closely accompanied by coverage in the print media. Leading environmental activists (Sunderlal

Bahuguna and Baba Amte come immediately to mind) sometimes write signed articles in newspapers, drawing attention to the struggle they are engaged in. More often, sympathetic journalists write on these struggles and their wider implications. Since the mid-1970s, there has been a virtual explosion of environmental writing in English and Indian language newspapers and magazines. With radio and television controlled by the state, the print media has played an important role in reporting, interpreting and publicizing nature-based conflicts in modern India.

In understanding the spread of environmental consciousness, however, one must not underestimate oral communication. For example, the popular science group, the KSSP, has performed plays and rendered folk songs in all parts of Kerala in order to increase popular awareness of deforestation and pollution. In the neighbouring state of Karnataka, themes of environmental abuse and renewal have figured in the traditional dance-drama of the west coast, Yakshagana. An activity which combines discussion and practical action is the 'eco-development' camp, widely used by action groups to promote afforestation and other forms of environmental restoration (Bhaskaran, 1990). But in the sphere of communication too, the most innovative technique of the environmental movement recalls its acknowledged patron saint, Mahatma Gandhi. This is the *padayatra* or walking tour. Used by Gandhi to spread the message of communal harmony and by his disciple Vinoba Bhave to persuade landlords to donate land to the landless, the *padayatra* has been enthusiastically revived by environmental activists. The first environmental *padayatra* was the trans-Himalayan march from Kashmir to Kohima, covering 4000 km., by Sunderlal Bahuguna (one of Bhave's disciples) and a group of his associates in 1982–3.

The most notable *padayatra* of this ilk was the Save the Western Ghats March of 1987–8, along the 2500 km. long mountain chain. After seven months of preparation involving over 150 voluntary organizations (from the states of Kerala, Tamil Nadu, Karnataka, Goa and Maharashtra), on 1 November 1987 the march commenced from the two extremities simultaneously—Kanyakumari in Tamil Nadu and Navapur in the Dhulia district of Maharashtra. Three months later, marchers from the north and south converged at Ponda in Goa, for the meeting, which marked the march's conclusion. By then they had collectively covered 4000 km of hill terrain, making contact with over 600 villages en route. The marchers themselves came from a variety of backgrounds and age groups. Their aim was threefold: to study at first hand environmental degradation and its consequences for communities living along the Ghats; to try to activate local groups in playing a watchdog role to prevent further ecological deterioration; and to canvass public opinion in general (Hiremath 1988; Vijaypurkar 1988).

One of the objectives of the Western Ghats March, in which it largely succeeded, was to draw attention to threatened mountain ecosystems other than the Himalaya, whose plight had hitherto dominated the Indian environment debate. As a haven of biological diversity (nearly 150 endemic species) and the source of many rivers, the Ghats are as crucial to the ecological stability of peninsular India as the Himalaya are to the Indo-Gangetic plain. The Western Ghats march inspired *padayatras* across other vulnerable mountain systems. A 'Save the Sivaliks' march was undertaken across 200 km of the Sivalik range in Jammu and Kashmir the following winter, while in early 1991 a 50-day march was undertaken through the Eastern Ghats of Andhra Pradesh and Orissa. The latter effort, the Vanya Prant Chaitanya Yatra (Tribal Areas Awareness March), focused on the interconnections between environmental degradation and tribal poverty, as exemplified by deforestation, pollution, land alienation and displacement (Saraf 1989; Vinayak 1990).

Our final illustration of an environmental *padayatra* highlights not a region but a threatened resource—water. This was the Kanyakumari march, organized by the National Fisherfolk Forum in April 1989 under the slogan 'Protect Waters, Protect Life'. As in the Western Ghats, two teams started independently—one from a fishing village in Bengal on the east coast, the other from a village near Bombay on the west coast. Making their way on foot and by van, the marchers organized a variety of meetings and seminars in villages along the way. Although initiated by organizations working among fisherfolk, the march had a wider ambit. As well as declining fish yields, the marchers studied the pollution of coastal waters by industry and urban sewage, and the destruction of key ecosystems like mangrove swamps and estuaries. The objectives of the march as enumerated by its organizers, were: (*a*) to widen people's awareness of the link between water and life and to encourage popular initiatives to protect water; (*b*) to form a network of all those concerned with these issues; (*c*) to pressurize the government into evolving a sustainable water utilization policy, and to democratize and strengthen the existing water management agencies; (*d*) to assess the damage already done, identify problem areas for detailed study, and evolve practises for rejuvenating water resources; (*e*) to revive and propagate traditional water conservation practises and regenerative fishing technologies (NFF 1989).

The marchers from the two coasts converged in Kanyakumari, on the southernmost tip of India, on May Day 1989 (this culminating date reflecting the trade union locus of the organizers). An exhibition on water pollution and conservation, held at a local high school, was followed by a march to the sea. Here the participants, led by 100 women, took a pledge to 'Protect Waters, Protect Life'. Finally, a crowd of nearly 10,000, at least half of whom were women, wound their way to the public meeting that was to mark the culmination of the march. Sadly, an incident provoked by a government bus disrupting the marchers led to a police firing in which several people were killed, and the rally was called off. Despite its unhappy ending, the Kanyakumari march had fulfilled its aim of highlighting the threats to a liquid resource, which, in the Indian context, must be reckoned to be as important as oil (Dietrich 1989; Kumar 1989).

## Ecological Restoration

As tactics of struggle and consciousness raising, the *satyagraha* and *padayatra* have received generous media coverage. Less visible, but equally significant, are the programmes of ecological restoration that various social action groups have undertaken. With the state's manifest inability to restore degraded ecosystems, many voluntary organizations have taken it upon themselves to organize villagers in programmes of afforestation, soil and water conservation, and the adoption of environmentally sound technologies.

In focusing on environmental rehabilitation in preference to struggle or publicity, some groups have been influenced by the Gandhian tradition of constructive work, others by religious reform movements, and yet others by the example of international relief organizations. Often, groups with a background of work in health care, education or women's issues have turned in recent 'years to promoting sound natural resource management. Three brief case' studies are presented below to illustrate the variety of groups engaged in ecological restoration.

We start with the group that pioneered the Chipko movement, the Dashauli Gram Swarajya Mandal (DGSM). While one wing of Chipko, identified with

Sunderlal Bahuguna, has preferred to connect Himalayan deforestation with national and global environmental concerns, the DGSM, under the leadership of Chandi Prasad Bhatt, has turned from struggle to reconstruction work at the grassroots. Over the last decade, the DGSM has concentrated chiefly on afforestation work in the villages of the upper Alakananda valley. Two notable features of these plantations have been the lead taken by women, and the high survival rate of saplings—an average of 75 per cent in contrast to the 14 per cent average rate in Forest Department plantations. In addition, in heavily eroded landscapes, volunteers have taken up appropriate soil conservation measures like the plugging of gullies, construction of small check-dams and the plantation of fast growing grass species. Finally, the DGSM has enthusiastically promoted energy saving devices such as fuel-efficient cooking stoves and biogas plants (CSE 1985; S.N. Prasad, personal communication).

A second example of successful eco-restoration work also originated in a process of struggle. In the Sangli district of Maharashtra, where socialist workers have long been active, peasants have been faced with persistent drought. In this context, two villages of Khanapur taluk, encouraged by socialist and popular science activists, decided to build on a cooperative basis, a small dam across a river that sporadically contained water. To finance the dam they requested the state government to allow, them to sell sand from a nearby river bed. The administration, however, preferred to auction the sand to private merchants: it was even reluctant to sanction the dam. A series of hunger strikes, processions and *gheraos* forced the government to abandon the auction system, although it permitted local sale to a limited extent. Helped by voluntary contributions, and under the technical guidance of a Bombay engineer, the villagers finally succeeded in building the 'Bali Raja Memorial Dam' by the end of 1988. The water thus stored is used to provide one irrigated crop to each family of the two villages, and for nursery and forestry work (*Indian Post* 1988b; Joy and Rao 1988; Omvedt 1987).

Our final case study originated not in a movement but in a remarkable individual, Anna Saheb Hazare of the village of Ralegaon Siddhi in the Ahmednagar district of Maharashtra. Ahmednagar too is drought prone—speaking of the scarcity of water, the *Bombay Chronicle* of 2 March 1913 had termed it 'the most unfortunate and heavily tried district in India'. Thus when Anna Hazare returned to the village on retirement from the army in the mid-1970s, he found that food production reached barely 30 per cent of its requirements. Quickly locating the problem as insufficient retention of rainwater, he organized villagers into building a series of storage ponds and embankments (*nallah bandhs*) along the low hills surrounding the village. Very quickly, run off was reduced and aquifers recharged, and the ground-water table rose considerably. There is now sufficient water for household use and irrigation, and crop yields have increased dramatically (the village has begun to export food). Besides this, Hazare has mobilized villagers to plant 400,000 saplings. With his village now acknowledged as a model of ecorestoration through self help, Hazare is training volunteers to work in other villages. He has simultaneously launched a movement against corruption in state forestry and drinking water programmes (Rai *et al*. 1991).

As these examples show, reconstruction work can proceed hand in hand with struggle. Yet in other instances, groups temperamentally unsuited to confrontation have done admirable work in promoting environmentally benign technologies and in rehabilitating degraded lands. Reconstruction work constitutes a valuable third front of the environmental movement, complementing the activities of consciousness building and popular resistance to state policies.

Individual groups working in the environmental field are typically confined to a small area. In the last decade, various attempts have been made to develop a macro-level organization to coordinate these various groups and activities. This process got a considerable boost with the rally against 'destructive development' held in Harsud in September 1989. In a follow-up meeting held in Bhopal in December— to coincide with the fifth anniversary of the gas tragedy in that city—groups that participated in the Harsud rally initiated the formation of the Jan Vikas Andolan (Peoples' Development Movement; JVA), a loosely knit national level organization to coordinate local struggles. Over the past. three years, the JVA has had meetings in different parts of the country, involving a wide range of groups and individuals. In defining itself as a movement against the existing pattern of development, the JVA's own objectives are fourfold: (*a*) to coordinate collective action against environmentally destructive policies and practises; (*b*) to provide national solidarity to these struggles; (*c*) to mobilize wider public opinion on the need for a new development path; and (*d*) to work towards an alternative vision, ecologically sustainable and socially just, for India's future (JVA 1990).

## Ideological Trends in Indian Environmentalism

Social action in the three generic modes outlined above (i.e. struggle, publicity and restoration) constitutes the bedrock of the Indian environmental movement. While such activism has characteristically been localized—with most groups working within one district—the links between the micro and macro spheres have been made most explicit (recent initiatives like the JVA excepted) through the environmentalists' critique of the ruling ideology of Indian democracy, that of imitative industrialization. For environmentalists have insistently claimed that the intensification of natural resources conflict is a direct consequence of the resource and capital-intensive pattern of economic development, modeled on the Western experience, followed since Independence. The resource illiteracy of development planning, they claim, is directly responsible for the impoverishment of the resource base and of the millions of rural people who depend on it (JVA 1990).

While there is widespread agreement within the environmental movement as regards the failures of the present development model, there is little consensus on plausible alternatives, all responding to the range of conflicts we have analysed above, but advocating widely varying proposals for mitigating these conflicts. It is, however, possible to identify three distinct ideological perspectives within the movement. It is of course entirely possible that none of the ideologies so identified is present in a particular struggle, or indeed that adherents of all three viewpoints might participate in unison in a specific initiative. However, close study and discussions with groups spread all over India does suggest that the three strands analysed below are the dominant ideologies of Indian environmentalism.

The first, which we may call *Crusading Gandhian,* relies heavily on a moral/religious idiom in its rejection of the modern way of life. Here, environmental degradation and social conflict are viewed above all as a *moral* problem, their origins lying in the wider acceptance of the ideology of materialism and consumerism, which draws humans away from nature even as it encourages wasteful lifestyles. Crusading Gandhians argue that the essence of 'Eastern' cultures is their indifference, even hostility to economic gain: thus, if India were to abandon its pursuit of Western models of economic development, it would only be returning to its cultural roots. These environmentalists call, therefore, for a return to precolonial (and pre-capitalist) village society, which they uphold as the exemplar of social and ecological harmony. Gandhi's own invocation of *Ram Rajya* (the mythical but benign rule of King Rama) is here being taken literally, rather than metaphorically. In this regard Crusading Gandhians frequently cite Hindu scriptures as exemplifying a 'traditional' reverence for nature and life forms.

Crusading Gandhians have worked hard in carrying their message of moral regeneration across the country and indeed across the globe. They have sharply attacked the stranglehold of modernist philosophies—particularly those upholding rationalism and economic growth—on the Indian intelligentsia; through the written and spoken word, they propagate an alternative, non-modern philosophy whose roots lie in Indian tradition (Bahuguna 1983; Nandy 1987, 1989; Shiva 1988).

The second trend, in many ways the polar opposite of the first, is Marxist in inspiration. Marxists see the problem in political and economic terms, arguing that it is unequal access to resources, rather than the question of values, which better explains the patterns and processes of environmental degradation and social conflict. In this sharply stratified society, the rich destroy nature in the pursuit of profit, while the poor do so simply to survive (the Crusading Gandhians would tend to deny altogether that the poor also contribute to environmental degradation). For Ecological Marxists, therefore, the creation of an economically just society is a logical precondition of special and ecological harmony. In their practical emphasis, socialist activists concentrate on organizing the poor for collective action, working towards their larger goal of the redistribution of economic and political power. While including various Naxalite and radical Christian groupings, Ecological Marxists in the Indian context are perhaps most closely identified with People's Science Movements (PSMs)—the best known of which is the KSSP—whose initial concern with taking 'science to the people' has been widened to include environmental protection. Ecological Marxists can be distinguished from Gandhians in two significant respects: their unremitting hostility to tradition (and corresponding faith in modernity and modern science) and in their relatively greater emphasis on confrontational movements (KSSP 1984).

Crusading Gandhians and Ecological Marxists can be seen as representing the 'ideological' and 'political' extremes of the Indian environmental movement, respectively. Because of their ideological purity and consistency, their arguments are often compelling, albeit to different sets of people. In between these two extremes, and occupying the vast middle ground, lies a third tendency, which may be termed (less controversially) 'Appropriate Technology'. Less strident than the Gandhian in its opposition to industrial society, this strand of the environmental movement strives for a working synthesis of agriculture and industry, big and small units, and Western and Eastern (or modern and traditional) technological traditions. Both in its ambivalence about religion and in its criticism of traditional social hierarchies it is markedly influenced by Western socialism. Yet in its practical emphasis on constructive work, it taps another vein in the Gandhian tradition. Thus Appropriate Technologists have done pioneering work in the generation and diffusion of resource conserving, labour intensive and socially liberating technologies. Their emphasis is not so much, *pace* the Marxists, on challenging the 'system'—or, *pace* the Gandhians, the system's ideological underpinnings—as in demonstrating in practise a set of socio-technical alternatives to the centralizing and degrading technologies presently in operation (Bhatt 1992; Reddy 1982).

All three tendencies are represented in that most celebrated of environmental initiatives, the Chipko movement (Guha 1989a). The Gandhian trend, associated above all with the figure of Sunderlal Bahuguna, is best known outside the Himalaya. The Marxist trend within Chipko has been represented by the Uttarakhand Sangarsh Vahini, a youth organization that has organized popular movements against commercial forestry, unregulated mining and the illegal liquor trade. Finally, the Appropriate Technologists are represented by the organization under whose auspices the movement began, the Dashauli Gram Swarajya Mandal, whose fine work in ecological restoration has already been alluded to.

These contrasting perspectives may be further clarified by examining each strand's attitudes towards equity and science, as well as their style and scale of activism. Most Crusading Gandhians reject socialism as a Western concept: this leads some among them to gloss over inequalities in traditional Indian society, and others even to justify them. Clearly the Marxists have been most forthright in their denunciations of inequality across the triple axes of class, caste and gender. The Appropriate Technologists have been sufficiently influenced by Marxism to acknowledge the presence and pervasiveness of inequality, but have rarely shown the will to challenge social hierarchies in practise. Attitudes towards modern science and technology also vary widely. The Gandhians consider science to be a brick in the edifice of industrial society, and responsible for some of its worst excesses. Marxists yield to no one in their admiration of modern science and technology, viewing science and the 'scientific temper' as an indispensable ally in the construction of a new social order. Here, the Appropriate Technologists are the most judicious, calling for a pragmatic reconciliation between modern and traditional knowledge and technique, to fulfill the needs of social equity, local self-reliance and environmental sustainability.

On the scale of activism, Appropriate Technologists prefer to work on a micro scale—a group of contiguous villages at best—in demonstrating the viability of an alternative model of economic development. The Gandhians have the largest attempted reach, carrying their crusade on worldwide lecture tours: they have often tended to think globally and act globally, even as the Appropriate Technologists have acted locally and occasionally thought locally too. The Marxist groupings work in the intermediate range, at the level of a district perhaps, or (as in the case of the KSSP) the level of a state. Finally, the three strands also differ in their preferred sectors of activism. Their rural romanticism has led the Gandhians to exclusively emphasize agrarian environmental problems, a preference reinforced by their well-known hostility to modern industry. While Appropriate Technologists do recognize that some degree of industrialization (though not of the present resource-intensive kind) is inevitable, in practise they too have worked largely on technologies aimed at relieving the drudgery of work in the village. Here it is the Ecological Marxists, with their natural constituency among miners and workers, who have been most alert to questions of industrial pollution and workplace safety.

Crusading Gandhians, Appropriate Technologists and Ecological Marxists represent the three most forceful strands in the Indian environmental debate; but we should also take account, however briefly, of two other points of view. First, we have the Indian variant of that vibrant strand in global environmentalism—the wilderness movement. Indian naturalists have provided abundant documentation of the decline of natural forests and their plant and animal species, urging the government to take remedial action (Krishnan 1975). Although their earlier efforts were directed almost exclusively towards the protection of large mammals, more recently wildlife preservationists have used the scientific rhetoric of biological diversity and the moral arguments in favour of 'species equality' in pursuit of a more Extensive system of parks and sanctuaries and a total ban on human activity in protected areas (Guha 1989b).

So we come, finally, to an influential strand of thinking within the state and state agencies, which might be termed scientific conservation. Pre-eminent here is the work of B.B. Vohra, a senior bureaucrat who was one of the first to draw public attention to land and water degradation. In a pioneering and impressively thorough paper (Vohra 1973), he documented the extent of erosion, waterlogging and other forms of land degradation. There was, he noted, no countrywide organization or policy to deal with these problems; nor was there coordination between concerned government departments. For Vohra, as for the early scientific conservationists (Hays 1957), the solution lies in the creation of new ministries and departments to deal with problems of environmental degradation. The central fovernment, he has written, 'has no option but to obtain a commanding position for itself in the field of land and soil management through financial and administrative measures' (Vohra 1973; see also Vohra 1980, 1982).

Neither wilderness protection nor scientific conservation command a popular following, yet each has had a considerable influence on government policy. Both tendencies look upon the state as the ultimate guarantor of environmental protection, and their energetic lobbying has informed stringent legislation in pursuit of this ideal, such as the Wildlife Protection Act of 1972 (modified in 1991), the Forest Conservation Act of 1980, and the Environment Protection Act of 1986. However, insofar as neither group is cognizant of the social roots of environmental use and abuse, they tend to be dismissed as 'elite' conservationists by environmentalists owing allegiance to Gandhian or Marxist traditions.

## First World and Third World Environmentalism

While there is a vigorous environmental debate and environmental movement in India, it should be noted that its very existence challenges the conventional wisdom of Western (and especially American) social science. Thus, a decade ago, a leading American economist confidently asserted:

> If you look at the countries that are interested in environmentalism, or at the individuals who support environmentalism within each country, one is struck by the extent to which environmentalism is an interest of the upper middle class. *Poor countries and poor individuals simply aren't interested* (Thurow 1980: 104–5, emphasis added).

As a social phenomenon, the economist went on to explain, environmentalism is

> a natural product of a rising rent standard of living. We have simply reached the point where, for many Americans, the next item on their acquisitive agenda is a cleaner environment. If they achieve it, it would make all of the other goods and services (boats, summer homes, and so forth) more enjoyable (ibid: 104–5).

This interpretation of environmentalism is in fact widespread in the West. Historians of American environmentalism are unanimous that environmentalism is a 'full stomach' phenomenon, a direct consequence of economic affluence by which wilderness areas and clean air come to be cherished once basic material needs have been fulfilled (Nash 1982). As a leading historian has remarked, the emergence of popular environmentalism in the US was 'not a throwback to the primitive, but an integral part of the modern standard of living as people sought to add new "amenity" and "aesthetic" goals and desires to their earlier preoccupation with necessities and conveniences' (Hays 1982: 21; see also, Hays 1987). Or to quote a leading British journalist, it is 'safe to assume that when everyone turns environmental, prosperity has truly arrived. Greenness is the ultimate luxury of the consumer society' (Moore 1989: ix).

In this perspective, environmentalism is organically related to the expansion of leisure opportunities in a 'postindustrial' society—it is itself an expression of a 'postmaterial' world view (Inglehart, 1977). Yet, contrary to what one might expect from this theory, poor countries and, even more strikingly, poor individuals and poor communities within them have shown a strong interest in environmental issues. India is not an exception in this regard, for Brazil, Malaysia and Kenya all have growing environmental movements with markedly lower class constituencies.

A detailed contrast between First World and Third World environmentalism would take us too far afield, into the realm of comparative and global environmental history: that is not possible here. Given the bias in the literature towards the study of North Atlantic environmentalism, and indeed the equation in many minds of environmental concern with economic prosperity, it will help to locate the Indian environmental movement by contrasting it with what one might call the 'ecology of affluence'. Just as we take the Indian case as a paradigm of Third World environmentalism, we use the American movement as a paradigm of First World environmentalism.

The histories of environmentalism in these two great and vibrant democracies have, inevitably, been very different. In the one case, environmentalism as a popular movement is, indeed, an unmistakable product of a postindustrial economy and a postmaterial society. India, however, is still a dominantly agrarian country—here the environmental movement has emerged at a relatively early stage in the industrialization process.

This is, of course, related to the very different trajectories of economic development in the two countries. The countries that pioneered industrial development in Western Europe and North America did face environmental problems relating to the degradation of land and forests. However, with technological substitution and scientific resource management, problems I such as timber scarcities and dust bowls, once faced by countries like the US, have disappeared. In the second phase of Western industrialization (i.e., after World War II) other forms of environmental degradation—especially air and water pollution and the destruction of wilderness—have come to occupy centrestage. In other words, with the maturing of the industrialization process, public attention has shifted from problems of environmental 'sustainability, such as the steady supply of forest produce, or the protection of soils to issues of environmental quality like cleanliness of air and water, or he protection of pristine habitats'.

On the other hand, in India's industrialization experience—and here it is typical of the Third World more generally—it has simultaneously faced problems of land and resource depletion, pollution and the decimation of biological diversity. The history of colonial exploitation and the process of planned development after Indian independence are both germane here. Moreover, unlike in the West, there is little hope of a large scale shift in consumption patterns—from fuel wood to oil, for example—to overcome the problems caused by deforestation, soil erosion and so forth. Consequently, at least in the immediate future, resource depletion and destruction are likely to persist (Gadgil and Guha 1992).

A second major difference, flowing logically from the first, concerns the social origins of the environmental impulse. Clearly, in the Indian case environmental degradation and the ensuing resource shortages directly threaten survival and livelihood options. Here, as we have documented at some length, environmentalism has its origins in conflicts between competing groups—typically peasants and industry—over productive resources. By contrast, environmental conflicts in the West have characteristically emerged out of threats to health and leisure options. The forces for environmental destruction are, in both cases, overwhelmingly state agencies and private enterprise. In one scenario, intensification of resource use undermines existing but subsistence-oriented economic activities, while in the other it poses a threat to the health or amenities of local communities. In advanced industrial societies, quality of life issues such as environmental protection, have somewhat displaced economic conflicts as the motivating factor behind collective action; while in the 'developing' world, environmental conflict is, for the most part, only another form of economic conflict.

These different motivations closely influence the tactics of protest. In India, direct action—tree hugging, demonstrations, attacks on official property—have from the beginning been a vital component of environmental action. Here there is a marked similarity in idiom and action to the archetypal peasant movement. In the US, environmental groups have relied to a greater degree on litigation, skilful use of the media and lobbying politicians—tactics with a greater chance of success in a more formal and mature democratic political system. The experience of recent years, however, somewhat qualifies this sharp contrast between direct action on the one hand and lobbying and litigation on the other. Environmental groups in India are turning increasingly to the courts as a supplement to popular protest, while in America, militant environmentalists disgusted with the incremental lobbying of mainstream groups have taken to direct action—the spiking of trees, for instance—to protect threatened wilderness.

A fourth important difference concerns the role of science and scientists. In the US, scientists have played a key role: indeed, the beginnings of modern American environmentalism are conventionally assigned to the writing of and reaction to the book *Silent Spring* by the biologist Rachel Carson (1962). In subsequent decades, the work of scientists such as Barry Commoner, Paul Ehrlich, Garret Hardin and the co-authors of the 'Limits to Growth' report have all helped bring ecological concerns to a wide public audience. In India, scientists (and social scientists) have played a severely circumscribed role in the environment debate. Rather, journalists, Gandhians and environmental activists themselves have been in the forefront. Comparative rates of literacy are relevant here, as is the attitude to science: unlike the US situation, science does not enjoy a high public profile in India, nor do scientists command moral authority.

The last difference is the most crucial of all. This is that environmental degradation has been, in terms of its human consequences, a far more serious issue in India, as in most of the Third World generally. For in the Western world, the destruction of the environment has had an adverse impact primarily on health and on natural habitats valued for reasons of science, aesthetics or leisure, whereas in the poorer countries it has in addition gravely undermined the life chances of millions of rural (and urban) households. This key distinction has meant that in the US, for example, the environmental movement has by and large run parallel to the consumer society without questioning its socio-ecological basis (cf. Guha 1989b). The sharper edge to environmental conflict in the Third World, and its close connections to questions of subsistence and survival, has prompted a more thorough-going critique both of consumerism and of uncontrolled economic development. This has been a critique primarily directed at the iniquitous and unsustainable patterns of economic growth that characterize most Third World countries; yet it is also a critique with much relevance to Western lifestyles and economic preferences, themselves the cause of massive environmental degradation worldwide.

It is thus that the environmentalism of the poor has a very different agenda from the environmentalism of the rich. The conflict between these two agendas came briefly to the fore at the Rio Conference in June 1992. It was brushed aside then, but will assuredly resurface at regular intervals. With the environment becoming a major theme in global politics, there is more need than ever for a fuller understanding of the social roots of environmental concern: of its origins, motivations and forms of expression in different countries and social systems.

## Note

1. In March 1993, the Indian government itself decided not to ask for further World Bank support for the project, indicating its inability to meet the Bank's criteria for resettlement and environmental rehabilitation.

## References

Alvares, Claude, 1989, 'No!', *The Illustrated Weekly of India,* 15 October.

Anklesaria, Shahnaz Aiyar, 1988, 'Narmada Project: Government-opposition alliance against Amte', *Indian Express,* 30 October.

Anonymous, 1991, 'The Sangharsh Yatra: A First Hand Account', in *Narmada: A Campaign Newsletter* 7 & 8, April.

Bahuguna, S., 1983, *Walking with the Chipko Message,* Silyara, Navjivan Ashram, Tehri Garhwal District.

Bandyopadhyay, J., 1989, *Natural Resource Management in the Mountain Environment: Experiences from the Doon Valley, India,* ICIMOD, Katmandu.

Bhaskaran, S.T., 1990, 'The Rise of the Environmental Movement in India', *Media Journal* 37(2).

Bhatt, C.P., 1992, *The Future of Large Projects in the Himalaya,* Peoples Association for Himalaya Area Research, Naini Tal.

Bhuskute, V.M., 1968, *Mulshi Satyagraha,* Pune: Dastane (in Marathi).

*Bombay Chronicle,* 2 March 1913.

Calman, L., 1985, *Protest in Democratic India,* Westview Press, Boulder.

Carson, R., 1962, *Silent Spring,* Fawcett Publications Inc., Greenwich.

Concerned Scholars, 1986, *Bharat Aluminium Company: Gandhamardan Hills aid Peoples Agitation,* Concerned Scholars, Sambalpur.

CSE, 1985, *The State of India's Environment 1982: A Citizens' Report,* Centre for Science and Environment, New Delhi.

Dalal, Nergis, 1983, 'Bring back my Valley', *The Times of India* 10 July.

Dietrich, Gabriele, 1989, 'Kanyakumari March: Breakthrough despite Breakup', *Economic and Political Weekly* XXIV (20): 1087-8 (20 May).

D'Monte, D., 1981, 'Time up for Tehri', *Indian Express* 30 May.

———, 1985, *Temples or Tombs? Industry versus Environment: Three Controversies,* Centre for Science and Environment, New Delhi.

Dogra, B., N. Nautiyal and K. Prasun, 1983, *Victims of Ecological Ruin,* from the authors, Dehra Dun.

Elwin, V., 1964, *The Tribal World of Verrier Elwin: An Autobiography,* OUP, Bombay.

*EPW*, 1991, 'Gujarat: An Advertiser's Supplement', *Economic and Political Weekly* XX VI(1/2): AS-1-27 (5–12 January).

Fernandes, W., and S. Kulkarni (eds), 1983, *Towards a New Forest Policy,* Indian Social Institute, New Delhi.

Fernandes, W., and E. Ganguly-Thukral (eds), 1988, *Development and Rehabilitation,* Indian Social Institute, New Delhi.

Gadgil, Madhav and Ramachandra Guha, 1992, *This Fissured Land: An Ecological History of India,* OUP, New Delhi, University of California Press, Berkeley, CA.

———, 1994, *Ecology and Equity: Steps towards an Economy of Permanence,* UNRISD, Geneva.

Ganguly-Thukral, E. (ed.), 1992, *Big Dams, Displaced People,* Sage Publications, New Delhi.

Goldsmith, E., and N. Hildyard, 1984, *The Social and Environmental Effects of Large Dams,* Wade bridge Ecological Centre, Camel ford, Cornwall, UK.

Guha, Ramachandra, 1989a, *The UnquietWoods: Ecological Change and Peasant Resistance in the Himalaya,* OUP, New Delhi; University of California Press, Berkeley, CA:

———, 1989b, 'Radical American environmentalism and Wilderness Preservation: A Third World Critique', *Environmental Ethics* 11(1): 71–6.

Hart, H.C., 1956, *New India's Rivers,* Orient Longman, Bombay.

Hays, S.P., 1957, *Conservation as the Gospel of Efficiency: The Progressive Conservation Movement,* 1880–1920, Harvard University Press, Cambridge, Massachusetts.

———, 1982, 'From Conservation to Environment: Environmental Politics in the United States since World War Two', *Environmental Review* 6(1).

———, 1987, *Beauty, Health and Permanence: Environmental Politics in the United Stales, 1955–85,* Cambridge University Press, New York.

Hiremath, S.R., 1987, 'On Fighting a Corporate Giant', in A. Agarwal, D. D'Monte and U. Samant (eds), *The Fight for Survival,* Centre for Science and Environment, New Delhi.

———, 1988, 'Western Ghats March was an Education', *Deccan Herald* 14 February; *Indian Post* (1988a) 8 August; *Indian Post* (1988b) 21 September.

Inglehart, R., 1977, *The Silent Revolution.* Princeton University Press, Princeton, NJ.

ISST, 1991, *Mining in the Himalayas: Report on a Field Study in the Almora and Pithoragarh Districts,* Institute of Social Studies Trust, New Delhi.

Joshi, Dinesh, 1983a, 'Magnesite Udyog: Vi.aash ki Ankahi Kahani' (Magnesite Mining: Destruction's Untold Story), *Parvatiya Times* (New Delhi) 30 November.

———, 1983b, 'Magnesite Kaman se Himalaya ko Bhaari Kshati' (Grievous Injuries lo the Himalaya caused by Magnesite Mining), *Parvatiya Times* 31 November.

Joy, K.J., and Nagmani Rao, 1988, 'The Great Sand Robbery and Impending Ecological Disaster', *Economic and PoliticalWeekly* XXIII (33): 1169–71 (13 August).

JVA, 1990, *Jan Vikas Andolan: A Working Perspective,* mimeo.

Kalpavriksh, 1988, *The Narmada Valley Project: A Critique,* Kalpavriksh, New Delhi.

Kanvalli, S., 1991, *Quest for Justice,* Samaj Parivartan Samudaya, Dharwad.

Kothari, R., 1984, 'The Non Party Political Process', *Economic and Political Weekly* XIX (5): 216–24 (4 February).

Krishnan, M., 1975, *India's Wildlife in 1959–70: An Ecological Survey of the Larger Mammals of Peninsular India,* Bombay Natural History Society, Bombay.

KSSP, 1984, *Science as Social Activism: Reports and Papers on the People's Science Movement in India,* Kerala Sastra Sahitya Parishad, Trivandrum.

Kumar, K.G., 1989, 'Police Brutality Besieges Ecology: National Fishermen's March', *Economic and Political Weekly* XXIV (21): 1155 (27 May).

Kurien, J., and Thankappan Achari, 1990, 'Overfishing along Kerala Coast: Causes and Consequences', *Economic and Political Weekly,* 1–8 September.

Modi, L.N., 1988, 'Whose forests are they anyway', *Blitz* 10 December.

Moore, C., 1989, 'Foreword', in Philip Marsden-Smedley (ed.), *Britain in the Eighties: The Spectator View of the Thatcher Decade,* Grafton Books, London.

Nandy, A., 1987, *Traditions, Tyrannies and Utopias,* OUP, New Delhi.

———. (ed.), 1989, *Science, Hegemony and Violence,* OUP, New Delhi.

Narain, Sunita, 1983, 'Brutal Oppression of Fisherfolk', *The Times of India*, 11 October.

*Narmada: A Campaign Newsletter* (1989-90, various issues), Narmada Bachao Andolan, New Delhi.

Nash, R., 1982, *Wilderness and the American Mind,* Yale University Press, New Haven.

NFF, 1989, 'Kanyakumari March: Breakthrough despite Breakup', National Fisherfolk Forum, Cochin.

Omvedt, Gail, 1987, 'Of Sand and King Bali', *Economic and Political Weekly* XXII (9): 365–66 '(28 February).

Paranjpye, V., 1989, *Evaluating the Tehri Dam: An Extended Cost-Benefit Appraisal,* Indian National Trust for Art and Cultural Heritage, New Delhi.

Pathak, S., 1987, Uttarakhand *mein Kuli Begat Pratha* (*The Forced Labour System in: Uttarakhand*), Radhakrishnan Prakashan, New Delhi.

PUCL, 1985, *Bastar: Ek Mutbhed ki Jaanch (Bastar: An Enquiry Report),* Peoples Union for Civil Liberties, Bhopal.

PUDR, 1982, *Undeclared Civil War,* Peoples Union for Democratic Rights, New Delhi.

———, 1986, *Gandhamardan Mines: A Report on Environment and People,* Peoples Union for Democratic Rights, New Delhi.

Rai, Usha Mukul, Sarosh Bana and Dinesh Kumar, 1991, *Call of the Commons: People versus Corruption,* Centre for Science and Environment, New Delhi.

Rangarajan, M., 1992, 'Forest Policy in the Central Provinces, 1860–1914', Unpublished D. Phil thesis, University of Oxford.

Reddy, A.K.N., 1982, 'An Alternative Pattern of Indian Industrialization', in A.K. Bagchee and Normal Bannerjee (eds), *Change and Choice in Indian Industry,* K.P. Bagchee, Calcutta.

Saraf, Suraj, 1989, 'An Odyssey to Save the Sivaliks', *The Hindu* 26 February.

Scott, J.C., 1985, *Weapons of the Weak: Everyday Forms of Peasant Resistance,* Yale University Press, New Haven.

Sengupta, N. (ed.), 1982, *Jharkhand: Fourth World Dynamics,* Authors' Guild, New Delhi.

Sharma, L.T., and R. Sharma (eds), 1981, *Major Dams: A Second Look,* Gandhi Peace Foundation, New Delhi.

Shiva, V., 1988, *Staying Alive: Women, Ecology and Survival in India,* Kali for Women, New Delhi.

23

# Monumental Folly

Darryl D'Monte

The Taj Mahal has been in the public eye for the wrong reasons in the early years of the twenty-first century. Former Chief Minister of Uttar Pradesh, Mayawati, launched an ill-conceived 'Taj Corridor' plan along the banks of the Yamuna river, across the monument. Fortunately, before too much damage was done, the Rs 175 crore project was nipped in the bud: the Chief Minister had to resign and the threat was averted. Not long after, there was a feeble attempt by the Waqf Board to claim the 'miracle in marble' as its property—no doubt with an eye on the enormous revenues that this most-visited tourist site in India earns.

But it is not as if the Taj has not had its share of threats even much earlier. The earliest such record is a letter from the Mughal Emperor Aurangzeb himself to his father Shah Jahan, who built the mausoleum in memory of his beloved wife Mumtaz. Mumtaz died giving birth to their fourteenth child in 1631. Aurangzeb had imprisoned his father in the Agra Fort, where he could see the monument from a distance. He wrote after visiting the 'illustrious sepulchers and gained the blessing of paying homage to that holy shrine' on a Friday.

Aurangzeb complained:

> The dome is leaking in two places, from the arches and galleries. The buildings in the sacred enclosure stand exactly as they were completed in your Majesty's august presence, except that the master builders state that if the roof of the second storey is opened out and treated with concrete, over which half a yard

* Darryl D'Monte wrote this paper especially for this volume. His book *Temples or Tombs?* (CSE, Delhi, 1985) is still the best source for a sound background on the Taj controversy.

> of mortar grout is laid, it is probable that the semi-domed arches, the galleries and the smaller domes will be set right, but they confess their inability to prescribe any corrective measures in respect of the main dome.

'My revered guide and benefactor', intoned the hypocritical and tyrannical son, 'may you live long! This great edifice has curiously sustained some injuries but it is believed that if the pious disposition of the Emperor inclines to its repairs, the defects will forthwith be set right.' Shah Jahan died in custody after eight long years in 1666. It should be mentioned that Aurangzeb was appalled, among other things, by the extravagance displayed by his father in his attempt to commemorate for all time to come the memory of his wife. He had spent an estimated Rs 6 crore, an enormous sum at the time, over 22 years, with the labour of 20,000 workers.

A century later, the Jats captured Agra and sacked Agra Fort. They camped in the Taj and burnt hay to keep themselves warm in winter, which must have harmed the marble mausoleum. They also helped themselves to the finest gems and silver gates. By the nineteenth century, the British had taken over and used the grounds for open-air 'frolics'. Outdoor balls were held on the marble terrace in front of the main door; the mosques on either side of the main monument were rented out to honeymooning couples. Lord Curzon, who is regarded as the doyen of Indian archaeology, records:

> At an earlier date, when picnic parties were held in the garden of the Taj, it was not an uncommon thing for the revelers to arm themselves with hammer and chisel, with which they whiled away the afternoon by chipping out fragments of agate and carnelian from the cenotaphs of the Emperor and his lamented Queen.

One of Curzon's predecessors, however, could have written the epitaph for this monument, which is regarded as one of the Wonders of the World. In the 1830s, Lord William Bentinck, the first Governor General of India, announced that he was going to demolish the best Mughal monuments in Agra and Delhi to retrieve the marble from the facades. Some of this marble would be shipped to London, to be disposed of piecemeal. Several of Shah Jahan's pavilions in the Red Fort in Delhi were stripped to the brick and the marble shipped to London; some of the shipment included treasures for King George IV himself.

As one author writes:

> Plans were then made to dismantle the Taj Mahal and wrecking machinery was moved into the garden grounds. Just as the demolition crew was setting to work, word came from London that the first auction had been a failure and that all further sales were cancelled—it would not be worth the money to tear down the Taj Mahal.

Saved by the gavel, in this case!

However, a century and a half later, in the 1970s, there was a serious threat to the monument. The Indian Oil Corporation (IOC) planned to set up an oil refinery in Mathura, which is 40 km as the crow flies from Agra. At the time, such was the innocence regarding the environment that no one could have guessed that the fumes from this refinery could corrode the marble façade of the Taj. When oil is refined, it gives off sulphur dioxide, which can combine with moisture in the atmosphere to form sulphuric acid. Indeed, one of the earlier locations sought for the refinery was Agra itself. Mathura was chosen because it was located 'centrally within the com mand area': the Green Revolution was in full swing at the time and there was obviously a huge demand for oil and its byproducts in north-west India.

IOC received permission from the central government in 1973 for a Rs 97 crore refinery to process six million tonnes of crude a year. It should have been built within five years but the oil price hike by OPEC that very year upset the government's plans and there was 'negative demand' for these products for some months. One of the first to raise questions was Dr Asad Rahmani, now the Director of the prestigious Bombay Natural History Society (BNHS), who had been studying the migration of cranes to the nearby Bharatpur bird sanctuary and was concerned about the impact of this industrial project on the winged visitors from Siberia and elsewhere.

Prof J.M. Dave, who later headed the School of Environmental Sciences at Jawaharlal Nehru University in Delhi, was among those early critics too. He then headed the air pollution division of the National Environmental Engineering Research Institute (NEERI) in Nagpur. He wrote to Prime Minister Indira Gandhi, who referred the letter to the Petroleum Ministry. The ministry formed an informal committee, which realized that it had first to establish how much sulphur dioxide would reach Agra from the refinery. Another strong critic was Prof. T. Shivaji Rao, a civil engineering academic from Visakhapatnam who, like Dave, tended to exaggerate the adverse effect of the project, even to the extent of magnifying figures.

Experts started worrying about the safety of other monuments around Agra—the Agra Fort, Itmad-ud-Daulah's tomb and Akbar's tomb at Sikandara (on the outskirts of the town), as well as Fatehpur Sikri (which lies on the way to Bharatpur). Soon, there was a flurry of letters in the press and questions raised in parliament. The late S.N. Chib, from the Indian Heritage Society in Delhi, was very active in drumming up opposition to the refinery. D.K. Barooah, the Petroleum Minister, directed the IOC to 'take the necessary precautionary measures to ensure that the effluents discharged into the atmosphere will not have any adverse effect on the monuments'. People were also worried about temples at Mathura: Vrindavan is revered as the birthplace of Krishna.

Somewhat belatedly, the Archaeological Society of India (ASI), which is a venerable institution in its own right, formed in 1861, also started stirring. R. Sengupta, later Director of Conservation, who was involved in restoring the Bamiyan statues in Afghanistan (blown up a few years ago by the Taliban), took an active interest. It commissioned its own study from NEERI. With pressure mounting from several quarters, the government commissioned a major study by a committee headed by Dr S. Varadarajan. This writer was somewhat skeptical about his bona fides, since he was a petrochemical technocrat par excellence—he headed the Indian Petrochemicals Corporation Ltd in Baroda.

I wrote editorials in *The Times of India* and, later, *Indian Express* questioning how someone in such a key position in the oil industry could possibly take an objective view of this controversy. Indeed, half the committee represented oil interests. However, appearances can prove deceptive. When I met Dr Varadarajan when I was writing a book which dealt with the Taj case, among two others (*Temples or Tombs? Industry versus Environment: Three Controversies*, CSE, New Delhi, 1985), he enthralled me by describing how he once insisted that all members of the committee wake up at 5 am when they met in Agra to witness the glory of the Taj at sunrise. Armed with a lamp, he also took them on a visit to the top of the dome, which can only be reached by a steep flight of winding steps.

When the committee met for the first time in 1974, it learned how Barooah had happened to pay a visit to Italy, where he was impressed by the precautions taken against pollution from a refinery near Venice, which has a very large number of monuments. Tecneco, a subsidiary of the state-owned oil conglomerate, ENI, had built one of the plants built to control the emissions. S.G. Nayak, the general manager of the refinery, travelled to Italy and met Tecneco to see whether the plumes of sulphur dioxide could be scrubbed clean. He learned that they had no method of doing so, but relied instead on limiting concentrations at different points by raising the height of smoke stacks, so that the pollutants were dispersed more widely. Such scientific uncertainly about methods continued for a whole year.

Dr B.B. Lal, who formerly headed the ASI's chemistry department, which is based in Dehra Dun, was commissioned to ascertain the capability of indigenous institutions to carryout the task of curbing the emissions. Dr Lal wisely assessed that there was a great deal of expertise. For instance, the Indian Institute of Science in Bangalore was measuring the impact on helicopter blades of hot, dusty winds in the Rajasthan desert for the defence forces. The Taj was also being assailed by such winds from the same desert, albeit at a much greater distance; the sand contained particles which could pit the sandstone façade of the monument. However, as Dr Lal concluded, although there was expertise in isolated institutions, there was no single agency which could take a holistic view of the situation. Ultimately, therefore, Tecneco was appointed as consultants to conduct the study.

By the end of 1975, wildlife experts were also concerned about the world-famous Bharatpur sanctuary, 40 km west of Agra. It had literally seen hundreds of migratory birds butchered on a single shoot by its erstwhile Maharaja and visiting dignitaries, including Prince Phillip. The renowned Dr Salim Ali, the 'bird man' who was the presiding deity of the BNHS, raised the spectre of this 30 sq km expanse, which harboured 2,50,000 birds at the height of the season, being contaminated by fumes from the refinery. His view was that the 'Taj Mahal can be built again, but birds from Siberia can't'. The Varadarajan committee wrote to British bird societies to seek their opinion; they replied that the threat would be minimal.

It was during the following year that studies conducted by Tecneco, NEERI and the India Meteorological Department identified the baseline air quality data in Agra. This revealed that more than the refinery, the monument was imperiled by two thermal power stations as well as the brass foundries which dotted the town. The studies involved a combination of mathematics to establish how a smoke plume 40 km away would descend on a monument, given the prevailing winds and other factors; chemistry to examine the effect of sulphuric acid on marble and archaeology to find out what the state of the Taj was. Samples of marble were painstakingly obtained from Markana in Rajasthan—identical to the material that was used for the monument in the mid-seventeenth century.

The Varadarajan committee submitted its report at the end of 1977: it was, at the time, the most comprehensive study ever conducted in the world of the impact on an ancient monument of industrial pollutants. It concluded that there were dangerous levels of sulphur dioxide and suspended particulate matter in Agra, which is why it recommended closing down the two thermal power stations and shifting the small-scale industries, particularly the foundries, to the outskirts of the town. It then called for a ban on the setting up of new industries if the Taj lay downwind of these and asserted that any large project should first be examined for its environmental impact.

As may be expected, such a report was not accepted by the critics. Shivaji Rao cited the damage caused to other monuments around the world: to the Acropolis in Athens, Cleopatra's Needle in New York and Cologne Cathedral in Germany, all of which had been virtually defaced by pollutants in the air, mainly from cars. In retrospect, their apprehensions were misplaced and the precautions taken, which later included the use of low-sulphur crude for the Mathura refinery and the creation of a protective tree belt around the monument and the ban on industries around it, appear to have done the trick. The Italian restorers, who lovingly surveyed the Taj, also confirmed that the monument was basically in good shape after 350 years, and local pollutants were a bigger threat than the plumes from the refinery.

It is interesting to see how this controversy raised issues which have plagued the environmental debate ever since. The first was that perennial problem: which is more important—preserving the environment or providing development? As a letter to a newspaper put it in 1979: 'It should be remembered that the Taj is basically a mausoleum in memory of the consort of a bygone era…Shah Jahan wasted crores of rupees just to satiate his fanciful desire and thereby engrave his own grandeur in the niche of history.' This crass view was echoed by Nikita Khrushchev when he visited the Taj in 1955. He declared that had Shah Jahan been a Marxist, he would have spent the funds on building a hospital instead. Even the Agra Chamber of Commerce asserted that the Taj could be pulled down, brick by brick: it wanted the refinery for the economic growth it would bring to the region.

This was short-sighted, to say the least: the cultural value of such a monument is 'priceless'. As Mary Sleeman, wife of the official who banned *thugee* in India early in the nineteenth century wrote to her friend in England: 'I would happily die to have one such monument over me!' It provides an iconic image of India—witness the Indian restaurants throughout the world which have miniature Taj replicas within their premises, if they are not actually named after it. The same is true of the pyramids of Egypt, Machu Picchu in Peru and Angkor Wat in Cambodia.

What is more, the earnings which the Taj brings to India can never be underestimated. It is in many cases the only reason people visit the country in the first place. By the 1980s, 40,000 people visited the Taj every day; the figure has risen several folds since. Tourism is now the world's biggest industry and growing faster than most. It is basically a service industry and one which the country badly needs. All monuments—some 15,000 were listed in 1966, and only 5,000 were under the supervision of the ASI even two decades later—deserve to be protected as much for their intrinsic value as their economic worth.

The Taj, which is recognized by UNESCO as a World Heritage Site, is relatively well looked after but even so, eternal vigilance is the price which any monument has to pay for its safety. Another UNESCO site, the temples at Hampi were once listed as endangered because the authorities built a bridge too close to the complex and there were other instances of insensitive construction near it. Another example of such world-recognized sites is the Ajanta cave paintings, which ASI attendants have been cleaning and in the process possibly defacing. A renowned British restorer, Sir Bernard Feilden [sic] once threatened to write to UNESCO, asking it to withdraw its recognition of the Elephanta caves in Mumbai because he saw youths clambering over the Shiva statues to get their photographs taken. By causing such a huge controversy, the Taj case has attracted the attention of the public to the need to preserve our ancient heritage.

The controversy also highlighted the role of scientists. The ASI was, and presumably is, not too competent to take a holistic view of the protection of a site. This was made abundantly clear when this writer discovered that the chemistry branch, based in Dehra Dun, was habitually cold-shouldered by the more prestigious conservation department, ensconced at headquarters in Delhi and thereby close to the seat of power. Indeed, one ASI chemist informed me that he had not been able to obtain a copy of the Tecneco report till he went on a trip to Italy, where he procured a copy. The ASI has applied the wrong chemical in attempting to preserve Angkor Wat.

There was also the tendency, already referred to, of resorting to misinformation on the part of scientists and experts. The problem required an inter-disciplinary approach, which was conspicuous by its absence, with many scientists preferring to harp on their narrow specialization and trying to score points over others. Perhaps the scientists should have taken their cue from the punishment that a Greek court once imposed on a prominent conservation expert who had exaggerated the impact on the Acropolis of traffic fumes in Athens: he was sentenced to a day's imprisonment for his sins of commission!

India has seen its shade of wrong technical solutions to preserve monuments. In the 1930s, the Ajanta caves fell under the purview of the Nizam of Hyderabad. He summoned Italian restorers to try and protect the famous paintings from the toll taken by time—and the excess of visitors. The Italians had no experience of dealing with monuments in a tropical country, where it rains only three months in the year, after which there is searing heat. In their country, it rains less but throughout the year and the moisture is trapped within stone: when this freezes, it cracks the façade. They applied shellac to the paintings, which blackened with time: the 'solution' turned out to be the problem. It has since had to be painstakingly removed—at considerable risk of obliterating the figures which have also given the country its most potent iconic images.

Today, while the Taj Mahal has been preserved for posterity, the same can hardly be said of thousands of other monuments, big and small, dotted across the length and breadth of the country. If nothing else, it requires much greater public opinion to exert pressure on the authorities to take proper precautions. Otherwise, as was the case of the Taj over different periods in history, there is every possibility that people insensitive to the demands of conservation may make a truly monumental mistake and cause irretrievable damage.

24

# The Tragedy of Displacement

Sanjay Sangvai

The Sardar Sarovar Project would cause the most massive displacement by any one project in India till date. Though the government has always underestimated the number of affected persons, it is certain that at least 41,500 families from 245 villages—in Maharashtra (33 villages, 5,000 families), Gujarat (19 villages, 4,500 families) and Madhya Pradesh (193 villages and over 33,000 families)—will be affected by the reservoir alone, as per the announcement of the Union Water Resource Minister in Parliament in July, 1994. That is a steep increase from Gujarat's claim of only 6,700 families to be affected. Only the reservoir affected persons are considered project affected persons (PAPs) by the government.

The Narmada Bachao Andolan (NBA) has been raising the issue of the definition of 'oustees' and the extent of actual displacement. In the reservoir affected villages alone, thousands of families of boats-people, fisher-people, labouring castes, riverbed farmers and the so-called 'encroachers' in Akrani-Akkalkua and Alirajpur tehsils, were and are still not counted as oustees. In Akrani tehsil, land settlement in the so-called forest land was not done since Independence. The *Adivasis* who have been cultivating the land for generations are declared as 'encroachers', denying them their right over the land and the consequent benefits as 'oustees'. NBA has been raising the issue of such legitimate tillers for years. From 1998, there has been a semblance of action on the part of the Maharashtra government.

In addition, the non-reservoir affected people are not at all considered to be PAPs. Among them are 1,70,000 landholding families to be affected by the canal network, out of whom at least 24,000 landholders (*khatedars*)

---

* Reprinted from Sanjay Sangvai, 2000, *The River and Life: People's Struggle in the Narmada Valley,* Earthcare Books, Mumbai and Kolkata.

would be rendered small and marginal farmers with unviable holdings.[1] Moreover, there are over 950 families affected by the dam colony, and thousands of families affected by secondary displacement—for making way for the resettlement of the Narmada oustees—as witnessed in Gujarat and Taloda in Maharashtra. More than 4,200 hectares of forest in Taloda-Akkalkua area was cut to resettle Narmada oustees from Maharashtra. Over 50 villages are dependent on these forests. There are earlier Adivasi settlers on this land, who also are called 'encroachers'.

They are to be ousted to resettle the Maharashtra oustees. There have been clashes between the early settlers ('encroachers') in this forest and the Narmada oustees. The government resorted to firing in Taloda forest on 13 July 1992 resulting in the death of Dhanibai Padvi, from the earlier settlers. In 1998, there was one murder due to these clashes. In Gujarat, absentee landlords are selling their lands for resettlements at a price of unto Rs 25,000 per hectare, depriving the actual tillers and landless labourers (*chakars*) working on those lands for their livelihood.

Catchments area treatment activities and compensatory afforestation for the dam have been usurping the lands of hundreds of tribals in Maharashtra, Gujarat and M.P. About 42,000 people from over 108 villages will lose their right over the forest, due to the Shoolpaneshwar sanctury that is being planned to compensate the loss of forests submerged by the dam. Now, the villages are getting evacuation notices for the power sub-station, Garudeshwar Weir Dam. Over 10,000 fisher people and other villagers dependent on the downstream flow of the river would be severely affected as the dam gets completed and the downstream river water is depleted. There has been little cognizance of these categories of dam affected people. Besides, a number of boats people, crafts people, artisans, shopkeepers, hawkers, etc. dependent on these villages would be affected. Thus, there are waves of displacement in this one single project, amounting to 10 lakh (1 million) people.

The resources that the tribals and peasants of Nimad possess have also not been calculated properly. Neither the methodology nor the sensitivity exists to evaluate the way the people live on the basis of common natural resources for their houses, water, food, fodder, fuel and nutrition, right from the cradle to cremation. The resource base and its usufruct rights, common property rights in the villages have not been considered. Nor has there been any appreciation of the prosperity, even in the conventional economic parlance, of the Nimad villages. The Andolan evaluated the resource base of the first nine villages way back in 1987, and found serious gaps between the evaluation of the people's resources by the government and that by the people themselves.

## Vain Attempts

Processes of civil dialogue through government, non-government, or academic initiative have proved fruitless. There is no common policy among the concerned states regarding resettlement, thus denying the oustees the chance to resettle in the state they want. According to the Narmada Water Disputes Tribunal (NWDT), a complete master plan of resettlement was to be ready by 1981; this is not yet in sight in 2000. Even now, the government does not have the exact number of oustees or an evaluation of their resources. The definition of a PAP is fraught with serious lapses. The government lack the resources—land, money and political will—to resettle such a large number of people. This has been regularly pointed out in the meetings of the Narmada Control Authority (NCA), many a time by the representatives of the Ministry of Environment and Forests (MoEF) and the Union Ministry of Social Welfare, the latter being responsible for the resettlement aspect.

Government functionaries, independent experts, institutions, journalists and fact-finding teams have repeatedly brought to light the impossibility of resettlement. The NBA had constantly pointed out the basic shortcomings in displacement and resettlement aspects to the World Bank and to the state governments and NCA. Thayer Scudder, a sociologist and consultant to the Bank on resettlement, pointed out the discrepancies in the early stages itself. The Commissioner for SC/ST of India, in his report in 1989–90, opined that the right to life of the tribals was being snatched away in the name of public purpose, violating all their constitutional rights. According to him, it would constitute contempt of the law and the Constitution to push ahead the work on the dam while important issues regarding its costs and benefits remained unanswered.

The Tata Institute of Social Sciences (TISS), the Monitoring and Evaluation agency for Maharashtra oustees, made it clear in its reports in June 1988 and 1993 that the resettlement of the Maharashtra tribal oustees in Parveta in Gujarat and Somaval in Maharashtra was far from satisfactory, and that it violated their right to life. The agency objected to the government policy of treating a large number of the oustees from Akrani tehsil as encroachers and landless. While many of the oustees were organizing to resist the unjust displacement, the government forced several oustees to migrate from their original villages. TISS opposed these government actions which amounted to a policy of 'divide-and-rule', and advised the Government of Maharashtra to explore alternatives to the big dams in order to minimize displacement. The Morse Committee, in its report in 1992, has termed the full and just resettlement of even the reservoir oustees as impossible, while recommending that the canal and colony oustees should also be considered as PAPs.

In December 1994, a Committee appointed by the Maharashtra government, comprising social activist Vivek Pandit and Manikrao Gavit, Member of Parliament from Nandurbar, highlighted the lack of land, housing, fuel, fodder, and food at the resettlement sites and the denial of rights to major sons and 'encroachers', who do not have *pattas*. It recommended that until all the vexed issues are solved with the consent of the oustees, the Maharashtra government must demand stoppage of the work on the dam and other activities causing displacement. In addition, a number of other independent agencies reviewed the situation. In 1992, the local BJP unit of Shahda, Taloda came out against the forced resettlement of the Narmada oustees on the Taloda and Akkalkua lands that were being cultivated by earlier 'encroachers'.

In M.P., a number of government-appointed and other committees highlighted the plight of the M.P. oustees resettled in Gujarat. A committee of legislators in M.P., headed by the former (and now acting) irrigation minister Dr. Ramchandra Singhdeo, toured the resettlement sites in December 1994 and came out with a scathing report. Hundreds of oustees had left their resettlement sites and returned to their original villages. His report called upon the Madhya Pradesh government to halt the project and have a complete review regarding resettlement and other aspects. The government also appointed the committee of Members of Parliament from the state to look into the conditions of oustees resettled in Gujarat. This Committee, headed by Dilip Singh Bhuria, came out with similar conclusions and recommendations. The government accepted in the state Assembly that hundreds of oustees returned to their original villages, abandoning the Gujarat resettlement sites.

Of all the oustees, those from the Gujarat villages have faced the worst conditions. They have been organizing against the injustice and cheating by the government and the NGOs that made them resettle some 10 to 17 years back. Most have been either without land or without cultivable land and most of them still live in tin sheds without water supply, fodder, fuel and other minimum requirements of life. They have been struggling for their rights from 1994 onwards, in cooperation with the NBA. According to the surveys of the resettlement sites carried out by the Gujarat Visthapit Sankalan Samiti (Gujarat Displaced People's Coordination Committee) in 1995, and later by the Gujarat People's Union of Civil Liberties in 1996, the oustees have been impoverished, and a majority of them could not survive in the resettlement sites.

In 1992, 24 families abandoned the Malu resettlement site to return to their original village, Vadgam. Over 500 oustees' families of Gadher, Makadkheda, Kadadha, Antras villages in Gujarat abandoned their resettlement sites in Dediapada and Dabhoi-Sankheda tehsils and returned to their original villages. The Gujarat government tried to obstruct and arrest the oustees who were returning along with their households.[2] All this exposes the claims made by the Gujarat government and the servile NGOs regarding the 'model resettlement' of the Gujarat oustees. ARCH-Vahini, one of the NGOs which had taken the lead in shifting the people with the confidence that resettlement was possible, admitted in 1996 that the resettlement of the Gujarat oustees appeared impossible, and the only way was not to leave the villages. 'Hundreds of Madhya Pradesh oustees have been cheated into accepting problematic, uncultivable land, hundreds of others remain without decent amenities. We have been pushed into *Koi nahin hatega, baandh nahin banega*[3] (no one will move out—dam shall not be built).'

## Unacceptable Means

Removing even this miniscule proportion of oustees from their original village required threats, police action, beating and jail. Every step relied on tactics such as siege of the villages, corrupt practises, allurements to the village strongmen, dividing the village and, above all, the constant threat of submergence by predicating the dam as inevitable and irreversible. Instead of giving full information to the villagers and proceeding with their participation and consent, the displacement was brought about by devious means, through considerations other than those of just resettlement, and recreating the economy and social milieu.

The unprecedented police action in Manibeli became routine from 1991 onwards. The government resorted to beatings, jail, rape of tribal women and forcible survey. These atrocities were accompanied by police firing near Chichkedi village in Akrani tehsil, killing Rehmal Vasave on 18 November 1993. Men and women were harassed and under constant threat and tension; life itself became insecure. In Gujarat, hundreds of police forcibly removed the Vadgam people, who had returned from their resettlement site. Similar was the treatment meted out to the Turkheda people, of whom the police raped Budhibehn Rathwa.

In Madhya Pradesh, apart from continuous repression, the state government resorted to giving cash compensation to the villagers in lieu of land, which was a direct violation of the provisions of resettlement. In Maharashtra, the government tried to lure away some of the villagers, using them as agents to get other people to leave their villages. As if this was not enough, large-scale corruption and fraud prevails in the resettlement process. To show resettlement quotas full, the names of minors are included as majors and the names of outsiders are included as villagers by the lower and middle-range officials. Thus, the governments of Maharashtra and other states can make the fraudulent claims of 'Resettlement Completed'. The state of Maharashtra used repression, constant insecurity and the threat of submergence to make people move out. The politics of inevitability and irreversibility was used to its fullest by the government. Displacement and resettlement are very far from the clean process that it is made out to be on paper by the dam builders. Instead of being a natural, just and rational process, it has been an extension of the exploitation and corruption of the system.

The Gujarat government employed over 25 NGOs to lure the people away from their villages, do contract jobs like installing hand-pumps, wells, building roads etc, on the resettlement sites. Whatever short-term and marginal improvements this may have brought, it has been used to the maximum by the government to lend false legitimacy to the resettlement. Without understanding the process and trauma of displacement, some worthy members of the academia in Gujarat glorified the displacement by means of such smooth phrases as 'Emergence, Not Submergence', 'Displacement Is Development', *'Doobi Jamin, Taryu Jeevan* (Land submerged, Life Afloat).

Withstanding this onslaught, the people fought valiantly. Many did not move out. Through Manibeli village experienced submergence thrice, the adivasis did not accept the government resettlement to this date. Even in the monsoons of 1994 and 1999, the villages in Akrani and Akkalkus tehsils did not leave their land and villages. Adivasi villages of M.P. are in equally strong spirits. New notices of submergence slammed by the government have worked up the villages of Nimad and put them in a fighting mood.

The long drawn and tedious struggle has also exhausted some of the villagers; some have accepted resettlement or compensation. As the years went by and the insecurity increased, accompanied by repression, and cajoling and bullying, some people in Maharashtra and M.P. shifted to the resettlement sites. However, many of them could not survive there and shifted back to the organized villages. The 'resettled' oustees of Gujarat and Maharashtra organized themselves. The government tried to lure away the landless labourers and the lower castes from the Nimad villages, as if it cared for them. The people in these villages knew the importance of social amity and have been trying to put up a united fight against the displacement. It is the adivasis, the small and medium peasants and the backward castes in Nimad that have been in the forefront of the fights against destitution. The appointment of the Grievance Redressal Authority (GRA) by the Gujarat government in February, 1999, which was endorsed by the Supreme Court, did not help to improve the basic situation. The government showcased some 'model' sites. But even here the issues regarding land and livelihood remained unsolved.

## Notes

1. Various surveys for Gujarat government (IIM 1991), House Committee (1992) and World Bank.
2. 'Displacement and Resettlement of SSP oustees in Gujarat', a note by Nandini Oza; various reports about the abysmal conditions of Gujarat oustees in newspaper—'Narmada Oustees Leave Rehabilitation Camps', *Indian Express*, Baroda, 29 July 1996; 'SSP Oustees Declare War on Government', *Indian Express*, Baroda, 31 May 1996.
3. Anil Patel, 1995, 'A Leap in the Dark', *ARCH Vahini*, *Times of India*, Ahmedabad, 22 December, preceded by the two-three letters by ARCH about the complaints regarding the rehabilitation.

25

# Community, Place and Citizenship

**Ajantha Subramanian**

Recent work on the politics of the environment has highlighted the centrality of 'community' to the assertion of rights claims. Too often, however, 'community',[1] like its correlated 'culture', is understood in terms of an uninterrupted continuity with the past, as a kind of permanence that does not allow for historicity or dynamism. By contrast, I suggest that the real power of community lies in its becoming, not in its being.

In this paper, I turn to artisanal fisher activism in the district of Kanyakumari to make two related arguments: first, for a processual understanding of 'community'; and second, for a consideration of ecological activism as a politics, not of cultural autonomy but of equal citizenship.

Within the environmental literature, the opposition of 'state' and 'community' is used to critique state developmental practises that displace the poor and deplete natural resources. More broadly, this opposition has come to stand in for the opposition between modernity and tradition. Many critics of Indian state developmentalism tend to seek ecological alternatives to the state and other institutions of 'modern' power in forms of 'non-modern' or community authority and consciousness that they claim either predate or exist outside modernity's sweep.

I find this binary between modernity and its Other, with the 'state' standing in for the modern and the 'community' for the non-modern, both ahistorical and politically unhelpful. It is ahistorical because it imprisons concepts in an original context of production and disregards the history of their changing uses and meanings. It is politically disempowering because it imposes a regime of cultural authenticity on social actors that limits their

* Reprinted with permission from Ajantha Subramanian, 'Community, Place and Citizenship', *Seminar,* 516, August, 2002.

range of political expression. By contrast, I consider it more historically and politically meaningful to document the production of community within the framework of state power. Only by acknowledging the mutual implication of state and community can we account for how bureaucratic categories and mechanisms of rule are incorporated into the self-representations and political strategies of subaltern groups.

For me, work on place-making has offered a useful way of highlighting the dynamism of community. Donald Moore (1998), for instance, has proposed a rethinking of locality and community, not as pristine receptacles of the non-state and the non-modern, but as products of ongoing struggles knitting together diverse histories and geographies. Localities, he insists, are not simply the backdrop *of* history but are made and remade *through* history. Finally, Moore points out that, while struggles over territory can be highly localized, they are never simply local, sealed off from the outside.

Taking his cue, I argue that Kanyakumari's artisans responded to their displacement by capitalist restructuring of the district's fishery with a politics of place-making that produced both new forms of locality and new understandings of community. Artisanal politics generated a category of local citizenship forged, not through the separation of state and community identities and spaces, but through their mutual implication.

Let me turn now to Kanyakumari. The district has a 68 kilometre coastline that is dotted with 44 coastal villages and inhabited by a low caste, Catholic fishing population numbering approximately 150,000. The social geography of the coast is at once religious and civil: the boundaries of fishing villages overlap with parish boundaries, and the parish priest is the moral authority of the village council. However, this mutual implication of the religious and civil is not without its tensions. Fisher struggles for greater caste rights within the church, or for greater lay authority on the coast, have occurred with frequency over the course of three centuries (Narchison 1983; Villavarayan 1956; Ballhatchet 1998).

It was into this dynamic cultural context that the developmental state entered in the 1950s. Mechanization of the Indian fishery was one strand of the national drive towards industrialization that took off during the decade after independence. The National Planning Commission proposed a radical transformation of capture fisheries that paralleled India's Green Revolution in agriculture: new mechanized fishing technologies would boost catches to levels commensurate with the postulated wealth of the oceans, contribute to the economic development of the country, and help feed its burgeoning population. This 'Blue Revolution' was to be an all-India affair, promoted by the central government and adopted with variation in every coastal state.

The commission's recommendation of rapid technological change was justified by perceptions of the coastal population as socially backward. The incorporation of the coast into a national framework of development was to help undermine those aspects of coastal culture that were inimical to social progress. At the same time and in accordance with Gandhian notions of the decentralized, self-governing village republic, the NPC identified the need to sustain the organic solidarity of the fishing village as a foundation for development.

It finally determined that community development, which would retain the fishing village as the basic unit of the development process, would be the ideal approach to ensuring the smooth transformation of the coast. In its final incarnation, community development was a peculiar blend of goals: it invoked the 'village community' as an organic space of 'moral economy' that would provide a foundation for the nation *and* it sought to restructure the village to suit the needs of nation-building.

In the context of the Kanyakumari coast with its Catholic population, the Tamilnadu state government sought legitimacy for fisheries development by soliciting the support of the Catholic Church and framing community development as religious minority uplift. Then Tamil Nadu Chief Minister K. Kamaraj courted the church as the moral authority of the coast, both for winning fisher votes and for endorsing fisheries development. By choosing the Catholic Church as *the* authority of the coast and disregarding the authority of village fishing councils, the state reduced the complex cultural history of Kanyakumari's fishers to a single referent of identity easily accommodated to developmental priorities. Through the political process, then, state developmental and secular understandings of 'community' came together, creating an overlap between the fisher collective of the development agenda and the religious collective of the state's secular imagination.

In the 1960s, the Blue Revolution's original goals of poverty alleviation and self-sufficiency in food were subverted by the rise in price of prawn in the world fisheries market. In Tamil Nadu, the 'pink gold rush' signalled the subordination of cooperative development for domestic consumption to the export trade in prawn. Accordingly, the Tamil Nadu Fisheries Department shifted emphasis to the rapid distribution of subsidized trawling boats for prawn harvest.

The pink gold rush restructured domestic fishing for monocrop, export oriented production. Apart from increasing levels of fish harvest, the pink gold rush radically altered social dynamics across the Indian coastal belt. Since prawn are most abundant in shallow waters, trawler owners equipped with the capital-intensive technology to take them to offshore fishing grounds now preferred to remain in the area closest to shore to avail of this valuable commodity. The crowding of the inshore sea has led to violent confrontations between trawler and artisanal fishers over access and use of the coastal waters (Kurien 1978, 1985, 1993; Kurien and Achari 1990; Mathew 1986).

On other parts of the Indian and Tamil Nadu coast, the prawn rush attracted outside entrepreneurs to fishing and created a class of non-operating merchant capitalists, most of whom had no previous connection to the sea. In Kanyakumari, however, a different pattern emerged. Here, the class of mechanized trawler owners arose from *within* the Catholic fishing population and as a result generated a unique politics of place and community around the access and use of natural resources.

Conflict between the two warring factions of Catholic fishers set in motion a triangular relationship between state, church and fishers. As artisanal militancy against trawler aggression increased, the state began to collaborate with the church to defuse the power of artisanal village councils. Both state and church have deployed the rhetoric of 'community' to present the upward mobility of one section of the Catholic fishing population as the advance of the community as a whole, and to link community uplift in turn with national progress. Both have promoted notions of religious minority solidarity and participation in the nation to present the material advance of Kanyakumari's trawler owners as the creation of a representative fisher middle class, and to condemn artisanal opposition to trawling as the reactive isolationism of a population resistant to progress.

Through fisheries development, then, the Indian state has exercised its power to incorporate coastal Catholics into nation-building frameworks, in the process producing them as 'difference' from a national mainstream. In its capacity as a developmental actor, the Indian state has identified artisanal fishers as an economic minority standing apart from the industrializing nation. And in its capacity as a secular force, it has identified them as a Catholic minority standing apart from the Hindu mainstream.

These two overlapping forms of 'community', each distinguished by its difference from a posited economic or cultural 'mainstream', have circumscribed the relationship of fisher artisans to the state and operated as limits to full citizenship. However, Kanyakumari's coastal artisans have responded, not by rejecting the state and its categories outright, but by appropriating and reworking them in unexpected ways.

In response to the coastal crisis, and their marginalization by both state and church, Kanyakumari's artisanal fishers have turned to a politics of place-making that maps identity onto territory. They have tapped recent developmental and political initiatives to constitute themselves as a collective of 'traditional practitioners'. This new understanding of community has three key elements—territory, technology and ecology. As I detail below, each of these elements has a longer history. However, over the last two decades artisanal fishers have redefined these elements and combined them to construct an artisanal community consciousness.

The reworked understanding of territory that grounded artisanal community consciousness reflected a spatial shift from village to zone. Previously, fishers asserted their right to shore space and the marine resource through the village. By the late 1980s, however, the village was supplanted by the zone as the primary basis for territorial identity. Interestingly, this shift was catalyzed by a state initiative. In response to widespread artisanal attacks on trawlers across the Tamil Nadu coast in the late 1970s, the Tamil Nadu state government instituted the 1983 Marine Fisheries Regulation Act that reserved the 3-mile inshore zone for artisanal fishing.

The Act was mainly compelled by 'law and order' concerns: its primary purpose was to separate fisher antagonists into distinct zones to stave off conflict while continuing to promote development through mechanization. In effect, however, the Act exacerbated tensions between warring fishers. In Kanyakumari, artisanal fishers took full advantage of the new Act. The line in the sea substituted a horizontal boundary for the vertical ones separating villages and became a territorial marker for the divisive hostility between mechanized and artisanal villages. Now, trawlers were attacked not only when they damaged artisanal craft and gear, or harvested large catches, but also if they transgressed the 3-mile inshore boundary. With every clash, the 3-mile zone became an even more potent symbol of artisanal identity.

Artisanal fishers did not only claim the 3-mile zone; the zone became symbolic of their link to the sea, a connection that they claimed the trawler fishers had lost. Take for instance this explanation provided by a fisherman who participated in the fire-bombing of a trawler owned by a friend's relative. When I asked him how a population sharing caste and faith came to be so divided, he explained,

> It's because the trawler owners have forgotten who they are and what they know about the sea. You see, anyone can use a trawl net—a farmer, a teacher, even a bureaucrat! But when we go out to sea, we have an instinctive sense of where the fish are. We can read the water like others read the land. It's this shared sense of the sea that makes us a community.

By this logic, the very markers of civilization adopted by trawler owners, such as land ownership and absentee boat ownership, are reinterpreted by artisanal fishers as markers of deracination, of an uprooting from the sea and, by extension, from community.

Like their redefinition of territory, artisanal fishers' redefinition of technology was also compelled by another development initiative, this time by the church. With the expansion of the development arena in the 1970s to include non-state actors, the Catholic Church too entered the fray. A decade after the onset of the prawn rush and frequent clashes, a section of the Kanyakumari clergy began to question the liberatory potential of the state's development agenda and rethink their own role as moral custodians of the coast.

Drawing inspiration from Latin American liberation theology and the Indian communist movement, they began talking about the economic and cultural rights of the poor and how to extend the church's 'natural authority' to fill a development gap left by the state. The ensuing 'option for the poor' was manifest in a church project to motorize artisanal crafts. The aim of the project was to create an intermediate technology that would in turn create an intermediate category of motorized fishers and help undercut the polarization of artisanal and mechanized fishers.

After much trial and error, a motorized canoe with a speed equal to the trawler became operational in the late 1980s. Instead of undercutting sectoral tensions however, the spread of canoes increased the militancy of artisanal politics. With trawling identified as the only real enemy, the new motorized technology was assimilated into the original antagonism between sectors. The inclusion of motors into the category of 'artisanal fisher' reflected its increased flexibility *and* specificity. Now, artisanal could include new forms of technology as long as they weren't trawlers. Not only were they assimilated, the motorized canoes became the policing arm of the artisanal sector. The speed of the canoes enabled head-on confrontation with trawlers at sea and the frequency of clashes increased sharply. In addition, artisanal village councils whose legislative authority had been undermined by their inability to restrict trawling were now revitalized through the deployment of vigilante canoes.

The political resonance of the opposition between artisanal and mechanized technologies is especially evident in the attitudes of fisher youth. Many young fishermen did not grow up learning their fathers' skills, opting instead to spend part of the time at school and part of it working as labour on mechanized craft. For most, this experience as seasonal labour on trawlers is inherently conflictual: they are attracted to trawling for its wealth and status, and repelled by the humiliation of working on the boats.

A number of the young fishermen I spoke to recounted tales of oppressive working conditions and poor pay. The experience of a labourer's humiliation has given them new appreciation for the relatively more egalitarian character of artisanal harvest. A 21-year-old fisherman who had worked successively on five trawl boats before turning to work on his uncle's artisanal craft spoke of the trawler with both hostility and fear: 'It changes people,' he said, 'it makes them arrogant and cruel. Before I worked on them, I used to pray for the money to buy a boat. But now that I see how it changes a person, I don't want one.' By associating trawling with a breakdown of social relations, artisanal fishers constructed a new moral economy that made the very ownership of trawlers a transgression of community.

Finally, artisanal fishers redefined ecology to reflect a new concern with sustainability. The lives of artisanal fishermen are marked by the unpredictability of harvest. While seasonal variation and individual skill do contribute to the outcome of fishing trips, there is also a great deal left to chance. On any given day, two groups of fishermen operating in the same area using the same craft and gear may be either blessed with a full net or cursed with an empty one. Artisanal fishers often contrast the unfathomable nature of the sea with the farmer's mastery over land.

An elderly fisherman and village councillor explained to me the integral role played by *kadal matha*, the goddess of the sea and a local incarnation of the Virgin Mary, in the lives of fishers: 'The land can be owned and farmers plant seeds knowing exactly what crop they'll harvest. But the sea isn't anyone's property. We never really know what our kadal matha will give us.' Although it causes bitterness, divine providence as a reason for empty nets is accommodated within the moral universe of artisanal fishers. This makes it all the more unacceptable that mere human beings would usurp this divine right through technological capability.

Artisanal outrage at such hubris on the part of the trawlers has found new expression through the language of 'sustainability'. Sustainability as a concept entered the political lexicon of local artisans through the mobilization work of the National Fishworkers Forum (NFF). The NFF's work in Kanyakumari began in the early 1990s when, in the name of economic liberalization, the state deregulated India's 200-mile exclusive economic zone and licensed the entry of over 2,000 foreign industrial fishing vessels.

In its anti-liberalization campaign, the NFF stood development on its head by equating trawling with destruction not production, and by identifying artisanal fishing as the only means to a sustainable future. This initiative drew Kanyakumari's artisans into a global political arena that linked local struggles over use and access of marine resources. But even as they were incorporated into a global politics of opposition, artisanal fishers increasingly used the language of fate and faith to counter trawler aggression. They began to speak of trawling, not simply as an expression of greed and unequal distribution, but as hubris against divinity.

Resource depletion was a warning from above not to disrespect the gift of nature. Significantly, 'nature' also included the god-given skill of artisanal fishing which made the deskilling effect of mechanized trawling an added affront to nature and divinity. The link between artisanal fishing, divine will, and the sustainable future of the resource produced a new sense of religiosity around their special connection to the sea, displacing moral authority from the church to artisanal fishers and making them the custodians of the sea and the moral arbiters of local conflict.

Together, territory, technology and ecology crystallized a place-based community consciousness that challenged the framework of state developmentalism. It did so by reconstituting community to exclude trawlers and reframing authority to exclude the church. These reworked forms of community and authority have anchored a sense of local belonging. However, I would argue that for Kanyakumari's artisans, the claim to local identity and rights was also a claim to citizenship. They endorsed so-called local identities and priorities and rejected their subordination to national concerns. But they did not demand local *autonomy*. Rather, artisanal fishers demanded *greater* state intervention to protect their rights as artisanal producers. It was this equation between the critical role of state power and local rights that made artisanal politics one of citizenship.

In articulating a sense of 'ecological citizenship', artisanal fisher politics offers the following challenges. First, it points to a rethinking of citizenship in terms other than that of national belonging. Most importantly, it suggests a new perspective on subalternity. On the Kanyakumari coast, there was no unitary insurgent consciousness rooted in premodern culture. In the process of reconstituting themselves as a community, artisanal fishers redefined culture, sometimes in old and sometimes in new terms. They invoked the protection of the Virgin and the state. They demanded their rights as locals and as citizens. And they reconstituted community in cosmological as well as in class terms. In their self representations and political strategies, Kanyakumari's artisans combined elements that are commonly separated into categories of 'traditional' and 'modern' to create a new politics that reveals the compatibility of subalternity with citizenship.

## NOTE

1. The essay in Arun Agrawal and K. Sivaramakrishnan (eds), 2000, *Agrarian Environments: Resources, Representations and Rule in India*, Duke University Press, offers empirically grounded and theoretically forceful arguments for rethinking 'community' in processual terms.

## REFERENCES

Ballhatchet, Kenneth, 1998, *Caste, Class and Catholicism in India, 1789–1914*, Curzon Press, London.

Kurien, John, 1993, 'Ruining the Commons: Overfishing and Fishworkers' Actions in South India', *The Ecologist* 23(1), 5–12.

———, 1985, 'Technical Assistance Projects and Socio-economic Change: Norwegian Intervention in Kerala's Fisheries Development', *Economic and Political Weekly* XX(25,26), A70–A87.

Kurien, John, 1978, 'Entry of Big Business into Fishing: its Impact on Fish Economy', *Economic and Political Weekly*, 8 September, 1557–1564.

———, and Thankappan Achari, 1990, 'Overfishing Along Kerala Coast: Causes and Consequences', *Economic and Political Weekly*, 1–8 September.

Mathew, Sebastian, 1986, 'Growth and Changing Structure of the Prawn Export Industry in Kerala, 1953–83', M.Phil. dissertation, Centre for Development Studies, Trivandrum, Kerala.

Moore, Donald, 1998, 'Subaltern Struggles and the Politics of Place: Remapping Resistance in Zimbabwe's Eastern Highlands', *Cultural Anthropology* 13(3), 344–381.

Narchison, J.R., V. Paul Leon, E. Francis and F. Wilfred, 1983, *Called to Serve: A Profile of the Diocese of Kottar*, Assisi Press, Nagercoil.

Villavarayan, J.M., 1956, *The Diocese of Kottar: A Review of its Growth*, Assisi Press, Nagercoil.

26

# Lessons from the Deluge: Priorities for Multi-Hazard Risk Mitigation

Aromar Revi

> Some of these (flood concentration areas in Mumbai[1]) have a tendency to disrupt traffic and paralyse city life. A number of steps such as de-silting of drainage and clearing of nallahs are taken up by Brihan Mumbai Corporation (BMC) and Railways to avoid such flooding. However, a combination of heavy precipitation and high tide may make such flooding unavoidable.
>
> —*Mumbai Disaster Management Plan*, Government of Maharashtra, 1999.

## A Watershed Event

Business as usual, irrespective of drought, plague, famine or riot has been Mumbai's hallmark through the great cotton boom of the mid-nineteenth century to becoming one of the great megacities of the early twenty-first century. The city's confident stride was broken slightly by the textile strikes of the 1980s and civil strife in the early 1990s, as it struggled to transform itself from an industrial and commercial powerhouse to a global player in a networked world. This led to some speculation about whether it would go the way of the genteel decline of old industrial centres like Manchester, Kolkata or even parts of east London; be slowly torn apart by disparities of class, communal strife and rapid institutional decline or possibly become the Hong Kong or Shanghai of south Asia (Bombay First-McKinsey 2003; Mukhopadhyay 2001). At the surface, feel-good rhetoric has ruled the day but to some the rot seemed to grow deeper with every year.

* Printed with permission from Aromar Revi, 'Lessons from the Deluge: Priorities for Multi-Hazard Risk Mitigation', *Economic and Political Weekly,* September 3–10, 2005.

Mumbaikers, pragmatic, vital people that they are, typically brushed aside these abstract concerns of a small clutch of academics, administrators and risk professionals. They went about life as usual, all the way into the early evening of 26 July 2005. But by nightfall, it was clear to many million people in Mumbai that life may never be quite the same again.[2] An exceptional rainstorm finally put to rest the long prevailing myth of Mumbai's indestructible resilience to all kinds of shocks, including that of the partition.

Over a week of incessant rainfall brought the world's second largest municipality to its knees. Over a million people were seriously affected; lives of many millions were dislocated. Lifeline infrastructure and services that are taken for granted in city life: water, sewerage, drainage, road, rail and air transport, power and telecommunications stopped functioning across much of the city. Civic and political institutions of all hues were paralysed. Emergency services were naturally overwhelmed. Even the much-touted private sector was caught napping with ATM banking services based in Mumbai coming to a halt across much of India. Schoolchildren, blue collar workers, CEOs and film stars alike waded home for hours through chest deep water.

The shock was not so much the ferocity of the worst rainstorm in the city's recorded history, but the breakdown of Mumbai's 150-year old institutions of civic management and Governance and the apparent lack of effective emergency response. The government of Maharashtra developed India's first urban Disaster Management Plan (DMP) for Mumbai[3] in the late-1990s, which identified flooding as a significant risk, pinpointed bottleneck locations in each ward, and vulnerable slums and settlements (GoM 1999; Vatsa and Joseph 2003). The sadness is that in spite of this, there was no systematic action taken over half a decade to mitigate this risk.

It is not that all support systems failed to deliver in Mumbai. Lakhs of ordinary people, the armed forces and in some areas the police and civic administration along with community based organizations (CBOs) and non-governmental organizations (NGOs) rose bravely to the challenge.

This seems to point in only one direction: the lack of a clear appreciation of the implications of Mumbai's hazard exposure; the considerable vulnerabilities of its people, infrastructure and institutions; and the lack of coordinated interventions by a series of governments to mitigate the multiple risks the city is exposed to. Yet, in spite of its tardiness, the responsibility of this systemic failure cannot be exclusively laid at the door of the government of the day. Successfully building a culture of risk mitigation and emergency preparedness, as Japan and even Bangladesh have demonstrated is often decades in the making and necessarily involves communities, the private sector, civil society, the state and national government. Therefore, interventions by each of these stakeholders at all these levels will be necessary if Mumbai has to live down the trauma of July 2005 and the international concern about it being a less than adequate place to work from and do serious business in.

## Driving Forces

Systemic failures have deep-roots that cannot be explained by sweeping generalizations like Mumbai's failure to respect nature or plan adequately, which are correct but inactionable. Similarly ad hoc solutions ranging from finding an appropriate CEO for the city, creating a separate Mumbai disaster management agency to building flood defences often create more problems than those that they seek to solve.[4] We need to delve deeper to uncover the driving forces that seem to have captured the city in a vicious and negative downward spiral. The irony is that these forces are often driven by successful past urban management practises that now feed on Mumbai's strategic vulnerabilities.

First, few know that Mumbai is moderately or highly exposed to a plethora of natural and human-made hazards: earthquakes (both on and offshore), landslides, cyclonic storms and possibly storm surge, sea-level rise, rainstorms and local and regional flooding, drought, chemical, industrial and nuclear accidents and civil strife (GoM 1999). In short, its strategic location that drew everybody from the East India Company to migrants from eastern India comes with inherent constraints that need to be respected. Mumbai, like other megacities, is such a large agglomeration of people, infrastructure and capital that it constitutes a huge concentration of risk that needs to be addressed both systemically and systematically through a process of strategic long-range urban planning (Moffat 2003).

Mumbai lies in the Bureau of Indian Standards (BIS) Seismic Zone III with a most severe historical earthquake of MM Intensity IX (in 1618) with a more recent (1951) event at intensity VIII. This led an IIT-Mumbai research team to ascribe a moderate to severe level of earthquake risk to the city. Conservative earthquake risk scenarios estimate up to 40,000 dead and 1,18,000 seriously injured people in the city, apart from many lakh buildings being damaged (Sinha and Adarsh 1999).

Mumbai was originally reclaimed from a series of seven islands, a dramatic tale of human ingenuity and the power of commerce and engineering in transforming and managing difficult landscape. The Mumbai DMP identifies 10 sections along the Central Railway and 12 along the Western Railway prone to serious flooding along with 235 flooding points within the city (GoM 1999). This was however, before climate change induced sea-level rise was seriously considered. It would be difficult to imagine an adequate and equitable response to a moderate 0.3 m sea-level rise from the residents of slums and middle class housing in Mumbai, within its current structure of governance. This may well become a reality over the next 50 years (Aggarwal and Lal 2001)—something that most city residents and enterprises are blissfully unaware of.

Fire and industrial accidents have been part of the industrial ecology of the city. Fortunately, except for the Victoria dock explosion due to the SS Fort Stikine in 1944, (which killed up to 6,000 and devastated an area of 1.2 sq km in the heart of the city) no serious Bhopal-scale incidents have taken place in the last half-century. This is however, no insurance against serious industrial or chemical accidents in the future, especially with the increasing financial pressure on over 1,000 hazardous major 'old-economy' industrial units, where safety is easier to compromise than bottom lines along with the fact that most industries have no capability to address off-site impacts (GoM 1999).

Civil strife was a common feature of many of Mumbai's suburban towns, till it swept through the heart of the city like a scourge in 1993 and is yet to be eliminated. Institutionalization of disparity, communalisation of politics and civic life, a decline in the credibility and quality of law enforcing and justice institutions and the ingress of organised crime and terror networks into the heart of urban life make this a moderate but high impact risk, something the city has often to rediscover after cyclic periods of complacency.

Mumbai is one of the few megacities that has a large and expanding nuclear power and research facility within its urban limits. Therefore, in spite of the efforts of both BARC and the AERB to maintain high levels of nuclear facility safety, most urban managers and re-insurance agencies would classify this as a significant medium-term hazard.

Second, Mumbai's greatest strategic vulnerability emerges from one of its most significant assets—its land market. The city has one of the most skewed land markets in the world, which makes it dollar-for-dollar the eighth most expensive city in the world (RE 2000). The primary outcome of this is that more than 55 per cent of its residents, who keep the city functioning and economically vibrant cannot afford to live in anything other than squatter settlements, slums and chawls, even though their per capita incomes are significantly higher than many other regions in India. Successive governments and Mumbai's multiple planning agencies have attempted to address this central question, over 30 years, but have failed.

This failure cannot last much longer, if Mumbai is to make yet another attempt at reinventing itself as a world class city. The textile strike of the 1980s marked a watershed in the life of the city, with outmigration exceeding immigration for a period. Similarly, the floods of 2005 may mark the decline of Mumbai as many know it—if the working underclasses, the newly mobile service sector enterprises choose to vote with their feet and move to more attractive locations like Chennai, Bangalore and Delhi or smaller urban centres along the western coast with a better quality of life.

City-scale mitigation measures are necessary and possible. These will need to be resource-intensive (in tens of thousands of crore) to have impact on aggregate risk levels. Given this, there would be two temptations, within the establishment. First, to not attempt mitigation because it is perceived and projected to be unaffordable. This is a false argument, as mitigation is one of the most cost-effective investments that can be made in the housing and infrastructure sectors. Second, to choose high-cost options that ensure only some mitigation elements are executed (with advantage to particular interests) as a result of which overall risk levels do not decline substantially. Third, the megacities of the early twenty-first century (e.g., Tokyo, Mumbai, Mexico) are remnants of a global industrial metabolism fed by carbon-rich fossil fuels: coal and oil. The beginning of the end of the age of oil is nearing. Dramatic increases in coal use will probably be constrained by climate change concerns. Recent studies have shown that the metabolism of megacities like Greater Mumbai is unsustainable in the medium-run on account of its huge demand for water energy and food, apart from the untreated waste that finally finds its way into its estuaries and the Arabian sea (Goa 2100 Team 2003).

Mumbai's sheer scale, the basis of its historical economic and financial power may become a liability in a more uncertain future populated by extreme climate events (of which the 900 mm July 2005 rainstorm was only a moderate example), sea level rise and possible cyclic droughts that would threaten the water security of the city. More important, the economies of scale of large industrial megacities are becoming less attractive in a world of flexible manufacturing, global supply chains, increasing mobility of capital and intense competition for better services and quality of life in smaller, more environmentally benign cities (Moffat 2003). The risk of future disruption of some of these critical supply chains, including power networks, water, oil and gas pipelines also needs to be considered.

If Manuel Castells and Kenichi Ohmae are correct, the future of the twenty-first century lies with a network of regions centred on small Tier-II million cities, with low resource footprints supported by distributed power, transportation and IT infrastructure (Castells 2003; Omhae 1991; Goa 2110 Team 2003). Mumbai may therefore have an opportunity to reinvent itself as a flexible network of smaller urban centres from Colaba to Kalyan with appropriate changes in economic structure, land use, urban ecology and metabolism. Part of the vision of Navi Mumbai of the 1970s attempted to speak to these concerns. Most important, in an increasingly uncertain future, the city would need to be designed so that a shock could ripple through the landscape without bringing the entire system to a halt.

In short, Mumbai, like its aspired alter ego Shanghai, is one of the more hazard prone megacities in the world. In the past, the high risk-return profile of the city helped it sail through with aplomb. If full external costs are counted, net returns are probably declining and possibly negative in bad years. The city must therefore immediately start on a scientific probabilistic assessment of multiple short, medium and long-term risks (both due to natural and human-made hazards) on a neighbourhood-by-neighbourhood basis.[5] This will when completed in a couple of years time, provide a systematic basis for building and enabling community and private enterprise-led resilience and the slow and difficult task of overhauling public infrastructure and services. This could give Mumbai some increased resilience during major emergencies that both New York and London exhibited over the last few years.

## Strategic Vulnerabilities and Risks

Mumbai's greatest strength and vulnerability is its people, especially the 6.5 million who live in slums, chawls and squatter settlements. The residents of these settlements because of their location, poor infrastructure and accessibility are often many times more vulnerable than those in more permanent housing.[6] The greatest burden of the recent floods has been borne by the poor, who not only have little or no security of tenure and depend on an insecure informal sector for their livelihood, but also have close to no insurance cover. Mumbai cannot reduce its overall risk profile in the future without addressing the needs of a majority of its residents.[7] This will require a strong series of interventions in its land and housing markets (Burra 2005; Patel 2005), which in turn will require considerable political courage and administrative competence to execute.

A large proportion of Mumbai's building stock is aged, dilapidated and partially engineered, implying that it does not meet contemporary standards of building safety (Sinha and Adarsh 1999). Technical measures to strengthen and retrofit these buildings are moderately well known, but have never been implemented for a city of the scale of Mumbai. It would take considerable institutional and financial innovation, a well thought through set of incentives and regulatory mechanisms and the development of a large new offshoot of the current building industry to enable this at a meaningful scale over a 10 to 15-year time frame. Gujarat's earthquake repair and retrofitting programme provides an indication that roughly 10 per cent of the capital investment in buildings will be required to support this. Nevertheless, given similar initiatives in California and a few other cities in Asia this could well be a realistic medium-term goal.

The unchecked growth of built-up areas in Mumbai that defy most planning, zoning and environmental regulations is one of the major causes of the structural vulnerability of Greater Mumbai. Added to this are weak and vitiated planning practises that are less driven by infrastructure and services availability and more by defining notional exclusive land use categories that have little relationship to the metabolism and fabric of a twenty-first century city. The central tenets of long-range urban planning have been systematically violated in Mumbai. This has choked off natural areas that provide its core ecological services: water, food, clean air, waste absorption, and protection against the tide and weather, in favour of more profit for its developers and realtors and greater rents for its regulators and property owners.[8]

Enough experience of managing urban flooding exists in Maharashtra (and nearby Gujarat) to have designed appropriate urban systems to mitigate the impact of the late July 2005 flooding. The city of Surat has cleaned up its solid waste management systems after the 1994 flood-induced plague outbreak. It has also attempted to develop a system of flood management that addresses the challenge of sudden upstream deluges that reach the city during a period of high tide causing the kind of regional flooding that was experienced in Mumbai.[9] The design and maintenance of Mumbai's strategic and local drainage infrastructure should have been largely within the technical capacity of the skilled and experienced cadre of urban engineers at the BMC. The failure may therefore be more with the institutional bottlenecks, incentives and checks and balances within India's most well-resourced municipality, rather than with technical competence.[10]

Given this debacle, there is a clear need to rationalise land cover and land use in Greater Mumbai in keeping with rational ecological and equitable economic considerations, within the framework of the law of the land. This in some cases will require people-sensitive relocation, using standards already developed under the MUTP and some areas, the acquisition of existing illegal and even legal developments.[11] The key concern here is that developers' interests do not overpower 'public interest', that the rights of the poor are upheld; else displacement from one location will force them to relocate to another often more risk-prone location.[12]

This will be a politically difficult task, but experience the world over shows that the long-term human, social and economic risks of not correcting these distortions will hugely outweigh the short-term political and financial gains of the status quo. Too many great cities have been abandoned, from Nineveh and imperial Rome to Dholavira and Hampi, once they lost their function and legitimacy as crossroads of commerce, culture or administration. It is an illusion that Mumbai is less mortal and a prey to decay than these great cities of the past.

Both public and privately owned infrastructure collapsed post-inundation in much of Mumbai and will in many areas take weeks and sometimes months to restore. It is an irony that apart from Mumbai's public water, sewerage and drainage infrastructure, the privately-owned power networks and hi-tech twenty-first century mobile telecommunication systems were all severely affected. This is contrary to the initial government claim that the failure was because of century-old infrastructure. The flooding was not so much a failure of a particular type or generation of infrastructure but the outcome of a long process of planning, regulatory and implementation failures.[13] Even basic common sense principles of having power transformers located over 50-year flood levels and ensuring fail-safe backup power for telecom, water and sewerage infrastructure were not thought through by both public and private agencies. It is hoped that at least the private sector can learn how to mitigate risk effectively and efficiently, in the future.

The once well-regarded BMC municipal services have degraded so much (even in non-slum areas) that adequate drainage and solid waste clearance operations are reported to have not been fully undertaken for years. The cumulative impact of this on local and regional flooding is more than apparent. The systematic restructuring and revamping of the BMC urban service delivery and land-use regulation functions are clearly in order. This will be necessary not only to strengthen the political and user accountability of these public service providers but also to ensure equitable service delivery to all areas, leveraging on existing programmes like the Dattak Vastee Yojana. This is a resource-intensive task that needs to be initiated in the next few months, if the credibility of the BMC as a professionally managed institution has to be restored.

Urban planners, administrators and political leaders in Maharashtra will need to understand that contemporary cities are as much about infrastructure networks and service access as about real estate values and conforming land uses. Building the energy, water, waste water, transportation, telecom and IT infrastructure for large cities takes many decades. These are very lumpy investments and require massive annual expenditures on operations and maintenance, especially if they are not adequately designed or executed. Some of these services are best provided by public providers in an appropriate regulatory frame, others given quantum leaps in technology and distributed network development are better managed privately. Yet, all of these forms of management will need to assess and mitigate risk adequately. This will require a strengthened municipality and a tenacious but flexible regulatory framework that enables public-private-community partnerships without permitting the externalisation of risk on to citizens at large and the poor in particular. If effectively executed, this is one experience that Mumbai could sell across the world.

A most serious vulnerability is the decay of Mumbai's rich civic culture and the displacement of a culture of citizen commitment to the city into short-time horizon rent seeking, sectarian political gain and bureaucratic quick fixes that miss key systemic challenges by focusing on near-term solutions: Bandra-Kurla complexes, expressways and sealinks.

Risk mitigation is difficult to implement because it requires long institutional development horizons. A political framework for long-range urban infrastructure development and risk unbundling will need to be constructed in Mumbai for the first time in the country. Simultaneously, incentives need to be intelligently designed to facilitate a stream of benefits to politicians, officials and the private sector so that their short- and medium-term interests do not compromise the goal of rebuilding a better, less-risk prone and more equitable city to live in.

## Mitigation Priorities

If Mumbai is to be renewed as a world-class city, based on the above analysis, a multi-level agenda of action will be required to mitigate multiple risks, as presented below.

### National Level

(i) Linking-up special elements of the National Seismic and Cyclone Risk mitigation programmes to create a national urban vulnerability reduction mission. This in turn will need to be linked with the National Urban Renewal Programme (NURP) at the city level to address mitigation of both natural and man-made hazards.

(ii) Revision of the Vulnerability Atlas of India (BMTPC 1997) to include estimates of risk to economic activity and capital stock due to natural and man-made hazards and hence identification of priority districts and cities for intervention.

(iii) Developing a series of new insurance instruments to provide risk coverage to multiple categories of households and enterprises in cities and incentives for public-community-private partnerships to mitigate risk.

(iv) Enactment and implementation of laws regulating professional activities of real estate developers, planning, engineering and building design professions, providing for liability for not meeting established national safety standards.

(v) Establishment of national lifeline infrastructure standards for water supply, sewerage and sanitation, power, telecommunications, transportation and IT and guidelines to upgrade legacy infrastructure.

## State Level

(i) Upgrading the existing departmental functions through the creation of a Maharashtra state disaster management authority (MSDMA) on the lines of the model bill and the experience of the Gujarat State Disaster Management Authority (www.gsdma.org) with appropriate powers and responsibilities to intervene in urban areas.

(ii) Notifying changes in relevant housing and urban development, town planning and infrastructure related legislations to integrate disaster mitigation concerns into urban planning and development drawing upon best practises in India and Asia.

## City Level

(i) Re-examining the urban governance, planning and service delivery framework and institutional arrangements for Greater Mumbai and the Mumbai Metropolitan Region (MMR) with a focus on linking urban renewal and development with risk mitigation. The overarching frame would need to be that of the 74th Constitutional Amendment and the need to correct distortions in the real estate and housing markets, through a mix of strong public and private intervention (Patel 2005).

(ii) Developing a structure plan for the MMR that links strategic urban services (transportation, energy, water supply, sewerage, sanitation, drainage and solid waste management), land-use planning and strategic risk mitigation.

(iii) Developing a long-range (30-year) vision and action plan for strategic risk sharing and mitigation for the MMR, through a multi stakeholder engagement and appropriate fiscal and financial incentives for risk mitigation.

(iv) Updating the Mumbai Disaster Management Plan and the ward DMPs (GoM 1999) within this, linking it to zonal development planning and building regulation via a common publicly accessible GIS database.

(v) Launching a Greater Mumbai multi-agency disaster mitigation mission with a chief secretary rank officer at the head that has sweeping powers of mobilizing public, private and civil society organizations and intervening in the land market, infrastructure planning and development, urban planning and regulation. Within this mandate:

(a) Strengthening emergency response and mitigation functions through a process of devolution of functions to the ward level and greater local political participation and community-based disaster management initiatives

(b) Developing and placing in the public domain a GIS-based probabilistic composite risk assessment and mitigation framework for earthquake, cyclone, flooding, storm surge, sea-level rise, civil strife, fire, chemical accidents at neighbourhood level (e.g., based on the experiences of GSDMA (2005)). This could be linked to a public database that would record all building permissions given and public investment in infrastructure and hence, could provide the overall framework within which ward and neighbourhood-level risk mitigation and planning would operate.

(c) Establishing a credible urban disaster response and recovery capacity in the city, built around existing specialised agencies (e.g., police, fire and trauma care services). Strengthening of existing central control rooms, ward emergency operations centres, protocols for emergency response and management and disaster communication via the public, electronic media and web-based interfaces.

(d) Development of a comprehensive multi-hazard, techno-legal regime for Mumbai: that covers land cover and use, zoning regulations, building and infrastructure.

(e) Establishing a public-private-community/owner partnership to finance, build and retrofit housing to disaster resistant standards.

(f) Establishing a public-private partnership to finance, build and retrofit infrastructure to disaster resistant standards, including the development of strategic flood, cyclone, storm surge and sea defences.

## Neighbourhood Level

(i) Initiating a city-wide network of community-based disaster management initiatives, especially in vulnerable locations, slums, chawls and squatter settlements.

(ii) Focus on information sharing, technical support and mobilisation to access entitlements from public agencies and mutual aid during emergency situations. Resourcing from MP, MLA and corporator local area development funds.

## Private Sector

(i) Initiation of appropriate risk assessment and disaster and recovery plans for key private service delivery agencies in the city especially in the power, transport and telecommunications sector.
(ii) Sensitisation of the electronic and print media to risk, disaster response, management and mitigation and establishment of standard protocols to enable effective emergency communication and mobilization.
(iii) Developing a viable set of financial instruments to enable the upgrading and retrofitting of buildings at risk. Developing a cadre of specialist service providers that could cater to this potential market.
(iv) Sensitization of private sector agencies to the specific risks that their infrastructure, operations and employees may be exposed to and developing a market for agencies that will provide technical support to enable risk mitigation.

## Civil Society

(i) Establishing a network of NGOs and civil society organizations that lead community centred emergency response in particular wards/neighbourhoods, sectors or interest groups under the coordination of a city-wide disaster management agency.
(ii) Advocacy and mobilization around risk mitigation for vulnerable groups.
(iii) Executing pilot projects to test new methods of community based disaster mitigation and management.
(iv) Providing feedback and checks and balances on the functioning of public agencies and private sector organisations in service delivery and risk mitigation using instruments like the right to information.

These and other interventions need to be encoded into city and ward-wise disaster management, response and mitigation plans that are openly debated, updated every three to five years, made publicly available in print and digital form and most important—implemented in a systematic, timely manner. Else, 'business-as-usual' will only bring more dramatic disruptions, loss of life and property and business in the future. The choice before the city, its residents and establishment is clear.

## Notes

1. Author's explanation. The original text refers to flooding points.
2. Multiple web reports and blogs. As an example see: http://en.wikipedia.org/wiki/2005_Mumbai_floods.
3. Following the Latur (1993) earthquake and the $ 221 million World Bank funded Maharashtra Emergency Earthquake Reconstruction Programme (MEERP), the government of Maharashtra took a number of positive steps to institutionalise disaster management and mitigation in the state, including the preparation of a State Disaster Management Plan (DMP) and a Mumbai Disaster Management Plan with subsidiary plans at ward level (for details see Vatsa and Joseph 2003). There has unfortunately been little convergence between the Mumbai DMP and the BMC and MMRDA planning, land use and building regulation functions.
4. A variety of opinions expressed in multiple newspaper, electronic media, web and other reports in July and August 2005.
5. A key technical intervention is the availability of critical building and infrastructure information as a web-enabled public domain GIS. The Mumbai DMP suggested in 1999 that 'All the infrastructural facilities and utilities in Greater Mumbai need to be mapped on to a GIS application on a multi-user basis. There is therefore a need to develop a GIS on a scale of 1:1000. This would help the planners, administrators, emergency services and utility providers'. It would also help address information asymmetry in the land and housing markets.
6. The Mumbai DMP (GoM 1999) recognized that 'All the slum colonies (whether authorised or unauthorised) are vulnerable to floods, health hazards, fires and cyclones' (p. 7). This differential vulnerability of slum areas is the basis of much of the 'structural' vulnerability of Greater Mumbai. Addressing this will require going beyond cosmetic measures like forced demolition and

relocation of slums and squatter settlements, which are short-term palliatives that do not address the core issue of Mumbai's dysfunctional land and housing market.

7. The Mumbai DMP was ahead of its time in recognising the need to integrate disaster mitigation with slum upgrading via a recognition of the need for tenure security; 'Providing collective tenure (leasehold rights) to co-operative housing societies of the (slum) settlements and encouraging them to upgrade the quality of shelter, demonstrated the possibilities of involving communities in the self-help process…It therefore becomes essential, that such strategies for shelter improvement take disaster mitigation into consideration, thereby reducing the vulnerabilities of these settlements' (GoM 1999: 25).
8. In keeping with good disaster mitigation practise the Mumbai DMP recommended, 'Control on land reclamation. All existing water bodies and storm water holding ponds will have to be protected under strict development control rules. Clauses providing for any exceptions should be deleted from the development control rules' (GoM 1999: 26).
9. This is a well-known risk which many cities (e g, Amsterdam and London) have evolved appropriate water management practises and flood defences for. But none have the unprecedented scale of land tenure problems that Mumbai has. Hence, technocratic solutions will only accentuate the problem, increasing rather than reducing vulnerability over time.
10. The responsibility and priorities for drainage/flood management in Mumbai are clearly defined by the Mumbai DMP: 'In the absence of training, soling and regular de-silting (cleaning), most nallahs have a tendency of flooding and choking. It is necessary that a programme of nallah training, soling and cleaning is undertaken rigorously through the Storm-water drainage department of the BMC' (GoM 1999: 24). A diagnosis of the malaise within urban governance institutions in Mumbai, is a necessary condition before a concrete set of solutions can be identified.
11. The Mumbai DMP states 'Settlements along the nallahs are vulnerable to floods…This may require shifting of some of the settlements along nallahs' (GoM 1999: 24).
12. Recognising the severe limitation of current planning practise, and its ability to be hijacked by short-term political and real estate interests, the Mumbai DMP recommends, 'The current typology of settlements only looks at the ownership and eligibility for regularisation. A detailed analysis of the existing settlements in terms of typology, of vulnerability would facilitate the preparation of a master plan for safe siting of such vulnerable settlements' (GoM 1999: 26).

13. Post-flooding, the government of Maharashtra has made a claim for Rs 1,200 crore from the government of India to upgrade Mumbai's drainage, which it claims is over 100 years old. It should be noted that much of Mumbai's colonial era infrastructure is located in the island city. This was less affected in this event, due to lower rainfall in south Mumbai. The bulk of the drainage challenges were experienced in areas extensively developed in the post-independence period. Large sums of international assistance including World Bank funding have gone into planning and upgrading the infrastructure for these areas, since the 1970s. While this has clearly not been enough to deliver adequate services, there is a crying need for an integrated structure plan for the city that links water supply, sewerage, sanitation, drainage and solid waste management, transportation, energy infrastructure and land-use planning for the metropolitan region.

## References

Aggarwal, D. and M. Lal, 2001, *Vulnerability of Indian Coastline to Sea Level Rise,* Centre for Atmospheric Sciences, Indian Institute of Technology, New Delhi.

BMTPC, 1997, *Vulnerability Atlas of India*, Building Materials and Technology Promotion Council, New Delhi.

Bombay First-McKinsey, 2003, *Vision Mumbai: Transforming Mumbai into a World-Class City: A Summary of Recommendations,* Mumbai, September.

Burra, S., 2005, 'Towards a Pro-Poor Framework for Slum Upgrading in Mumbai, India', *Environment and Urbanisation* 17(1), April, London.

Castells, M., 2003, *The Rise of Information Society, 2nd edition*, Blackwell, London.

Goa 2100 Team, 2003, 'Goa 2100' in S. Itoh (ed.), *Urban Design for a Sustainable Future,* IGU, Tokyo.

GoM, 1999, *Mumbai Disaster Management Plan*, Government of Maharashtra, http://mdmu.maharashtra.gov.in/pages/Mumbai/

———, 2005, *Maharashtra Floods 2005,* Department of Relief and Rehabilitation, Government of Maharashtra, http://mdmu.maharashtra.gov.in/pdf/Flood/statusreport.pdf

GSDMA/TARU, 2005, *Gujarat Vulnerability and Risk Atlas*, Gandhinagar.

Moffat, S., 2003, 'Cities' Plus' in S. Itoh (ed.), *Urban Design for a Sustainable Future,* IGU, Tokyo.

Mukhopadhyay, T., 2001, *Shanghai and Mumbai: Sustainability of Development in a Globalising World,* Sanskriti, New Delhi.

Ohmae, K., 1991, *The Borderless World,* Harper and Row, New York.

Patel, S., 2005, 'Housing Policies for Mumbai', *Economic and Political Weekly,* 40(33), August 13, Mumbai.

RE, 2000, *Global Rent Review*, Richard Ellis.

Sinha, R., and N. Adarsh, 1999, 'A Postulated Earthquake Damage Scenario for Mumbai', *ISET Journal of Earthquake Technology,* 36(2–3), pp. 169–76. Vatsa, K., and J. Joseph, 2003, 'Disaster Management Plan for the State of Maharashtra, India: Evolutionary Process', *Natural Hazards Review,* 4(4), November, pp. 206–12.

27

# The Tiger and the Honeybee

**Savyasaachi**

What modes of thinking and codes of conduct underlie the phenomena of deforestation? Can deforestation be explained by inefficient management of forests or is it a result of the inappropriate use of forest materials? In order to meet the demand for industrial timber, monoculture plantations have destroyed forest diversity; and wildlife sanctuaries and biosphere reserves have enclosed a forest biodiversity to preserve genetic materials required for the manufacture of a variety of industrial products of which seeds and medicine are most valuable.

This mode of intervention attempts to arrest and, if possible, reverse the processes of deforestation. It preserves the abundance and diversity of the forest materials to make them available for industrial production alone. An important prerequisite of this mode is to marginalize forest dwellers and declare that their presence is destructive of forest materials. Forest-dwellers are then displaced and development programmes are implemented to facilitate their rehabilitation. The failure of the accompanying development programmes to provide a large number of people with resources to earn there livelihood, pushes them towards the forest. These people are landless labourers. Where they stand face-to-face with the forest, there is a threshold, the point of contact, the interface, the frontier from where there emerges a social foundation, which makes landless people the most deserving inhabitants of the universe of the forest.

The social and cultural practises they survive by are parallel to those of dominant development programmes. These parallel practises are informed by their experience and perceptions of the forest on a day-to-day basis.

---

* Reprinted from Savyasaachi, 'The Tiger and the Honey-bee', *Seminar* 423, November, 1994, pp. 30–35.

They are in search, as it were, for a mode of earning a livelihood and the dominant practises deprive them of that mode. The ground for parallel practises, the universe of the forest, is also a living space. The two arms of this mode of scientific intervention, forest policy and land reforms, prepare the ground for the preservation and conservation of forest materials.

The forest policy prepares way to displace forest-dwellers from the forest, and land reforms prepare the way for their resettlement. The forest space, thus freed of the social and cultural perceptions of its dwellers, is transformed into a biosphere reserve. Several plants and animal species are declared endangered, threatened and rare, and therefore become expensive. It is their preservation that directs the management strategy of the reserves.

The question therefore is how can an inquiry be opened up to examine whether this mode of creating biosphere reserves leads to the preservation of the variety and abundance of forest materials? Also, is this the most appropriate perception of the forest?

These questions attempt to explore the ways in which man and nature are interlinked. Can there be a meaning to the history of social formations when it does not attempt to grasp how the presence of man influences an understanding and a transformation of nature? Reciprocally, must not a meaningful history of nature attempt to throw light on an appropriate mode of man's social and cultural presence in nature (and not outside it), which is essential to its preservation in the light of these considerations.

The significance of Simlipal for the geographical region surrounding it is as follows. The Bombay Natural History Society Encyclopaedia reports:

> Simlipal is the meeting ground of the northern and southern flora of the country. There are found animals of the Himalayan as well as those inhabiting the southern reaches of the Nilgiri mountains.

The Orissa Environment Society reports that:

> Simlipal stands as a barrier to monsoon currents passing into the Gangetic plains. Orissa gets its rainfall from the Bay of Bengal. As low pressure over the Bay of Bengal passes north of Balasore and strikes Meghasan—the highest point in Simlipal—the monsoon current is diverted to the south-west and crosses Gonasika range in Keonjhar, Pal-Lahada, Madadaneha range of Bonai and passes westwards to the Barapahar-Gandharmardana range. Had there been no Simlipal Meghasan range, Orissa would have been a desert tract like Rajasthan.

> Simlipal is at the northern tip of the Eastern Ghats. These hill ranges give rise to several perennial rivers. Most of these rivers are used for irrigation in Mayurbhanj, Keonijhar and Balasore districts. (Bose 1985; Sahu 1985).

Emphasizing the overall impact Simlipal on the ecology and economy of the region, it is stated:

> Simlipal holds the life of eastern Orissa: if depleted of forest growth it would expose the rocky pleads to radiate immense heat and create a hot atmosphere which would heat up the monsoon-bearing winds to go higher and higher and thus deprive the entire eastern Orissa from monsoon rains (Bose 1985).

Reports suggest that the management of Simlipal under Project Tiger is neither commensurate with its significance for the region nor with its internal constitution.

In 1969, the General Assembly of the International Union for Conservation of Nature prepared the ground for setting up Project Tiger, which was launched in April 1973. Since then, several tiger reserves have been created, of which Simlipal is one. The Project Tiger sanctuary is organized into three concentric circles. At the centre is the core area; surrounding it are the buffer zone and the periphery. In the core area, wildlife is to be given complete protection. In the buffer area commercial forestry operations, modified to meet the requirements of wildfied conservation are permitted. The periphery is the transition zone between the reserve and the outside.

The National Commission on Agriculture (1976) prohibited grazing and collection of minor forest produce from the core area. The principal impediments for the creation of this 'core' were identified to be forest dwellers. A survey conducted in 1991 points out that the shifting of village from the tiger reserve under Project Tiger had been accepted by the central and state governments.

The objectives of this relocation were to provide better opportunities to the residents of these villages, since they were deprived of such basic facilities as communication, education and medicine. In Simlipal, there were 223 households distributed over nine villages. Of these, 93 were Ho, 68 Munda, 45 Kharia, 9 Bahudi, 7 Santal and 1 Makhud household (Alexander *et al*. 1991). These forest-dwellers are unwilling to be relocated. They have faced difficulties ever since Project Tiger designated the core area concept.

Tigers have become a nuisance; the frequency of their intrusion into villages has increased, as has the presence of poachers. And when they damage property or cause injury to human beings, the forest department protects the tigers, not the people!

There is no assessment of how the condition created by Project Tiger open the forest to destructive practises. Further, it is not known whether the quantity of forest produce needed by forest-dwellers for their subsistence harms the forest. Neither is it certain that education and infrastructure support for displacements of forest-dwellers will reduce their dependence on forest produce; nor whether severing their relation with forest and inducting outsiders to manage will lead to its preservation.

The core area where wildlife is to be protected is in fact open to reckless exploitation. In collaboration with forest officials, poachers kill animals for tusks, hide and other exotic foods. Also, their methods harm the forest. For instance, trees are sprayed with insecticides to kill elephants, and tigers are killed when they are young as their coats are in good condition. An inventory of such methods would enable us to estimate the actual harm done to the forest by poaching. As regards biodiversity, available information draws attention to the uneven distribution of wildlife and its concentration in the core area, which attracts poachers, creates conditions for a population imbalance, disrupts processes of self-regeneration in the forest and implicates forest-dwellers.

The buffer area, intended to provide ecological and environment support for the biodiversity in the core area, has in fact become the launching ground for poachers. The commercial forestry permitted in this area has not been regulated and modified to the needs of the core area. It has been reported in this regard that:

> The Simlipal Forest Development Corporation was set up in 1980 with the sole purpose of scientific management and commercial exploitation superceded scientific management beyond carrying capacity. There was pressure to ban tree cutting from 1982 (Patro and Misra, 1985).
>
> Interference by forest department with natural plant community and introduction of exotics on large scale will destroy the biosphere of Simlipal and usher in a new plant climax and wildlife community.... Indiscriminate cutting of Sal trees... by the Forest Development Corporation results in reduced production of biomass and timber wealth....

> Frequent floods in Burabalang, Salandi, Baitarani (rivers flowing through Simlipal) are indications of denudation of Simlipal forest and results in loss of life, property and sand casting of fertile agricultural land (Sahu 1985).

The commercial collection of minor forest produce has affected the food chain and the balance of population and this has impoverished the forest-dweller. It is pointed out that commercial forestry has inhibited wildlife from coming into the buffer area. On the contrary, it has created conditions for outsiders to get a strong foothold in the area and operate in the core area. For instance, the Sahu family living in Gurgudia is reported to monopolize the collection of minor forest produce from the core area. This family has encouraged settled rice cultivation and in the process has acquired land, which has marginalized forest-dwellers.

The living condition of forest-dwellers in the buffer area is exemplified by the Hill Kharias. The government wants them to begin rice cultivation in the hope that it will wean them away form the forest, but the Kharias refuse to be seduced by these efforts. They have no faith in government-sponsored development programmes, which provide them with subsidized homes and infrastructure equipment like ploughs and cattle. These implements are of no use to the Kharias because they are not taught methods for rice—cultivated rice cannot be a staple diet and even those who cultivate rice maintain that it cannot be a substitute for food-gathering.

The house is a brick and cement structure covered by a tin roof; two small rooms are not large enough to make a home. It looks more like a punishment cell and shows disrespect to the people for whom it is made. In the summer months, the interior of these houses become over-heated as a result of which the Kharia families live under trees. Those who live closer to the core area are affected by the road opened for poachers by such development programmes.

The competition between the private and public sectors of the forest industry has pushed out the forest dwellers. Further, the creation of a core area has invited the private sector to prepare grounds for deforestation. The two sectors collaborate to practises a method of resource extraction whose rhythm and speed are not commensurate with the rhythm and speed of forest self-regeneration. A faster rate of extraction impoverishes without allowing for replenishing. This is a particular example of how human intervention, over a passage of time, has accentuated discontinuities in nature, only to impair the process of self-regeneration and render irreversible the overall degradation of the forest.

There is, therefore, a need to recover and make intelligible a principle of interaction between forest and man together in one living space. What is this on account of? The forest-dwellers recognize a life-force which flows through the forest, through its food-chains and life-cycles, from which all the elements of the forest, including man, draw nourishment. They are careful observers of food chains and life-cycles, and of the origins, development and demise of the individual and collective social life of the varied materials in a forest. The forest-dwellers' social and cultural life makes them keen observers of aspects of the forest.

Development programmes, when they attempt to train forest-dwellers for the market, damage this habit of observation. In areas within the proximity of the market place, such detailed knowledge of plants and trees is known only to a few. In areas away from the market there exists a tradition of sharing information and observations of the forest, sitting around a fire at night, when people return home. The forest-dwellers' search for food is simultaneously a study of the forest.

Forest-dwellers, in whichever part of Simlipal they are situated, continue to rely on their skills of food gathering. Under conditions created by the management of Simlipal forest produce is not easily available for subsistence: a sign of over-extraction. The forest-dwellers resist collection before the product is fully mature. However, in their impoverished condition, created by Project Tiger, they now collect before it is time to do so: for this remains their only means for survival.

The circumstances created by Project Tiger interferes with the life-cycle of plants, disrupts the food chain and prepares the ground for deforestation. For instance, premature collection of resin from tree trunks and honey from hives disrupts their natural cycle, breaks food-chains and impairs the process of self-regeneration. In this context, it has been observed that there is a dissonance between natural and artificial divisions of the Simlipal reserve. There is thus 'a change in micro-climate. For instance, the Project Tiger reserve forest is getting moister while the buffer area managed by the Forest Development Corporation is getting drier' (Sahu 1985).

The forest-dwellers complain that over the past decades the condition of the forest has deteriorated. According to them a sure sign of deforestation is the depletion of subsistence forest produce, particularly tubers and honey, from the standpoint of which they construct a view of the forest. They are of the view that, among other things, timely production in substantial quantity is indicative of the health of the forest. This differs from the dominant view where the tiger is the core species. It is embedded in the forest-dwellers experience of living in the Simlipal forest.

According to the Hill Kharias, there are several honey reservoirs in Simlipal. These are called *mahu bhandars.* They are spread evenly, cutting across the core, buffer and the periphery areas. Traditionally, each of these reservoirs has been taken care of by Kharia families who collect honey form these *bhandars.* They point out that only a decade ago, Simlipal was famous for the quantity and variety of honey. Lamenting the loss of honey, an elder said, 'these days honey-bees live in the city'.

From this parallel standpoint, the forest is a work place. The work done here involves production of nectar and its plural aromas, all of which contribute to the being both food and medicine, honey is an essence. It is made from nectar which is filtered and absorbed through numerous food-chains and life-cycles. The survival of all species is essential to the production of honey. The work processes in nature (food chains and life-cycles) do not exist independent of each other. Accordingly, survival of the fittest does not mean the elimination of the less sturdy species.

On the contrary, it is in the nature of things in a forest to ensure the fitness of all as far as possible, and to ensure the production of honey. In this perception, the tiger has an important role to play—it protects the forest from intruders who disrupt the production of nectar, aromas and honey. The interdependence of living being in any ecosystem is such that no one species is dispensable. Thus, it may be stated that a forest rich in honey production must correspondingly be rich in plant diversity. The Kharias maintain that the absence of a honey-bee can affect plant diversity in a forest as well.

There are several reasons for the decline in honey production. Of these, commercial forestry is one. The most wanted commercial product is the sal (*Shorea robusta*) tree. A description of the deforestation due to the cutting of sal trees illustrates ecological processes that sustain a healthy forest: high in its canopy, dense in its undergrowth, rich in diversity and abundant in its plant and animal species, the sal tree ecology is protected by the religious tradition of the Hill Kharias.

The inclusion of the regenerative processes and products of the forest in the Kharia way of life is also a way of keeping an eye on the well being of the forest. The forest-dwellers integrate the forest in their way of life—social, cultural or religious. This is best seen in the case of the sal tree. Sal wood is known as *daru* meaning God. This nomenclature extends to wood from other trees as well. Sal is placed in the centre of a *zaheera,* a sacred grove situated at one end of a village where human intervention for livelihood is taboo. Here, forest-dwellers in Simlipal as in other forests of central India propitiate this tree when it begins to flower.

Sal is reported to be indigenous to Simlipal. It is a predominant species in this forest, although it is not found in large number elsewhere in Orissa. This is sufficient ground for forest-dwellers to believe this plant to be sacred. It is a naturally germinating plant and can survive by itself, depending on no other plant for growth. The seeds mature while they are attached to the mother plant and begin to germinate as soon as they fall on the ground. These two qualities make the sal tree auto ecological.

The gummy latex secreted from the trunk of the sal tree forms an aromatic resin, which is burnt to identify the atmosphere of any sacred event—thus making it a special occasion. The latex attracts the honey-bee, which finds the tall tree appropriate for housing its hive. The sal flower provides more nectar than other forest flower. Moreover, the sal provides shelter and is the primary host for the silkworm. These multiple uses of the sal are guarded within the boundaries of the *zaheera.* However, outside it, forest-dwellers eat sal seeds; use its wood for construction; its latex for driving away mosquitoes; its leaves to make cups and plates to serve and eat from; and its flower for curing diahorrea and dysentery.

Commercial forestry has destroyed the practical uses associated with sal. It has created a market for all the forest produce generated from the sal tree. The cutting of sal trees in large numbers (larger than the number reproduced) has forced honey collectors to take beehives from short trees. In these hives, honey production is low as they are prematurely taken and because the bees are disturbed by the close proximity of human beings.

The collection of sal seeds, it is reported, reduces the quantity of food material for herbivores and saprophytes. Accordingly, the population of rodents, squirrels, mole rats, porcupine, hare and rabbits begins to decline. This in turn, impacts the population of reptiles that feed on these herbivores. Further, the collection of sal and *tendu* leaves affects the activities of saprophytes. In the forest, the principal sources of organic matter are the leaves, stems, fruits and seeds of trees. Reduction in the quantity of leaves affects the humus layer which is getting thinner in Simlipal. This, in turn, affects the growth of mushroom and toadstool. With a decrease in mushrooms, people in Simlipal look for wild animals for their protein nutrition (Sahu 1985). These, the Kharias say, are also on the decline.

In the overall situation that emerges, the detritus food-chain, which orders the processes of self-regeneration, is disturbed. These processes operate on the ground at the level of the flower. The ecological processes that are ordered around the sal tree begin with the detritus food-chain. The organic matter which falls on the ground is food for micro and other organisms, which are food for predators (small carnivores). The process of decomposition contributes to the self-regenerating quality of the forest. This is the beginning of a detritus food-chain. At the ground level, its operation is less dependent on direct solar energy and more on the influx of organic matter produced in other systems—the life-cycle of living species.

In a very basic way it creates conditions for the production and reproduction of the green vegetable plant world. This is the base for grazing food-chains—herbivores and the carnivores eating these herbivores. This is one development of a detritus food-chain—through it the carnivores eating these herbivores. This is one development of a detritus food-chain—through it the self-regeneration of different species in the forest ecosystems. The detritus food-chain also generates active reproductive materials (flowers), which ensure the production of nectar and honey under the guidance of the Queen bee. The Queen is fertile and all males of its community are drones and worker bees. The nectar is a self-regenerative material, worthy of propitiation, being both food and medicine.

In contrast to the process of decomposition, which is operative at the ground level of the height of a flowering plant, where direct solar energy is necessary for growth. The honey-bee is at the top of this ecological pyramid which begins with humus and continues to the production of nectar, which is processed into honey. This ecological process stands in contrast to those structured by the herbivores and carnivores of a grazing food-chain. Their honey-bee is part of a process that produces the materials on which the diversity of animals in a forest depend. Therefore, to place any other species at the top of a food-chain is considered to open the doors to a misunderstanding and mismanagement of the forest. This wrongly positions a detritus food-chain in its method and strategy for preservation and conservation of a forest.

The dependence of the insect population on plant diversity is both food and habitat specific. The honey-bee is placed at an important noble centre of this food-chain. It interacts with all the tiers of the forest and collects nectar from a variety of plants. The honey-bee can therefore be used to monitor the health of a forest. With its help it is possible to identify deforestation before it gets too late. A forest with abundant honey must therefore have the following properties: a high canopy, having grown through a 3-4 storey structure of a forest; dense undergrowth; and a rich diversity of flowering plants providing an important source of nectar. Finally, there must be continuous water supply.

In a healthy forest, honey-bees, of which there are several kinds, gather nectar from flowers distributed over different stories of a forest. This is their territory of operation. They prefer not to be disturbed in their work of honey production. Some are more sensitive to human sounds, such as tiger-bees. Accordingly, they make their hives in high places, like tall trees, on rocks and stones on top of hills, between crevices, or inside tree trunks. These living spaces are normally shaded and in proximity of water sources.

Under the dominating Project Tiger regime, the programme to preserve the biodiversity of Simlipal forest aims to make raw material available for the production of a variety of industrial goods. The processes in nature that produce and reproduce these raw materials are understood to centre around a species—the tiger—and in an area designated as the core.

This mode is centralized and undemocratic, for not only does it marginalise people who are familiar with the forest on a day-to-day basis, it also marginalises plant and animal species which are not directly needed for industrial production. The biodiversity in a forest is arranged in a hierarchical order depending on its requirement. The most privileged species are labeled 'endangered', or 'rare'. From the standpoint of industrial production they are 'rare' and therefore, the most expensive. Second, in order are the minor forest products. The rest of the plant and animal kingdom is part of the general forest until such time as their use is discovered and their need felt. This mode, as we have seen, sets in processes of deforestation. The greatest disadvantage of this mode is that it does not have an index internal to it to identify the onset of this process.

A parallel mode of preservation and conservation for biodiversity is suggested by the Hill Kharias. According to this mode, the programme to preserve biodiversity aims to make available raw material for the production of honey and to sustain processes that will maintain and sustain a healthy forest. Only subsequently can demands for raw materials for industrial production become important. In this mode, the material is ensured alongside the processes for the production of honey.

This mode is decentralized and segmentary in so far as different *mahu bhandars* in different places in Simlipal are taken care of by different Kharia families. Their knowledge of the plant and animal kingdom derives from their day-to-day experience of living in a forest and from their mode of gathering food. They are acquainted with its seasonal variations. In their view, endangered species are those that are important for their subsistence and for the health of a forest. Some of these species die last when deforestation sets in. They are auto ecological. This mode enables a close watch over the process of self-regeneration of a forest. For instance, depletion of honey production immediately shows that there is a disturbance in life-cycles and food-chain processes in a forest.

According to the dominant mode of preservation, the tiger stands in opposition to the honey-bee, such that the preservation of one is at the expense of the other. A parallel mode on preservation, on the other hand, places the tiger and the honey-bee within a larger ecological framework; the emphasis rests on the symbiosis of their relationship, which in turn determines the overall preservation of the forest.

To better understand the phenomenon of arresting it, it is necessary to revise perceptions of biodiversity to accommodate a broader understanding of the impact of a particular species on the impact of a particular species on the totality of the forest. There is need to strengthen the processes which encourage a day-to-day interaction between plant and animal. The dominant mode needs to be dissolved not only because it distances the plant from the animal world, but also because it prevents the understanding of the forest as a dynamic, interactive ecosystem.

It is of little value to attempt to free forests of human intervention. The protectionism that this mode propounds only creates an illusion of non-interference. The self-defeating character of this enterprise is given in the unalienable relation of man to nature: man is as much a part of nature as nature is of man. Under protectionism, man only transforms his relation to the forest, and in the direction of destruction of both. The question that remains is: What mode of relation to the forest, in particular, and to nature as a whole is appropriate to the well-being and endurance of both?

## NOTE

* The discussion in this paper is based on field investigations undertaken in Simlipal in 1992–93. This was part of a larger study on changing relations of 'Man and Forest'. It was undertaken in collaboration with The Universe, a non-government organization based in Cuttack, Orissa.

I thank the Kharia people for willingly sharing their experiences of living in the Simlipal forest. My thanks also to Abha Mishra and Mihir K. Jena for helping to collect and record research findings. I am grateful to Shanti Ranjan Behra, Basant Mohanta and Bijoy Lal Mohanta for actively encouraging this research. Finally, I thank Dr. Klaus Seeland of the Swiss Federal Institute of Technology for creating an opportunity for this research and for his sustained interest in intellectual problems associated with man's relation to the forest in the context of deforestation.

## REFERENCES

Alexander, A.C., R.R. Prasad, and M.P. Jagirdar, 1995, *Tribals, Rehabilitation and Development,* Rawat Publications, Jaipur.

Bose, N.C., 1985, 'Simlipal Holds Life of Eastern Orissa: Save it or Perish', in Patro and Misra (ed.), *Conservation of Simlipal in its Wilderness* (CSW), Proceedings of National Conference on Biosphere, Simlipal, Baripada. May 9–12, 1985, Orissa Environmental Society, Orissa.

Patro, S.N., M.K. Misra (eds), 1985, *Conservation of Simlipal in its Wilderness,* Proceedings of National Conference on Biosphere, Simlipal, Baripada, May 19–21, 1985, Orissa Environmental Society, Orissa.

Sahu, B.N., 1985, 'Simlipal as a Biosphere in the Natural Context', in Patro and Misra (ed.), *Conservation of Simlipal in its Wilderness* (CSW), Proceedings of National Conference on Biosphere, Simlipal, Baripada. May 9–12, 1985, Orissa Environmental Society, Orissa.

# Empowerment and Disempowerment of Forest Women in Uttarakhand, India

Madhu Sarin

What factors trigger changes in women's position in a patriarchal forest-based society? How effective are devolution policies that are rhetorically committed to empowering women to participate in forest management? This paper explores these questions with the help of case studies spread over nine districts of Uttarakhand (now the major part of the newly constituted state of Uttaranchal in November 2000) in North India. Beginning from colonial rule in the nineteenth century, 67 per cent of the region's uncultivated commons, critical for sustaining its subsistence based agro-pastoral livelihood systems, have progressively been appropriated by the state as 'forests'. The accompanying reduction in villagers' forest access in a context where the use of forest resources has traditionally been considered 'women's work' has had a profound impact on reshaping traditional gender relations. More recently, a host of state policies have attempted to promote women's socioeconomic and political empowerment. These have included normative interventions such as the reservation of one-third seats for women in panchayati raj institutions for local self-government and promotion of all women *van* panchayats (elected councils for managing village forests), combined with more innovative government programmes such as Mahila Samakhya (women's empowerment) focused on mobilizing women for self-empowerment.

---

* Reprinted with permission from Madhu Sarin, 2003, 'Empowerment and Disempowerment of Forest Women in Uttarakhand, India', in Govind Kelkar, Dev Nathan and Pierre Walter (eds), *Gender Relations in Forest Societies in South Asia: Patriarchy at Odds,* Sage Publications, New Delhi, pp. 287–309.

The recent policy intervention for devolution of forest management through village forest joint management (VFJM) introduced through a World Bank-funded forestry project has been overlaid on the existing and ongoing state initiatives. The forestry project also has a stated commitment to specifically target women and the poor as project 'beneficiaries'. In addition, there are several other government and donor-funded projects (such as watershed development, water resource development, diversification of agriculture, etc.), each with its own component and strategy for enhancing women's participation. All these interventions have been introduced in a regional context which has a long history of people's own struggles for rights over land, forests, and water of which the Chipko Movement of the 1970s is the most famous. More significantly, village women themselves have been organizing their own struggles within their households and communities to gain greater control over their forests, as it is their daily lives which are most intimately impacted by forest quality, their proximity to the village, and institutional structures determining forest access.

The first section of this provides a brief overview of the historical evolution of the complex diversity of tenurial regimes for common lands and community institutions for their management in Uttarakhand. The second section examines local struggles to gain greater control over forest use and management, and of the forces aiding and hindering such struggles within existing formal as well as informal community institutions. In the third section, case studies undertaken to examine the impact of devolution policies for forest management on creating space for local control over decision-making, forest quality and management for enhancing livelihoods are discussed. Due to women's continuing centrality in the use and management of forest resources in Uttarakhand, the case studies focus on the impact of devolution policies on gender relations and women's spaces for forest management. The final section examines the impact of the recent forest devolution policy of VFJM, and the promotion of women's representation in *van* panchayats.

## The Context for Forest Management in Uttarakhand

Nestled in the Central Himalayas, Uttarakhand is an area with rich forests and unique biodiversity. According to the 1991 census, Uttarakhand's population was 5.93 million, 78 per cent of which lived in rural areas. The rural population of the hill areas is even higher at over 90 per cent. While the area had an economy of abundance at the time of colonial occupation in the early nineteenth century (Guha 1989; Nanda 1999), today an estimated 45 per cent of its economically productive workforce is employed outside the region for lack of local employment. While only 12.6 per cent of the hill region's area is officially under cultivation (Saxena 1995), rural livelihoods are sustained by authorized as well as unauthorized use of about 60 per cent of the total geographic area, comprising both cultivated and uncultivated lands. Around 67 per cent of Uttarakhand's total geographic area, representing most of the uncultivated commons, stands legally notified as 'forests' (Ghildyal and Banerjee 1998). Unlike many other states, all forestland is not under the jurisdiction of the forest department. About 69 per cent is reserved forest under the forest department's control, about 16.8 per cent is protected forest[1] under the revenue department's jurisdiction, and 13.6 per cent consists of legally notified village forests managed by elected *van* panchayats. Although socio-economic differentiation has increased over the years, village communities are still relatively homogenous compared to the high social stratification existing in the plains. Land distribution is relatively equal with rare cases of land holdings of over 2 ha, and landlessness is low.

Women own or control little of the privately held land, and their forest rights are mediated through the male head of the household in whose name they are recorded. The agro-pastoral economy of the region is still predominantly subsistence based with about 50 per cent of rural households, including the rural elite, having high dependence on village commons and forestlands. High migration of men in search of employment makes the women effective managers of the rural household economy. About 40 per cent of the households are estimated to be headed by women (CECI 1998, quoted in Ecotech 1999). The gender-based division of labour is highly skewed, with women responsible for most agricultural work (barring ploughing and marketing), livestock care (excepting during seasonal migration and marketing) and collection of firewood, fodder, leaf litter and non-timber forest produce from village commons and forestlands.

The history of state appropriation of the uncultivated commons has had a profound impact on reshaping traditional gender relations in the area. Prior to the British conquest in 1815, community institutions of the hill peasantry effectively exercised direct control over the use and management of both cultivated lands and the uncultivated commons within customary village boundaries, with little interference from earlier rulers (Agarwal 1996; Guha 1989; Nanda 1999; Somanathan 1991). Agriculture and animal husbandry comprised inseparable components of hill farming systems, dependent on spatially and temporally integrated use of cultivated and uncultivated lands. Seasonal migration to alpine pastures and grasslands prevented resource degradation by dispensing with the need for uninterrupted use. High dependence on the forests generated conservation values embedded in cultural and religious traditions, such as the maintenance of sacred groves. Traditional village panchayats dealt with community affairs and inter and intra-village disputes (Guha 1989). Despite women being the primary users of forest resources, tradition excluded them from political decision-making at the community level.

A number of interventions during colonial rule permanently altered this landscape of integrated local resource use and management, initially in Kumaon and British Garhwal, which were under direct rule, and then subsequently even in the adjoining princely state of Tehri Garhwal. In 1823, the colonial regime undertook the first land revenue settlement in the Kumaon region, categorizing the land within customary village boundaries as cultivated *naap* (measured) and uncultivated *benaap* (unmeasured) lands. In 1893, all unmeasured 'waste' *(benaap)* lands in Kumaon were declared 'district protected forests' under the control of the district commissioners. From 1910–17, the colonial government attempted to appropriate further control over forest resources by notifying over 7,500 sq km of the commons in British territory as reserved forests, thereby severely restricting people's use rights. The 'rights and concessions' of people, such as cattle grazing and lopping for fuelwood and fodder, so essential for sustaining the agro-pastoral economy, were severely curtailed. The annual practise of burning the forest floor for increasing grass yields was also banned within 1 mile of reserve forests.

As few settlements remained at such distance from the new reserves, it virtually made the practise illegal (Guha 1989).

## Struggles for Local Forest Rights

Forest reservation led to large-scale rebellions and incendiarism (literally setting the reserved forests ablaze in protest against denial of traditional access to them) as it played havoc with the customary patterns of resource use, dislocating existing agrarian practises. The report of the Kumaon Grievances Committee,[2] which was set up by an alarmed administration to look into the causes of unrest among the people, contains perceptive insights into the prevalent gendered nature of resource use and gender relations. It identified the 'employment of forest guards to enforce numerous rules and regulations and their constant interference with women and children who, under the customs in vogue in Kumaon, are the chief people to exercise on behalf of the villagers such rights as lopping, collection of minor forest produce, grazing, etc' (GOUFJ 1922) as a major grievance. Lopping restrictions imposed in the new reserves accounted for 75 per cent of the recorded offences. Although there was no *purdah* in the hills, the men bitterly resented women being summoned to court for such 'offences'. Men's sentiments against the forest guards' daily interference with women's use of the forest caused the Grievances Committee to recommend that 'the forest guard be removed so as to do away with the real grievance in all classes of forests where it can possibly be effected' (ibid.). As a consequence, 4,460 sq km of the less commercially valuable forests out of 7,500 sq km of the new reserves were taken away from the forest department and handed back to the revenue department with full restoration of people's forest rights. However, rights in these Class I reserves were given to 'all bonafide residents of Kumaon,' thereby converting customary common property resources into open access areas. Provision was made for *van* panchayats to exercise community control over legally constituted 'village forests' demarcated from within the Class I reserves and civil forests, though applicable only in those villages which applied for them. This enabled sections of the peasantry to retrieve some space for local forest management. Thus, by the early twentieth century, the uncultivated commons had been divided into three legal categories of forests: commercially valuable Class II reserves under the forest department; commercially less valuable Class I reserves; and civil/*soyam* forests, under the civil administration. This, however, did not appease the villagers' discontent over curtailed resource rights, and protests continued.

The independent state of India continued commercial forest exploitation with even greater vigour. The reach of the forest department and its contractors spread to the remotest corners with the expansion of the road network. Local livelihoods received even less attention than under colonial rule as state policy consistently favoured export of raw timber and resin for processing by large industry in the plains. By the 1970s, the Chipko Movement had emerged to demand that priority be given to local employment in the extraction and processing of forest produce (Guha 1989). Increasing incidents of landslides and floods, and the declining availability of biomass for subsistence needs propelled even the hill women into the movement, broadening the popular base of Chipko protests. Ironically, the Kumaon *van* panchayat rules of 1931 were revised in 1976 at the height of the Chipko Movement, substantially reducing *van* panchayat authority and entitlements even over village forests.

The issue of local forest rights, however, was soon subsumed within the new national and global ideology of environmental conservation. Instead of giving priority to local forest-based livelihoods and employment, Chipko was used to justify a spate of centralized environmental policies and laws. The Forest Conservation Act of 1980 empowered the central government to make decisions related to the alienation of even the smallest patch of forestland. The Uttar Pradesh Resin and Forest Produce Act, 1976, made tapping, sale and purchase of all resin a state monopoly. In 1981, a 15-year ban was imposed (and has since been extended) on all commercial felling in the Uttar Pradesh Himalayas above 1,000 m.[3] Today the only permitted felling are for the villagers' timber rights (*haq haquque*) based on consumption recorded in 1911–17, which have not been revised since.

Despite extensive state appropriation and continuing centralization of control over forests, Uttarakhand today provides extensive examples of a unique combination of officially constituted and informal community forest management (CFM) institutions, the latter increasingly developed by women's groups. Autonomous community management of legally demarcated 'village forests' (on forest and revenue department land) by *van* panchayats has existed in Uttarakhand for over seven decades. Traditionally practicing restricted democracy due to their almost exclusive control by men, some *van* panchayats have been undergoing internal transformation with women appropriating control over their management. Unofficial community management, with diverse institutional arrangements on all legal categories of forestlands, has co-existed with formally constituted *van* panchayats, and in fact predates them from pre-colonial times.

As the largest custodian of state property,[4] the forest department has been unable to maintain the forests in good condition or meet people's forest-based livelihood needs. Its responsibility for enforcing the Forest Conservation Act and Wildlife Protection Act has reinforced its image as an anti-people agency. Thus, in 1988–89, some of the Chipko activists who had spearheaded the movement against commercial forest felling, felt compelled to start yet another, relatively less known *ped kato andolan* (cut the trees movement). They argued that the Forest Conservation Act 'was being used to hold up basic development schemes for the hill villages while the builders' mafia continued to flout it brazenly under the guise of promoting tourism' (Rawat 1998: 128). More recently, resource displacement and loss of livelihoods caused by expansion of the protected area network over 20 per cent of the geographic area of Uttarakhand, has produced the *jhapto cheeno andolan* (snatch and grab movement) reflecting the intense feelings of alienation and disempowerment. Women who earned international fame for stopping contractors from felling their forests during Chipko have come to hate the word *paryavaran* (environment). As one of these women from Reni village complained, 'They have put this entire [surrounding forest] area under the Nanda Devi National Park. I can't even pick herbs to treat a stomach ache any more' (Mitra 1993: 35).

Both the *van* panchayat and unofficial CFM systems now confront challenges from new systems being imposed on villagers by the state. Under a World Bank funded forestry project, the Uttar Pradesh forest department has promoted village forest joint management (VTJM) with autonomous *van* panchayats. This is in contrast to the joint forest management (JFM) practised on degraded reserve and protected forests under the forest department jurisdiction in other states. It is misleading to refer to the UP VFJM approach as JFM. This hides the crucial difference from JFM in other states—that *van* panchayats had enjoyed autonomous authority over forests prior to VFJM. Reserve forest areas are also to be included in VFJM but till the time of this research (2000), it had primarily focused on *van* panchayat and civil/*soyam* lands (CDS, 2000). VFJM is thus creating space for the forest department to intrude on village forests managed by community institutions instead of creating space for villagers to participate in management of reserve forests under departmental jurisdiction. The decision-making autonomy of van panchayats participating in VFJM is now 'subject to the supervision, direction, control and concurrence of the Divisional Forest Officer' (FDUR 1997: 3.1). A functionary of the forest department is being made the joint account holder and the proposed member secretary of van panchayats (GOU 2001), after having no role for seven decades.

At the same time, informal community management is under pressure from the state-directed and target-driven new *van* panchayat formation. The revenue department is demarcating civil forestlands under its jurisdiction as village forests to be managed by officially constituted *van* panchayats. The department is also dividing existing multi-village *van* panchayats without consulting them into single village ones, often generating inter-village inequity and conflict in the process. 'Eco-development committees' promoted by the wildlife wing of the forest department solicit villagers' 'participation' in replacing their existing forest dependent livelihoods (against their will) with new, non-forest-based alternatives. Decisions earlier taken by villagers through negotiation and consensus-building, such as whether to take up community management at all, whether to do so officially or unofficially, and whether to do so at a hamlet, village or multiple village level, are now being taken by the state on their behalf. Through the formation of new *van* panchayats, non-governmental organizations (NGOs) and civil society groups, which earlier played an important role in policy advocacy and spearheading social movements, have largely been co-opted to work as 'private service providers' for the many donor-funded projects in the region.

## Women in Informal Community Forest Management

Community forest management outside any formal legal framework is widespread in Uttarakhand in all categories of forestlands, within or near villages. Traditional informal community institutions (*lath* panchayats), informal forest committees (*van samiti*s) and more recently, increasing numbers of Mahila Mangal Dal*s* (women's welfare associations) are regenerating and regulating use of reserve and civil/*soyam* forestlands, often compelling unofficial cooperation by forest and revenue department staff. These informal community management systems represent another important form of appropriating space for local forest management.

Women's ability to appropriate such forest management space outside any formal policy framework is shaped by diverse influences and factors. These include the degree of local scarcity of forest resources, and therefore the conflicts and competition governing access to essential forest products; the extent of migration of men from the area; women's exposure to and participation in social movements such as Chipko or empowerment programmes such as Mahila Samakhya; support of progressive local male leaders and the indirect impact of positive discrimination policies such as the reservation of one-third seats for women in panchayati raj institutions or the promotion of all-women *van* panchayats. Out of our 16 case studies, we found village women actively engaged in informal CFM in eight cases, either exclusively through women's groups or together with men.

Holta, one of our case study villages, is a large village of 400 households in Tehri Garhwal district, which has a rich tradition of informal CFM. In part, this is because the *van* panchayat rules were not made applicable to the area which was under the erstwhile state of Tehri Garhwal till 1991. Due to not having a *van* panchayat, Holta village initiated protection of its *sanjaiti* (communal) land around 1986 entirely on its own. The area has a high dependence on agriculture with limited migration of men. Literacy rates are low and most men work in the village itself or for wages, or in small shops on the roadside in the vicinity. Although there is visible dominance of men in the village, the local self government institution (*gram sabha*), is headed by a woman. Interestingly, Holta's informal forest protection committee had no women representatives when it was first constituted 13 years ago. However, despite the *samiti* employing a guard for protection, women of neighbouring villages as well as those of Holta itself continued stealing firewood and grass from the forest. According to the men, their failure in stopping the women resulted in them inducting four women as members of the committee about five years ago to 'get thieves to nab other thieves' (Gairola 1999a).

Interestingly, the village young men had applied for forming a *van* panchayat six years ago, but had received no response from the administration. Asked why they wanted a *van* panchayat when their informal system was working so well, the men felt that a *van* panchayat had greater access to government funds for plantations. They had heard about generous budgets for VFJM. The women, in contrast, did not want any funds or government scheme. They were proud of their regenerated forest from which they could meet their biomass needs.

On being told that the government was promoting the formation of all women *van* panchayats and that if they submitted such an application it was likely to be cleared quickly, the women's faces lit up. However, the men were taken aback. They said that the village women were illiterate and knew little, and that they would not be able to handle the work. It was then pointed out that the *pradhan* (village head) of Holta *gram sabha* was a woman who was also non-literate. If she could be a panchayat *pradhan,* why couldn't the women manage a *van* panchayat?

According to Holta villagers, water sources had dried up and firewood and fodder had become scarce because of unregulated forest use by the surrounding villages, and encroachment on the common land by local families. Some village youth successfully persuaded the encroachers to vacate the commons, setting an example by giving up their own encroachments. Letters were sent to the *pradhans* of surrounding villages that anyone entering the forest would be fined. Major conflicts followed with one village going to court against Holta due to unclear boundaries of their respective common lands. However, with improvement in forest conditions and availability of water, resistance declined.

At the time of our research in 1999–2000, the entire village's biomass needs, except timber, were met from the regenerated forest. Vegetable cultivation had become feasible with regeneration of three natural water sources. Rules were framed and strictly enforced for grass, tree leaf fodder and firewood collection, with all households contributing to pay the guard. The informal forest protection committee has representation from all the hamlets and castes, and four women representatives of the village Mahila Mangal Dal, empowered by the Mahila Samakhya programme, had also been able to wedge their way in. Community relations with the forest department, however, were extremely sour. In the words of the village women, 'the forest department has made us into thieves. The women were protecting their forest like their own children' (Gairola 1999a). Women and men had different perspectives on gender-specific changes in the forest protection committee (*samiti*).

According to the women, after gaining exposure and self-confidence through the government's Mahila Samakhya programme, they had had to fight hard to gain representation in the protection committee. Due to the limited migration of men from the area the women had few opportunities to occupy leadership roles. The men had resisted their demands saying that they were protecting the forest for the benefit of the women and that there was no need for the women to become members themselves. The men also considered it difficult to decide which caste groups the women should represent as the men of all castes considered themselves indispensable.

The women insisted that if seats were now being reserved for them even in panchayats, the same could be done for them in the village forest protection committee (*samiti*). Once two of the women were elected as ward members during the panchayat elections, they were made members of the forest protection *samiti* (FPS) also. Significantly, most members of the FPS are also ward members of the village panchayat.

According to the women, now all their grass, firewood and bamboo requirements are met from the forest. Similarly, all the villagers' small timber needs for agricultural implements, such as sickles, ploughs and axes are met from the forest. Everyone is satisfied with the work of the forest protection *samiti*. Daily life has become much easier for their daughters and daughters-in-law. Although a few troublesome individuals do keep creating problems now and then, it is easy to keep them quiet as the women of their families too are benefiting from easy availability of firewood and fodder. When asked whether the villagers had helped put out fires in government forests during the summer, the women's response was:

> Why should we help when those are not our forests? The Forest Department does not permit us to lift even a dry branch from them. After the government forests are burnt, we are able to fetch firewood from them. So we actually benefit from fires in government forests. We have not taken a contract for environmental protection! If there is environmental damage, it should equally affect the Forest Department. They are also citizens of this country but behave as if they are the descendants of the British! (Gairola 1999a)

Such antipathy towards the forest department staff among village women is widespread. Once, the women of Dhar Kot village in Pratapnagar block of Tehri district tied the forester and members of the forest patrol to trees and continued lopping green oak leaf fodder for their cattle. The forest staff kept shouting the whole day in the thick forest, but no one was there to hear them. While most village men are fully aware of their rights and the procedures for exercising them, the women knew next to nothing about such matters.

In several hamlets of the three villages under the Makku *van* panchayat, Mahila Mangal Dal*s* had asserted informal control over patches of civil or communal land closer to their settlements for day to day management of firewood and fodder. According to the former *sarpanch* of the Makku *van* panchayat, the scarcity of firewood and fodder is increasing conflicts over forests to such an extent that women have started resorting to physical fights among themselves. He had encouraged the village women to enclose patches of communal *gram* panchayat land for meeting their needs, while also saving it from encroachment by the elite. Both he and the women faced a lot of resistance from powerful vested interests and husbands who had to do housework while women patrolled. However, effective protection by the women led to dramatic regeneration of the women's forests/*mahila bans* (Bhatt 1999).

In Bareth village, which also has a *van* panchayat, while the men control the pine dominated *van* panchayat forest, the women effectively manage and control the mixed species forest on the village common land. The forest is closer to the village and useful for meeting their daily requirements of fuelwood and fodder. In both Makku and Bareth, women perceived local *van* panchayat councils to be dominated by men. Panchayat forests were also far from the villages, and therefore not convenient for daily fuelwood and fodder collection. The formal and informal CFM arrangements here complemented each other with the women occupying informally carved out space. They could access such space with the mediation of the *gram* panchayat without having to deal with cumbersome bureaucratic procedures. In Arakot village, the Mahila Mangal Dal has been protecting the village *soyam* land for the past 20 years, paying a guard with voluntary contributions. In Naurakh and Resal, civil land was being protected by individual families through private enclosures with day to day management under the women's control. Officially viewed as 'encroachment' on government lands, such informal systems are fairly widespread owing to their low transaction costs (Singh 1997a).

## Women Appropriating Space within Van Panchayats

Till the 1970s, *van* panchayat councils seem to have been almost exclusively male due to the traditional exclusion of women from community institutions. The Chipko Movement for the first time drew women out from their homes exposing them to a new world of assertion and articulating their demands. The movement also had a significant impact on traditional gender relations, with the men accepting women's participation in non-domestic affairs, particularly those related to management of forest resources. The subsequent period has also seen specific government policies promoting women's participation in forest management, including promotion of women's representation on *van* panchayat councils. Our case studies included one case where government officials had compelled a woman to be elected as a *van* panchayat *sarpanch* (in Bareth village) simply on the grounds of it being government policy. In the absence of any capacity building support, or any previous exposure to such a role, her forced election to the position of sarpanch was a disaster. She resigned within a month.

In two other cases, however, women had appropriated space for themselves within their respective *van* panchayats to have a decisive say in their day to day functioning. In the case of the Pakhi *van* panchayat, the Mahila Mangal Dal had gained informal control over the *van* panchayat whereas in the case of Dungri Chopra *van* panchayat, the women had succeeded in getting an all women *van* panchayat council elected. Pakhi *van* panchayat had subsequently been brought under VFJM under the ongoing World Bank funded forestry project. We examine here how women appropriated space for themselves in the two *van* panchayats and their functioning under the women's leadership.

The Pakhi *van* panchayat was formed in 1958 and it lies in Chamoli district, in the vicinity of Gopeshwar from where Chandi Prasad Bhatt spearheaded the Chipko Movement in the early 1970s. The village became well known for the active participation of women and men in the movement, which stopped the felling of their forests, by the forest department. Pakhi and Jalgwad, the two villages falling under the same *gram* panchayat, have a common *van* panchayat. Out of about 180 total households only eight are Scheduled Castes and the rest are upper caste Hindus (Thakur or Brahmin). The *van* panchayat forest area of 240 ha is in good condition and has mixed species dominated by oak and rhododendron, with a sprinkling of deodar (blue pine). Fuelwood, fodder, animal bedding and some non-timber forest produce, rather than cash income, are the primary benefits the villagers get from the forest. There is limited male migration from the area.

The two villages have an active Mahila Mangal Dal, whose leaders have received considerable exposure since their involvement with Chipko. The Mahila Mangal Dal continues to interact with several NGOs, participate in government-sponsored camps for women and is also a member of Himvanti, a multi-country federation of mountain women's organizations for the Hindukush–Himalayan region. In 1999–2000, although the elected council had two women members, they did not participate actively in *van* panchayat meetings. Instead, the Mahila Mangal Dal as a whole influenced the *van* panchayat council by the women's collective decisions. One of the elected women *van* panchayat members had received training and marketing support from Himalayan Environmental Studies and Conservation Organisation (HESCO), an NGO promoting economic development in villages, and has initiated processing of locally grown fruit into jams, pickles, and juices. This has increased the villagers' returns from fruit significantly.

Participation in the Chipko Movement enabled the Mahila Mangal Dal to effectively wrest control over the day to day management of the village forest from the male dominated *van* panchayat council. Prior to the initiation of VFJM with the *van* panchayat in August 1999, decisions about when to open the forest for grass, leaf and firewood collection, the rules for collection, the fines for violation, etc., were taken by the Mahila Mangal Dal and communicated to the *van* panchayat *sarpanch*. His primary responsibility was to publicly announce these decisions in both the villages. As no external funds were available, the women also used to repair the forest boundary wall with voluntary labour. They had employed a woman as forest guard to whom they paid Rs 300 per month also raised through voluntary contributions. Fines from those violating the protection rules were collected by the Mahila Mangal Dal and deposited in its own account. At the time of the research, the Mahila Mangal Dal had Rs 3,100 in its account.

The women's control over forest-use decisions enables them to ensure that forest product collection does not conflict with periods of heavy agricultural work. Soon after harvesting the monsoon *mandwa* crop in October, they open the forest closest to the village for grass collection. The furthest forest area is opened in December when all agricultural work is finished and the women can devote most of their time to stocking up firewood and grass before the snow falls. Cutting of bushes and pruning tree branches is done from April to May.

Although pleased with having appropriated control over the village forest, the women expressed resentment over the men leaving all the forest protection work to them. They had attempted to coax the men to assist with voluntary patrolling but the men had refused saying that it was only the women who needed the forest. The women also reported that when outsiders came to the village (and there are many visitors due to the contacts established during Chipko) the men pushed the women forward to talk to them. However, when important village related decisions are made, the women are often kept in the dark. This was evident with the introduction of the VFJM with the *van* panchayat in August 1999. Despite the World Bank project document's emphasis on specific targeting of women and the poor in 'participatory' forest management, neither the forest department-NGO 'spearhead team' nor the village men provided the women much information about VFJM. The men only told them that a budget of Rs two million was being approved for the *van* panchayat which would be very beneficial for the village.

The sudden availability of a generous budget for the village forest, however, led to a rapid gender-based shift in power and control. The same men, whom the women had complained that they had left all forest protection work to them, suddenly became overenthusiastic for it. Three watchmen, at salaries of Rs 1,000 per month, were employed together with one watchwoman for forest protection. After three months of working without a salary, the *sarpanch* started paying the woman Rs 200 per month. Knowing that the men were being paid Rs 1,000 per month, she refused to accept such a payment. After a lot of arguing, she was finally paid Rs 700 for the previous month's work and then laid off on the grounds that it was difficult for a woman to protect the far ends of a large forest. Similarly, initially the men monopolized wage work in the nursery. Only after strong protests by the women were some women also employed. When no funds were available for forest development, women were left to take care of it with voluntary labour. As soon as money came in, the women were labeled ineffective for undertaking the task.

The men too, however, were not outright winners in the subtle shifts in the balance of power and control within the village. The *van* panchayat council and the *sarpanch* experienced a similar loss in local decision-making control to the forest department. Maintenance of the muster roll for wage work and preparation of the monthly progress report was now done by the guard or forester instead of the *sarpanch* as earlier. According to the *sarpanch,* the villagers' role in VFJM was reduced to providing information for preparation of the micro-plan and working as paid labour for forestry operations. The villagers could no longer do anything on their own without prior approval of the forest staff. Neither the women nor the men were clear about the new VFJM rules, or the legal agreement they were supposed to have signed. There was no copy of the agreement in the *van* panchayat records, and the *sarpanch* did not have a copy of the micro-plan with him. He said that years of experience had made him familiar with the rules governing wan panchayats. But he knew little about the VFJM rules.

In the words of one of the worried women, in their greed for money, the men had made a deal over their village forest with the forest department. Since conducting the fieldwork, an all women *van* panchayat council has been elected in Pakhi. However, the husband of the woman *sarpanch* is the forest guard who is the joint signatory for the *van* panchayat's VFJM account with his wife as the *sarpanch*. Due to this, her questioning voice related to VFJM has become muted. It is not known how this has impacted women's unity within the Mahila Mangal Dal.

The Dungri Chopra *van* panchayat in Pauri Garhwal district was formed in 1939 and is the oldest *van* panchayat among our research case studies. The district has a high literacy rate combined with high rates of migration of men. Due to the men getting good jobs, their interest in managing the village forest has declined which has left the village women with problems resulting from poor satisfaction of their biomass needs. Under the Mahila Samakhya programme the women have gained self-confidence, are better organized, and better informed about government policies for women's empowerment. They made a concerted bid to take over the management of the village forest from the disinterested men's leadership. Many men disapproved of the women's initiative, but some of the village elders encouraged the women to take over. As a consequence, around 1997–98, the village women succeeded in getting an all-woman *van* panchayat council elected. The woman *sarpanch,* however, faces daunting challenges.

Today, government schemes worth millions come to the villages and there is rampant corruption. No government official visits the village without negotiating a commission in advance. In 1999, the District Rural Development Agency sanctioned Rs 60,000 for undertaking plantation in the village forest. When Dwarka Devi, the woman *sarpanch,* went to collect the first installment of Rs 30,000, the *van* panchayat inspector made her sign a receipt for the full amount but gave her only Rs 24,000. She went to Dilip Singh, an elder former *sarpanch,* to seek advice on what to do. He told her that in future, whenever any such payment had to be collected, she should always take other women *sarpanch*s with her, and on returning to the village, place the entire amount in front of the general house to prevent anyone from suspecting her. The villagers would themselves help her work out how to deal with the situation (Gairola 1999c).

Seemingly, Dwarka Devi had internalized this valuable lesson in transparent governance. It has enabled her to maintain collective responsibility for managing the village forest and also to evolve coping strategies for dealing with the increasingly unsavory and dramatically changing world outside the village. The panchayat forest is one of the best in the district and the women meet almost all their forest needs from it.

Rather than strengthening such transparent governance mechanisms within *van* panchayats, VFJM, which is being promoted under the World Bank funded forestry project, assumes that misuse of funds for implementing micro-plans as well as the misuse of the villagers own funds can be prevented by making a forest department functionary as the member secretary-cum-joint account holder of the *van* panchayats. While perverting the tradition of the leadership's accountability to the general body of villagers, the arrangement has created yet another avenue for lower level forest department staff forging alliances with a class of elite village men to misappropriate micro-plan funds.

## Women-Specific Impacts of VFJM: Livelihoods and Equality

Discussions on the merits and demerits of VFJM rules often centre round the percentage of the share of income that the villagers would get from their forests. Women forest users, however, have been driven to physically attack other women competing for increasingly scarce fuel and fodder resources, some even resorting to suicides to end their daily drudgery (Nanda 1999). Their priorities are to increase the direct use values of their forests. An improved quality of life and ecological security for them precedes considerations of income from sale of forest products, although additional income is never unwelcome. The project document claims to target women and the poor but provides no analysis of how a shift in management priorities for increasing income would impact their access to requirements for daily subsistence or their work burdens. VFJM micro-plans in the case studies had few provisions for supporting livestock, a critical local livelihood requirement, instead emphasizing closure of forests to grazing.

A key assumption underlying support for VFJM is that the major problem plaguing *van* panchayats was the lack of funds, and therefore the best incentive for increasing the villagers' stake in forest protection was offering them attractive shares of income from the sale of forest products. Yet, a survey of 644 *van* panchayats in Ranikhet subdivision in Almora district found that as many as 433 did not have any income and only 45 could boast of a balance of at least Rs 25,000 in their accounts (Singh and Ballabh 1991, quoted in Singh, 1997b). A large number of forests had been managed well by villagers without any source of income. It has also been observed that the income of a *van* panchayat has no bearing on what the villagers consider to be a 'good' *van* panchayat. A *van* panchayat is regarded as good if it meets the needs of fuel and fodder and helps recharge water sources. Therefore, oak forests were generally preferred to pine forests even though they provided less revenue and employment (through resin tapping). In contrast, the revenue department considered *van* panchayats with bigger bank balances to be performing better. The World Bank forestry project subscribed to the same assumption.

The World Bank funded forestry project has provided an average of Rs 1.5–2 million for implementing a micro-plan in each village brought under VFJM under the project. Besides promoting inequity between neighbouring villages, the sudden offer of large sums of money to selected villages with high unemployment and limited opportunities for cash incomes, however, had led to the eruption of major conflicts to gain control over the funds. Sustainable voluntary protection, often by women's groups, had been replaced by patrols of externally-funded guards. In our case study villages, selection of the paid guards had been done by elite village men providing them a new avenue for patronage. Women's groups, in particular, had been able to negotiate/assert their authority to manage civil/*soyam* lands by negotiating with their *gram* panchayats. In at least three out of 10 case studies (Pakhi, Arakot, and Chora) where VFJM was introduced, village women were actively protecting the *van* panchayat/*soyam* forests. In all three cases, no effort was made to build upon, and strengthen the women's efforts. In Pakhi village, a poor watchwoman paid by women's voluntary contributions was replaced by four watchmen paid much higher salaries with World Bank loan funds.

While overlooking existing systems of voluntary contributions, the project demands that villagers contribute 20 per cent of micro-plan costs. This was being collected through compulsory deductions from wages, thereby transferring the costs to those doing the wage work. The majority of these workers were women or poorer villagers. They had been forced to contribute on behalf of the whole community, often without even being aware that there were deductions and as to why they were being paid less than the minimum wage. In none of the case study villages had any open discussion been held on how the mandatory contribution could be shared more equally by all those theoretically gaining entitlements to the specified benefits. In Kharag Karki village, women thought they could at least take the firewood from cleaning operations as compensation for accepting lower wages, but were not permitted to do so, leaving them alienated and bitter. Organized and acutely forest-dependent women have borne disproportionate costs of (in)voluntary contributions or unpaid protection duties in order to build up panchayat and forest committee funds controlled by elite men and the forest department staff. This situation is highlighted by recent developments in the much publicized case of the Parvera *van* panchayat which has had an active woman *sarpanch*. The department had encouraged the women to take up voluntary protection to build up the *van* panchayat fund with the project money available for protection. This is considered an indicator of good panchayat functioning under VFJM (FDUP n.d.). However, with the transfer of a committed senior forest officer who had been taking personal interest in supporting the woman *sarpanch* and the Mahila Mangal Dal, the good performance of the Parvera van panchayat has proved to be short lived. Elite village men have made a bid to appropriate control over the women's savings in the VFJM account by getting a new *van* panchayat elected in cahoots with the *van* panchayat inspector of the revenue department. The woman *sarpanch* has challenged the new election in court on the grounds that a new council cannot be elected during the five-year agreement for VFJM signed by the existing van panchayat with the forest department. While the old and the new *sarpanch*s *are* fighting the battle in court, the biggest casualty has been the well protected village forest due to the disputed authority over its management (Author's interview, 22 June 2001).

## Conclusion

In Uttarakhand, NGOs and civil society groups have historically played a strong advocacy role. Chipko, for example, was triggered by protests led by Dasholi Gram Swaraj Mandal. Today, the NGO movement is split into different camps and factions. The vast majority have been co-opted to work as 'private service providers' for the several donor funded projects in the region, including the forestry project. Once they have accepted working on project terms, they effectively lose their critical and questioning voice. Among civil society groups perturbed over such impacts of donor funded projects, there are different sub-sets and worldviews. The overall impact is that today the NGO and civil society movements have been considerably weakened with hardly any concerted public action for protecting people's forest rights. Consequently, no unified voice has been raised against the potential damage to the region's unique institution of the *van* panchayat from the VFJM. A large number of concerned individuals and advocacy groups, however, have been articulating such concerns at different forums (SKS 1999; SPWD, 2000).

A series of government actions has weakened the existing local forest management systems substantially over the course of the last century. Progressive restrictions on local use of forest resources through the Forest Conservation Act, the felling ban, and recent Supreme Court judgments, combined with large-scale livelihood and resource displacement caused by expansion of the Protected Area Network, are changing peoples' attitudes towards forests and undermining the primary incentives for the CFM.

However, despite the imposition of crippling bureaucratic controls on their functioning, a large number of Uttarakhand's *van* panchayats have survived as vibrant self-governing community forestry institutions. A large number of diverse and informal institutional arrangements for community management on all legal categories of forestlands, several of them led by acutely forest-dependent women, co-exist with, and even within, the formal *van* panchayats. Such informal arrangements, often with negotiated support of elected *gram* panchayats, many now headed by women, provide more accessible and flexible space for CFM by poor women and marginalized groups outside the ambit of bureaucratic procedures and controls. These formal and informal community institutions have traditionally functioned on principles of restricted participatory democracy by men. The larger context of socio-cultural transformation and change, however, is creating space for women and other marginalized groups to broaden their democratic base. Women have started asserting their rights to participate in community decision-making and defining forest use and management priorities through organized action and struggle within their households and communities.

In the name of devolution, VFJM is empowering the forest department to reassert control over both *van* panchayat forests and civil/*soyam* lands, the only surviving village commons. Instead of revalidating the rich and diverse base of indigenous knowledge of local women and men, and the diverse management systems they have developed for supporting livelihoods and maintaining ecological services, VFJM reinforces the forest department's claim to be the monopoly holder of technical forest knowledge, despite its historical lack of experience with forest livelihoods and biodiversity conservation. Externally imposed micro-planning teams are insensitive to the internal dynamics of existing self-governing institutions and women's ongoing struggles within them for gaining greater voice and control over livelihoods and decision-making. Instead of empowering women, such top-down interventions do the opposite by disrupting and marginalizing women's own struggles and initiatives. Placing a forest department functionary as the joint account holder and proposed member secretary inside *van* panchayats shifts institutional accountability to the forest department and away from forest users, of whom a great majority are women

Nurturing democratic, self-governing CFM institutions requires a framework ensuring tenurial security over community forests, clear boundaries defining communal property rights and empowerment of forest-dependent women and men within communities to make real choices for enhancing sustainable livelihoods in accordance with their gender-specific priorities. State interventions need to build upon and further facilitate gender equal democratization of existing local initiatives and institutional arrangements. These need to take into account women's traditional lack of independent access and control over communal forest resources and decision-making in forest management institutions, instead of seeking to replace them wholesale with standardized state-engineered institutional frameworks such as VFJM.

Further support for policy interventions such as VFJM, target-driven new *van* panchayat and Eco-Development Committee formation in Uttarakhand needs to be abandoned. Alternatives to these state-sponsored institutional structures need to be developed through broad based consultative processes in which the least powerful and most forest-dependent, women in particular, are assured prominent opportunities for discussion and decision-making.

## Notes

* This paper is based on research funded by the Centre for International Forestry Research (C1FOR) and International Fund for Agricultural Development (IFAD). I am grateful to Geeta Gairola, Tarun Joshi, other *Sainion Ka Sangathan* team members and Seema Bhatt for their field work and to Govind Kelkar for comments and editorial inputs. The article has drawn considerably on two of my earlier papers—a draft overview paper based on the research in Uttarakhand titled 'From Right Holders to 'Beneficiaries'? Community Forest Management, *Van* Panchayats and Village Forest Joint Management in Uttarakhand', January 2001, and 'Disempowerment in the Name of "Participatory" Forestry?—Village Forests Joint Management in Uttarakhand', *Forests, Trees and People Newsletter,* 44, April 2001, Uppsala, Sweden.

1. In the adjoining Tehri state, non-reserve forestlands under the civil administration were called *soyam* lands. The protected forests are called 'civil land' in the Kumaon region. Together these lands are managed by the revenue department in collaboration with elected village panchayats.
2. The Forest Grievances Committee for Kumaon set up in 1921 examined some 5,040 witnesses from all grades of society in British Garhwal, Almora, and Nainital.
3. In 1986, the ban was made applicable above an altitude of 2,500 m. At lower altitudes, green felling of only pine in areas specified in forest working plans is permitted (Saxena 1995).
4. Out of 67 per cent of Uttarakhand's area classified as forests, about 69 per cent is reserved forests exclusively under the jurisdiction of the forest department. The rest, comprising of civil/*soyam* and VP forests, respectively falls under the revenue department and *van* panchayat jurisdiction with the forest department responsible for technical supervision.

## References

Agarwal, Chetan, 1996, 'Boundary and Property Rights in Uttarakhand Forests', *Wastelands News,* February–April, New Delhi.

Bhatt, Seema, 1999, 'A Case Study of Some People's Institutions in the Akash Kamini Valley', Garhwal, Mimeo.

Centre for Development Studies (CDS), 2000, *Sanyukt Van Prabandhan,* 3(1), Nainital.

Canadian Centre for International Studies and Cooperation (CECI), 1998, 'Community Based Economic Development Project, India: Preliminary Proposal', CECI, Montreal, Mimeo.

Ecotech , 1999, 'Study on Management of Community Funds and Local Institutions, State Perspectives and Case Studies', Volume II, Draft for Discussion, Ecotech Services, New Delhi, Mimeo.

Forest Department of Uttar Pradesh (FDUP) , 1997, Village Forest Joint Management Rules, August 30, Lucknow.

———, n.d., 'Empowerment of People through Forestry—A Status Paper on JFM in UP,' Lucknow.

Gairola, Geeta, 1999a, Field case study of CFM in Holta Village, District Tehri Garhwal.

———, 1999b, Field case study of VTJM with Pakhi Van Panchayat, District Chamoli.

———, 1999c, Field case study of Dungri Chopra Van Panchayat, District Pauri Garhwal.

Ghildyal, M.C., and A. Banerjee, 1998, 'Status of Participatory Management in Uttarakhand Himalayas', Paper presented at regional workshop on participatory forest management implications for policy and human resource development, Kunming, People's Republic of China.

Government of the United Provinces (GoUP), 1922, Kumaon Forest Grievances Committee Report, Government of the United Provinces, Lucknow.

Government of Uttaranchal (GoU), 2001, The Uttaranchal Panchayati Forest Rules, 2001, Final draft prepared by the UPFD in May 2001, Dehradun.

Guha, Ramachandra , 1989, *The Unquiet Woods: Ecological Change and Peasant Resistance in the Himalaya,* Oxford University Press, Delhi.

Mitra, A., 1993, 'Chipko, An Unfinished Mission', *Down to Earth,* April 30, New Delhi.

Nanda, N., 1999, *Forests for Whom? Destruction and Restoration in the UP Himalayas,* Har Anand Publications, New Delhi.

Rawat, A.S., 1998, 'Biodiversity Conservation in U.P. Hills: A People's Viewpoint', *Studies,* 3(4), Centre for Sustainable Development, UP Academy of Administration, Nainital.

Sainion Ka Sangathan (SKS), 1999, Proceedings of NGO discussion held in July, 1999.

Saxena, N.C. , 1995, *Towards Sustainable Forestry in the UP Hills,* Centre for Sustainable Development, LBS National Academy of Administration, Mussoorie, UP.

Singh, K., and V. Ballabh , 1991, 'People's Participation in Forest Management: Experience of Van Panchayats in UP Hills', *Wasteland News,* Aug-Oct, pp. 5–13.

Singh, Satyajit, 1997a, 'Diverse Property Rights and Diverse Institutions: Forest Management by Village Forest Councils in the U.R Hills', IDS, Sussex.

———, 1997b, 'Collective Action for Forest Management in the UP Hills,' IDS, Sussex, Mimeo.

Somanathan, E., 1991, Deforestation, Property Rights and Incentives in Central Himalaya,' *Economic and Political Weekly,* 26(4): PE37–PE46.

SPWD, 2000, Proceedings of the Seminar on Sustainable Forest Management in Uttarakhand held on February 22–23 at Dehradun.

29

# The Dreams of a Water Warrior

Roopali Phadke

Vilasrao Salunkhe, best known for his founding of the Pani Panchayat model of equitable water distribution in Maharashtra, passed away on 23 April 2002 of a massive heart attack. One by one, his friends around the world are coming to terms with his loss. When environmentalists are eulogized, it is often by those who have known them longest, fought beside them and revelled in well-earned victories. I, honestly, did not have that privilege. On these pages, I commemorate a man whom I knew for the briefest of times. Yet, his vision dominates my understanding of how circumstances can and do change, and why we must invest in our shared human capacity for justice.

Within our present focus on pathways of 'new environmentalism', Vilasrao Salunkhe's affirmations that technology design can embed principles of social equity, resource redistribution and participatory planning presents us with an interesting set of ideas to reflect upon and rally around. Moreover, the broader Maharashtra water equity movement, with its roots in the Pani Panchayat model, opens up a new frontier in Indian environmentalism where collective resistance and alterative development strategies are hand-in-hand compelling bureaucratic reform.

In this paper, I reflect upon the lessons that can be learnt from Vilasrao's efforts at changing the rural landscape toward ecological and social sustainability. I also extend these ideas to the theoretical scale by suggesting methods by which social movements can better engage in the process and prospect of democratic technology design.

The world renowned Pani Panchayat model of water distribution began with Vilasrao Salunkhe's efforts at combating drought in Naigon village in

---

* Reprinted with permission from Roopali Phadke, 'Dreams of a Water Warrior', *Seminar* 516, August, 2002.

Purandhar tehsil after the devastating Maharashtra drought of 1972. Evolving from Salunkhe's work at Naigon, five main principles guide water rights and access in Pani Panchayat schemes: water is granted on a per capita basis equivalent to half an acre per person; irrigation is intended for seasonal water conserving crops, irrigation rights are extended to landless persons; water is held as an unalienable community right unattached to a parcel of land; and lastly, 20 per cent of irrigation costs are borne by community members as cash contributions. Beginning with one project in 1972, the Pani Panchayat model currently operates in 52 irrigation projects in 25 villages in Maharashtra. Pani Panchayat organizing is coordinated by an NGO trust called the Gram Gaurav Pratisthan.

When Vilasrao welcomed me into his home in Pune for the first time, I was nervous to meet the man whose development efforts had heralded such international repute. As we sat down to discuss his life's work and visions for the future, he said to me with a modest, yet enthusiastic, grin, 'It is so interesting that you want to study the equitable distribution of water. How did this become your interest?' Having faced deeply entrenched male chauvinism and technical elitism in my interviewers on irrigation development, I found it so refreshing to meet an engineer, especially of Salunkhe's stature, who intently reflected on my ideas about participatory development and bureaucratic reform.

In addition to providing moral support to a young researcher, Vilasrao opened his work to my ethnographic gaze. In 2001, he invited me on several occasions to join his meetings and tours with government and NGO officials interested in learning about Pani Panchayat successes. One afternoon, I travelled with a coterie of government officers to Mahur village, one of Pani Panchayat's crowning achievements, to a dialogue with villagers about cooperative water harvesting efforts.

Looking out onto an irrigation tank *with water* in the dry month of March, we listened to farmers explain how their lift irrigation scheme had been designed through community input and the social and economic transformations they had experienced as a result of adopting the Pani Panchayat model. On our return to Pune, several engineers in the belied privacy of their Tata Sumo, argued that Mahur was an isolated and nonreplicable example of equity in action.

The rounds of interviews I conducted with irrigation agency executives yielded similar results. In my notes, I recorded emphatic claims that Vilasrao was a fanatic, a dreamer and most unreasonable concerning his position on water equity. One high ranking irrigation engineer said tome, 'nowhere in the world is water shared equitably—why should Maharashtra be the exception'.

To Vilasrao's credit, his achievements on the ground have impressed some politicians—though often outside of Maharashtra. Before his untimely death, Vilasrao Salunkhe served as an official adviser to the Andhra Pradesh chief minister's water conservation mission. In this capacity, he was helping to design a government-sponsored version of the Pani Panchayat system to irrigate 900 hectares for 400 families in Khammam district. Vilasrao had been facilitating information exchanges by bringing Pani Pachayat farmers form Maharashtra to visit villages in Andhra.

While bureaucrats often cite faults with the utopian idealism of the Pani Panchayat model, through my interactions with Vilasrao and his colleagues, I have been particularly impressed by the cognitive links between technology and equity that are manifest in the Maharashtrian water redistribution movement. Recent examples from these Maharashtra efforts illustrate that a pivotal mode for environmental organizing in India will be to invectively mobilize social movement theory to address the democratic design of technology. An important part of this project is to bring the academic fields of political ecology and science and technology studies into constructive engagement.

The field of political ecology has been home to social movement scholars interestingly studying the structural dynamics of poverty, underdevelopment and environmental equity through the examination of the government/civil society interface. While social movement theory aims to provide detailed understandings about the how, why and what of environmental resistance, the politics around technical decision-making, such as the design of infrastructure projects, is often simply expressed in terms of societal impacts, rather than the conceptual potential for sustainable and participatory design.

The field of science and technology studies (STS), with its intellectual roots in the sociology of scientific knowledge tradition (SSK) in Europe, provides some interesting insights around the design and negotiation of technology. Analytical concerns in STS theory include the study of local and traditional knowledge systems, accountable scientific expertise and the mobilization of citizen science. While STS research efforts have predominantly focused on American and European examples, there is great potential to bring these theoretical insights to bear on natural resource struggles in the developing world.

While the topical concerns of STS and political ecology are often disparate, there are many methodological and analytical areas of overlap. By integrating these fields, a STS approach for examining micro-level technical decision-making can be combined with the macro-institutional concerns of political ecology to study how participatory technology development can impact livelihood struggles. Environmental movements in India can productively deploy the broader scholarship that is developing in American and European STS programmers on democratic technology design. By embracing this integrated perspective, alternative development models can be explored that transcend expert dominated technological development in favour of more community based approaches at enabling environmental equity.

The water redistribution movement in Maharashtra, based in NGO initiatives for participatory technology development, offers an opportunity to develop a richer understanding of how an expanded STS programme can come together with political ecology concerns. Before his death, Vilasrao Salunkhe was involved in orchestrating the Maharashtra State Equitable Water Distribution Council. As a broad coalition of technical NGOs social movement organizations and legal and political actors, this council has been opposing technocratic elitism by reconstituting a role for community-based research in the design and planning of irrigation projects.

Working simultaneously at the scales of collective resistance and the sculpting of technical altematives, NGOs affiliated with this movement have opened up the black box of water technology to map out the cognitive disconnects between how irrigation is and can be designed. One important example of this work has been the redesign of the Chikotra irrigation project in Kohlapur district.

Since 2000, Vilasrao Salunkhe had invested his energies on reforming an irrigation project in Maharashtra that would make a mark. Building on his two decades of research and activism on water equity in Maharashtra, Salunkhe believed that the government's Chikotra Valley irrigation project could be redesigned into India's first example of river basin scale water equity. A closer look at this project demonstrates how alterative technical designs can be guided by a sophisticated tool box of participatory research techniques, like on farm agroecological research and community resource mapping, to embed locally derived needs and demands into watershed development planning.

In 2001, a medium size dam was constructed by the Maharashtra Krishna Valley Development Corporation on the Chikotra river, a tributary of the Vedganga, to increase irrigation infrastructure in Kohlapur district. With a 60 metre tall main dam storing 38 million cubic metres of water, this irrigation project was to serve a command area of 27 villages. The water distribution system for this project consisted of 25 intermittent weirs downstream of the main dam. In theory, farmers in the benefit zone would be granted permits by the agency to sink pumps at water sites and lift water to their fields. The project has submergence zones of 317 hectares from five villages.

As is the case with almost every dam under construction in India, the Chikotra project elicited vehement public opposition. In addition to resettlement issues, local objection to this project focused on the poor technical design of the irrigation system and limited project benefits. The social movement, advocating for a project redesign, was led by Anandrao Patil of the Shram Shakti Pratisthan in Kohlapur and Vilasrao Salunkhe of the Gram Gaurav Pratisthan. Technical assistance was also provided by several engineers form the Pune based NGO, the Society for Promoting Participative Ecosystem Management (SOPPECOM). These NGOs argued that, by design, the government project benefited only half of the water scarce villages in the basin. In addition, the Chikotra scheme narrowly focused on surface water for irrigation, rather than integrating ground water and domestic needs into water demand calculations.

Under the banner 'Water for All', these NGOs have led the design of an alterative basin development plan which provides all 52 villages in the Chikotra Valley with equal access to water. After two years of agricultural and hydrological investigations, formation of water users' cooperative associations in every village and detailed community consultations, the final alternative plan provides an assured supply of water for every family in the basin for irrigation and domestic needs. This plan combines reservoir storage from the main dam with water conservation projects and lift irrigation schemes throughout the extended command area. Water distribution will be managed by users' cooperatives, with participatory resource mapping and water shed planning giddied ecological regeneration and sustainable agricultural development.

In January 2002, in response to the alternative proposal that was submitted to the irrigation agency, the Maharashtra chief minister pledged his support for the redesigned Chikotra project. The irrigation agency was directed by the CM to conduct a feasibility study regarding project implementation. The agency reported back that while in theory there was enough water available in the Chikotra Valley to provide every family with water, the government should not set a precedent in Maharashtra by financing state subsidization of lift irrigation or agreeing to a principle of equitable access where even landless people are entitled to water.

As a result of this administrative footdragging, formal affidavits were presented to the agency from each gram panchayat in the Chikotra Valley, signaling widespread support for a Pani Panchayat based notion of water equity. In addition, since March 2002, Chikotra Valley residents have taken to the streets of Mumbai with sit-ins and protests to draw attention to their proposal. The lack of administrative action on the Chikotra alternative plan led Salunkhe to argue to me that 'the whole mindset of politicians and bureaucrats in the irrigation department is such that they have no understanding of what has been achieved through Pani Panchayat social equity projects'.

Based on Pani Panchayat efforts over the last 20 years, as well as other well-known examples from drought prone rural Maharashtra such as Ralegaon Siddhi and Hiware Bazar, it has been well established that participatory watershed planning can yield transformative social and economic effects for rural societies. To scale up these experiences beyond a single village, it is crucial to distil specific technical, design principles from these projects. Herein, it will be vital for social movement scholars to integrate approaches from science and technology studies towards analysing democratic technology design.

This integrated perspective will help articulate the answers to three key questions: how are values, like equity and access, embedded in technical design? Which social mobilization techniques enable knowledge-making and knowledge-sharing? And lastly, what role should institutions of mediation, like technical NGOs, play in negotiating alternlfives? In the end, nothing will do more justice to Vilasrao Salunkhe's memory than to actualize a basic level of water equity in India. Beyond the scope of the Chikotra project, this means reinvesting in the importance of social equity and valuing water as the lifeblood of agrarian society.

## NOTE

* Author's acknowledgements in the original paper: I would like to thank Rohan D'Souza and Ravi Rajan for encouraging me to write this piece. The research I reference in this paper is part of a larger effort to understand the historical and contemporary features of the Maharashtra water equity movement. I am particularly indebted to K. J. Joy; Anandrao Petal; and Vilasrao Salunkhe for their sustained help. This research has received generous support from the American Institute, the US National Science Foundation and University of California.

# Section V

# Global Issues

*Photo credit Pallava Bagla*

# Introduction

India is part of a larger world and more often than not, issues and concerns that exist within its borders as a nation state, have close linkages with the lands and peoples beyond. Part of this is inevitable. The world may be divided into nearly 200 states that vary greatly in military and economic capability, population size and resource endowment. Ecological phenomena do not respect the bounds of territory drawn or contested so fiercely by men and women. The oceanic tides and prevailing winds, the flow of river systems and the migration of birds with the cycle of seasons are but small ways in which nature defies human-made boundaries.

But more than that, India is also a major arena of human-nature interactions, with consequences that could be of global significance. Not only does more than one out of every six of the six billion people on the planet live here, the billion people share living space with a myriad other organisms, making this one of the 12 countries that has the largest number of living species. Yet, with one of four people living in poverty and with the world's largest middle class, it also faces deep internal rifts.

Economic changes are often at the centre of great debates not only over poverty and wealth of different sets of people but also about what do with the wealth of nature.

Over the last quarter century, India moved from being one of the 12 fastest growing economies in the world to one of the nine fastest. In the process, there are huge, sometimes irreversible changes in the way land, water, soil are used, or as is often the case, abused. How far these issues of economy and politics can be approached in a manner sensitive to the environment is a complex subject. Dunu Roy, a graduate of the Indian Institute of Technology, Bombay, has long worked on such issues, with special attention to how better environmental quality and entitlement of the poor can go together. His experiences are of great value in the on going debate (see Chapter 30).

Energy is a must for more rapid economic growth. For long, the expansion of fossil fuel use was equated with progress. Coal, oil and natural gas were the key driving forces of the industrial revolution from the eighteenth century and of the petrochemical revolution in the twentieth century. India was a latecomer to industrial growth, with over 80 per cent of the work force engaged in agriculture in 1947. This meant the rate of use of fossil fuels lagged far behind that of industrialized economies. It also means that as the economy in India modernizes, the levels of use will increase even if not to the levels of North America, Japan and Europe.

The green house gases (methane, ozone and carbon dioxide) make the planet habitable for humans. But the present levels may well be changing faster than in recorded human history. Such a rapid increase in these 'green-house gases' as these compounds are known can have far reaching impacts on climates worldwide. Since they trap heat, temperatures will rise, with significant implications for growing cycles of crops, polar ice-caps and sea levels. Even thoughtful, serious science-based scenarios warn that these effects can be deeply damaging to human settlement and safety.

It is normal to say we ought to think of what is good for the planet or the future of the land. It is much more difficult to specify who will define quite what those interests are. Much hinges on 'whose interests one rates higher than others' (McNeill 2000: xxiii, 114).

How one assesses the role of fossil fuel use in the developed countries (that use much more per capita) and of the developing countries like India and China (which together make up nearly four of every 10 people on earth). Much depends on whether we accept the huge disparity between levels of energy and resource use between nations, say between India and the US, or on a different scale, India and Bangladesh. These are often related to uneven rates of access and use within countries. A middle class person in a metropolis or even a small town draws more water and fossil fuels than a poor villager in the mountains or on sea shore. For now, developing countries have resisted bids to 'cap' their growth and asked for transfer of cleaner technologies to them by the developed countries. These efforts have met with some success on the ozone depleting refrigerant chemicals. Global use of Chloral Fluoro Carbons (CFCs) dropped by 80 per cent in eight years ending 1994.

Negotiators have not made much headway on issues of global warming. The rise in carbon dioxide and other such gases that will radically alter global climates is a common threat. But there is little accord on what to do about it and how.[1] India's position is very significant: it is both a developing country and one that aspires all the more so since reforms began in 1991 to 'catch up with the West'. Whether this can be done in a manner that does not replicate the wasteful impacts so evident in the developed world remains a challenge (see Chapter 33; also see Abbasi 1999).

There has been much more progress on the issue of conserving biodiversity with a major treaty signed by most countries, even the US, dating to 1992. But even here, there are major divisions. Much of the biological wealth is in poor countries and many international NGOs as well as the UN are funded by the richer nations.

Equally significant are hazards and benefits from choices of technology made by national leaderships. India witnessed in 1984 one of the world's worst industrial disasters at Bhopal where the Union Carbide pesticide plant had a gas leak that caused not less than 2,500 deaths. Since then many more who were maimed have had a wait for justice, medication and hopes of help to rebuild their work, health and lives (see Chapter 31, also see Rajan 2001). From Minamata in Japan to Silveso in Italy, chemical pollution has extracted a heavy cost in human lives, with chemical compounds lingering long in ecosystems and in the human body, causing unintended and adverse impact on health.

The genie in the atom gives rise among critics to similar fears. Atomic energy through fission has never quite gained in the developing world the centrality it has acquired in countries like France. The Three Mile Island accident in the US in 1978 and Chernobyl in the USSR eight years later, gave rise to serious criticism. In addition to the threat of radiation, they point to the issue of disposal of nuclear waste and to the risks in mining uranium. Over the last decade, fears of global warming due to fossil fuel use have given fresh impetus to a renewal of the nuclear energy industry worldwide. The genie in the atom cannot, it appears, be put back. Whether it can be used peacefully and safely is the question to consider (see Chapter 32).

In recent years, the intransigence of the US has made difficult any progress on international cooperation on environmental issues. Yet, civil society initiatives and citizen pressures have helped bring about action in the past as with the agreements on ozone. When industrial nations promised to cut carbon dioxide to 1990 levels by 2012, the US alone stayed away.

There is also evidence that some corporate groups are more responsive to consumer and investor pressure. Either way, the role played by countries like India will be critical. Much will hinge on whether the internal choices made here about technology or equity show the rest of the world there are better ways of striking a balance (Speth 2004; also see *The Economist* 10 December 2005).

## Note

1. On the Montreal Protocol on ozone and how it was secured see Rajan (1992); Agarwal and Narain (1992).

## References


Abbasi, S.A., P Krishnakumari and F.I. Khan, 1999, *Hot Topics: Everyday Environmental Concerns*, OUP, Delhi.

Agarwal, Anil and Sunita Narain, 1992, *Global Warming in an Unequal World*, CSE, Delhi.

Govind Rajan, Mukund, 1992, *India and the North-South Politics of Global Environmental Issues*, OUP, Delhi.

McNeill, J.R., 2000, *Something New under the Sun: An Environmental History of the Twentieth Century*, Penguin Allen lane, London.

Rajan, S Ravi, 2001, 'Towards a Metaphysic of Environmental Violence: The Case of the Bhopal Gas Disaster', in Nancy Lee Pelusso and Michael Watts (eds), 2001, *Violent Environments*, Cornell University Press, Ithaca, pp. 380–98.

Speth, James Gustave, 2004, *Global Environmental Challenges: Transitions to a Sustainable World*, Yale University Press, Ithaca; Indian edition, Orient Longman, Delhi, 2005.

*The Economist*, 'Don't Despair, Grounds for Hope on Global Warming', 10 Dec. 2005, pp. 11–12.

# 30

## Environmentalism and Political Economy

**Dunu Roy**

In tracing the journey of ideas, some landmarks stand out. In the sphere of what has come to be known as the 'green' discourse, one of the early markers of the impact of developmental products on nature was Rachel Carson's 1962 classic, *Silent Spring*. Ten years later, the Club of Rome's *Limits to Growth* sparked off a debate on whether these impacts would constrain development. The year 1972 was also the time when Barbara Ward and Rene Dubos wrote *Only One Earth*, which explored the nature of the constraints on development. It served as background material for the Stockholm Conference on the Human Environment. In 1982, Erik Eckholm came out with *Down to Earth*, a review of the efforts to protect the world's environment. And in 1992, the Rio Conference discussed the linkages between environment and development.

What will 2002 bring us? And where do we place the Indian debate within this global context? These are questions that may interest both academics and activists concerned with environmental issues. An appropriate way of reviewing environmentalism may be to ask what has been the debate over the last 20 years and what can we learn from it; what have been the ebb and flow of the ideas that have affected us all? Environment itself is a totality that encompasses many parts and wholes and such an essay provides an opportunity to locate various actors within the trajectory of what might be called the dynamics of political ecology, the science of the interaction between society and nature.

---

* Reprinted with permission from Dunu Roy, 'Environmentalism and Political Economy', *Seminar* 516 August, 2002, pp. 65–69.

As an example of events in this trajectory, the Kerala Shastra Sahitya Parishad (KSSP) held a Convention in 1978 at Trivandrum at which it passed a resolution opposing the construction of a dam in the Silent Valley. That represented a significant environmental step, because the convention had gathered groups from all over India to discuss the promotion of KSSP's technique of science popularization at a national level. Many of the participants were greatly impressed by KSSP's work and ideas and several wanted to immediately undertake similar work in their regions. However, there was a small but significant section that pointed to the socio-economic differences between Kerala and other states and felt these differences were important enough to affect the nature of science and science popularization.

In 1980, *Science Today* published a small piece questioning science (Roy 1980), which summarized the collective experiences with the application of science and technology in several areas since 1969. Beginning with earthquake relief at Koyna to experiments in drought proofing in Marathawada and pollution control and planning in Shahdol, this article chronicled a growing awareness of how, 'having built something for use, we neglected to examine how useful it was'. Thus, 'there were categories not only within what was observed, but also amongst those who did the observation, and these categories were often in conflict with each other'. This did not quite harmonize with the notion of popular science being promoted by KSSP and echoed what, perhaps, was a step on the road to political ecology.

Two years later, in 1982, the Centre for Science and Environment (CSE) published the first *State of India's Environment* (SoE-1). This was, essentially, a compilation of the work by many researchers and activists over the previous two decades. It also included some brief sections on people's protests because, by then, several peasants' and workers' agitations and (what were then known as) non-party political formations had made an appearance in civil society. The editors' achievement was to pull together all these contributions into a comprehensive whole that described the enormous degradation of the nation's environment. However, CSE itself did not advance a political perspective on the causes for this degradation, except to speak on behalf of women and forest dwellers who bore the brunt of the degradation.

During the same year, a small group in Shahdol authored a text on environmental planning (Ghotge *et al.* 1982). Based on eight years of praxis; this publication observed that 'not only were there hierarchical functions in planning theory, but that people too are hierarchical.' In contrast to SoE-1's general tenor, the Shahdol group argued that there was concern for the environment either when profit rates were affected or due to popular discontent. It further stipulated that people's planning should be, 'that political exercise in allocation and management of resources which improves the well-being of those engaged in production, prevents the harmful byproducts of industrialization, and conserves the natural resources.' Hence, the group attempted to argue that all man–man and man–nature conflicts were related to the central contradiction over control of resources.

In 1984, the second *State of India's Environment* (SoE-2) was published. The editors by then had recognized that there was a 'politics' involved in thinking about environment. Consequently, they included a chapter on the 'Politics of Environment', and Anil Agarwal, then Director of CSE, wrote one essay while I was asked to write the other. (It may be pertinent at this stage to observe that CSE has never again asked me to contribute!) Agarwal situated his discussion within existing systems of governance, foregrounding the venality of corrupt leaders, and argued forcefully for equity and community ownership and management of resources. But, having put the issue of political economy on the canvas, he brushed it over with a coat of democratic paint. How communities would actually take over and manage resources, in the face of a rapacious state, was left unattended.

My article began with the observation that 'different persons give different answers to the same question', and located that difference within an understanding of 'ecobalances and interpenetration of different systems in a generalised world market'. It built upon the magnificent environmental critique of the Damodar Valley projects by the civil engineer, Kapil Bhattacharya, and tried to show the ecological as well as socio-economic linkages between the impacts of dams, mining, forest laws and occupational hazards. While centralization of authority failed to protect the environment, it was also clear that, 'despite the existence of enough evidence of damage, authorities refuse to pay attention.' Thus, environmental politics needed to be understood in terms of 'who benefits', and 'whose interests are being protected'.

In 1984, an inhuman accident occurred at the Bhopal factory of Union Carbide that impinged upon the understanding of politics of several environmental groups. By then, it may be remembered, Gail Omvedt and Prakash Karat (both with affiliations to Left parties) had trenchantly criticised non-party groups, accusing them of being part of an imperialist design of pitting environmental concerns against working class interests. This was allegedly due to the kind of development work these groups were involved in and the funding they received from foreign (meaning western) agencies. Many political groups were (and continue to be) greatly influenced by these arguments and it marked a schism between political and apolitical environmentalists. The subsequent betrayal of the Bhopal victims by the government only sharpened the divide.

However, there were a few attempts to address the divide. For instance, there was the case of retrenchment in Hindustan Aluminum—allegedly sparked off by 'environmentalists'. In an article published in 1986 (Roy 1986), I observed that Hindalco was itself 'a prime example of exploitation both of labour power as well as natural resources.' In effect, what had happened was that there was 'pressure by a labour union on government to condone anti-people and anti-nature policies,' and this pointed to a 'difference of aspirations between leaders and led'. The environmental issue was used 'to obfuscate relations between labour and capital and to keep the labour movement within the narrow confines of the economic arena'. This was equally relevant for the Union Carbide factory, where workers' protests against repeated accidents in the plant had largely been ignored by both management as well as the unions.

This attempt to break through a narrow concept of 'environmentalism' and to link up issues of despoliation and pollution with those of occupational safety and job-centred development, continued in an article on the production of power (Roy 1988). The article noted:

> The environmental question is not merely confined to pollution of air, water and land. Nor is it just a slogan for afforestation. It is organically related to the drive for generation of greater and greater amounts of energy in the shortest possible time.

Such a thrust inevitably leads to higher hazards at the workplace, reduced maintenance, and lowered safety. Ironically enough, in 1988 Ramachandra Guha presented a typology for different ideologies of environmentalism, but his category of 'ecological Marxism' did not admit to any linkage between labour and environment.

By 1990, the first 'green' judgements were delivered by the Supreme Court. It was held that the Right to Clean Environment was superior to the Right to Livelihood, both emanating from Article 21 of the Constitution. This, naturally, led to a lively interest in the law and I made a foray into the area of jurisprudence with a fictional piece (Roy 1990). It explored various legislative codes (such as the Factories Act, the Workman's Compensation Act and the Environment Protection Act) and the loopholes implicit in them. While the tale, as it unfolded, demonstrated that the 'dangers inside were related to dangers outside the factory', it also attempted to show that justice was being subverted. Government agencies, ministries and boards were set up to protect private interests and so could hardly be expected to serve the public. As one of the characters in the story states, 'While we, as lawyers, have to work within the framework of the law, the law itself rests on shaky foundations'.

The year 1990 also, marked the 'opening up' of the economy with the first phase of liberalisation sponsored by the World Bank and the International Monetary Fund. By 1992, the 73rd and 74th Amendments to the Constitution had been passed, the Indian Science Congress session at Baroda was seized of the debate on population versus environment, the United Nations Conference on Environment and Development at Rio had been concluded, and the Babri Masjid had been razed.

With this in the background, I contributed an essay to a volume on the attitude of indigenous peoples to the environment. The essay (Roy 1992) argued that the sub-divisional officer sitting in his Rs 270,000 house shared nothing in common with the tribal for whom he was constructing a Rs 2,500 hut under the rural housing scheme. It pointed out that myths and fables, whether of the nineteenth or the twenty-first century, were born out of a 'convenient marriage between revenue and duty'. It concluded, 'Refusal to oblige government is not born out of a dullness of mind but is a deliberate and intelligent response to an unwanted situation—as deliberate a political act as civil disobedience'.

In 1993, the World Bank withdrew from the Sardar Sarovar Project, a major victory for those trying to advance an alternative developmental paradigm based on the mobilization of indigenous peoples. In 1994, the All India People's Science Network organized a daylong debate on the various themes highlighted by the Narmada Bachao Andolan. Several senior officers and engineers from the establishment participated in the debate, which ranged over issues of irrigation, power, agricultural production and rehabilitation. During the course of the discussion I made the somewhat controversial claim that 'displacement is as much one of the design objectives of the dam as power or irrigation' (Roy 1994). While the Narmada Water Disputes Tribunal was the first to propose land for land in 1979, there was no rehabilitation policy prior to 1985, and there is still no national resettlement policy. This, despite the fact that sufficient documentary evidence of the impact of displacement caused by hundreds of projects, affecting millions of people, existed from 1864 onwards. Hence, the paper asked the political question, 'When will there be sufficient evidence?'

The next two years witnessed many minor and major upheavals in different parts of the country on issues of race, caste, ethnicity, region, religion and class. Thus, in the run-up to the general elections of 1996, the possibility of dramatic changes in governance appeared strong. Several 'people's agendas' were prepared by different non-party groups as part of a campaign to further democratize the electoral process. These included the agendas of INREP (Initiative for National Renewal and Empowerment of the People) and INSAF (Indian National Social Action Forum), the resolve of the NAPM (National Alliance of People's Movements), and the manifesto of the SJP (Samajwadi Jan Parishad). The notion of 'people'—so attractive to environmental discourse—lay at the core of these political statements.

It, therefore, appeared useful to take this as background material to write on 'people's politics' (Roy 1996) as a commentary on an alternative vision of society. The analysis showed that, 'all these non-party declarations, while representing a break from traditional electoral practise, conformed to a notion of a democratic, secular, egalitarian, and sustainable society'—all of which had been 'mentioned in the Janata Dal manifesto before the elections, but ignored by the United Front government after the elections! The various manifestos agreed that the major hurdles were communalism and liberalization, gender inequity and violence. But they did not have a common definition of people', apart from a vague unspecified consensus on 'dalits, women, tribal, and backwards'. And, compared to the resolutions passed at a National Convention of Rural Workers, 'there was a gap between people's demands and "people's agendas"'.

This gap between perception and reality formed the core of an exposition on child labourers published in 1998 (Roy 1998). The essay posed three theoretical questions: (*a*) What is the ethical basis for deciding the rights of child labourers? (*b*) Who decides these rights? (*c*) How do child labourers make their views known? It further illustrated the paradox of governance that 'those who preside over exploitative and unjust systems are also those who prescribe its remedies!' The state was that apparatus, in a position of authority, whose task it was to conceal exploitation in order to protect production. It, therefore, determined the categories of knowledge and the contours of prevailing logic.

In challenging this worldview, within which the ethics of children at work was located, the paper attempted to bridge the gap between generations: 'Children desire both work and play. Working men and women too want meaningful work and play.' By taking examples of forest regeneration, children's education, and gender-driven transportation in Himachal, it tried to show that people's rights, which had been appropriated, could been reappropriated. 'The system cannot be changed by taking away the world's children. The struggle is not only for what the society and state are willing to concede but to change society and state.'

Working people's struggles demand a theoretical basis for sustaining themselves as well as for changing the system which forces them to struggle. But the production of theory does not necessarily follow this imperative. Thus, for the Fifth World Conference on Injury Prevention and Control in 2000, I tentatively chose to revisit an earlier theme on occupational safety (Roy 2000). The context had been set by the closure of 'polluting units' in Delhi and the loss of thousands of jobs. The courts were beginning to take a meddlesome lead in initiating actions that lay in the realm of the executive and legislature. Much of this initiative sprang from concerns with the environment and the propaganda blitz launched by 'environmentalists' with special interests. Hence, it was necessary, once again, 'to expose the ideological roots of the careless worker'.

The paper traced the growth of ideas, from the compensation-safety apparatus of the late nineteenth-century to the merger of principles of pollution prevention with those of occupational safety in the 1990s. It dwelt on the contemporary relationships between 'international finance capital and working conditions, with the direct impact of globalisation on industrial development'. Thus, new hazards were emerging for workers and consumers and, therefore, new principles were required for the evolution of safety measures. The paper brought together elements from diverse campaigns on the right to information, right to organise, right to resources, right to health, right to participate, and added a right to vulnerability, to suggest a fundamental right to safe livelihood.

For an essay such as this, there really is no conclusion because the learning experience builds upon the past to construct the future. But perhaps it may be possible to set up some temporary landmarks. I obviously belong to one school in the environmental university. But, as this essay demonstrates, this school exists in conjunction with the others, conflicting with them, searching out fallibilities, asserting its own dynamics, and constantly, perpetually, unceasingly, learning about the world outside the university. So what have we, who belong to this particular school, learnt over the last two decades?

We have, I think, re-learnt the old lesson, but in a new context, that the nature of knowledge, of learning, is determined by class interests and the conflicts between them. Thus, our university is as much a prey to the political economy of production for greed (while millions teeter on the brink of survival) as is any other institution in civil society. We have to see our university as part of the world and, hence, the environment is everything, and we have to grasp the linkages between all phenomena. In particular, we have to comprehend the linkage between environment and work, between the food-gatherer and the tool-producer.

The challenge that such an exploration throws up is not merely a change in lifestyles, but a revolution in state and society; not just the right to be human, but in defending and enlarging human rights. This is the heart and the brain and the muscle of political ecology—that we hauntingly wage war against all oppression and exploitation, against all injustice and violence, against war itself.

## References

Ghotge, S., A. Gupta, D. Roy and A. Deshpande, 1982, *Planning the Environment,* Gandhi Gram Press, Madurai.

Roy, Dunu, 1980, 'What is Science For?' *Science Today*, October, Bombay.

———, 1986, 'Environment and Politics', *Economic and Political Weekly,* 15 November.

———, 1988, 'Energy, Work and Danger', paper presented at the seminar on Energy and Environment, Institute for Cultural Research and Action, Bangalore, August 1988.

———, 1990, 'The Legal Eye', *Economic and Political Weekly,* 1–8 December and 15 December.

Roy, Dunu, 1992, 'Myth, Fable, and Guile: The Art of Survival in Shahdol', in *Indigenous Vision, Peoples of India, Attitudes to Environment,* Sage Publications, New Delhi.

———, 1994, 'The Accountability Principle' (Unpublished) People's Science Institute, September, Dehradun.

———, 1996, 'People's Politics: The Third Force?' *Lokayan Bulletin* 12(6).

———, 1998, 'Rights of Child Labour: Ethics, Production and Nation-State', *Economic and Political Weekly,* 31 January.

———, 2000, 'Labour and Safety: Science, Technology, and Society', Paper presented at the Fifth World Conference on Injury Prevention and Control, March, New Delhi.

(SoE-2) *State of India's Environment*, 1984–85, 'Politics of Environment-II', Anil Agrawal and Sunita Narain (eds), Centre for Science and Environment, New Delhi.

# Are We Prepared for Another Bhopal?

Vijay K. Nagaraj and Nithya V. Raman

The Bhopal gas leak and the disaster that ensued raises a series of questions regarding industrial safety, risk, compensation and relief for victims of industrial disasters, multinational enterprises and liability, regulation of the transfer and use of hazardous technology and so on. The leak prompted a range of responses—statutory, policy and the judicial—both in India and the USA which have now played out over two decades. The twentieth anniversary is an appropriate occasion to consider some of these responses and assess whether we are prepared for another Bhopal.

Internal documents that have come to light during the discovery process in the US courts in the contamination case over the last few years clearly indicate that UCC (*a*) transferred unproven technology to UCIL; (*b*) did everything it could to ensure that it maintained a majority stake of over 50per cent in UCIL; (*c*) was aware of the possibility of a potential runaway reaction that triggered the MIC leak in Bhopal; (*d*) had far lower safety standards in place in Bhopal than it did in the USA; and (*e*) was aware right from 1982 that the Bhopal plant suffered from serious safety problems.[1]

Even as UCC's responsibility for the accident is well established, there is no doubt that the Governments of India and Madhya Pradesh too have to accept their share of responsibility for not regulating the safety of the plant. Between the late 1970s and 1984 there were several accidents at the Bhopal plant, including one which resulted in the death of a worker due to a phosgene gas leak. The inspectors of the Industrial Safety and Health Department, Government of Madhya Pradesh recorded at least six accidents at the plant before 1984, but although they made recommendations and

---

* Reprinted with permission from Vijay K. Nagaraj and Nithya V. Raman, 'Are We Prepared for Another Bhopal?' *Seminar*, December, 2004, pp. 52–58.

instructions the inspections did not lead to a 'deeper probe or stricter follow-up of the action Carbide was required to take' (Ramaseshan 1984).

It is also impossible for the Government of India to plead innocence regarding the potential threat from the Bhopal plant. For instance, the 1982 application for approval of foreign collaboration submitted by UCIL clearly states that the manufacture and storage of MIC involved extremely hazardous substances and technology.[2] Prior to the 1984 leak several media reports and protests by the UCIL workers' union at the plant also drew attention to the serious safety hazards, but both the state and central governments overlooked the issue (Everest 1985: 132).

It is also significant to note that while the GOI was allowing industries employing hazardous substances, technology and processes to operate, no attempt was made to develop an appropriate regulatory framework to govern safety and risk of such industries. Further, little or no attention was paid to enhancing capacities of bodies responsible for industrial safety to actually monitor hazardous industries. The lack of legislative frameworks and corresponding institutional preparedness notwithstanding, it seems that what the state really lacked 'was the will and the intent to come down strongly on Union Carbide' (Ramaseshan 1984).

The state made no attempt to ensure that the community was informed regarding the potential threat from the Carbide plant. In fact attempts made by some in the local administration to shift the plant to a safer site were overruled in 1975 and just months before the fatal leak a large number of settlers living illegally around the plant were given legal titles to the land (Cassels 1993: 15–16).

Following the Bhopal gas leak the Factories Act of 1948 was amended and a new chapter on Hazardous Industries added in 1987. This amendment also incorporated some of the Supreme Court pronouncements on industrial safety made in the context of an oleum gas leak in New Delhi in 1986.

The amendments essentially focused on ensuring that information regarding potential risks and hazards are made available to local authority and to communities in the vicinity of the plant, and that workers have a right to participate in safety management and regulation of the location and expansion of hazardous industries. The 1987 amendments also redefined the 'occupier' (the person designated to be responsible for the affairs of the factory—specifically safety in the present context) to be one of the Directors and explicitly laid down that the occupier has an obligation to show, in the event of an accident, that due diligence had been exercised to enforce the safety obligations laid down in the Act.

One of the most critical lessons from Bhopal is the importance of transparency and public participation in decisions relating to the location and operation of hazardous industries. It needs no emphasis that secrecy breeds a lack of accountability. Recognizing the right to know and enforcing transparency, i.e., the obligation to inform, is critical also because knowledge brings with it a greater sense of responsibility on all sides—the public, the regulatory authorities and of course the corporation itself.

The law has, for some time now, been protective of the right against disclosure about matters connected with industry. Unfortunately the 1987 amendments to the Factories Act rather than change this only reinforced non-transparency and secrecy. While Chapter IV A of the Act emphasized transparency the punishment for any 'unauthorized' disclosure was actually enhanced substantially. Similarly S.118 which places further restrictions on disclosure was also allowed to remain unchanged.

What is, however, most startling is that there were no amendments to the Factories Act or any other statute that made it mandatory for industry to disclose all information that may help mitigate the effects of the disaster. 'The emphasis on industrial secrecy, and the enforced silences, rest uneasily with the dire need for disclosure and of information sharing witnessed in the days, months, and years following the Bhopal gas disaster.'

For the victims of the leak the real disaster began after the leak. In 1985 the Government of India enacted the Bhopal Gas Claims Act and appropriated to itself the sole privilege of representing the victims of the disaster. The Act also laid down an elaborate mechanism of processing victims' claims for compensation including medical categorization. Apart from an initial ex-gratia, the state made no attempt to grant any interim compensation to victims until the Supreme Court so directed years after the leak. In 1989 the Government of India agreed to a full and final settlement with UCC (without consulting the victims) and signed off all of UCC's civil and criminal liabilities in exchange for a paltry sum of US$ 450 million even though it had begun by claiming damages to the tune of over US$ 3 billion. And it was only three years later that the Claims Courts began adjudication of compensation claims.

For the victims of the leak the compensation mechanism proved to be another nail in the coffin. It pitted thousands of poor, illiterate and powerless gas victims against a ruthless bureaucracy that reduced victims to claimants. The compensation mechanism became an unholy nexus between petty bureaucrats, doctors, lawyers and, in some cases, even judges. Characterized by heavy bureaucratic procedure, arbitrariness and corruption, the victims of Bhopal ended up receiving too little compensation too late and that too at a high cost. Of course these were in fact the fortunate; thousands had their claims rejected altogether or were even refused registration.

One of the most pertinent lessons of Bhopal was the need to ensure prompt relief to victims without them having to establish individualized fault and causation. The plight of the Bhopal gas victims has prompted a consideration of public no-fault compensation models that separate the issue of victim compensation from the question of liability and deterrence, allowing victims to be compensated quickly without removing the deterrent effects of traditional tort law from the system (ibid.: 265).

In 1991, India enacted the Public Liability Insurance Act (PLIA) to provide for interim compensation on a no-fault basis. In 1992 this was amended because insurance companies were unwilling to insure hazardous companies for a sum without an overall ceiling. This, although the PLIA already prescribed limits on the amounts to be paid to each affected person where death, serious injury, loss of work, or damage to property occurs. The PLIA was an attempt to use insurance as a risk spreading exercise which would enable the immediate payment of minimal amounts as an interim measure. This would cover not only Bhopal-like incidents but the multitude of mini-Bhopals that are a regular occurrence. There is little evidence, however, that this account under the PLIA is being drawn upon—not very good news for present or future victims of industrial disasters.

In 1995 the National Environment Tribunal Act was enacted to set up tribunals to deal exclusively with the determination and disbursement of compensation. The law, however, is yet to come into force. There is also a proposal under discussion to merge these tribunals with the Environment Appellate Authority (EAA) which was set up by a 1997 law to 'hear appeals with respect to restriction of areas in which any industry's operations or processes... shall not be carried out...' The merger apparently is being mooted because both these forums are being underutilized—a paradox given the ever-increasing multitude of conflicts around location of industrial projects and increasing number of accidents involving hazardous substances.

Victims of the Bhopal gas leak attempted and failed to access justice through the tort system in both the USA and in India. In theory, the tort system is a powerful tool to obtain justice, compensation and remediation and to act as a deterrent in cases of environmental damage (Anderson 2002). The Bhopal case revealed some of the weaknesses of using tort law.

In Judge Keenan's decision to send the Bhopal case to the Indian courts, he noted that he was 'firmly convinced that the Indian legal system is in a far better position... to determine the cause of the tragic event and thereby fix liability' and, because of access to greater information than the American courts, to fix the appropriate amount of compensation (Baxi and Dhanda 1992: ii). Yet, tort litigation in India did none of this.

A number of cases have recently been filed by affected communities which have sought to use national law to tackle cases of personal injury or environmental damage claims against MNCs.

However, tort cases hold many obstacles for plaintiffs. Cases are generally slow and expensive. Toxic tort cases often have very high standards for proving certain legal elements such as causation and liability, which make it difficult for plaintiffs to meet evidentiary burdens (Rolle 2003). While tort cases are reasonably good at assessing personal injury and property damage, tort cases are 'clumsy and inflexible' in assessing, evaluating, and quantifying environmental goods and processes outside the market (ibid.). For cross-border torts, questions arise about the appropriate forum for the case and the applicable law. The application of the doctrine of *forum non conveniens*, like in the Bhopal case, can often determine whether a case will succeed or fail (ibid.).

Even if cases are allowed, awards can be small if a court decides it is proper to use the more limited law of the state in which the tort occurred (ibid.). Courts may also choose to apply the doctrine of limited liability which allows corporations to hide behind the corporate veil, effectively prohibiting the enforcement of any damages awarded. MNCs are also problematic for torts because they defy assumptions about 'the mapping of legal persons to territorial jurisdiction', the basis of traditional tort law (ibid.).

The increasing proliferation of tort cases against MNCs is a symptom of the failure of other regulatory systems or, in fact, the lack of them—despite Bhopal—which have left victims/plaintiffs with no option but to turn to tort law for redress. Affected workers and communities have formed alliances with NGOs and public interest lawyers to attack a perceived 'governance deficit' in the regulation of MNCs (Anderson 2002). The big question, however, is whether the judiciary should and can make up this deficit, especially within the framework of tort?

In a December 1986 written statement filed in the Bhopal district court UCC claimed in its written statement that 'there is no concept known to law as "multinational corporation"'.[3] UCC further argued that 'the phrase "multinational corporation" or "monolithic multinational" has no relevance, significance, or legal consequence in the context of the present suit' (ibid.: 62). UCC therefore reduced its own corporate identity to a 'phrase'.

In a written statement submitted at a combined hearing of two Congressional Sub-Committees in March 1985, the CEO of UCC, Warren Anderson, noted that among the important public policy issues raised by the Bhopal tragedy include 'Third World questions, such as the whole relationship between multinationals and developing countries'.[4] In fact, he went on to say that 'without the technologies and capital that multi-nationals help to introduce, developing countries would have little hope of eradicating hunger and poverty' (ibid.: 204).

Multinational corporations today are more powerful, organised, complex and certainly as elusive to the law as they were in 1984. Jamie Cassels observed: 'Indeed the multinational company, though the most important nongovernmental entity in the daily life of citizens around the world, is not even recognized to exist by most legal systems' (Cassels 2000/2001). Warren Anderson can look back over the past 20 years with some satisfaction—the relationship between multinational corporations and developing countries is just where he left it.

In recent years, there has been a shift of hazardous or polluting industries, so-called 'dirty' industries, to low-income nations (McCallion and Sharma 2001). This shift is prompted by what Lawrence Summers while at the World Bank, called the 'impeccable economic logic of dumping a load of pollution on the lowest wage country' (ibid.). Potentially hazardous industries from the developed world will move to host states in the developing world that offer the lowest levels of environmental regulation and compliance costs, and the least liability for international investment (ibid.). Governments in developing nations face a contradictory situation: they are responsible for the health and safety of their citizens, but over-regulation of multinational corporations can drive away investment, reducing the wealth of the nation and the number of available jobs (Cassels 1993). States, particularly low-income ones, are unable to adequately regulate hazardous industries that move into their nations for fear of driving investment elsewhere.

Such 'impeccable' economic logic ignores the fact that market pricing theories have yet to find a quantifiable economic cost for environmental harm. This ignorance is compounded by the fact that long term costs of environmental damage are uncalculated and largely borne by the host state whereas benefits of the movement of hazardous technology to developing nations are discrete and quantifiable, in the form of increased money for investors, and higher rates of economic growth, GDP, income or productivity for the home state.

While international law has addressed the problem of transboundary pollution, it has not directly addressed the responsibilities of corporate actors within this context. Since 1963, the International Law Commission has debated the question of transboundary environmental harm. In 2001, the Commission adopted a draft preamble and 10 draft articles on the Prevention of Transboundary Harm from Hazardous Activities.

The question of *liability* for transboundary environmental harm is still only a matter for debate. There was an order to the International Law Commission from the General Assembly in General Comment Operative paragraph 3 of resolution 56/82 to continue its consideration of international liability for injurious consequences arising out of acts not prohibited by international law. Despite a great deal of work from the Special Rapporteur on this question, there were some in the Commission who still expressed a great deal of dissent on the appropriateness of the topic (ILC 2003). The topic of liability, according to this view, did not easily lend itself to codification or progressive development, and there existed no agreement on the matter in doctrine, jurisprudence, or practise (ibid.). However, there was general support for the statement that the innocent victim should not be 'as far as possible, left to bear the loss resulting from transboundary harm arising from hazardous activity'.

This limited progress on the question of international liability has had almost no impact on the question of multinational corporation liability. The Commission stated in its Annual Report for 2003 that

> [w]hile issues concerning damage by transnational corporations in the territory of a host country and their liability were critical, some members viewed any consideration of such issues within the context of the topic, or at any rate by the Commission, with reticence. Moreover, it was noted that questions

concerning civil liability such as those on proper jurisdiction, in particular the consideration of cases such as… the 1984 Bhopal disaster litigation went beyond the general scope of the topic.

So who is going to bell the cat? In 2001 the OECD adopted its Guidelines for Multinational Corporations, which are essentially agreements and guidelines for member home governments of multinational corporations. And in 2003 the UN Sub-Commission adopted the UN Norms on the Responsibilities of Transnational Corporations and Other Business Enterprises with regard to Human Rights. Both are often claimed by human rights groups in particular as significant achievements but neither deal with the issue of corporate liability—at the heart of tragedies like Bhopal.

The issue of transnational liability is obviously not as widely accepted as transnational profit. The Damocles sword of *forum* still hangs above every move to bring multinationals to the courts of home countries. As Judge Doggett of the Supreme Court of Texas observed in Dow Chemical v. Castro Alfaro, in reality *forum non conveniens* is nothing 'but connivance to avoid corporate accountability' (Cassels 1993: 277). Judge Doggett went on to observe that comity to ensure corporate accountability cannot be achieved 'when the United States allows its multinational corporations to adhere to a double standard when operating abroad and subsequently refuses to hold them accountable for these actions.' (Ibid.)

Indeed even the deaths of thousands of people did not prompt any steps by the US government to ensure that the multinationals that brought in millions by way of profits from their overseas operations actually adhered to the same safety standards abroad as they did at home. It is significant that while no attempt was made by the US government to get their multinationals to improve safety standards abroad, many steps were taken within the US to enhance safety and protect potential victims of industrial hazards.

In the US the Bhopal disaster also resulted in the passing of the Emergency Planning and Community Right to Know Act, 1986. This act made it mandatory for industries to disclose the names of hazardous chemicals they were dealing with, quantities of such chemicals they were releasing into the atmosphere, and the possible dangers. In 1990 the Clean Air Act was amended and required a range of industries to disclose worst case scenarios and the preventive measures in place. The US Environment Protection Agency (EPA) moved to post all the information on the internet in order to make it more widely accessible (Davis 2002: 1).[5]

In 1986, the Comprehensive Environmental Response, Compensation, and Liability Act (CERCLA),[6] more commonly known as Superfund was amended by the Superfund Amendments and Reauthorization Act (SARA) to improve on standards of remedies as well as enhance enforcement and expand the size of the trust fund. The amendments also stressed the importance of permanent remedies and innovative treatment technologies in cleaning up hazardous waste sites and also provided new enforcement authorities and settlement tools. Further, it increased the focus on human health problems posed by hazardous waste sites and required EPA to revise the Hazard Ranking System (HRS) to ensure that it accurately assessed the relative degree of risk to human health and the environment posed by uncontrolled hazardous waste sites that may be placed on the National Priorities List (NPL) (ibid.).

Thus while the US sought to tighten its own regulatory capacities and control over 'dirty' industries it has steadfastly refused to take any action to get its multinational corporations to adopt higher safety standards abroad. Indeed, by continually enhancing corporate regulation at home and turning a blind eye to the actions of its multinationals abroad the US government (like many other northern governments) is encouraging corporate double standards in safety but acting very much in keeping with the 'impeccable' economic logic of dumping pollution and hazardous technology on poorer countries—a logic that believes in simply relocating risk rather than enhancing safety.

The Government of India argued before the American court in the Bhopal case, the concept of *multinational enterprise liability*. Essentially, this involved marrying strict liability for hazardous technology and enterprise liability for MNCs. The Indian lawyers argued that MNCs,

> by virtue of their global purpose, structure, organization, technology, finances and resources have it within their power to make decisions and take actions that can result in industrial disasters of catastrophic proportion and magnitude. This is particularly true with respect to those activities of the multinationals which are ultrahazardous or inherently dangerous (Baxi and Dhanda 1990: 187).[7]

The 'complex corporate structure' of MNCs is characterized by close intertwining coupled with fine distinctions within a network of subsidiaries, affiliates and divisions operating in a maze of 'interlocking directors, common operating systems, global distribution and marketing systems, design development and technology worldwide, and financial and other controls'. (Ibid.) This 'complex corporate structure' of the MNC makes it extremely difficult for victims to 'pinpoint responsibility for the damage' or precisely 'isolate which unit of the enterprise caused the harm'. (Ibid.)

Only the MNC in question has 'the means to know and guard against the hazards likely to be caused by the operation of the said plant'. (Ibid.) As a result, 'the multinational enterprise which caused the harm is liable for such harm'. (Ibid.)

A year after the Bhopal gas leak, a major leakage of oleum gas from a plant owned by Shriram Industries in New Delhi resulted in a large number of persons being affected and at least one death (IELRC). In this case (Shriram) the Supreme Court made a pronouncement of absolute liability and enterprise liability that changed the way in which Indian companies were held liable.

The Supreme Court said:

> We are of the view that an enterprise which is engaged in an hazardous or inherently dangerous industry which poses a potential threat to the health and safety of the persons working in the factory and residing in the surrounding areas owes an absolute and non-delegable duty to the community to ensure that no harm results to any one on account of hazardous or inherently dangerous nature of the activity which it has undertaken (M.C. Mehta vs Union of India, (1987) 1 SCC 395).

The Supreme Court reasoned that,

> Since the persons harmed on account of the hazardous or inherently dangerous activity carried on by the enterprise would not be in a position to isolate the process of operation from the hazardous preparation of substance or any other element that caused the harm the enterprise must be held strictly liable for causing such harm as part of the social cost for carrying on the hazardous or inherently dangerous activity. [...] Such hazardous or inherently dangerous activity for private profit can be tolerated only on the condition that the enterprise engaged in such hazardous or inherently dangerous activity indemnifies all those who suffer on account of the carrying on of such hazardous or inherently dangerous activity regardless of whether it is carried on carefully or not (ibid.).

The court concluded by pronouncing that,

> Where an enterprise is engaged in a hazardous or inherently dangerous activity and harm results to anyone on account of an accident in the operation of such hazardous or inherently dangerous activity resulting, for example, in escape of toxic gas the enterprise is strictly and absolutely liable to compensate all those who are affected by the accident and such liability is not subject to any of the exceptions which operate vis-à-vis the tortuous principle of strict liability under the rule in Rylands v. Fletcher. (Ibid.)

The case, though limited to Indian companies, sets out a model of liability that can be extremely useful as the international community moves forward in its consideration of how MNCs, especially those engaged with hazardous technology, can be held more accountable.

Underlying both absolute liability and enterprise responsibility is that:

1. The enterprise is obliged to take steps to anticipate all risks, and plan and prevent them from materializing. This corresponds on the one hand to the principle of due diligence in human rights law and on other also echoes the now well established principle of strict liability in environmental law.
2. The enterprise internalizes the cost of the personal or environmental harm due to its activities. This is very much in line with the 'polluter pays' and 'precautionary principles', now well established in international and domestic law in almost all jurisdictions.
3. In the context of 1 and 2 above, the victims are not burdened with the responsibility of pinpointing fault. In a departure from traditional tort law but responding to the realities of the increasingly sophisticated nature of technology and multinational business, it is considered fair to shift the onus of proof on the tortfeasor, i.e., the enterprise/multinational corporation.

Despite judicial pronouncements on the question, the statutory law in India clearly reveals the state's ambivalence on the issue of multinational corporate liability. Amongst the amendments to the Factories Act, 1948 was one in 1987 which absolves the designer, manufacturer, importer or seller of plant and machinery once the user to whom the plant and machinery were handed over, gave an undertaking that, 'if properly used', no harm would ensue. It seems this provision was in the nature of an assurance to the high priests of capital and technology—from absolute liability to absolute prejudgement of liability.

In sum, 20 years after the world's worst industrial disaster hundreds of thousands of people are still at grave risk. Whatever little gains the victims of Bhopal have made is primarily due to their own tireless struggle for justice and redress. Unfortunately it seems the disaster that is Bhopal and the struggle of the victims has not been enough to prompt the executive, legislature and judiciary to act together to meet the challenge posed by a combination of political economy of neoliberal globalization, hazardous technology and the power of transnational capital.

## NOTES

1. See (*a*) Memorandum from B.T. Burgoyne, Union Carbide Eastern addressed to the Management Committee dated 2 December 1973 enclosing 'Sevin Project – India, Finance Plan' and Capital Budget Proposal 73-8 dated 12.2.73, 'Union Carbide India Limited Methyl-Isocayante Based Agricultural Chemical Project; (*b*) Operational Safety Survey CO/MIC/SEVIN Units, Union Carbide India Ltd Bhopal Plant, UCC, July 1982. The members of the team were J.M.Poulson, Steve Tyson, and Leonard Kail; (c) Operational Safety/ Health Survey- MIC II Unit, Institute Plant', 10 September 1984. In fact this survey was also led by J.M Poulson, who led the 1982 Operational Safety Survey of the Bhopal Plant; (*d*) Testimonies of Bhopal plant workers in T.R. Chouhan *et. al.*; (*e*) Inspection Report of the Union Carbide Corporation Institute West Virginia, Occupational Safety and Health Administration (OSHA), USA, February 1985; (*f*) Testimony of Ronald Wishart, Vice President, Government Relations, UCC in Hearing on the Bhopal Gas Disaster and its Implications before the House Sub-Committee on Asian and Pacific Affairs, 98th Congress, Second Session, 12 December 1984; (*g*) Bhopal Methyl Isocyanate Incident Investigation Team Report, March 1985; Union Carbide Corporation, Danbury, Connecticut.
2. Exhibit 8 to Ghosh Affidavit, Exhibit 11, Memorandum of Law, Michael V. Ciresi, Stanley M. Chesley and F. Lee Bailey, In Re: Union Carbide Corporation Gas Plant Disaster at Bhopal, India in December 1984. MDL Docket No. 626, Misc. No. 21-38, 85 Civ. 2696 (JFK), US District Court, Southern District Court of New York.
3. Written Statement, Counter Claim and Set-Off of Union Carbide Corporation, 10 December 1986 in Regular Civil Suit No. 1113 of 86 in

Court of the District Judge, Bhopal, reproduced in Upendra Baxi and Amita Dhanda, *op.cit.*, p. 61.

4. Written Statement of Warren M. Anderson, 26 March 1985, in Release of Poison Gases and Other Hazardous Air Pollutants From Chemical Plants; Joint Hearing Before The Subcommittee on Health and the Environment and The Subcommittee on Commerce, Transportation and Tourism of The House Committee on Energy and Commerce House of Representatives, 99 Congress, first session, 26 March 1985, Serial No 99-18, US Government Printing Office, Washington 1985, p. 204.
5. The move towards disclosure however suffered a setback when in the wake of the World Trade Centre bombing the 'Chemical Manufacturers Association—now the American Chemistry Council—issued a report in 1998 warning of "the dark side of the Internet". It accused the EPA of allying with "professional environmentalists" to provide "one-stop shopping" for terrorists.' The industry managed to secure the support of law enforcement agencies and this resulted in the Congress legislating a restriction on keeping detailed information relating to toxic gas dispersion models, casualty figures etc. off the internet and allowing only the publication of summaries of such information.
6. CERCLA was enacted by the US Congress in 1980. It created a tax on the chemical and petroleum industries and established that Federal authority can respond 'directly to releases or threatened releases of hazardous substances that may endanger public health or the environment.' The taxes collected (amounting to US $1.6 billion in the first five years) went to a trust fund for cleaning up abandoned or uncontrolled hazardous waste sites. <http://www.epa.gov/superfund/action/law/sara.htm accessed on 31/08/2004>
7. Amended Plaint of Union of India filed in Union of India versus Union Carbide Corporation in Case 1113 of 1986, Court of the District Judge, Bhopal reproduced in Baxi and Dhanda (1990: 187).

## References


Anderson, Michael, 2002, 'Transnational Corporations and Environmental Damage: Is Tort Law the Answer?', *Washburn Law Journal* 41(399), Spring.

Baxi, Upendra and Amita Dhanda, 1990, Keenan's 1986 decision quoted in, *Valiant Victims and Lethal Litigation: The Bhopal Case*, Indian Law Institute, N.M. Tripathi, Bombay.

Cassels, Jamie, 1993, *The Uncertain Promise of Law: Lessons From Bhopal*, University of Toronto Press, Toronto.

Cassels, Jamie, 2000/2001, 'Outlaws: Multinational Corporations and Catastrophic Law', *Cumberland Law Review* 31(311).

Davis, Ann, 2002, 'New Alarms Heat up Debate on Publicizing Chemical Risks', *The Wall Street Journal*, 30 May.

Everest, Larry, 1985, *Behind the Poison Cloud*, Banner Press, Chicago.

(ILC) International Law Commission, *Annual Report 2003*, available on <http://www.un.org/law/ilc/ reports/2003/2003report.htm (accessed on August 30, 2004).

McCallion, Kenneth F., and H. Rajan Sharma, 2001, *International Resolution of Environmental Disputes and the Bhopal Catastrophe*, The Permanent Court of Arbitration/Peace Palace Papers, Volume 2, May 2001, The Hague.

Ramaseshan, Radhika, 1984, 'Government Responsibility for Bhopal Gas Tragedy', *Economic and Political Weekly* 50, 15 December.

Rolle, Mary Elliott, 2003, 'Unravelling Accountability: Contesting Legal and Procedural Barriers in International Toxic Tort Cases', *Georgetown International Environmental Law Review* 15(135).

# 32

## Is the Friendly Atom Poised for a Comeback?

Eliot Marshal

'Nuclear power faces stagnation and decline.' So warned a group of scientists in a sweeping review published two years ago by the Massachusetts Institute of Technology (MIT) in Cambridge.[1] Led by chemist John Deutch and physicist Ernest Moniz, both of MIT, the study concluded that nuclear power was in trouble and deserved a helping hand from government. Despite high construction costs, the authors argued that the United States should triple the number of unclear power plants by mid-century because they can deliver electricity without emitting green-house gases such as carbon dioxide ($CO_2$). The MIT group proposed a hefty tax on carbon emission to help get this cleaner energy source moving.

The political and economic environment has changed dramatically since that report came out. On 8 August, President George W. Bush signed into law the first major U.S. energy bill in a decade. Although it does not tax carbon, it promises subsidies across the board for new investments in renewable energy, such as wind and solar power, and a grab bag of more than $6 billion in benefits narrowly tailored for builders of new nuclear reactors (*Science*, 5 August, p.863). The bill was a plum for the nuclear power industry—one of several events that have got people talking about a 'nuclear renaissance'. Indeed, that's the title of a book published earlier this year by physicist and energy policy analyst William Nuttall of the University of Cambridge, U.K. One reason for optimism, Nuttall points out, is that oil and natural gas process have shot up since 2003, making non-fossil fuel energy more attractive. Meanwhile, some public leaders have cited nuclear power as a way to reduce the impact of global warming—and even some environmental advocates seem to agree.

---

* Reprinted with permission from Eliot Marshal, 'Is the Friendly Atom Poised for a Comeback?', *Science*, Vol. 309, 19 August, 2005, pp. 1168–69.

Although a few Asian countries never got off the nuclear bandwagon, new ones are now climbing aboard to meet rapidly growing electricity demand. India, with the most reactors under construction in the world, is planning a unique system that relies mainly on thorium rather than uranium fuel (see inset box). Japan continues work on fast neutron reactors that can 'breed' plutonium. And China announced in April that it will more than quadruple its nuclear electric capacity by 2020, buying among other designs a new 'pebble bed' reactor that shuts down if it over-heats. Nuclear advocates in the West also hope that advanced reactor designs can help overcome the lingering memories of Three Mile Island and Chernobyl.

### India's Homegrown Thorium Reactor

For more than five decades, India has followed its own path on nuclear power. After refusing to join the Nuclear Non-proliferation Treaty and detonating a nuclear device in 1974, it was excluded from the international group that shares fission technology. In isolation, it launched an ambitious nuclear electric programme that relies heavily on homegrown technology.

What makes India's strategy unique is its plan to build commercial reactors that run not on uranium but on a lighter element, thorium-232, India has one of the world's largest reserves of thorium—about 225,000 metric tones—but little uranium ore. Thorium does not fission; when irradiated with neutrons from a source material such as uranium-235, however, some of the thorium becomes uranium-233 (U233), which does fission and can sustain a nuclear reaction.

In 1958, India announced that it was embarking on an ambitious three-stage plan to exploit its thorium reserves. The first stage required building pressurized heavy-water reactors powered by natural uranium. They yield plutonium as a byproduct. Twelve are now operational. The plan called for stage two to kick in after sufficient plutonium had been extracted from spent cores; it would be used as a fuel in future fast-neutron reactors, which can irradiate thorium and produce u-233 as a byproduct. In the third stage, Advanced Heavy Water Reactors will burn a mixture of U-233 and thorium, generating about two-third of their power from thorium. Other nations—including the United States, Russia, Gemany, and Israel—have studied the route but have not attempted to use it to generate electricity.

Stage two of this grand strategy began officially last October. In the sleepy southern township of Kalpakam, a government-owned company began building a 500-magawatts of electricity (MWE) fast-breeder reactor that will use fast neutrons to produce U-233. In its core, the reactor will use a 'seed' fuel containing uranium and plutonium oxide; this source will send neutrons into a surrounding thorium blanket.

India atomic energy officials are confident that this exotic fuel system can be scaled up from a smaller, 40-megawatt Fast Breeder Test Reactor (FBTR) that has been running in Kalpakkam without major problems since 1985. This reactor and other research projects at the Indira Gandhi Centre for Atomic Research in Kalpakkam have demonstrated, IGCAR officials say, that India has mastered the new technology. In a 'bold step forward,' says Anil Kakodkar, Chair of the Atomic Energy Commission (AEC) in Mumbai, researchers at IGCAR in May of this year successfully extracted plutonium in high purity from the unique plutonium-rich mixed carbide fuel discharged from FBTR.

AEC anticipates that the fast breeder at Kalpakkam will cost about 5,700 million and produce 500 MW today to 'around 275 gigawatts' by the middle of this century.

Construction at Kalpkkam ran into trouble early this year; The 26 December 2004 tsunami flooded building and set the schedule back by four months, says Baldev Raj, IGCAR's director. But he says that the work is the reactor will go critical as planned in September 2010.

Mujid Kazimi, a nuclear engineer who studies thorium fuels at the Massachusetts Institute of Technology in Cambridge, says India's approach to breeding nuclear fuel from thorium is 'slightly more complicated' than fuel breeding planned elsewhere in the world. But he adds, 'everything they have reported to date indicates they are on track.'

India cannot go it entirely alone, however. It still requires uranium, including for two boiling water reactors it bought from General Electric in the 1960s, and that may be one reason it is interested in opening nuclear trade with other countries. At a meeting last month with Prime Minister Manmohan Singh, President George W. Bush called India 'a responsible state' with 'advanced nuclear technology'. The opening could lead of future exchanges of personnel and technology—and possibly fuel. Singh reassured Parliament however, that the deal would not undermine India's nuclear self-sufficiency.

(Pallava Bagla, *Science*, Vol. 309, 19 August 2005, pp. 1174–1175)

Does all of this amount to a nuclear renaissance? Skeptics point out that it would take a huge leap in the pace of plant construction simply to maintain nuclear power's current global share of electric output—about 17 per cent—let alone increase it. Many aging U.S. and European reactors will have to be dismantled in the next couple of decades. Even new ones remain more expensive than coal-or-gas-fired systems. And governments are not imposing stiff taxes on carbon emissions, the one strategy the MIT report said would tip investment decisions toward nuclear. Moreover, even if the economics were to favour nuclear power, two issues will continue to dog the industry: fears of nuclear weapons proliferation; and disputes about how to dispose of high-level wastes.

Optimists still think that the problems can be fixed. Reiterating his view of two year ago, Deutch says: 'If nuclear power can get its costs down and address the important issues of waste management and proliferation, its future will be very bright.' The next few years may reveal just how bright.

## Apocalypse Pending

The threat of global warming is perhaps the key factor in the rethinking of nuclear power. The nuclear industry, in particular, has seized on it as a reason to switch from fossil fuel to atom. For example, John Ritch executive director of the London-based World Nuclear Association (WNA), an advocacy group backed by power supply companies, told an audience in Idaho last month that unless the world cuts greenhouse gases, it will 'face catastrophic climate change, with the severest consequences for sea levels, species extinction, epidemic disease, drought, and extreme weather events that could combine to disrupt all civilization'.

WNA suggests that the best solution would be to raise the number of nuclear electric plants in the world from 441 today to 5,000 by the end of the century. That is the most ambitious scheme anyone has proposed, but so far, it has few takers. A more modest proposal—to maintain the nuclear share of electrify at the current level as a 'bridge' to future clean energy technologies—has struck a chord, however.

David King, science adviser to the U.K. government, has spoken publicly about the need to keep nuclear power as a clean energy option. Britain, the world's most visible campaigner for action on global warming, faces a common dilemma, as King explained to the *Independent* newspaper in May.

He described a looming 'gap' in clean energy production. About 27 per cent of U.K. electricity now comes from nuclear power, he noted, but without a 'new build', only one reactor unit (Sizewell B) will still be running in 2025, producing an estimated 4 per cent of the needed electricity. King said he was 'not a great fan of nuclear' but was willing to consider it because 'the climate change issue is so important'.

A recent U.K. government forecast lends weight to king's analysis: Solar panels, windmills, and wave-driven generators cannot pick up the slack anytime soon. An electricity strategy issued in May by the U.K. Council of Science and Technology, which reports to King, notes that 'the existing policy to reduce $CO_2$ will not be sufficient ... since the unclear stations are likely to be replaced by carbon-based technology (e.g., gas) in the short term'.

And even the United Kingdom, which has championed the international effort to curb $CO_2$ is failing to meet its self-imposed $CO_2$ reduction goals. Physicist David Wallace, vice president of the Royal Society in London, warned in May that 'our emissions are clearly going in the wrong direction', and that U.K. government forecasts of achievable $CO_2$ reductions have been 'frankly unrealistic'. Royal Society President Robert May has written that 'it is difficult to see how we can reduce our dependence on fossil fuels without the help of nuclear power'.

A few leaders in the green movement have endorsed the idea of using nuclear power as a bridge to cleaner systems in the future—including U.K. ecologist James Lovelock. Creator of the 'Gaia' metaphor that describes Earth as a living organism, Lovelock published a broad appeal last year. 'Our one immediately available source [of energy] does not cause global warming, and that is nuclear energy,' he wrote. 'I entreat my friends in the movement to drop their wrong-headed opposition [to it].' A few others, such have made similar statements. But environmental advocacy groups are not following.

Stephen Tindale, executive director of Greenpeace International in London, says it's 'misleading' to suggest that 'the green movement is suddenly embracing nuclear power on the back of Lovelock's statement'. He sees nuclear revival talk as 'a big distraction' from the need to invest in moderate scale, renewable energy systems. He adds that Moore is 'vehemently opposed to everything that Greenpeace stands for' and now makes his living 'by being anti-Greenpeace'.

Likewise, the head of Friends of the Earth in London, Tony Jouniper, says, 'we have reviewed our position on nuclear power'. In part because of the urgency of the climate change issue, and concluded that it is a 'false solution' pushed as part of 'a cleaver public relations campaign' by 'nuclear industrial interests'.

The Natural Resources Defenses Council has also reviewed its policy recently, says NRDC physicist Thomas Cocharan in the Washington, D.C., office, and concluded that nuclear couldn't survive without massive subsidies. As a June NRDC issue paper says, nuclear 'suffers from too many security, safety, and environmental exposure problems and excessive costs to qualify as a leading means to combat global warming pollution'.

Cochran offers a scenario to illustrate why he doesn't see nuclear as a good option. He begins with a modest goal: avoiding a small amount (0.2°C) of global warming at the end of this century. He calculates that relying on nuclear electricity, would require increasing the number of reactors in the world from the current 441 to at least 700 by mid-century and holding that number steady for 50 years. Allowing for retirement of obsolete equipment, he suggest, this will require building 1,200 new plants in all, at a rate of about 17 per year. The support requirements, he argues, would be staggering: a dozen new fuel-enrichment plants for reprocessing, the same number of Yucca Mountain-sized waste repositories if there were no reprocessing. Because just 8 kilograms of diverted plutonium would be enough to 'take out lower Manhattan', a nuclear renaissance isn't worth the risk, Cochran says.

The MIT review two years ago acknowledged that 'shortcomings' in the international safeguards on nuclear materials 'raised significant questions about the wisdom for a global growth scenario' for nuclear power. It did offer a fix: Tighten up the management of nuclear materials by the International Atomic Energy Agency (IAEA) and persuade France, Japan, Russia, and the United Kingdom to cut down the traffic in plutonium by shutting their reprocessing factories. But those changes have not occurred.

The threat of global warming may not have sparked a nuclear renaissance yet, but it is breathing new life into a debate over nuclear power that, in many countries, had been quiescent for the past few years.

## Note

1. 'The Future of Nuclear Power', funded by MIT and the Alfred P. Sloan Foundation, MIT, Cambridge, Massachusetts, 2003.

33

# Global Warming and India

Anand Patwardhan

Global warming has emerged as one of the most important environmental issues ever to confront humanity. This concern arises from the fact that our everyday activities may be leading to changes in the earth's atmosphere that have the potential to significantly alter the planet's heat and radiation balance. It could lead to a warmer climate in the next century and thereafter, portending a potpourri of possible effects, mostly adverse.

International efforts to address this problem have been going on for the last decade, with the Earth Summit at Rio in 1992 as an important launching point, and the Conference of Parties in Buenos Aires in 1998 as the most recent step. Although India as a developing country does not have any commitments or responsibilities at present for reducing the emissions of greenhouse gases such as carbon dioxide that lead to global warming, pressure is increasing on India and other large, rapidly developing countries such as China and Brazil to adopt a more proactive role.

At the same time, the developed countries of the North are trying to limit the extent of their commitments for emission reduction. In this situation, the public and policy-makers need to be aware of the ramifications and implications of the global warming problem, even if it is a problem that may manifest itself only sometime in the next century.

What is climate change? Climate change is a newcomer to the international political and environmental agenda, having emerged as a major policy issue only in the late 1980s and thereafter. But scientists have been working on the subject for decades. They have known since the nineteenth century that carbon dioxide in the atmosphere is a 'green house gas', that is, its

* Reprinted with permission from Anand Patwardhan, 'Global Warming and India', *Seminar* 486, February, 2000, pp. 63–69.

presence in the atmosphere helps to retain the incoming heat energy from the sun, thereby increasing the earth's surface temperature. Of course, carbon dioxide is only one of several such greenhouse gases in the atmosphere. Others include methane, nitrous oxide and water vapour. However, carbon dioxide is the most important greenhouse gas that is being affected by human activities. Carbon dioxide is generated by a multitude of processes ranging from animal and plant respiration to the burning of any kind of fuel containing carbon, including coal, oil, wood and cow dung. For a long time, human activities that generated carbon dioxide caused only a small perturbation in the natural cycle of the gas. However, since the Industrial Revolution, when our usage of fossil fuels increased dramatically, the contribution of carbon dioxide generated from human activities has grown large enough to constitute a significant perturbation of the natural carbon cycle.[1] Since the early 1950s, when regular measurements of the atmospheric concentrations of carbon dioxide were started, it has been conclusively established that these concentrations are increasing rapidly, driven by human activities.

The concentration of carbon dioxide in the earth's atmosphere was about 280 parts per million by volume (ppmv) in 1750, before the Industrial Revolution began. By 1994 it was 358 ppmv and rising by about 1.5 ppmv per year. If emissions continue at the 1994 rate, the concentration will be around 500 ppmv, nearly double the pre-industrial level, by the end of the twenty-first century.

The concentrations of other greenhouse gases such as methane and nitrous oxide have also been rising at a fairly rapid rate. The effect is that the atmosphere retains more of the sun's heat, warming the earth's surface. Of course, not all man-made additions to the atmosphere increase warming. For example, aerosols, tiny particles of solid or liquid suspended in the air, which result from the emissions of soot and sulphur dioxide from power plants tend to reflect heat and diminish warming. But aerosols are mostly short-lived while the carbon dioxide released into the atmosphere will stay there for decades.

At the same time, concern about local air quality is driving many countries to impose stringent controls on emissions of substances such as sulphur dioxide. As a result, many scientists feel that even as these emissions decrease in the future, the full effect of the greenhouse gases will be unmasked, leading to an even more rapid warming pattern.

While the pattern of future warming is open to debate, it is indisputable that the surface of the earth has warmed, on average, 0.3 to 0.6 degrees Celsius since the late nineteenth century when reliable temperature measurements began. Recent decades appear to be the warmest since at least 14000 BCE, according to the fragmentary information available.

It is against this backdrop of knowledge that Inter-Governmental Panel on Climate Change (IPCC) concluded in its second assessment report in 1995 that the current state of knowledge *'now points towards a discernible human influence on global climate'*. In this assessment report, the IPCC also concluded that under the existing scenarios of economic growth and development leading to greenhouse gas emissions, on a worldwide average, temperatures would rise by 1 to 3.5 degrees Celsius by the year 2100, and global mean sea level by about 15 to 95 cm. It is likely that changes of this magnitude and rapidity could pose severe problems for many natural and managed ecosystems, as well as important economic sectors such as agriculture and water resources. Indeed, for many low-lying and deltaic areas and small islands, a sea level rise of one meter could threaten complete loss of land and extinction of habitation.

Scenarios of future climate change are usually developed using complex three-dimensional models of the earth's atmosphere and oceans. However, while we have some degree of confidence in the gross or aggregate estimates for climate parameters (such as globally averaged surface temperature) from these models, there is a great deal of uncertainty with regard to regional details. In addition, most of the ill-effects of *climate change* are linked to extreme weather events, such as hot or cold spells of temperature, or wet or dry spells of rainfall, or cyclones and floods. Predictions of the nature and distribution of such events in a changed climate are even more uncertain, to the extent that virtually no authoritative predictions exist at all. Despite these uncertainties, it is clear that even the *possibility* of changes in, such extreme events is quite alarming.

Global warming has often been described as one of the most serious environmental problems ever to confront humanity, as this problem is inextricably linked to the process of development and economic growth itself. Since greenhouse gases are generated by burning fossil fuels, for example, in power plants, factories and automobiles, it is not easy to reduce emissions, since virtually every facet of our lives is intimately tied to the consumption of energy. Climate change is an unusually difficult issue for the people who make the decisions in democratic governments. First of all, the science is uncertain while governments have to make firm policy decisions, if only the decision to do nothing, long before these uncertainties can be resolved.

Political leaders are already beginning to overstate the clarity of the science in order to attract public support. A lot of money is now going into climate research, and new findings with varying political implications will continue to appear. Any serious attempt to cut emissions will have clear and immediate costs, but the benefits may not appear for a long time. To the extent that the benefits may be disasters that didn't happen, they may never be obvious. But the costs will be. As the debate develops, much of it is being cast in terms of the restraint that the present generation owes to future generations.

Unlike many other environmental issues, such as local air or water pollution, or even stratospheric ozone depletion caused by chlorofluorocarbons (CFCs), global warming poses special challenges due to the spatial and temporal extent of the problem—covering the globe and with decades to centuries of time scale. Again, in this particular issue, science has played, and continues to play, a critical role in defining the structure and basis of the debate. The following three dimensions of the issue illustrate the vexing features of the science underlying the problem: (*a*) Cumulative effect of the historical emissions. The climate system acts as a large integrator, that is, the response of the system is a result of the *entire history* of the forcing being applied; (*b*) Lags in the system: The response of the ocean-atmosphere system occurs several decades to centuries after the changes in the atmospheric greenhouse gas concentrations. As a result, even if emissions of greenhouse gases were stabilized immediately, it would take many years for the climate system to reach a new quasi-steady state, and some changes (such as sea level rise) would continue to happen; (*c*) The actual consequences of *climate change* are likely to exhibit considerable spatial and temporal variability—thus some regions may actually experience a transition to a milder, warmer, wetter, and overall better climate regime. As a result, there are costs as well as benefits associated with climate change; although the scientific consensus is clearly that the overall effects are likely to pose a significant burden.

How have we tried to respond to climate change? Negotiations began in 1991 under the auspices of the United Nations to formulate an international treaty on global climate protection. Those negotiations resulted in the completion by May 1992 of a Framework Convention on Climate Change (FCCC). The Convention was opened for signature at the Earth Summit in Rio de Janeiro in June 1992, and it came into force in March 1994.

The Convention has few binding requirements. It calls for nations to limit carbon dioxide and other greenhouse emissions, by 'addressing anthropogenic emissions by sources and removals through sinks of greenhouse gases....' It does not set out specific targets or timetables for reducing emissions. It only requires the developed country signatories to formulate and adopt policies that aim at stabilizing greenhouse gas emissions at 1990 emission levels, recognizing that the return

> by the end of the present decade to earlier levels of anthropogenic emissions... would contribute to... modifying longer term trends in anthropogenic emissions consistent with the objective of the Convention... to achieve... stabilization of greenhouse gas concentrations in the atmosphere at a level that would prevent dangerous anthropogenic interference with the climate system.

The Convention adopted the notion *of common but differentiated responsibility,* recognizing that the global climate was a common resource and responsibility, but that there were clear asymmetries between the developed and the developing countries in terms of both the past and present contributions to the problem as well as the resources to respond to it. That is, the developed countries are, by far the largest emitters of carbon dioxide and other greenhouse gases. At the same time, they also have the technical and financial resources to try and reduce their emissions. Two broad groupings of countries emerged after the Convention, the countries listed in Annex-1 of the Convention, or the developed countries, and the others. Countries such as Russia or Ukraine (parts of the former Soviet Union) although a part of the Annex-1 countries are placed in a special category as Economies in Transition.

At the time of the Rio Summit, proponents of more specific, legally binding targets and timetables for reducing greenhouse gas emissions successfully urged follow-on talks leading to future negotiation of a protocol or other legal instrument in order to strengthen the Framework Convention. In 1995, the Parties to the Framework Convention at their first meeting in Berlin, Germany, declared that commitments made in 1992 to reduce greenhouse gas emissions were inadequate to meet the objective of the

Convention. So-called 'next steps' were needed to confront the potential of global warming in the post-2000 time frame. Consequently, the Parties agreed to a process, set forth in their 'Berlin Mandate', of analysis and assessment of just what next steps might be taken to limit greenhouse gas emissions.

This process resulted in the negotiation of a protocol, the final details of which were completed at the third meeting of the Conference of the Parties to the Framework Convention held 1–12 December 1997, in Kyoto, Japan. The Kyoto Protocol to the United Nations Framework Convention on Climate Change commits industrialized nations to specific, legally binding emission reduction targets for six greenhouse gases: carbon dioxide, methane, nitrous oxide, hydrofluorocarbons, perfluorinated compounds, and sulphur hexafluoride. The protocol was opened for signatures on 16 March 1998.

International political implications have proven significant. By far the majority of greenhouse gases are emitted by sources in industrial and transportation sectors (especially automobiles) that are concentrated in developed countries. These countries have shown concern not only about their own emissions, but about increased emissions from poorer countries as they expand their economies. Friction has been evident in the debates over which actions by developed and developing countries should be undertaken, on what schedules, and which parties should pay incremental costs for mitigation measures. Developing countries generally have argued that the financial burden of change should be borne by developed countries, which are mainly responsible for current atmospheric change due to human activity. While some of the developed countries such as China and India do appear to have large emissions, on a per capita basis they are still negligible as compared to any of the developed countries.

As the Framework Convention (FCCC) states, the basic goal of the negotiation process is to return the concentrations of greenhouse gases to a level that prevents dangerous anthropogenic interference with the climate system. The simplest way of conceptualizing this goal is to consider a target or limit for the atmospheric concentrations of the greenhouse gases set at a level that does not lead to unacceptable climate change.[2] Of course, since our ability to predict future *climate change* is very limited, the notion of what is 'unacceptable' is itself quite imprecise and fuzzy. In this conceptualization, the economic activities in different countries that lead to greenhouse gas emissions correspond to this limit or resource being used up.

The entire negotiation process then may be regarded as an effort to address the following three questions: (*a*) What exactly is the limit, and how should it be defined? (*b*) What is the basis that ought to be used for the manner in which different countries can use up this resource? (*c*) What are the instruments that could be used to divide up and actually distribute this resource to the different countries once the allocation basis has been determined?

The first question centres around the level of atmospheric concentrations that would be considered acceptable in view of the possible consequences of climate change. A related issue is whether the limit would be specified individually for each greenhouse gas, or whether some sort of a 'basket' approach could be used where countries could trade-off amongst the different gases. This issue depends critically on whether the effects of the different gases could be made commensurate with each other through a set of equivalences[3] and if greater flexibility or economy would be obtained. It has also been suggested that rather than concentrate on the greenhouse gas concentrations, it may be better to focus on the sinks for these gases—which is primarily the terrestrial biosphere and the oceans.

The second question centres on the basis for the allocation and is currently the subject of much debate. Large, populous developing countries like India and China would clearly favour a per capita basis, as it gives them the greatest scope for increasing emissions further in their development processes.

The final question deals with the approach to be followed once the allocations have been determined. A large variety of market based instruments such as taxes and tradable permits have been deployed for conventional pollutants such as sulphur dioxide and there is much research on their applicability in the climate context; However, the key issue to recognize here is that any instrument will necessarily have to address large-scale technology and monetary transfers since developing countries could, in principle, 'sell' their allocations to the developed countries.

For India, the *climate change* issue has several ramifications: First, although India does not currently have any obligations under the Convention to reduce its greenhouse gas emissions, international pressure will keep increasing in this regard. It is therefore important for us to develop a clear understanding of our emission inventory. We also need to document and analyse our efforts in areas such as renewable energy, wasteland development and afforestation—all of which contribute towards either reducing carbon dioxide emissions or increasing carbon dioxide removal from the atmosphere. Considering that these efforts may often be undertaken for a variety, of reasons not directly related to global warming, but yet have benefits as far as *climate change* is concerned, we may be able to leverage such efforts in the international context.

Second, we need to develop a clear and well articulated position on each of the three basic questions indicated earlier. This position needs to be supported by appropriate analysis. The Indian research community could contribute substantially in this regard.

Finally, we need to recognize that even if countries do undertake immediate and rapid action to reduce their emissions, some degree of *climate change* is inevitable. If we consider the fact that we have very limited abilities to deal with weather extremes in the present day, the situation may get worse in the future. Therefore, we need to significantly improve our ability to plan and adapt to extreme events such as floods, droughts, cyclones and other meteorological hazards. Any robustness that we build into the system in this regard will always stand us in good stead, no matter what *climate change* actually transpires.

## Notes

1. For example, over 700 billion tons of carbon dioxide cycle annually through the biosphere. The anthropogenic contribution in this cycle is around 24 billion tons. Though the natural cycles are finely balanced, this is still a significant perturbation as it leads to an accumulation of carbon dioxide in the atmosphere.
2. For example, a value for long term atmospheric carbon dioxide concentration of 500–550 ppmv has often been used in the discussions. This value then defines the size of the resource that can be 'used up'.
3. Equivalences have indeed been suggested in the form of 'global warming potentials', an index that attempts to capture the ability of each gas to cause changes in the climate system. However, since different gases have different lifetimes in the atmosphere, and since the entire history of forcing is important, this becomes a fairly complex problem.

## References

Authoritative reviews of the science underlying the climate change issue are provided by the Inter-Governmental Panel on Climate Change (IPCC). See, for example:

J.T. Houghton *et. al*. (eds), 1996, *Climate Change 1995: The Science of Climate Change,* Cambridge University Press, Cambridge; and. James P. Bruce *et. al*. (eds), 1996, *Climate Change 1995: Economic and Social Dimensions of Climate Change,* Cambridge University Press, Cambridge. Also see the IPCC Website at <www.ipcc.ch>

A good introduction to the costs of reducing emissions is provided by Robert Repetto and Duncan Austin, 1997, *The Costs of Climate Protection: A Guide for the Perplexed,* World Resources Institute, Washington. Also see, the Framework Convention secretariat Website at <www.unfccc.de>, which provides very complete information on the entire negotiation process, as well as the actual Convention and Protocol documents.

# Further Readings

Detailed footnotes are given in the general introduction as well as in the sectional introductions. This brief essay merely points the way forward and identifies books or journals that will add considerably to one's learning about a complex, growing field. Several of the works listed here will give further leads and references. Readers are encouraged to explore these and also follow leads given in footnotes to the chapters.

The five Citizen's Reports on India's environment remain an indispensable source of information and insight. The first two were Anil Agarwal, Ravi Chopra and Kalpana Sharma (ed.), *The State of India's Environment, the First Citizen's Report, 1982* (Delhi: Centre for Science and Environment [CSE], 1982) and Anil Agarwal and Sunita Narain (ed.), *The State of India's Environment, the Second Citizen's Report, 1984-85* (Delhi: CSE, 1985). Subsequent reports and issue specific studies retain a lucidity of writing with a wealth of detail. The first report is available in an elegant Hindi version done by Anupam Mishra of the Gandhi Peace Foundation and published by the CSE.

Dating to the late 1980s and early 1990s are two overviews that are still indispensable. These are Madhav Gadgil and Ramachandra Guha, *This Fissured Land, An ecological history of India* (Delhi: OUP, 1989) and *Ecology and Equity* (Delhi: Penguin, 1995). Both have been published in a combined volume titled *The Use and Abuse of Nature* by Oxford University Press (OUP).

The pre-colonial era is better served for the ancient period than for centuries more proximate to our own times. A vivid sense of changes in the land emerges from Jean Deloche, *Transport and Communication in India prior to Steam Locomotion,Volume I* (Delhi: OUP, 1993) and Irfan Habib, *Atlas of Mughal India* (OUP, 1983). Sumit Guha, *Environment and Ethnicity in India, 1200-1991* (Cambridge University Press, 1999) is a useful corrective to versions that romanticize India's past ecology.

Anil Agarwal and Sunita Narain provide insights into linkages of past and present in their *Dying Wisdom* (Delhi: CSE, 1997). Anupam Mishra's Hindi works, *Rajasthan ki rajat boonde* and *Abhi bhi Khare hain talaab* published by the Gandhi Peace Foundation are fine grained works on water harvesting. The former is available in an English translation from the Research Foundation for Science and Technology (Delhi, 1991)

Excellent edited collections strong on the imperial impact are David Arnold and Ramachandra Guha (ed.), *Nature, Culture, Imperialism* (Delhi: OUP, 1995); Richard Grove, Vinita Damodaran and Satpal Sangwan (ed.), *Nature and the Orient* (Delhi: OUP, 1998); and Paul Greenough and Anna Tsing (ed.), *Southern Projects in Nature* (Delhi: Orient Longman, 2003).

Post-1947 changes are traced in Roger Jeffery and Nandini Sundar (ed.), *A New Moral Economy for India's Forests?* (Delhi: Sage Publications, 1999) and V.K. Saberwal and M. Rangarajan (ed.), *Battles over Nature* (Delhi: Permanent Black, 2003). On water see the special section of *Economic and Political Weekly*, 18 February 2005 edited by Biksham Gujja, K.J. Joy, Suhas Paranjpe, Vinod Gupte and Shruti Vispute. Also see A. Vaidyanathan, *India's Water Resources* (Delhi: OUP, 2006).

Popular works by scientists and activists, professionals and journalists can be informative and enjoyable. See Madhav Gadgil, *Ecological Journeys* (Delhi: Permanent Black, 2001); K. Ullas Karanth, *A View from the Machan* (Delhi: Permanent Black, 2005); Zai Whitaker, *Salim Ali for Schools* (Delhi: Permanent Black, 2003). Salim Ali's autobiography, *The Fall of a Sparrow*, (Delhi: OUP, 1987) is a charming and candid account. Also see, D.C. Kala, *A life of Jim Corbett* (Delhi: Ravi Dayal, 1979, Revised edition, 1999). Jim Corbett's *The Man Eaters of Kumaon* (Bombay: OUP, 1944, Reprint, 1995) is a classic. For more detailed listings and short extracts see M. Rangarajan (ed.), *The Oxford Anthology of Indian Wildlife: Volume II, Watching and Conserving* (Delhi: OUP, 1999) and also Valmik Thapar (ed.), *Saving Wild Tigers, The Essential Writings* (Delhi: Permanent Black, 2001).

Environmental movements have evoked several studies. On Chipko and its antecedents, see Ramachandra Guha, *The Unquiet Woods* (Delhi: OUP, 1989). On Kerala Shastra Sahitya Parishad, see M. Zachariah and S. Sooryamurthy, *Science for Social Revolution* (Delhi: Sage Publications, 1994).

Other significant works on environmental conflicts and policies include Amita Baviskar, *In the Belly of the River* (Delhi: OUP, 1995); Mahesh Rangarajan, *Fencing the Forest* (Delhi: OUP, 1996); Marlene Buchy, *Teak and Arecanut* (Delhi: INTACH, 1996), Satyajit Singh, *Taming the Waters* (Delhi: OUP, 1997); Purnendu S. Kavoori, *Pastoralism in Expansion* (Delhi: OUP, 1999); Arup Jyoti Saikia, *Jungles, Reserves, Wildlife, A History of Forests in Assam* (Guwahati: Wildlife Areas and Development Trust, 2005); Rohan D'Souza, *Drowned and Damned* (Delhi: OUP, 2006). On science see Michael Lewis, *Inventing Global Ecology* (Delhi: Orient Longman, 2003).

For contemporary memoirs and accounts see Anupam Mishra and Satyendra Tripathi, *Chipko* (Delhi: Gandhi Peace Foundation, 1987); Jasjit Mansingh, *Oona, Mountain Wind* (Delhi: Bluejay, 2002) and Sanjay Sangvai, *The River and Life* (Mumbai: Earthcare Books, 2000).

For views critical of privatization and globalization see the various works of Vandana Shiva, *Water Wars* (Delhi: India Research Press, 2002) and *Staying Alive* (Delhi: Kali for Women, 1987). Also see Arundhati Roy, *The Greater Common Good* (Delhi: India Book Distributors, 1999).

Gender has been a focal point of not only activism and debate but also of scholarship. See Govind Kelkar, Dev Nathan and Pierre Walter (ed.), *Gender Relations and Forest Societies in Asia: Gender and Patriarchy at Odds* (Sage Publications, 2003); Kuntala Lahiri-Dutt (ed.), *Fluid Bonds* (Kolkata: Stree, 2006); and Sumi Krishna (ed.), *Livelihood and Gender, Equity in Community Resource Management* (Delhi: Sage Publications, 2004). Bina Agarwal, *A Field of Her Own, Gender and Land Rights in South Asia* (Cambridge, 1995) remains a vital source.

The journals *Seminar* and *Economic and Political Weekly* are excellent sources. In particular see the *Seminar* issues on *The Politics of Ecology* (February 1987); *Our Natural Resources* (June 1993); *People and Parks* (February 1995); *Grassroots Governance* (February 1996); *Environment, Myth and Reality* (February 2000); *Protecting Nature* (August 2000); *Wastelands* (March 2001) and *Shades of Green* (August 2002).

*Down to Earth* founded in 1992 is published from Delhi and is a fortnightly with first rate ground level reportage, opinion pieces and debates. In addition to the editorials, it has the unusually good reporting on neglected but ecologically critical regions, such as the North-East which is covered by Nitin Sethi.

On the North-East, see P.S. Ramakrishnan, *Shifting Agriculture and Sustainable Development, An Inter-Disciplinary Study from North East India* (Paris: UNESCO, 1992) and Daman Singh, *The Last Frontier, People and Forests in Mizoram* (Delhi: TERI, 1995).

Works of imaginative and creative literature in English and other Indian languages often provide a window on landscapes, peoples and cultures. For a sample, try Bibhutibhushan Bandyopadhyaya, *Aranyak of the Forest*, Translated by Rimli Bhattacharya (Kolkata: Seagull, 2002); Anita Desai, *A Village by the Sea* (Delhi: Puffin, 1982); Amitav Ghosh, *The Hungry Tide* (Delhi: Penguin, 2004). Excellent anthologies include Ravina Aggarwal, *Into the High Ranges, The Penguin Book of Mountain Writings* (Delhi: Penguin, 2002); Amita Baviskar, *Waterlines* (Delhi: Penguin, 2005); and Savyasaachi, *Between Earth and Sky, The Penguin Book of Forest Writings* (Delhi: Penguin, 2006).

The incomparable collection of poetry deeply evocative of seasons, lands, and multiple meanings is of course the late A.K. Ramanujan's *An Interior Landscape, Poems from a Classical Tamil Anthology,* first published in 1967 and available in several reprints by Oxford University Press.

Similarly, in a class of her own is Mahasweta Devi, *The Book of the Hunter* translated by Sagaree and Mandira Sengupta (Kolkata: Seagull Books, 2002). Also worth a look is her collection, *Dust on the Road, The Activist Writings of Mahasweta Devi.*

Recent reviews are *Global Environmental Challenges* by James Gustave Speth (Delhi: Orient Longman, 2005) and James Lovelock, *The Revenge of Gaia* (London: Penguin, Allen Lane, 2006). Dev Nathan *et al.* (ed.), *Globalization and Indigenous peoples in Asia* (Delhi: Sage Publications, 2004). For a perceptive work see Joan Martinez-Alier, *The Environmentalism of the Poor* (Delhi: OUP, 2006).

There is as yet little to equal the magisterial Michael Williams, *Deforesting the Earth, From Pre-history to Global Crisis* (Chicago: University of Chicago

Press, 2000). A radically provocative and Trans-regional view of early environmentalism is to be found in Richard Grove, *Green Imperialism* (Delhi: OUP, 1995).

One book that is indispensable for its global sweep and a masterly tying together of detail is J.R. McNeill, *Something New Under the Sun, An Environmental History of the Twentieth Century* (London: Penguin, Allen Lane, 2000). The evolution of environmental concerns and ideologies is provided by Ramachandra Guha, *Environmentalism, A Global History* (Delhi: OUP, 2000). Both books are especially good on a range of experiences across cultures and continents.

# The Editor and Contributors

## The Editor

**Mahesh Rangarajan** is a historian of environmental change and a well known political commentator. A former Assistant Editor of *The Telegraph*, he is presently a Visiting Professor at Jadavpur University, Kolkata. He did his BA, History (Honours) at Hindu College, University of Delhi and was a Rhodes Scholar at Oxford University. He completed a second BA in Modern History at Balliol College and his doctorate at Nuffield College. He was Visiting Faculty at Cornell University, Ithaca, during 2001–04 and has also taught at the University of Delhi. He has also been a visitor to the Centre for Ecological Sciences, Indian Institute of Science, Bangalore and taught at the National Centre for Biological Sciences, Bangalore. His books include *Fencing the Forest* (1996) and *India's Wildlife History* (2001) and a two volume, edited collection, *The Oxford Anthology of Indian Wildlife* (1999). He is the co-author of *Towards Coexistence, People, Parks and Wildlife* (2000). He has co-edited two inter-disciplinary volumes, *Battles over Nature* with Vasant Saberwal (2003), and *Making Conservation Work, Securing Biodiversity in The New Century* with Ghazala Shahabuddin (2007). Dr Rangarajan was Corresponding Editor of the journal, *Environment and History* (1996–2001) and is associated with the journal *Conservation and Society* (founded 2003). He is a member of the Advisory Board of the Mysore-based Nature Conservation Foundation.

He is currently working on a co-edited volume on India's environmental history (with K. Sivaramakrishnan) and a full length study of human–wildlife relations in colonial and independent India.

## The Contributors

**Bina Agarwal** is Professor of Economics at the Institute of Economic Growth, Delhi. She is the author of several works including *Cold Hearths* *and Barren Slopes: The Wood Fuel Crisis in the Third World* (1986) and *A Field of Her Own: Gender and Land Rights in South Asia* (1994). She studied at the universities of Delhi and Cambridge and is a founder member of the Indian Society of Ecological Economics.

**Arun Agrawal** is a Professor at the University of Michigan at Ann Arbor. His works include *Greener Pastures* and a monograph on the Western Himalayas, *Environmentality* (2006). Agrawal graduated with a Bachelor's in History from Hindu College, Delhi University, did an MBA from Indian Institute of Management, Ahmedabad and went on to become a political scientist.

**Pallava Bagla** has been a photojournalist for two decades, writing for *Science*; *The Indian Express* and *The National Geographic News* among others. A graduate of Botany from Hansraj College, Delhi University, he also contributes to *Orbis*, one of the world's largest photo agencies. In 2006, he was the first recipient of the National Award for Science and Technology Communication in the print medium instituted by the Ministry of Science and Technology, Government of India.

**Divyabhanusinh** is Vice President of World Wide Fund for Nature (India) and a B. Litt. from Pune University. He has authored two books, *The End of a Trail: The Cheetah in India* (1995) and *The Story of Asia's Lions* (2005). He is a member, Cat Specialist Group, Species Survival Commission of the World Conservation Union.

**Darryl D'Monte** is one of Asia's leading environmental journalists and is based in Mumbai. His *Temples or Tombs? Industry versus Environment* (Centre for Science and Environment, 1985) remains a classic. Formerly Resident Editor in Mumbai of *The Times of India* and *The Indian Express*, he chairs the Forum of Environmental Journalists. His most recent work is *Ripping the Fabric: The Decline of Mumbai and its Mills* (2002).

**Madhav Gadgil** is a leading ecologist and a polymath. He studied at Pune, received his doctorate from Harvard, worked with the Maharashtra Association of Cultivation of Science and then moved to Bangalore, where he was at the Indian Institute of Science till 2004. In 1982, he founded the Centre for Ecological Sciences. He has co-authored two landmark books on environmental history: *This Fissured Land* (1992) and *Ecology and Equity* (1995). An essay collection, *Ecological Journeys* was published in 2001.

Professor Gadgil writes extensively in Marathi and Kannada, and has served as Chair of the Science and Technology Committee of the Global Environment Facility. He continues to be closely associated with a range of initiatives, including the People's Biodiversity Registers and the Children's Science Congress.

**Ramachandra Guha** is a historian and writer based in Bangalore. An economist and sociologist by training, he is a historian by vocation. *The Unquiet Woods* (1989) was his first book. *Environmentalism: A Global History* (2000) and *How Much Should a Person Consume?* (2007) are among more recent works. He writes a fortnightly column in *The Hindu*.

**Sumit Guha** is Professor of History at Rutgers University, New Jersey, USA. His books include *Environment and Ethnicity in India, 1200–1991* (1999) and *Health and Population in South Asia from Earliest Times to the Present* (2001). He studies the medieval and early modern periods of Indian history with a particular focus on social and environmental history. He has previously taught at St Stephen's College, Delhi University, The Indian Institute of Management, Kolkata and at Brown University, USA.

**N.S. Jodha** was born and brought up in an arid area of Rajasthan. He did his Ph.D. in Agricultural and Resource Economics, covering arid areas. He worked for about two decades with institutions such as the Central Arid Zone Research Institute (CAZRI), the Agro-Economic Research Centre for Rajasthan and Gujarat, the Indian Agricultural Reasearch Institute (IARI), and the International Centre for Research in Semi Arid Tropics (ICRISAT) and IITA (Africa). He has focused on issues and problems of dry lands. In 1987, he joined ICIMOD, an institution located in Nepal working in eight Himalayan countries. He has worked with several agencies dealing on common property resources, and poverty–environment links particularly in the context of dry tropical areas and mountains. His six books include *Life on the Edge* (2001).

**D.C. Kala** is based in Ranikhet and the author of several works including *Jim Çorbett of Kumaon* published by Ravi Dayal Publishers.

**K. Ullas Karanth** is a carnivore ecologist who studies the ecology and conservation of large mammals in India. He is especially interested in tigers, methods for estimating animal numbers and in resolving human–animal conflicts. He pioneered the use of radio tracking for estimating tiger numbers in India. Dr Karanth is associated with the Wildlife Conservation Society (WCS), New York and directs his India Programme based in Bangalore. He has written extensively at the popular level in both Kannada and English. His books include *The Way of the Tiger* and most recently, *A View from the Macchan* (2006).

**M. Krishnan** who died in 1996 was the doyen of India's naturalists. Selections from his weekly nature columns have been collected together in two anthologies, *Jungle and Backyard* and more recently in an edited collection, *Nature's Spokesman* (1995). Krishnan was based in Chennai and was awarded a Jawaharlal Nehru Memorial Fellowship that resulted in his book, *India's Wildlife, 1958–70.* His outstanding black and white photographs have been recently published in *Eye in the Jungle, Photographs and Writings* edited by Ashish and Shanti Chandola (2006).

**Mayank Kumar** teaches history at Satyawati College (Evening), University of Delhi. Dr Kumar specializes on the historical ecology of medieval and early modern Rajasthan. He did his BA in History at MS University, Baroda, and his Masters and a doctorate at Jawaharlal Nehru University (JNU). Dr Kumar also has an M. Phil in History from the Delhi University.

**Makkhan Lal** has taught archaeology and history at Banaras Hindu University and Aligarh Muslim University. He has been a Charles Wallace Fellow at Cambridge University and was a founding member of the World Archaeological Congress.

**Eliot Marshal** is News Editor with the journal *Science.*

**J.R. McNeill** is Professor of History at Georgetown University, USA. His several books include *Something New Under the Sun: An Environmental History of the Twentieth Century* (2000).

**V.N. Misra** is one of India's most eminent archaeologists and was Professor at Deccan College, Pune.

**Vijay K. Nagaraj** is based in Rajsamand, Rajasthan and works with the Mazdoor Kisan Sangharsh Samiti (MKSS).

**Anand Patwardhan** heads the Techonology Information and Forecasting Assessment Council, an autonomous unit of the Ministry of Science and Technology. He holds a B. Tech from the Indian Institute of Technology (IIT), Mumbai and a Masters and a Doctorate from Carnegie Mellon

University, USA. His research is largely focused on the environment–climate–society interface. He has worked on areas such as: the analysis of climate variability and climate change, assessments of vulnerability and adaptation; industrial ecology, diffusion of cleaner technology and environmental management. Dr Patwardhan is also interested in issues of technological change, innovation and technology policy.

**Roopali Phadke** is Assistant Professor of Environmental Politics and Policy at the Environmental Studies Department at Macalester College, USA. Her research and teaching interests concern water development, sustainable agriculture and renewable energy promotion and implementation. She is completing a book on sustainable and equitable water systems in Maharashtra.

**Nithya V. Raman** is based in Rajsamand and works with the Mazdoor Kisan Sangharsh Samiti (MKSS).

**Aromar Revi** is a Founder-Director of TARU, an international consulting and research firm that has pioneered work on governance and environment related issues. He is also Fellow of the India-China Institute of the New School University, New York. He is a member of the *Goa 2001 Team* that developed the RURbanism framework for long term planning of sustainable regions and their embedded cities. Aromar is an alumnus of the Indian Institute of Technology (IIT), Delhi, and the schools of Law and Management at the University of Delhi.

**Dunu Roy** An alumnus of IIT Bombay, Dunu worked for over a decade in Shahdol, Madhya Pradesh on environmental planning issues. He is Director of the Hazards Centre, Delhi and Dean of the People's Science Institute, Dehra Dun. He describes himself as a 'political environmentalist' and works closely with mass organizations. He is the co-author of *Planning the Environment* (1982).

**Vasant Saberwal** is a film-maker and author. Presently a Programme Officer with the Ford Foundation, he has studied at the universities of Delhi and Yale and also at the Wildlife Institute of India, Dehra Dun. Dr. Saberwal's books include *Pastoral Politics* (1999) and a co-authored volume, *Towards Coexistence, People, Parks and Wildlife.* His most recent work is *Democratizing Nature*, a co-authored study of conservation in the Great Himalayan National Park.

**Sanjay Sangvai** is the author of *The River and Life, People's struggle in the Narmada Valley* (2000), an insider's account of the movements against big dams on the river Narmada in central India. Sangvai's extensive grassroots experience is combined with reflection on the larger challenges of equity and the environment.

**Madhu Sarin** is best known for her extensive work on gender and natural resource issues in community based resource management. She graduated as an architect from Chandigarh, followed by studying development planning in London. She is the author of *Urban Planning in the Third World: The Chandigarh Experience* (1982) and *Joint Forest Management: The Haryana Experience* (1996). She has co-authored and independently written several works of which *Laws, Lore and Logjams: Critical Issues in Indian Forest Conservation* (2005) is the most recent. She has written extensively on tribal forest rights. She also served as a member of the Technical Policy Group that drafted the National Biodiversity Strategy and Action Plan (NBSAP).

**Savyasaachi** is a Reader in Sociology at Jamia Millia Islamia University, New Delhi. He did a Bachelor's at Hindu College and a Master's in Sociology at the Delhi School of Economics. He also has an M.Phil in Chinese Studies at Delhi University. His doctoral work on the Abhujmarias of Bastar was published by Wider, Helsinki. He is the editor of *Between the Earth and Sky: The Penguin Book of Forest Writings* (2005).

**N.C. Saxena** was a member of the Indian Administrative Service and a former Secretary, Planning Commision of India. A graduate in mathematics, he opted to study farmers and forestry in Uttar Pradesh for a doctoral thesis at Oxford University. *The God that Failed: India's Eucalyptus Craze*, is one of his several works. He writes extensively on issues of forestry, displacement, decentralization, poverty and rural development.

**Mukul Sharma** is the author of *Landscapes and Lives* (2001). He is working on a full-length study of the relationship of revivalist social movements and Green issues. Mukul studied political science in Delhi University and worked for nearly two decades with the leading Hindi daily, *Navbharat Times*. He has written and reported extensively on environmental issues in India and elsewhere in the developing world. He has won several awards as a journalist including from *The Statesman* and from the Food and Agriculture Organization(FAO).

**Daman Singh** is the author of *The Last Frontier: People and Forests in Mizoram*, (1996), a pioneering study of conservation issues in India's North-East. She majored in Mathematics at St Stephen's College, studied Rural Management at the Institute of Rural Management, Anand (IRMA) and worked for several years with the Society for Wasteland Development (SPWD).

**Ajantha Subramanian** is Assistant Professor of Social Anthropology and Social Studies at Harvard University. She worked on citizenship, community and ecology among fishers on Tamil Nadu's Coromandel coast for her doctoral thesis.

**M.S. Swaminathan** is an eminent agricultural scientist. He is a former Director of the Indian Council for Agricultural Research (IARI) and the International Rice Research Institute (IRRI), Philippines. He studied at Travancore, Coimbatore and Cambridge, UK. A recipient of several awards including the Ramon Magsaysay (1971) and the Norman Borlaug Award (1979), his several books include *Plants and Society* (London, 1989).

**Himanshu Thakkar** works with the South Asia Network on Dams, Rivers and People, Delhi. His work on water-related issues brings together a deep knowledge of policy with extensive grassroots experience.

**Romila Thapar** is Professor Emeritus at the Centre for Historical Studies, Jawaharlal Nehru University, New Delhi. One of India's most eminent scholars, she is a graduate of Panjab University and a PhD from London University. She has been a Reader at the University of Delhi and a Professor at Jawaharlal Nehru University. Her several books include *Asoka and the Decline of the Mauryas* and *Sakuntala, Texts, Readings, Histories*. She is author of *The Penguin History of India, I* (1966) and *The Penguin History of Early India* (2002).